Cengiz Kahraman (Ed.)

Fuzzy Applications in Industrial Engineering

Studies in Fuzziness and Soft Computing, Volume 201

Further volumes of this series can be found on our homepage: springer.com

Vol. 185. Spiros Sirmakessis (Ed.)
Knowledge Mining, 2005
ISBN 3-540-25070-0

Vol. 186. Radim Bělohlávek, Vilém Vychodil
Fuzzy Equational Logic, 2005
ISBN 3-540-26254-7

Vol. 187. Zhong Li, Wolfgang A. Halang, Guanrong Chen (Eds.)
Integration of Fuzzy Logic and Chaos Theory, 2006
ISBN 3-540-26899-5

Vol. 188. James J. Buckley, Leonard J. Jowers
Simulating Continuous Fuzzy Systems, 2006
ISBN 3-540-28455-9

Vol. 189. Hans Bandemer
Mathematics of Uncertainty, 2006
ISBN 3-540-28457-5

Vol. 190. Ying-ping Chen
Extending the Scalability of Linkage Learning Genetic Algorithms, 2006
ISBN 3-540-28459-1

Vol. 191. Martin V. Butz
Rule-Based Evolutionary Online Learning Systems, 2006
ISBN 3-540-25379-3

Vol. 192. Jose A. Lozano, Pedro Larrañaga, Iñaki Inza, Endika Bengoetxea (Eds.)
Towards a New Evolutionary Computation, 2006
ISBN 3-540-29006-0

Vol. 193. Ingo Glöckner
Fuzzy Quantifiers: A Computational Theory, 2006
ISBN 3-540-29634-4

Vol. 194. Dawn E. Holmes, Lakhmi C. Jain (Eds.)
Innovations in Machine Learning, 2006
ISBN 3-540-30609-9

Vol. 195. Zongmin Ma
Fuzzy Database Modeling of Imprecise and Uncertain Engineering Information, 2006
ISBN 3-540-30675-7

Vol. 196. James J. Buckley
Fuzzy Probability and Statistics, 2006
ISBN 3-540-30841-5

Vol. 197. Enrique Herrera-Viedma, Gabriella Pasi, Fabio Crestani (Eds.)
Soft Computing in Web Information Retrieval, 2006
ISBN 3-540-31588-8

Vol. 198. Hung T. Nguyen, Berlin Wu
Fundamentals of Statistics with Fuzzy Data, 2006
ISBN 3-540-31695-7

Vol. 199. Zhong Li
Fuzzy Chaotic Systems, 2006
ISBN 3-540-33220-0

Vol. 200. Kai Michels, Frank Klawonn, Rudolf Kruse, Andreas Nürnberger
Fuzzy Control, 2006
ISBN 3-540-31765-1

Vol. 201. Cengiz Kahraman (Ed.)
Fuzzy Applications in Industrial Engineering, 2006
ISBN 3-540-33516-1

Cengiz Kahraman
(Ed.)

Fuzzy Applications in Industrial Engineering

Dr. Cengiz Kahraman
Istanbul Technical University
Industrial Department of Engineering
Faculty Management
Campus Macka, 80680
Istanbul, Turkey
E-mail: kahramanc@itu.edu.tr

ISSN print edition: 1434-9922
ISSN electronic edition: 1860-0808

ISBN 978-3-642-07011-2 e-ISBN 978-3-540-33517-7

Springer is a part of Springer Science+Business Media
springer.com

Softcover reprint of the hardcover 1st edition 2006

Cover design: Erich Kirchner, Heidelberg

Preface

The definition of Industrial engineering from the *American Institute of Industrial Engineers* is given as "Industrial Engineering is concerned with the design, improvement, and installation of integrated systems of people, material, equipment, and energy. It draws upon specialized knowledge and skills in the mathematical, physical and social sciences together with the principles and method of engineering analysis and design to specify, predict, and evaluate the results to be obtained from such systems." Fuzzy set approaches are suitable to use when the modeling of human knowledge is necessary and when human evaluations are needed. Fuzzy set theory is recognized as an important problem modeling and solution technique. Fuzzy set theory has been studied extensively over the past 40 years. Most of the early interest in fuzzy set theory pertained to representing uncertainty in human cognitive processes. Fuzzy set theory is now applied to problems in engineering, business, medical and related health sciences, and the natural sciences. Over the years there have been successful applications and implementations of fuzzy set theory in industrial engineering. Industrial engineering is one of the branches that fuzzy set theory found a wide application area. Industrial engineers face many problems with incomplete and vague information in these cases since the characteristics of these cases often require this kind of information. Fuzzy Sets Theory developed by L.A. Zadeh (1965) is an excellent tool to solve these problems. Many industrial engineering curriculums of undergraduate and graduate programs include many courses teaching how to use fuzzy sets when you face incomplete and vague information.

This book presents some application examples of fuzzy sets in industrial engineering. It contains 24 original research and application chapters from different perspectives, and covering different areas of Industrial Engineering. The book contains papers on the major seven areas of industrial engineering to which fuzzy set theory can contribute. These areas are fuzzy control and reliability, fuzzy engineering economics and investment analyses, fuzzy group and multi-criteria decision-making, human factors engineering and ergonomics, manufacturing systems and technology management, optimization, and statistical decision-making.

Many books and special issues of journals have been published on the fuzzy applications in the various topics of Industrial Engineering. This book aims at summarizing these works and presenting the future directions of the use of fuzzy sets in Industrial Engineering.

As the editor of this book, I very much thank the authors accepting to contribute with their invaluable chapters even though they are very busy with their other works. My special thanks go to Prof. H-J. Zimmermann, Prof. Waldemar Karwowski, Prof. Janusz Kacprzyk, Prof. P. Paul Wang, Prof. Hideo Tanaka, Prof. Da Ruan, Prof. Hung T. Nguyen, and the others.

I hope that this book would provide a useful resource of ideas, techniques, and methods for further research in the applications of fuzzy sets in Industrial Engineering. I am grateful to the referees whose valuable and highly appreciated works contributed to select the high quality of chapters published in this book. My sincere thanks go to Dr. Murat Gülbay for his helps in editing this book.

Istanbul, Turkey
March 2006

Prof. Cengiz Kahraman

Contents

Applications of Fuzzy Sets in Industrial Engineering: A Topical Classification

Cengiz Kahraman, Murat Gülbay and Özgür Kabak

Istanbul Technical University, Department of Industrial Engineering 34367 Maçka İstanbul Turkey

Summary: A rational approach toward decision-making should take into account human subjectivity, rather than employing only objective probability measures. This attitude towards the uncertainty of human behavior led to the study of a relatively new decision analysis field: Fuzzy decision-making. Fuzzy systems are suitable for uncertain or approximate reasoning, especially for the system with a mathematical model that is difficult to derive. Fuzzy logic allows decision-making with estimated values under incomplete or uncertain information. A major contribution of fuzzy set theory is its capability of representing vague data. Fuzzy set theory has been used to model systems that are hard to define precisely. As a methodology, fuzzy set theory incorporates imprecision and subjectivity into the model formulation and solution process. Fuzzy set theory represents an attractive tool to aid research in industrial engineering (IE) when the dynamics of the decision environment limit the specification of model objectives, constraints and the precise measurement of model parameters. This chapter provides a survey of the applications of fuzzy set theory in IE.

Key words: Industrial engineering, fuzzy sets, research areas, decision-making, multiple criteria, quality management, production planning, manufacturing, ergonomics, artificial intelligence, engineering economics.

1 Introduction

Fuzzy set theory, which was founded by Zadeh (1965), has emerged as a powerful way of representing quantitatively and manipulating the imprecision in decision-making problems. Fuzzy sets or fuzzy numbers can appropriately represent imprecise parameters, and can be manipulated through different operations on fuzzy sets or fuzzy numbers. Since imprecise parameters are treated as imprecise values instead of precise ones, the process will be more powerful and its results more credible. Fuzzy set theory has been studied extensively over the past 40 years. Fuzzy set theory is now applied to problems in engineering, business, medical and related

C. Kahraman et al.: *Applications of Fuzzy Sets in Industrial Engineering: A Topical Classification*, StudFuzz **201**, 1–55 (2006)
www.springerlink.com

health sciences, and the natural sciences. In an effort to gain a better understanding of the use of fuzzy set theory in industrial engineering (IE) and to provide a basis for future research, a literature review of fuzzy set theory in IE has been conducted. Over the years there have been successful applications and implementations of fuzzy set theory in IE. Fuzzy set theory is being recognized as an important problem modeling and solution technique. The use of fuzzy set theory as a methodology for modeling and analyzing decision systems is of particular interest to researchers in IE due to fuzzy set theory's ability to quantitatively and qualitatively model problems which involve vagueness and imprecision. Fuzzy set theory can be used to bridge modeling gaps in descriptive and prescriptive decision models in IE.

There are many national and international associations and societies to support the development of the fuzzy set theory. The *International Fuzzy Systems Association* (IFSA) is a worldwide organization dedicated to the support, development and promotion of the fundamental issues of fuzzy theory related to (a) sets, (b) logics, (c) relations, (d) natural languages, (e) concept formation, (f) linguistic modeling, (g) vagueness, (h) information granularity, etc. and their applications to (1) systems modeling, (2) system analysis, (3) diagnosis, (4) prediction and (5) control in decision support systems in management and administration of human organizations as well as electro-mechanical systems in manufacturing and process industries. IFSA organizes and encourages participation in open discussion session on the future directions and restructuring of fuzzy theory in IFSA congresses. Furthermore, IFSA publishes the *International Journal of Fuzzy Sets and Systems* and sponsors workshops and conferences on fuzzy theory. Japan Society for Fuzzy Theory and Systems (SOFT) was established in 1989. SOFT have 1,670 individual members and 74 company members, publishes an official bimonthly journal and organizes fuzzy systems symposiums. There are 8 regional branches and 8 research groups in SOFT. The *North American Fuzzy Information Processing Society* (NAFIPS) was established in 1981 as the premier fuzzy society in North America. Its purpose is to help guide and encourage the development of fuzzy sets and related technologies for the benefit of mankind. *Berkeley Initiative in Soft Computing (BISC) Program* is the world-leading center for basic and applied research in soft computing. The principal constituents of soft computing are fuzzy logic, neural network theory and probabilistic reasoning, with the latter subsuming belief networks, evolutionary computing including DNA computing, chaos theory and parts of learning theory. Some of the most striking achievements of BISC Program are: fuzzy reasoning (set and logic), new soft computing algorithms making intelligent, semi-unsupervised use of large quantities of complex data, uncertainty analysis, perception-based decision analysis and decision support systems for risk analysis and management, computing with words, computational theory of perception, and precisiated natural language. *Spanish Association of Fuzzy Logic and Technologies* promotes and disseminates the methods, techniques and developments of Fuzzy Logic and Technologies; Establish relations with other national or international Associations with similar aims; Organizes seminars and round tables on Fuzzy Logic and Technologies. The European Society for Fuzzy Logic and Technology (EUSFLAT) was established in 1998. The main goal of EUSFLAT is to represent the European fuzzy community of IFSA. Hungarian Fuzzy Society was established in 1998. EUROFUSE Working Group on Fuzzy Sets of EURO was established in 1975. The purpose of EUROFUSE is to communicate and promote the knowledge of the theory of fuzzy sets and related areas and their applications. In 1985, the Group constituted itself as the European Chapter of IFSA. *EUropean Network on Intelligent TEchnologies for Smart Adaptive Systems* (EUNITE) has started on 1 January 2001. It is funded by the *Information Society Technologies Programme* (IST) within the European Union's Fifth RTD Framework Programme.

Some bibliographies related to the interest areas of IE are given as follows. Gaines and Kohout (1977), Kandel and Yager (1979), Kandel (1986), and Kaufmann and Gupta (1988) show the applications of fuzzy set theory in general. The bibliographies by Zimmerman (1983), Chen and Hwang (1992) and Lai and Hwang (1994) review the literature on fuzzy sets in operations research, fuzzy multiple attribute decision making (Fuzzy MADM) and fuzzy multiple objective decision-making (Fuzzy MODM) respectively. Maiers and Sherif (1985) review the literature on fuzzy industrial controllers and provide the applications of fuzzy set theory to many subject areas including decision making, economics, engineering and operations research. Karwowski and Evans (1986) identify the potential applications of fuzzy set theory to the following areas of IE: new product development, facilities location and layout, production scheduling and control, inventory management, quality and cost benefit analysis. In their book, Evans et al. (1989) combined contributions from world renowned experts on the various topics covered such as: traditional IE techniques; ergonomics; safety engineering; human performance measurement; man-machine systems and fuzzy methodologies. Klir and Yuan (1995) and Zimmermann (1996) reviewed the applications of fuzzy set theory and fuzzy logic.

In this paper, we review the literature and consolidate the main results on the application of fuzzy set theory to IE. The purpose of this paper is to: (i) review the literature; (ii) classify the literature based on the application of fuzzy set theory to IE; and, (iii) identify future research directions. This paper is organized as follows. Section 2 introduces a classification scheme for fuzzy research in IE. Section 3 reviews previous research on fuzzy set theory and IE. The conclusions to this study are given in Section 4.

2 Classification Scheme for Fuzzy Set Theory Applications in IE Research

Table 1 illustrates a classification scheme for the literature on the application of fuzzy set theory in IE. In this table, many categories are defined and the frequency of citations in each category is identified. Artificial Intelligence resulted in the largest number of citations (50), followed by Multiple Criteria Decision-making (42), and statistical decision-making (32) and manufacturing (22). This survey is restricted to research on the application of fuzzy sets to IE decision problems. In our study, a total of 473 citations on the application of fuzzy set theory in IE were found (see Table 1 and Table 2). All of these studies come from the departments of IE of the universities. Eight journals, *Fuzzy Sets and Systems, Information Sciences, Computers & Industrial Engineering, Computers & Mathematics with Applications, European Journal of Operational Research, International Journal of Production Economics,* and *International Journal of Approximate Reasoning* accounted for 70 percent of the citations. Table 3 gives the main books published on fuzzy applications of IE topics.

Table 1. Classification Scheme for Fuzzy Set Research in IE (1972 – 2005)

Research topic	Number of citations
Artificial Intelligence	50
Multiple Criteria Decision-making	42
Statistical decision-making	32
Manufacturing	22
Classification / Clustering	19
Optimization	19
General Modeling	18
Expert systems	15
Decision-making	14
Control	13
Ergonomics	12
Linear programming	12
Mathematical Programming	10
Engineering economics	9
Decision support system	8

Research topic	Number of citations
Group decision-making	8
Information Systems	7
Production management	6
Production Systems	6
Project management	5
Technology Management	4
Quality management	3
Forecasting	2
Simulation	2
Risk assessment	1
Others	134
Total	473

Table 2. Summary of Journal Citations on Fuzzy Set Theory in IE (1972 – 2005)

Journal Title	Frequency
Accident Analysis & Prevention	1
Advances in Engineering Software and Power Systems: A Fuzzy Network Model	1
Applied Mathematics and Computation	9
Applied Mathematics Letters	2
Applied Soft Computing	4
Applied Thermal Engineering	1
Artificial Intelligence In Medicine	2
Automatica	2
Computers & Electrical Engineering	1
Computers & Industrial Engineering	33
Computers & Mathematics with Applications	35
Computers & Operations Research	8
Computers in Industry	7
Conservation and Recycling	1
Control Engineering Practice	3
Decision Support Systems	3
Displays	1
Ecological Modelling	1
Energy Conversion and Management	1
Engineering Applications of Artificial Intelligence	7
Ergonomics	3
European Journal of Operational Research	37
Expert Systems with Applications	9

Journal Title	Frequency
Fuzzy Group Decision-Making for Facility Location Selection	1
Fuzzy prediction of maize breakage	1
Fuzzy Sets and Systems	177
Information and Control	1
Information Sciences	22
International Journal of Approximate Reasoning	11
International Journal of Human-Computer Studies	2
International Journal of Industrial Ergonomics	5
International Journal of Machine Tools and Manufacture	2
International Journal of Production Economics	14
International Journal of Project Management	2
International Transactions in Operational Research	1
Journal of Agricultural Engineering Research	1
Journal of Computational and Applied Mathematics	1
Journal of Environmental Management	1
Journal of Human Ergology	1
Journal of Materials Processing Technology	3
Journal of Mathematical Analysis and Applications	8
Mathematical and Computer Modelling	3
Mathematical Modelling	1
Mathematical Models Used for Adaptive Control of Machine Tools	1
Mathematics and Computers in Simulation	1
Mechatronics	3
Methods of Information In Medicine	1
Microelectronics and Reliability	2
Neural Networks	3
Neural Networks: The Official Journal of the International Neural Network Society	2
Neurocomputing	1
Nuclear Engineering and Design	1
Omega	3
Pattern Recognition	4
Pattern Recognition Letters	4
Reduction and Axiomization of Covering Generalized Rough Sets	1
Reliability Engineering & System Safety	1
Renewable Energy	1

Journal Title	Frequency
Robot Selection Decision Support System: A Fuzzy Set Approach	1
Robotics and Autonomous Systems	6
Robotics and Computer-Integrated Manufacturing	4
Technological Forecasting and Social Change	1
Trends in Pharmacological Sciences	1
Total	472

Table 3. Some books published on fuzzy IE

Authors	Contents
Bector and Chandra (2005)	This book presents a systematic and focused study of the application of fuzzy sets to two basic areas of decision theory, namely Mathematical Programming and Matrix Game Theory. Most of the theoretical results and associated algorithms are illustrated through small numerical examples from actual applications.
Dompere (2004)	This monograph is devoted to the development of value theory of computable general prices in cost-benefit analysis under fuzzy rationality. The book demonstrates the use of fuzzy decision algorithms and logic to develop a comprehensive and multidisciplinary cost-benefit analysis by taking advantage of current scientific gains in fuzziness and soft computing.
Aliev et al. (2004)	This monograph provides a self-contained exposition of the foundations of soft computing, and presents a vast compendium of its applications to business, finance, decision analysis and economics. The applications range from transportation and health case systems to intelligent stock market prediction, risk management systems, and e-commerce.
Onwubolu and Babu (2004)	The book describes a variety of the novel optimization techniques which in most cases outperform the standard optimization techniques in many application areas. New Optimization Techniques in Engineering reports applications and results of the novel optimization techniques considering a multitude of practical problems in the different engineering disciplines.

Authors	Contents
Niskanen (2004)	This book includes an in-depth look at soft computing methods and their applications in the human sciences, such as the social and the behavioral sciences. The powerful application areas of these methods in the human sciences are demonstrated.
Verdegay (2003)	The aim of this book is to show how Fuzzy Sets and Systems can help to provide robust and adaptive heuristic optimization algorithms in a variety of situations. The book presents the state of the art and gives a broad overview on the real practical applications that Fuzzy Sets have, based on heuristic algorithms.
Yu and Kacprzyk (2003)	This book provides a comprehensive coverage of up-to-date conceptual frameworks in broadly perceived decision support systems and successful applications. Different from other existing books, this book predominately focuses on applied decision support with soft computing. Areas covered include planning, management finance and administration in both the private and public sectors.
Castillo and Melin (2003)	The book describes the application of soft computing techniques and fractal theory to intelligent manufacturing. The text covers the basics of fuzzy logic, neural networks, genetic algorithms, simulated annealing, chaos and fractal theory. It also describes in detail different hybrid architectures for developing intelligent manufacturing systems for applications in automated quality control, process monitoring and diagnostics, adaptive control of non-linear plants, and time series prediction.
Buckley et al. (2002)	The book aims at surveying results in the application of fuzzy sets and fuzzy logic to economics and engineering. New results include fuzzy non-linear regression, fully fuzzified linear programming, fuzzy multi-period control, fuzzy network analysis, each using an evolutionary algorithm; fuzzy queuing decision analysis using possibility theory; fuzzy hierarchical analysis using an evolutionary algorithm. Other important topics covered are fuzzy input-output analysis; fuzzy mathematics of finance; fuzzy PERT (project evaluation and review technique).

Authors	Contents
Carlsson and Fuller (2002)	This book starts with the basic concepts of fuzzy arithmetics and progresses through fuzzy reasoning approaches to fuzzy optimization. Four applications of (interdependent) fuzzy optimization and fuzzy reasoning to strategic planning, project management with real options, strategic management and supply chain management are presented.
Yoshida (2001)	The book focuses on the recent dynamical development in fuzzy decision making. First, fuzzy dynamic programming is reviewed from a viewpoint of its origin and we consider its development in theory and applications. Next, the structure of dynamics in systems is considered in relation to fuzzy mathematical programming. Furthermore, some topics potentially related to dynamics are presented: financial management, fuzzy differential for fuzzy optimization of continuous-time fuzzy systems, and fuzzy ordering in multi-objective fuzzy systems.
Leiviskä (2001)	Applications of Soft Computing have recently increased and methodological development has been strong. The book is a collection of new interesting industrial applications introduced by several research groups and industrial partners. It describes the principles and results of industrial applications of Soft Computing methods and introduces new possibilities to gain technical and economic benefits by using this methodology.
Fodor et al. (2000)	In this book, leading scientists in the field address various theoretical and practical aspects related to the handling of the incompleteness. The problems discussed are taken from multi-objective linear programming, rationality considerations in preference modeling, non-probabilistic utility theory, data fusion, group decision making and multicriteria decision aid.
Yen et al. (1995)	It brings a functional understanding of the industrial applications of fuzzy logic in one self-contained volume. Application areas covered range from sensors, motors, robotics, and autonomous vehicles to process control and consumer products.

Authors	Contents
Klir and Yuan (1995)	This book both details the theoretical advances and considers a broad variety of applications of fuzzy sets and fuzzy logic. Part I is devoted to basics of fuzzy sets. Part II is devoted to applications of fuzzy set theory and fuzzy logic, including: the use of fuzzy logic for approximate reasoning in expert systems; fuzzy systems and controllers; fuzzy databases; fuzzy decision making; and engineering applications.
Evans et al. (1989)	Traditional IE techniques; ergonomics; safety engineering; human performance measurement; man-machine systems and fuzzy methodologies; locational analysis; job shop scheduling; financial planning; project management; data analysis; and fuzzy control of industrial processes; human factors modeling and organizational design, and, fuzzy set applications in operations research and decision analysis.

3 Fuzzy Set Theory and Industrial Engineering

Extensive work has been done on applying fuzzy set theory to research problems in IE. Using the classification scheme developed in Section 2, recent research findings in each area of IE will be reviewed.

3.1 Statistical Decision-Making

Statistics is a science of making decisions with respect to the characteristics of a group of persons or objects on the basis of numerical information obtained from a randomly selected sample of the group.

Provided with plenty of data (experience), data mining techniques are widely used to extract suitable management skills from the data. Nevertheless, in the early stages of a manufacturing system, only rare data can be obtained, and built scheduling knowledge is usually fragile. Using small data sets, Li *et al.* (2006) improve the accuracy of machine learning for flexible manufacturing system scheduling. They develop a data trend estimation technique and combine it with mega-fuzzification and adaptive-network-based fuzzy inference systems (ANFIS).

Clustering multimodal datasets can be problematic when a conventional algorithm such as k-means is applied due to its implicit assumption of Gaussian distribution of the dataset. Cho *et al.* (2006) propose a tandem clustering process for multimodal data sets. The proposed method first divides the multimodal dataset into many small pre-clusters by applying k-means or fuzzy k-means algorithm. These pre-clusters are then clustered again by agglomerative hierarchical clustering method using Kullback-Leibler divergence as an initial measure of dissimilarity.

Modarres et al. (2005) develop a mathematical programming model to estimate the parameters of fuzzy linear regression models with fuzzy output and crisp inputs. The method is constructed on the basis of minimizing the square of the total difference between observed and estimated spread values or in other words minimizing the least square errors.

Fuzzy relation is a crucial connector in presenting fuzzy time series model. However, how to obtain a fuzzy relation matrix to represent a time-invariant relation is still a question. Based on the concept of fuzziness in Information Theory, Tsaur et al. (2005) apply the concept of entropy to measure the degrees of fuzziness when a time-invariant relation matrix is derived.

Wang and Kuo (2004) propose a method of factor analysis for a huge database so that not only the independence among the factors can be ensured and the levels of their importance can be measured, but also new factors can be discovered. To measure the independence of the factors, statistical relation analysis and the concept of fuzzy set theory are employed. A fuzzy set of 'the factors are almost dependent' is used to measure the degree of dependence between factors, and then through a hierarchical clustering procedure, the dependent factors are detected and removed. To measure the weights of importance for the independent factors, a supervised feedforward neural network is developed. In addition, they design a hierarchical structure to facilitate the extraction of new factors when the information of the system is not complete.

3.2 Manufacturing

The area of manufacturing systems is one of the main themes in industrial engineering. Many interest areas of industrial engineering such as job shops, machine efficiency, line balancing, job shop design, etc. are researched using fuzzy sets under vagueness.

Hop (2006) addresses the mixed-model line balancing problem with fuzzy processing time. A fuzzy binary linear programming model is formulated for the problem. This fuzzy model is then transformed to a mixed zero-one program. Due to the complexity nature in handling fuzzy computation, new approximated fuzzy arithmetic operation is presented. A fuzzy heuristic is developed to solve this problem based on the aggregating fuzzy numbers and combined precedence constraints. The general idea of this approach is to arrange the jobs in a sequence by a varying-section exchange procedure. Then jobs are allocated into workstations based on these aggregated fuzzy times with the considerations of technological constraint and cycle time limit.

Gholamian and Ghomi (2005) try to stimulate empirical research into the overall impacts of intelligent systems in manufacturing aspects. To reach this goal, a schema of intelligent applications is provided for each aspect as frame base structure, meaning the knowledge of intelligent applications in that specific aspect. Then, a semantic network is developed for intelligent manufacturing based on hierarchical structure of manufacturing systems to provide Meta knowledge of intelligent manufacturing applications. The analysis of semantic network indicates an increasing growth in application of soft computing in manufacturing aspects. Specially, fuzzy logic and its derivates have found more applications in recent years.

Chang and Liao (2005) present a novel approach by combining self-organizing map and fuzzy rule base for flow time prediction in semiconductor manufacturing factory. Flow time of a new order is highly related to the shop floor status; however, the semiconductor manufacturing processes are highly complicated and involve more than hundred of production steps. There is no governing function identified so far among the flow time of a new order and these shop flow status. Therefore, a simulation model which mimics the production process of a real wafer fab located in Hsin-Chu Science-based Park of Taiwan is built and flow time and related shop floor status are collected and fed into the self-organizing map for classification. Then, corresponding fuzzy rule base is selected and applied for flow time prediction. Genetic process is further applied to fine-tune the composition of the rule base.

Kahraman et al. (2004) propose some fuzzy models based on fuzzy present worth to measure manufacturing flexibility. The fuzzy models based on present worth are basically engineering economics decision models in which the uncertain cash flows and discount rates are specified as triangular fuzzy numbers. To build such a model, fuzzy present worth formulas of the manufacturing flexibility elements are formed. Flexibility for continuous improvement, flexibility for trouble control, flexibility for work force control, and flexibility for work-in-process control are quantified by using fuzzy present worth analysis. Formulas for both inflationary and non-inflationary conditions are derived. Using these formulas, more reliable results can be obtained especially for such a concept like flexibility that is described in many intangible dimensions. These models allow experts' linguistic predicates about computer integrated manufacturing systems.

Chen et al. (2003) introduce a new tool path generation method to automatically subdivide a complex sculptured surface into a number of easy-to-machine surface patches; identify the favorable machining set-up/orientation for each patch; and generate effective 3-axis CNC tool paths for each patch.

3.3 Single Criterion and Multiple Criteria Decision-Making

Single criterion decision-making takes only one criterion into account such as profitability. Multiple criteria decision-making is divided into two components: MADM and MODM. MADM problems are widespread in real life decision situations. A MADM problem is to find a best compromise solution from all feasible alternatives assessed on multiple attributes, both quantitative and qualitative. In case of MODM, you have more than one objective function and some constraints. In the following, some recent works on this topic are summarized.

The advent of multiagent systems, a branch of distributed artificial intelligence, introduced a new approach to problem solving through agents interacting in the problem solving process. Chen and Wang (2005) address a collaborative framework of a distributed agent-based intelligence system to control and resolve dynamic scheduling problem of distributed projects for practical purposes. If any delay event occurs, the self-interested activity agent, the major agent for the problem solving of dynamic scheduling in the framework, can automatically cooperate with other agents in real time to solve the problem through a two-stage decision-making process: the fuzzy decision-making process and the compensatory negotiation process. The first stage determines which behavior strategy will be taken by agents while delay event occurs, and prepares to next negotiation process; then the compensatory negotiations among agents are opened related with determination of compensations for respective decisions and strategies, to solve dynamic scheduling problem in the second stage.

A MADM problem deals with candidate priority alternatives with respect to various attributes. MADM techniques are popularly used in diverse fields such as engineering, economics, management science, transportation planning, etc. Several approaches have been developed for assessing the weights of MADM problems, e.g., the eigenvector method, ELECTRE, TOPSIS, etc. Chen and Larbani (2006) obtained weights of a MADM problem with a fuzzy decision matrix by formulating it as a two-person zero-sum game with an uncertain payoff matrix. Moreover, the equilibrium solution and the resolution method for the MADM game is also developed; these results are validated by a product development example of nano-materials.

The information axiom proposes the selection of the proper alternative that has minimum information. Selection of the convenient equipment under determined criteria (such as costs and technical characteristics) using the information axiom is realized by Kulak et al. (2005). The unweighted and weighted multi-attribute axiomatic design approaches developed in this paper include both crisp and fuzzy criteria. The fuzzy set theory has the capability of capturing the uncertainty under the conditions of incomplete, non-obtainable and unquantifiable information. These approaches are applied to the selection among punching machines while investing in a manufacturing system.

Beneficiation of coal refers to the production of wash coal from raw coal with the help of some suitable beneficiation technologies. The processed coal is used by the different steel plants to serve their purpose during the manufacturing process of steel. Chakraborty and Chandra (2005) deal with the optimal planning for blending raw coal of different grades used for beneficiation with a view to satisfy the requirements of the end users with desired specifications. The input specifications of coal samples are known whereas the output specifications are imprecise in nature. The aim of the work is to fix the level of the raw coal from different coal seams to be fed for beneficiation to meet the desired target of yield and ash percentage to maximum extent. Further, it is also desired by the decision-maker to restrict the input cost of raw coal to be fed for beneficiation. The problem is modeled as multicriteria decision-making problem with imprecise specifications. Fuzzy set theoretic approach is used and a corresponding model is developed.

MADM involves tradeoffs among alternative performances over multiple attributes. The accuracy of performance measures are usually assumed to be accurate. Most multiattribute models also assume given values for the relative importance of weights for attributes. However, there is usually some uncertainty involved in both of these model inputs. Outranking multiattribute methods have always provided fuzzy input for performance scores. Many analysts have also recognized that weight estimates also involve some imprecision, either through individual decision maker uncertainty, or through aggregation of diverging group member preferences. Many fuzzy multiattribute models have been proposed, but they have focused on identifying the expected value solution (or extreme solutions). Olson and Wu (2005) demonstrate how simulation can be used to reflect fuzzy inputs, which allows more complete probabilistic interpretation of model results.

Araz et al. (2005) propose a multi-objective covering-based emergency vehicle location model. The objectives considered in the model are maximization of the population covered by one vehicle, maximization of the population with backup coverage and increasing the service level by minimizing the total travel distance from locations at a distance bigger than a prespecified distance standard for all zones. Model applications with different solution approaches such as lexicographic linear programming and fuzzy goal programming are provided through numerical illustrations to demonstrate the applicability of the model.

Cochran and Chen (2005) develop a fuzzy set approach for multi-criteria selection of object-oriented simulation software for analysis of production system. The approach uses fuzzy set theory and algebraic operations of fuzzy numbers to characterize simulation software so that the strength and weakness of each alternative can be compared. The linguistic input from experts and decision makers are fuzzified and the results of the fuzzy inference are translated back to linguistic explanation. Further, by aggregating decision maker's preference (weighting) to the expert's rating, a single number for each candidate software can be obtained. One can use this number as an index to choose the most suitable simulation software for the specific application.

3.4 Information Systems

Information system (IS) is a system for managing and processing information, usually computer-based. Also, it is a functional group within a business that manages the development and operations of the business's information. There are many works on IS using fuzzy set theory.

Wang et al. (2005) give an overview of applying fuzzy measures and relevant nonlinear integrals in data mining, discussed in five application areas: set function identification, nonlinear multiregression, nonlinear classification, networks, and fuzzy data analysis. In these areas, fuzzy measures allow us to describe interactions among feature attributes towards a certain target (objective attribute), while nonlinear integrals serve as aggregation tools to combine information from feature attributes. Values of fuzzy measures in these applications are unknown and are optimally determined via a soft computing technique based on given data.

Imprecise or vague importance as well as satisfaction levels of criteria may characterize the evaluation of alternative network security systems and may be treated by the fuzzy set theory. As fuzzy numbers are adopted, the fuzzy weighted averages (FWAs) may become a suitable computation for the fuzzy criteria ratings and fuzzy weightings. Chang and Hung (2005) propose to use the FWA approach that operates on the fuzzy numbers and obtains the fuzzy weighted averages during the evaluations of network security systems. The algorithm constructs a benchmark adjustment solution approach for FWA.

Maimon et al. (2001) present a novel, information-theoretic fuzzy approach to discovering unreliable data in a relational database. A multilevel information-theoretic connectionist network is constructed to evaluate activation functions of partially reliable database values. The degree of value reliability is defined as a fuzzy measure of difference between the maximum attribute activation and the actual value activation. Unreliable values can be removed from the database or corrected to the values predicted by the network. The method is applied to a real-world relational database which is extended to a fuzzy relational database by adding fuzzy attributes representing reliability degrees of crisp attributes. The highest connection weights in the network are translated into meaningful if, then rules. This work aims at improving reliability of data in a relational database by developing a framework for discovering, accessing and correcting lowly reliable data.

Demirli and Türkşen (2000) identify the robot's location in a global map from solely sonar based information. This is achieved by using fuzzy sets to model sonar data and by using fuzzy triangulation to identify robot's position and orientation. They obtain a fuzzy position region where each point in the region has a degree of certainty of being the actual position of the robot.

Tanaka et al. (1986) formulate a fuzzy linear programming problem and the value of information is discussed. If the fuzziness of coefficients is decreased by using information about the coefficients, they expect to obtain a more satisfactory solution than without information. With this view, the prior value of information is discussed. Using the sensitivity analysis technique, a method of allocating the given investigation cost to each fuzzy coefficient is proposed.

3.5 Expert Systems

Fuzzy expert systems have been successfully applied to a wide range of problems from different areas presenting uncertainty and vagueness in different ways. These kinds of systems constitute an extension of classical rule-based systems because they deal with fuzzy rules instead of classical logic rules.

Barelli and Bidini (2005) developed a diagnostic methodology for the optimization of energy efficiency and the maximization of the operational time based on artificial intelligence techniques such as artificial neural network (ANN) and fuzzy logic. The first part of the study deals with the identifying the principal modules and the corresponding variables necessary to evaluate the module "health state". Also the consequent upgrade of the monitoring system is described. Moreover it describes the structure proposed for the diagnostic procedure, consisting of a procedure for measurement validation and a fuzzy logic-based inference system. The first reveals the presence of abnormal conditions and localizes their source distinguishing between system failure and instrumentation malfunctions. The second provides an evaluation of module health state and the classification of the failures which have possibly occurred.

Wang and Dai (2004) propose a fuzzy constraint satisfaction approach to help buyers find fully satisfactory or replacement products in electronic shopping. For the buyer who can give precise product requirements, the proposed approach can generate product-ranking lists based on the satisfaction degrees of each product to the given requirements. For the buyer who may not input accurate requirements, a similarity analysis approach is proposed to assess buyer requirements automatically during his browsing process. The proposed approach could help buyers find the preferred products on the top of the ranking list without further searching the remaining pages.

Fang et al. (2004) propose a fuzzy set based formulation of auctions. They define fuzzy sets to represent the seller and buyers' valuations, bid possibilities and win possibilities. Analyzing the properties of these fuzzy sets, they study fuzzy versions of discriminating and nondiscriminating bidding strategies for each bidder in a single-object auction by using Bellman and Zadeh's concept of confluence of fuzzy decisions instead of the game theoretic Nash equilibrium. The monotonically decreasing membership function replaces of the corresponding distribution density function in Milgrom and Weber's auction theory. In the second part of the paper, they develop a soft computing approach to maximize the seller's revenue in multiple-object auctions through the use of object sequencing. The proposed sequencing approach is based on "fuzzy rule quantification".

Fonseca and Knapp (2001) develop a fuzzy reasoning algorithm and implement via an expert system to evaluate and assess the likelihood of equipment failure mode precipitation and aggravation. The scheme is based upon the fuzzification of the effects of precipitating factors provoking the failure. It consists of a fuzzy mathematical formulation which linearly relates the presence of factors catalogued as critical, important or related to the incidence of machine failure modes. This fuzzy algorithm is created to enable the inference mechanism of a constructed knowledge-based system to screen industrial equipment failures according to their likelihood of occurrence.

3.6 Artificial Intelligence

In order to model compex and sophisticated models artificial intelligence has the ability of combining several tools from several sources thereby, reducing the uncertainty. The most common known artificial intelligence techniques are ANN, fuzzy ANN, genetic algorithm and neuro-fuzzy approaches. Some of new studies on artificial intelligence are summarized in the following.

Palmero et al. (2005) introduce a system for fault detection and classification in AC motors based on soft computing. The kernel of the system is a neuro-fuzzy system, FasArt (Fuzzy Adaptive System ART-based), that permits the detection of a fault if it is in progress and its classification, with very low detection and diagnosis times that allow decisions to be made, avoiding definitive damage or failure when possible. The system is tested on an AC motor in which 15 nondestructive fault types were generated, achieving a high level of detection and classification. The knowledge stored in the neuro-fuzzy system is extracted by a fuzzy rule set with an acceptable degree of interpretability and without incoherency amongst the extracted rules.

Chang et al. (2005) present a fuzzy extension of the economic lot-size scheduling problem (ELSP) for fuzzy demands, since perturbations often occur in products demand in the real world. The ELSP is formulated via the extended basic period approach and power-of-two policy with the demands in triangular membership function. The resulting problem thus consists of a fuzzy total cost function, fuzzy feasibility constraints and a fuzzy continuous variable (the basic period) in addition to a set of crisp binary variables corresponding to the cycle times and starts of the products' schedule. Therefore, membership functions for the fuzzy total cost function and constraints can be figured out, from which the optimal fuzzy basic period and cycle times can be determined in addition to the compromised crisp values in fuzzy sense. Also, a genetic algorithm governed by the fuzzy total cost function and fuzzy feasibility constraints is designed and assists the ELSP in search for the optimal or near-optimal solution of the binary variables. This formulation is tested and illustrated on several ELSPs with varying levels of machine utilization and products demand perturbations. The results obtained are also analyzed with the lower bound generated by the independent solution approach to the ELSPs.

Gholamian et al. (2005) study an application of hybrid systematic design in multiobjective market problems. The target problem is suggested as unstructured real world problem such that the objectives cannot be expressed mathematically and only a set of historical data is utilized. Obviously, traditional methods and even meta-heuristic methods are broken in such cases. Instead, a systematic design using the hybrid of intelligent systems, particularly fuzzy rule base and neural networks can guide the decision maker towards noninferior solutions. The system does not stay in search phase. It also supports the decision maker in selection phase (after the search) to analyze various noninferior points and select the best ones based on the desired goal levels.

Kuo et al. (2005) propose a hybrid Case-Based Reasoning system with the integration of fuzzy sets theory and Ant System-based Clustering Algorithm (ASCA) in order to enhance the accuracy and speed in case matching. The cases in the case base are fuzzified in advance, and then grouped into several clusters by their own similarity with fuzzified ASCA. When a new case occurs, the system will find the closest group for the new case. Then the new case is matched using the fuzzy matching technique only by cases in the closest group. Through these two steps, if the number of cases is very large for the case base, the searching time will be dramatically saved. In the practical application, there is a diagnostic system for vehicle maintaining and repairing, and the results show a dramatic increase in searching efficiency.

Chen et al. (2004) develop an effective learning of neural network by using random fuzzy back-propagation learning algorithm. Based on this new learning algorithm, neural network not only has an accurate learning capability, but also can increase the probability of escaping from the local minimum while neural network is training. For demonstrating the new algorithm they develop has its outperformance, the classifications of the non-convex in two dimensions problem are simulated. For comparison, the same simulations by using conventional back-propagation learning algorithm with constant pairs of learning rate (α = 0.1-0.9) and momentum (x_i=0.1-0.9) and stochastic back propagation learning are also performed.

3.7 Engineering Economics

Engineering economics is the process involving mathematical analysis for selecting (or making a choice) from a set of alternatives based on clearly defined economic criteria. Many subtitles of engineering economics have been examined under fuzziness such as fuzzy capital budgeting techniques, fuzzy replacement analysis, etc.

Tolga et al. (2005) aim at creating an operating system selection framework for decision makers. Since decision makers have to consider both economic and non-economic aspects of technology selection, both factors have been considered in the developed framework. The economic part of the decision process is developed by Fuzzy Replacement Analysis. Non-economic factors and financial figures are combined using a fuzzy analytic hierarchy process (Fuzzy AHP) approach. Since there exists incomplete and vague information of future cash flows and the crisp AHP cannot reflect the human thinking style in capturing the expert's knowledge, the fuzzy sets theory is applied to both AHP and replacement analysis, which compares two operating systems with and without license.

Chang (2005) presents a fuzzy methodology for replacement of equipment. Issues such as fuzzy modeling of degradation parameters and determining fuzzy strategic replacement and economic lives are extensively discussed. For the strategic purpose, addible market and cost effects from the replacements against the counterpart, i.e., the existing equipment cost and market obsolescence, are modeled fuzzily and interactively, in addition to the equipment deterioration. Both the standard fuzzy arithmetic and here re-termed requisite-constraint vertex fuzzy arithmetic (and the vertex method) are applied and investigated.

Kahraman et al. (2002) develop capital budgeting techniques using discounted fuzzy versus probabilistic cash flows the formulas for the analyses of fuzzy present value, fuzzy equivalent uniform annual value, fuzzy future value, fuzzy benefit-cost ratio, and fuzzy payback period are developed and given some numeric examples. Then the examined cash flows are expanded to geometric and trigonometric cash flows and using these cash flows fuzzy present value, fuzzy future value, and fuzzy annual value formulas are developed for both discrete compounding and continuous compounding.

Kuchta (2000) propose fuzzy equivalents of all the classical capital budgeting (investment choice) methods. These equivalents can be used to evaluate and compare projects in which the cash flows, duration time and required rate of return (cost of capital) are given imprecisely, in the form of a fuzzy number.

The application of discounted cash flow techniques for justifying manufacturing technologies is studied in many papers. State-price net present value and stochastic net present value are two examples of these applications. These applications are based on the data under certainty or risk. When we have vague data such as interest rate and cash flow to apply discounted cash flow techniques, the fuzzy set theory can be used to handle this vagueness. The fuzzy set theory has the capability of representing vague data and allows mathematical operators and programming to apply to the fuzzy domain. The theory is primarily concerned with quantifying the vagueness in human thoughts and perceptions. Assuming that they have vague data, Kahraman et al. (2000) use the fuzzy benefit-cost (B/C) ratio method to justify manufacturing technologies. After calculating the B/C ratio based on fuzzy equivalent uniform annual value, they compare two assembly manufacturing systems having different life cycles.

3.8 Ergonomics

Ergonomics is the science of adapting products and processes to human characteristics and capabilities in order to improve people's well-being and optimize productivity. The main topics of ergonomics in IE include anthropometry, human machine interaction and physical and mental health of workers. These topics are also studied by using the fuzzy logic techniques.

Kaya et al. (2004) take eighteen anthropometric measurements in standing and sitting positions, from 387 subjects between 15 and 17 years old. ANFIS is used to estimate anthropometric measurements as an alternative to stepwise regression analysis. Six outputs (shoulder width, hip width, knee height, buttock-popliteal height, popliteal height, and height) are selected for estimation purpose. The results show that the number of inputs required estimating outputs varied with sex difference. ANFIS perform better than stepwise regression method for both sex groups. Relevance to industry, more recently, as industry and marketing reach around the globe, body size has become a matter of practical interest to designers and engineers. In ergonomics anthropometric data are widely used to specify the physical dimension of workspaces, equipment, furniture and clothing. This is especially true for school children, which spend most of their time sitting at their chairs and desks and ought to be able to adopt comfortable body postures.

Park and Han (2004) propose a fuzzy rule-based approach to building models relating product design variables to affective user satisfaction. Affective user satisfaction such as luxuriousness, balance, and attractiveness are modeled for office chair designs. Regression models are also built on the same data to compare model performance. The results show that fuzzy rule-based models are better than regression models in terms of prediction performance and the number of variables included in the model. Methods for interpreting the fuzzy rules are discussed for practical applications in designing office chairs.

Bell and Crumpton (1997) present the development and evaluation of a fuzzy linguistic model designated to predict the risk of carpal tunnel syndrome (CTS) in an occupational setting. CTS is of the largest problems facing ergonomists and the medical community because it is developing in epidemic proportions within the occupational environment. In addition, practitioners are interested in identifying accurate methods for evaluating the risk of CTS in an occupational setting. It is hypothesized that many factors impact an individual's likelihood of developing CTS and the eventual development of CTS. This disparity in the occurrence of CTS for workers with similar backgrounds and work activities has confused researchers and has been a stumbling block in the development of a model for widespread use in evaluating the development of CTS. Thus this research is an attempt to develop a method that can be used to predict the likelihood of CTS risk in a variety of environments.

Work generally affects workers in terms of both physical and mental health. Workers must adapt their life pattern to match shift-work styles which can result in family problems, increased fatigue level, lower work efficiency, higher accident rate, illnesses, and lower productivity. Srithongchai and Intaranont (1996) obtain an entry permission to study the impact of shift work on fatigue level of workers in a sanitary-ware factory. The objectives are: 1) to evaluate fatigue levels of workers who work in the morning shift and night shift, and 2) to prioritize contributing factors affecting fatigue levels using a model of fuzzy set theory. Twelve male workers participate in the study. Four subjects are recruited from each of 3 departments, i.e., glazing, baking and quality inspection. The measurement is conducted before and after the shift for both shifts. Variables include heart rate monitoring throughout the work shift, critical flicker fusion frequency, reaction time response, hand-grip strength, and wet-bulb globe temperature. Results are analyzed using a computerized statistical package. It is concluded that mental fatigue from working in the morning shift is significantly higher than the one for working in the night shift. The same indication is also true in the case of physical fatigue, though it is not statistically significant. From the fuzzy set analysis, it is confirmed that working in the morning shift result in a higher fatigue level than working in the night shift and the temperature of the work environment is the most important factor contributing to the higher fatigue level.

Drastic changes in the nature of industrial work force, from manufacturing related industries to service and information industries, and increased automation have led to abundance of jobs for which there are no established performance standards. While work performance standards exist for physical tasks that are routinely performed in manufacturing industries, limits on acceptable levels of work are yet to be developed for mental tasks which are fast becoming common. Mital and Karwowski (1986) describe an approach to quantify qualitative human responses and describes a procedure to integrate them with other quantitative responses resulting from physical, mental, behavioral, or psycho-social factors. The theory of fuzzy sets and systems is utilized in developing the modeling concept which can be used in developing work-performance standards or optimizing overall man-machine systems effectiveness.

3.9 Production Planning and Control

Production systems form the base for building and improving the economic strength and vitality of a country. In order to manage production systems, production planning and control (PPC) techniques in various dimensions such as demand forecasting, aggregate planning, inventory management, material planning and operations scheduling, can be employed. In the literature very high number of papers on fuzzy PPC has been published. A summary of the direction of research on fuzzy production planning and control is found in the following.

Kasperski (2005) proposes a possibilistic approach to sequencing. For each parameter, whose value is not precisely known, a possibility distribution is given. The objective is to calculate a sequence of jobs, for which the possibility (necessity) of delays of jobs is minimal.

Chanas and Kasperski (2004) consider the single machine scheduling problem with parameters given in the form of fuzzy numbers is considered. It is assumed that the optimal schedule in such a problem cannot be determined precisely. They introduce the concepts of possible and necessary optimality of a given schedule. The degrees of possible and necessary optimality measure the possibility and the necessity of the event that a given schedule will be optimal. It is shown how to calculate the degrees of possible and necessary optimality of a given schedule in one of the special cases of the single machine scheduling problems.

Operation of complex shops needs specific, sophisticated procedures in order to guarantee competitive plant performance. Adenso-Díaz et al. (2004) present a hierarchy of models for roll shop departments in the steel industry, focusing on the calculation of the priority of the rolls to produce. A fuzzy-based model is developed and implemented in a real environment, allowing the simulation of expert behavior, considering the characteristics of an environment with imprecise information.

Majority of the products can be assembled in several ways that means the same final product can be realized by different sequences of assembly operations. Different degree of difficulty is associated with each sequence of assembly operation and such difficulties are caused by the different mechanical constraints forced by the different sequences of operations. In the past, few notable attempts have been made to represent and enumerate the degree of difficulty associated with an assembly sequence (in the form of triangular fuzzy number) by using the concept of assembly graph. However, such representation schemes do not possess the capabilities to model the user's reasoning and preferences. Ben-Arieh et al. (2004) present an intelligent Petri net model that combines the abilities of modeling, planning and performance evaluation for assembly operation. This modeling tool can represent the issues concerning degree of difficulty associated with assembly sequences. The proposed mechanism is enhanced expert high-level colored fuzzy Petri net that is a hybrid of knowledge-based system and colored Petri net.

Most of the studies on capacitated lot size problem are based on the assumption that the capacity is known exactly. However, in most practical applications, this is seldom the case. Fuzzy number theory is ideally suited to represent this vague and uncertain future capacity. Pai (2003) applies the fuzzy sets theory to solve this capacitated lot size problem.

Samanta and Al-Araimi (2001) propose a model based on fuzzy logic for inventory control. The periodic review model of inventory control with variable order quantity is considered. The model takes into account the dynamics of production-inventory system in a control theoretic approach. The control module combines fuzzy logic with proportional-integral-derivative (PID) control algorithm. It simulates the decision support system to maintain the inventory of the finished product at the desired level in spite of variations in demand.

Singh and Mohanty (1991) characterize the manufacturing process plan selection problem as a machine routing problem. The routing problem is formulated as a multiple objective network model. Each objective is defined by a fuzzy membership function as a means of capturing the imprecision that exists when defining objectives. A dynamic programming solution procedure identifies the network path representing the best process plan. A dual-objective example is demonstrated for a component part requiring three machining operations. Two of the machining operations can be performed at alternative machining centers, resulting in a network model consisting of six nodes and eight branches. Cost and processing time per component are each represented by triangular fuzzy numbers. Zhang and Huang (1994) use fuzzy logic to model the process plan selection problem when objectives are imprecise and conflicting. Fuzzy membership functions are used to evaluate the contribution of competing process plans to shopfloor performance objectives. The optimal process plan for each part is determined by the solution of a fuzzy integer programming model. A consolidation procedure, which uses a dissimilarity criterion, then selects the process plan that best utilizes manufacturing resources. The algorithm is demonstrated for a problem consisting of three parts and eight process plans. The algorithm was also tested against non-fuzzy algorithms found in the literature. In some circumstances, more reasonable solutions where achieved as a result of the algorithm's ability to deal with the fuzziness inherent in manufacturing process planning. Inuiguchi *et al.* (1994) compare possibilistic, flexible and goal programming approaches to solving a production planning problem. Unlike conventional methods, possibilistic programming allows ambiguous data and objectives to be included in the problem formulation. A production planning problem consisting of two manufacturing processes, two products and four structural constraints is considered. The problem is solved using possibilistic programming, flexible programming and goal programming. A comparison of the three solutions suggests that the possibilistic solution best reflects the decision maker's input, thereby emphasizing the importance of modeling ambiguity in production planning.

Lee *et al.* (1991) extend their 1990 treatment of the MRP lot-sizing to include fuzzy modifications to the Silver-Meal, Wagner-Whitin, and part-period balancing algorithms. The authors argue that when demands of the master schedule are truly fuzzy, demand should be modeled using membership functions. The performance of the three fuzzy lot-sizing algorithms is compared based on nine sample problems. Fuzzy set theory has been applied to problems in inventory management and production and process plan selection. The appeal of using fuzzy set theory in these IE problems echoes that of aggregate planning. Inventory management requires demand forecasts as well as parameters for inventory related costs such as carrying, replenishment, shortages and backorders. Precise estimates of each of these model attributes are often difficult. Similarly, in production and process plan selection problems imprecision exists in specifying demand forecasts, inventory and processing cost parameters, processing times, and routing preferences. Potential ambiguity is further increased when the problem is formulated with multiple objectives. The research studies reviewed in this section demonstrate the usefulness of fuzzy set theory in modeling and solving inventory, and production and process plan selection problems when data and objectives are subject to ambiguity.

3.10 Aggregate Planning

Aggregate planning, sometimes called intermediate-range planning, involves production planning activities for six months to two year with monthly or quarterly updates. Changes in the workforce, additional machines, subcontracting, and overtime are typical decisions in aggregate planning.

Wang and Liang (2004) develop a fuzzy multi-objective linear programming (FMOLP) model for solving the multi-product aggregate production planning (APP) decision problem in a fuzzy environment. The proposed model attempts to minimize total production costs, carrying and backordering costs and rates of changes in labor levels considering inventory level, labor levels, capacity, warehouse space and the time value of money.

Wang and Fang (2001) present a novel fuzzy linear programming method for solving the aggregate production planning problem with multiple objectives where the product price, unit cost to subcontract, work force level, production capacity and market demands are fuzzy in nature. An interactive solution procedure is developed to provide a compromise solution.

Tang et al. (2000) focus on a novel approach to modeling multi-product APP problems with fuzzy demands and fuzzy capacities, considering that the demand requirements are fuzzy demand in each period during the planning horizon, The objective of the problem considered is to minimize the total costs of quadratic production costs and linear inventory holding costs. By means of formulation of fuzzy demand, fuzzy addition and fuzzy equation, the production inventory balance equation in single stage and dynamic balance equation are formulated as soft equations in terms of a degree of truth, and interpreted as the levels of satisfaction with production and inventory plan in meeting market demands. As a result, the multi-product aggregate production planning problem with fuzzy demands and fuzzy capacities can be modeled into a fuzzy quadratic programming with fuzzy objective and fuzzy constraints.

Guiffrida and Nagi (1998) made a literature review on aggregate planning. Rinks (1981) cites a gap between aggregate planning theory and practice. Managers prefer to use their own heuristic decision rules over mathematical aggregate planning models. Using fuzzy conditional "if-then" statements, Rinks develops algorithms for fuzzy aggregate planning. The strengths of the fuzzy aggregate planning model over traditional mathematical aggregate planning models include its ability to capture the approximate reasoning capabilities of managers, and the ease of formulation and implementation. The robustness of the fuzzy aggregate planning model under varying cost structures is examined in Rinks (1982a). A detailed set of forty production rate and work force rules is found in Rinks (1982b). Turksen (1988a, 1988b) advocates using interval-valued membership functions over the point-valued membership functions found in Rinks (1981, 1982a, 1982b) when defining linguistic production rules for aggregate planning. Ward *et al.* (1992) develop a C language program based on Rinks' fuzzy aggregate planning framework. Gen *et al.* (1992) present a fuzzy multiple objective aggregate planning model. The model is formulated as a fuzzy multiple objective programming model with objective function coefficients, technological coefficients, and resource right-hand side values, represented by triangular fuzzy numbers. A transformation procedure is presented to transform the fuzzy multiple objective aggregate planning model into a crisp model. Aggregate planning involves the simultaneous determination of a company's production, inventory, and workforce levels over a finite planning horizon such that total relevant cost is minimized.

Many aspects of the aggregate planning problem and the solution procedures employed to solve aggregate planning problems lend themselves to the fuzzy set theory approach. Fuzzy aggregate planning allows the vagueness that exists in the determining forecasted demand and the parameters associated with carrying charges, backorder costs, and lost sales to be included in the problem formulation. Fuzzy linguistic "if-then" statements may be incorporated into the aggregate planning decision rules as means for introducing the judgment and past experience of the decision maker into the problem. In this fashion, fuzzy set theory increases the model realism and enhances the implementation of aggregate planning models in industry. The usefulness of fuzzy set theory also extends to multiple objective aggregate planning models where additional imprecision due to conflicting goals may enter into the problem.

3.11 Quality Management

Research on fuzzy quality management is broken down into three areas, acceptance sampling, statistical process control, and general quality management topics. Quality management is a set of actions of the general management function which determines the quality policy, aims and responsibilities and realizes them within the framework of quality by planning, control, ensuring and improving the quality. An overview of research on fuzzy quality management is found in the following:

3.11.1 Acceptance Sampling

Ohta and Ichihashi (1988) present a fuzzy design methodology for single stage, two-point attribute sampling plans. An algorithm is presented and example sampling plans are generated when producer's and consumer's risk are defined by triangular fuzzy numbers. The authors do not address how to derive the membership functions for consumer's and producer's risk. Chakraborty (1988, 1994a) examines the problem of determining the sample size and critical value of a single sample attribute sampling plan when imprecision exists in the declaration of producer's and consumer's risk. In the 1988 paper, a fuzzy goal programming model and solution procedure are described. Several numerical examples are provided and the sensitivity of the strength of the resulting sampling plans is evaluated. The 1994a paper details how possibility theory and triangular fuzzy numbers are used in the single sample plan design problem.

Problems in the design of sampling inspection plans have been studied for a long time as an important subject in quality control. Especially, in the case of a single sampling attribute plan, one would like the producer's risk to equal α and the consumer's risk to equal β exactly, but this is usually not possible since the sample size n and the acceptance number c must be integers. Kanagawa and Ohta (1990) present a new design procedure for the single sampling attribute plan based on fuzzy sets theory by means of formulating this problem as a fuzzy mathematical programming one. Chakraborty (1992, 1994b) addresses the problem of designing single stage, Dodge-Romig lot tolerance percent defective (LTPD) sampling plans when the lot tolerance percent defective, consumer's risk and incoming quality level are modeled using triangular fuzzy numbers. In the Dodge-Romig scheme, the design of an optimal LTPD sample plan involves solution to a nonlinear integer programming problem. The objective is to minimize average total inspection subject to a constraint based on the lot tolerance percent defective and the level of consumer's risk. When fuzzy parameters are introduced, the procedure becomes a possibilistic (fuzzy) programming problem. A solution algorithm employing alpha-cuts is used to design a compromise LTPD plan, and a sensitivity analysis is conducted on the fuzzy parameters used.

3.11.2 Statistical Process Control

Cheng (2005) presents the construction of fuzzy control charts for a process with fuzzy outcomes derived from the subjective quality ratings provided by a group of experts. The proposed fuzzy process control methodology comprises an off-line stage and an on-line stage. In the off-line stage, experts assign quality ratings to products based on a numerical scale. The individual numerical ratings are then aggregated to form collective opinions expressed in the form of fuzzy numbers. The collective knowledge applied by the experts when conducting the quality rating process is acquired through a process of fuzzy regression analysis performed by a neural network. In the on-line stage, the product dimensions are measured, and the fuzzy regression model is employed to automate the experts' judgments by mapping the measured dimensions to appropriate fuzzy quality ratings. The fuzzy quality ratings are then plotted on fuzzy control charts, whose construction and out-of-control conditions are developed using possibility theory. The developed control charts not only monitor the central tendency of the process, but also indicate its degree of fuzziness.

Gülbay et al.'s (2004) approach provides the ability of determining the tightness of the inspection by selecting a suitable α-level. The higher α, the tighter inspection. The article also presents a numerical example and interprets and compares other results with the approaches developed previously.

Wang and Chen (1995) present a fuzzy mathematical programming model and solution heuristic for the economic design of statistical control charts. The economic statistical design of an attribute np-chart is studied under the objective of minimizing the expected lost cost per hour of operation subject to satisfying constraints on the Type I and Type II errors. The authors argue that under the assumptions of the economic statistical model, the fuzzy set theory procedure presented improves the economic design of control charts by allowing more flexibility in the modeling of the imprecision that exist when satisfying Type I and Type II error constraints.

Wang and Raz (1990) illustrate two approaches for constructing variable control charts based on linguistic data. When product quality can be classified using terms such as 'perfect', 'good', 'poor', etc., membership functions can be used to quantify the linguistic quality descriptions. Representative (scalar) values for the fuzzy measures may be found using the fuzzy mode or the alpha-level fuzzy midrange or the fuzzy median or the fuzzy average. The representative values that result from any of these methods are then used to construct the control limits of the control chart. Raz and Wang (1990) present a continuation of previous work on the construction of control charts for linguistic data. Results based on simulated data suggest that, on the basis of sensitivity to process shifts, control charts for linguistic data outperform conventional percentage defective charts. The number of linguistic terms used to represent the observation is found to influence the sensitivity of the control chart. Kanagawa *et al.* (1993) develop control charts for linguistic variables based on probability density functions which exist behind the linguistic data in order to control process average and process variability. This approach differs from the procedure of Wang and Raz in that the control charts are targeted at directly controlling the underlying probability distributions of the linguistic data.

Bradshaw (1983) uses fuzzy set theory as a basis for interpreting the representation of a graded degree of product conformance with a quality standard. When the costs resulting from substandard quality are related to the extent of nonconformance, a compatibility function exists which describes the grade of nonconformance associated with any given value of that quality characteristic. This compatibility function can then be used to construct fuzzy economic control charts on an acceptance control chart. The author stresses that fuzzy economic control chart limits are advantageous over traditional acceptance charts in that fuzzy economic control charts provide information on the severity as well as the frequency of product nonconformance.

3.11.3 General Topics in Quality Management

In view of the compatibility between psychology and linguistic terms, Tsai and Lu (2006) generalize the standard Choquet integral, whose measurable evidence and fuzzy measures are real numbers. The proposed generalization can deal with fuzzy-number types of measurable evidence and fuzzy measures. In comparison with the previous research on service quality and fuzzy sets, this study integrates the three-column format SERVQUAL into the generalized Choquet integral. A numerical example evaluating the overall service quality of e-stores is provided to illustrate how the generalized Choquet integral evaluates service quality in the three-column format SERVQUAL. This integral is also compared with the possibility/necessity model when measuring service quality.

Chien and Tsai (2000) propose a new method of measuring perceived service quality based on triangular fuzzy numbers. It avoids using difference scores (perceptions minus expectations) which are applied by many marketing researchers but are criticized by some other researchers. From consumers' standpoints, they replace perceptions by satisfaction degree as well as expectations by importance degree. To evaluate the discrepancy between consumers' satisfaction degree and importance degree, they induce general solutions to compute the intersection area between two triangular fuzzy numbers, then the weak and/or strong attributes of retail stores are clarified.

Kim et al. (2000) present an integrated formulation and solution approach to Quality Function Deployment (QFD). Various models are developed by defining the major model components (namely, system parameters, objectives, and constraints) in a crisp or fuzzy way using multiattribute value theory combined with fuzzy regression and fuzzy optimization theory. The proposed approach would allow a design team to reconcile tradeoffs among the various performance characteristics representing customer satisfaction as well as the inherent fuzziness in the system. In addition, the modeling approach presented makes it possible to assess separately the effects of possibility and flexibility inherent or permitted in the design process on the overall design. Knowledge of the impact of the possibility and flexibility on customer satisfaction can also serve as a guideline for acquiring additional information to reduce fuzziness in the system parameters as well as determine how much flexibility is warranted or possible to improve a design.

Khoo and Ho (1996) present a framework for a fuzzy quality function deployment (FQFD) system in which the 'voice of the customer' can be expressed as both linguistic and crisp variables. The FQFD system is used to facilitate the documentation process and consists of four modules (planning, deployment, quality control, and operation) and five supporting databases linked via a coordinating control mechanism. The FQFD system is demonstrated for determining the basic design requirements of a flexible manufacturing system. Glushkovsky and Florescu (1996) describe how fuzzy set theory can be applied to quality improvement tools when linguistic data is available. The authors identify three general steps for formalizing linguistic quality characteristics: (i) universal set choosing; (ii) definition and adequate formalization of terms; and (iii) relevant linguistic description of the observation. Examples of the application of fuzzy set theory using linguistic characteristics to Pareto analysis, cause-and-effect diagrams, design of experiments, statistical control charts, and process capability studies are demonstrated. Gutierrez and Carmona (1995) note that decisions regarding quality are inherently ambiguous and must be resolved based on multiple criteria. Hence, fuzzy multicriteria decision theory provides a suitable framework for modeling quality decisions. The authors demonstrate the fuzzy multiple criteria framework in an automobile manufacturing example consisting of five decision alternatives (purchasing new machinery, workforce training, preventative maintenance, supplier quality, and inspection) and four evaluation criteria (reduction of total cost, flexibility, leadtime, and cost of quality). Yongting (1996) identifies that failure to deal with quality as a fuzzy concept is a fundamental shortcoming of traditional quality management. Ambiguity in customers' understanding of standards, the need for multicriteria appraisal, and the psychological aspects of quality in the mind of the customer, support the modeling of quality using fuzzy set theory. A procedure for fuzzy process capability analysis is defined and is illustrated using an example. The application of fuzzy set theory in acceptance sampling, statistical process control and quality topics such as quality improvement and QFD has been reviewed in this section. Each of these areas requires a measure of quality. Quality, by its very nature, is inherently subjective and may lead to a multiplicity of meanings since it is highly dependent on human cognition. Thus, it may be appropriate to consider quality in terms of grades of conformance as opposed to absolute conformance or nonconformance. Fuzzy set theory supports subjective natural language descriptors of quality and provides a methodology for allowing them to enter into the modeling process. This capability may prove to be extremely beneficial in the further development of quality function deployment, process improvement tools and statistical process control.

3.12 Project Scheduling

Project scheduling is the task of planning timetables and the establishment of dates during which resources such as equipment and personnel, will perform the activities required to complete the project. In the following some recent works on fuzzy project scheduling are given.

Kim et al.'s (2003) develop a hybrid genetic algorithm (hGA) with fuzzy logic controller (FLC) to solve the resource-constrained project scheduling problem (rcPSP) which is a well-known NP-hard problem. Their new approach is based on the design of genetic operators with FLC and the initialization with the serial method, which has been shown superior for large-scale rcPSP problems. For solving these rcPSP problems, they firstly demonstrate that their hGA with FLC (flc-hGA) yields better results than several heuristic procedures presented in the literature. Then they evaluate several genetic operators, which include compounded partially mapped crossover, position-based crossover, swap mutation, and local search-based mutation in order to construct the flc-hGA which has the better optimal makespan and several alternative schedules with optimal makespan.

Wang (2002) develop a fuzzy scheduling methodology to deal with this problem. Possibility theory is used to model the uncertain and flexible temporal information. The concept of schedule risk is proposed to evaluate the schedule performance. A fuzzy beam search algorithm is developed to determine a schedule with the minimum schedule risk and the start time of each activity is selected to maximize the minimum satisfaction degrees of all temporal constraints. In addition, the properties of schedule risk are also discussed. They show that the proposed methodology can assist project managers in selecting a schedule with the least possibility of being late in an uncertain scheduling environment.

Hapke and Slowinski (1996) present a generalization of the known priority heuristic method for solving resource-constrained project scheduling problems (RCPS) with uncertain time parameters. The generalization consists of handling fuzzy time parameters instead of crisp ones. In order to create priority lists, a fuzzy ordering procedure has been proposed. The serial and parallel scheduling procedures which usually operate on these lists have also been extended to handle fuzzy time parameters.

In the following Guiffrida and Nagi's (1998) literature survey on project scheduling is given. Notice that the majority of the research on this topic has been devoted to fuzzy PERT. Prade (1979) applies fuzzy set theory to the development of an academic quarter schedule at a French school. When data are not precisely known, fuzzy set theory is shown to be relevant to the exact nature of the problem rather than probabilistic PERT or

CPM. The aim of this work is to show how and when it is possible to use fuzzy concepts in a real world scheduling problem. An overview of a fuzzy modification to the classic Ford solution algorithm is presented along with a 17 node network representation for the academic scheduling problem. Calculations are demonstrated for a small portion of the overall scheduling problem. Chanas and Kamburowski (1981) argue the need for an improved version of PERT due to three circumstances: (i) the subjectivities of activity time estimates; (ii) the lack of repeatability in activity duration times; and (iii) calculation difficulties associated with using probabilistic methods. A fuzzy version of PERT (FPERT) is presented in which activity times are represented by triangular fuzzy numbers. Kaufmann and Gupta (1988) devote a chapter of their book to the critical path method in which activity times are represented by triangular fuzzy numbers. A six step procedure is summarized for developing activity estimates, determining activity float times, and identifying the critical path. A similar tutorial on fuzzy PERT involving trapezoidal fuzzy numbers may be found in Dubois and Prade (1985). McCahon and Lee (1988) note that PERT is best suited for project network applications when past experience exists to allow the adoption of the beta distribution for activity duration times and when the network contains approximately 30 or more activities. When activity times are vague, the project network should be modeled with fuzzy components. A detailed example demonstrates modeling and solving an eight activity project network when activity durations are represented as triangular fuzzy numbers. Lootsma (1989) identifies that human judgment plays a dominant role in PERT due to the estimation of activity durations and the requirement that the resulting plan be tight. This aspect of PERT exposes the conflict between normative and descriptive modeling approaches. Lootsma argues that vagueness is not properly modeled by probability theory, and rejects the use of stochastic models in PERT planning when activity durations are estimated by human experts. Despite some limitations inherent in the theory of fuzzy sets, fuzzy PERT, in many respects, is closer to reality and more workable than stochastic PERT. Buckley (1989) provides detailed definitions of the possibility distributions and solution algorithm required for using fuzzy PERT. A ten activity project network example in which activity durations are described by triangular fuzzy numbers, is used to demonstrate the development of the possibility distribution for the project duration. Possibility distributions for float, earliest start, and latest start times are defined, but not determined, due to their complexity. DePorter and Ellis (1990) present a project crashing model using fuzzy linear programming. Minimizing project completion time and project cost are highly sought yet conflicting project objectives. Linear programming allows the optimization of one objective (cost or time). Goal programming allows consideration of both time and cost objectives in the optimization scheme. When environmental factors present additional vagueness, fuzzy linear programming should be used. Linear programming, goal programming and fuzzy linear programming are applied to a ten activity project network. Project crashing costs and project durations are determined under each solution technique. McCahon (1993) compares the performance of fuzzy project network analysis (FPNA) and PERT. Four basic network configurations were used. The size of the networks ranged from four to eight activities. Based on these networks, a total of thirty-two path completion times were calculated using FPNA and PERT. The performance of FPNA and PERT was compared using: the expected project completion time, the identification of critical activities, the amount of activity slack, and the possibility of meeting a specified project completion time. The results ofthis study conclude that PERT estimates FPNA adequately. When estimating expected project completion time however, a generalization concerning compared performance with respect to the set of critical activities, slack times and possibility of project completion times cannot be made. When activity times are poorly defined, the performance of FPNA outweighs its cumbersomeness and should be used instead of PERT. Nasution (1994) argues that for a given alpha-cut level of the slack, the availability of the fuzzy slack in critical path models provides sufficient information to determine the critical path. A fuzzy procedure utilizing interactive fuzzy subtraction is used to compute the latest allowable time and slack for activities. The procedure is demonstrated for a ten event network where activity times are represented by trapezoidal fuzzy numbers. Hapke (1994) present a fuzzy project scheduling (FPS) decision support system. The FPS system is used to allocate resources among dependent activities in a software project scheduling environment. The FPS system uses L-R type flat fuzzy numbers to model uncertain activity durations. Expected project completion time and maximum lateness are identified as the project performance measures and a sample problem is demonstrated for a software engineering project involving 53 activities. The FPS system presented allows the estimation of Project completion times and the ability to analyze the risk associated with overstepping the required project completion time. Lorterapong (1994) introduces a resource-constrained project scheduling method that addresses three performance objectives: (i) expected project completion time; (ii) resource utilization; and (iii) resource interruption Fuzzy set theory is used to model the vagueness that is inherent with linguistic descriptions often used by people when describing activity durations. The analysis presented provides a framework for allocating resources in an uncertain project environment. Chang *et al.* (1995) combine the composite and comparison methods of analyzing fuzzy numbers into an efficient procedure for solving project scheduling problems. The comparison method first eliminates activities that are not on highly critical paths. The composite method then determines the path with the highest degree of criticality. The fuzzy Delphi method (see Kaufmann and Gupta (1988)) is used to determine the activity time estimates. The solution procedure is demonstrated in a 9 node, 14 activity project scheduling problem with activity times represented by triangular fuzzy numbers. Shipley *et al.* (1996) incorporate fuzzy logic, belief functions, extension principles and fuzzy probability distributions, and developed the fuzzy PERT algorithm, 'Belief in Fuzzy Probabilities of Estimate Time' (BIFPET). The algorithm is applied to a real world project consisting of eight activities involved in the selling and producing of a 30-second television commercial. Triangular fuzzy numbers are used to define activity durations. BIFPET is used to determine the project critical path and expected project completion time. The specification of activity duration times is crucial to both CPM and PERT project management applications. In CPM, historical data on the duration of activities in exact or very similar projects exists, and it is used to specify activity durations for similar future projects. In new projects where no historical data on activity durations exists, PERT is often used. Probabilistic-based PERT requires the specification of probability distribution (frequently the beta distribution) to represent activity durations. Estimates of the first two moments of the beta distribution provide the mean and variance of individual activity durations. Fuzzy set theory allows the human judgment that is required when estimating the behavior of activity durations to be incorporated into the modeling effort. The versatility of the fuzzy-theoretic approach is further championed in resource constrained and project crashing scenarios where additional uncertainty is introduced when estimating resource availability and cost parameters.

3.13 Facility Location and Layout

Facility location problems consist of choosing when and where to open facilities in order to minimize the associated cost of opening a facility and the cost of servicing customers. The design of the facility layout of a manufacturing system is critically important for its effective utilization: 20 to 50 percent of the total operating expenses in manufacturing are attributed to material handling and layout related costs. Use of effective methods for facilities planning can reduce these costs often by as much as 30 percent, and sometimes more. The problems of facility location and layout have been studied extensively in the industrial engineering literature.

3.13.1 Facility Location

Gen and Syarif (2005) deal with a production/distribution problem to determine an efficient integration of production, distribution and inventory system so that products are produced and distributed at the right quantities, to the right customers, and at the right time, in order to minimize system wide costs while satisfying all demand required. This problem can be viewed as an optimization model that integrates facility location decisions, distribution costs, and inventory management for multi-products and multi-time periods. To solve the problem, they propose a new technique called spanning tree-based genetic algorithm (hst-GA). In order to improve its efficiency, the proposed method is hybridized with the fuzzy logic controller (FLC) concept for auto-tuning the GA parameters.

Kahraman et al. (2003) solve facility location problems using different solution approaches of fuzzy multi-attribute group decision-making. Their paper includes four different fuzzy multi-attribute group decision-making approaches. The first one is a fuzzy model of group decision proposed by Blin. The second is the fuzzy synthetic evaluation. The third is Yager's weighted goals method and the last one is fuzzy analytic hierarchy process. Although four approaches have the same objective of selecting the best facility location alternative, they come from different theoretic backgrounds and relate differently to the discipline of multi-attribute group decision-making. These approaches are extended to select the best facility location alternative by taking into account quantitative and qualitative criteria. A short comparative analysis among the approaches and a numeric example to each approach are given.

Chu (2002) presents a fuzzy TOPSIS model under group decisions for solving the facility location selection problem, where the ratings of various alternative locations under different subjective attributes and the importance weights of all attributes are assessed in linguistic values represented by fuzzy numbers. The objective attributed are transformed into dimensionless indices to ensure compatibility with the linguistic ratings of the subjective attributes. Furthermore, the membership function of the aggregation of the ratings and weights for each alternative location versus each attribute can be developed by interval arithmetic and α-cuts of fuzzy numbers. The ranking method of the mean of the integral values is applied to help derive the ideal and negative-ideal fuzzy solutions to complete the proposed fuzzy TOPSIS model.

Guiffrida and Nagi (1998) made a literature review on facility location. Narasimhan (1979) presents an application of fuzzy set theory to the problem of locating gas stations. Fuzzy ratings are used to describe the relative importance of eleven attributes for a set of three location alternatives. A

Delphi-based procedure was applied, and the input of decision makers was used to constructmembership functions for three importance weights for judging attributes. Computations are summarized for the selection decision. The author concludes that the procedure presented is congruent to the way people make decisions. The procedure provides a structure for organizing information, and a systematic approach to the evaluation of imprecise and unreliable information. Darzentas (1987) formulates the facility location problem as a fuzzy set partitioning model using integer programming. This model is applicable when the potential facility points are not crisp and can best be described by fuzzy sets. Linear membership functions are employed in the objective function and constraints of the model. The model is illustrated with an example based on three location points and four covers. Mital *et al.* (1988) and Mital and Karwowski (1989) apply fuzzy set theory in quantifying eight subjective factors in a case study involving the location of a manufacturing plant. Linguistic descriptors are used to describe qualitative factors in the location decision, such as community attitude, quality of schools, climate, union attitude, nearness to market, police protection, fire protection, and closeness to port. Bhattacharya *et al.* (1992) present a fuzzy goal programming model for locating a single facility within a given convex region subject to the simultaneous consideration of three criteria: (i) maximizing the minimum distances from the facility to the demand points; (ii) minimizing the maximum distances from the facilities to the demand points; and (iii) minimizing the sum of all transportation costs. Rectilinear distances are used under the assumption that an urban scenario is under investigation. A numerical example consisting with three demand points is given to illustrate the solution procedure. Chung and Tcha (1992) address the location of public supply-demand distribution systems such as a water supply facility or a waste disposal facility. Typically, the location decision in these environments is made subject to the conflicting goals minimization of expenditures and the preference at each demand site to maximizing the amount supplied. A fuzzy mixed 0-1 mathematical programming model is formulated to study both uncapacitated and capacitated modeling scenarios. The objective function includes the cost of transportation and the fixed cost for satisfying demand at each site. Each cost is represented by a linear membership function. Computational results for twelve sample problems are demonstrated for a solution heuristic based on Erlenkotter's dual-based procedure for the uncapacitated facility location problem. Extension to the capacitated case is limited by issues of computational complexity and computational results are not presented. Bhattacharya *et al.* (1993) formulate a fuzzy goal programming model for locating a single facility within a given convex region subject to the simultaneous consideration of two criteria: (i) minimize the sum of all transportation costs; and (ii) minimize the maximum distances from the facilities to the demand points. Details and assumptions of the model are similar to Bhattacharya *et al.* (1992). A numerical example consisting of two facilities and three demand points is presented and solved using LINDO.

3.13.2 Facility Layout

Dweiri (1999) presents a distinct methodology to develop a crisp activity relationship charts based on fuzzy set theory and the pairwise comparison of Saaty's AHP, which ensures the consistency of the designer's assignments of importance of one factor over another to find the weight of each of the factors in every activity.

Deba and Bhattacharyyab (2005) present a distinct decision support system based on multifactor fuzzy inference system (FIS) for the development of facility layout with fixed pickup/drop-off points. The algorithm searches several candidate points with different orientation of incoming machine blocks in order to minimize flow cost, dead space and area required for the development of layout.

Ertay et al. (2006) presents a decision-making methodology based on data envelopment analysis (DEA), which uses both quantitative and qualitative criteria, for evaluating facility layout design. The criteria that are to be minimized are viewed as inputs.

Guiffrida and Nagi (1998) made a literature review on facility layout. Grobelny (1987a, 1987b) incorporates the use of 'linguistic patterns' in solving the facility layout problem. For example, if the flow of materials between departments is high, then the departments should be located close to each other. The linking between the departments and the distance between the departments represent linguistic (fuzzy) variables; the 'high' and 'close' qualifications represent values of the linguistic variables. The evaluation of a layout is measured as the grade of satisfaction as measured by the mean truth value, of each linguistic pattern by the final placement of departments. Evans *et al.* (1987) introduce a fuzzy set theory based construction heuristic for solving the block layout design problem. Qualitative layout design inputs of 'closeness' and 'importance' are modeled using linguistic variables. The solution algorithm selects the order of department placement which is manual. The algorithm is demonstrated by determining a layout for a six department metal fabrication shop. The authors identify the need for future research toward the development of a heuristic that address both the order and placement of departments, the selection of values for the linguistic variables, and the determination of membership functions. Raoot and Rakshit (1991) present a fuzzy layout construction algorithm to solve the facility layout problem. Linguistic variables are used in the heuristic to describe qualitative and quantitative factors that affect the layout decision. Linguistic variables capture information collected from experts for the following factors: flow relationships, control relationships, process and service relationships, organizational and personnel relationships, and environmental relationships. Distance is also modeled as a fuzzy variable and is used by the heuristic as the basis for placement of departments. Three test problems are used to compare the fuzzy heuristic with ALDEP and CORELAP. The authors note that the differences achieved by each of the three methods is a function of the different levels of reality that they use. Raoot and Rakshit (1993) formulate the problem of evaluating alternative facility layouts as a multiple criteria decision model (MCDM) employing fuzzy set theory. The formulation addresses the layout problem in which qualitative and quantitative factors are equally important. Linguistic variables are used to capture experts' opinions regarding the primary relationships between departments. Membership functions are selected based on consultation with layout experts. The multiple objectives and constraints of the formulation are expressed as linguistic patterns. The fuzzy MCDM layout algorithm is demonstrated for the layout of an eight department facility. Raoot and Rakshit (1994) present a fuzzy set theory-based heuristic for the multiple goal quadratic assignment problem The objective function in this formulation utilizes 'the mean truth value', which indicates the level of satisfaction of a layout arrangement to the requirements of the layout as dictated by a quantitative or qualitative goal. The basic inputs to the model are expert's opinions on the qualitative and quantitative relationships between pairs of facilities. The qualitative and quantitative relationships are captured by linguistic variables, membership functions are chosen arbitrarily. Three linguistic patterns (one quantitative; two qualitative) are employed by the heuristic to locate facilities. The performance of the heuristic is tested against a set of test problems taken from the open literature. The results of the comparison indicate that the proposed fuzzy heuristic performs well in terms of the quality of the solution. Dweiri and Meier (1996) define a fuzzy decision making system consisting of four principal components: (i) fuzzification of input and output variables; (ii) the experts' knowledge base; (iii) fuzzy decision making; and (iv) defuzzification of fuzzy output into crisp values. The analytical hierarchy process is used to weigh factors affecting closeness ratings between departments.

3.14 Forecasting

Forecasting methods are used to predict future events in order to make good decisions. Since the subject is about predicting and foreseeing, it is unavoidable to encounter uncertainty or vagueness. Therefore fuzzy set theory is intensively applied to forecasting issues. In the following, first Guiffrida and Nagi's (1998) literature review is given and then some recent works are summarized.

The first application of forecasting using fuzzy set theory to our knowledge appeared in Economakos (1979). A simulation based model was used to forecast the demand for electrical power when load components at various times of the day were described in linguistic terms. Interest in fuzzy forecasting has grown considerably since this initial article. Research on fuzzy forecasting is divided into three categories: qualitative forecasting employing the Delphi method, time series analysis and regression analysis. A summary of fuzzy forecasting applications is presented in the following.

Murray *et al.* (1985) identify fuzzy set theory as an attractive methodology for modeling ambiguity in the Delphi method. Ambiguity results due to differences in the meanings and interpretations that experts may attach to words. A pilot Delphi study using graduate business students (divided into control and test groups), as experts, was conducted to estimate the percentage of students attaining an "excellent" grade point average. Over four rounds of the Delphi, the control group received the typical feedback consisting of a summary of their responses. The test group received similar feedback but was also given feedback on a group average membership function. Methodological issues inherent in the study prevent a statistical analysis of the performance of the control and test groups with respect to the Delphi task. However, the authors demonstrate an appropriate approach for modeling ambiguity in the Delphi method, and provide insights on methodological issues for future research. Kaufmann and Gupta (1988) present a detailed tutorial on the fuzzy Delphi method in forecasting and decision making. The authors outline a four-step procedure for performing fuzzy Delphi when estimates are represented by triangular fuzzy numbers. Fuzzy Delphi is demonstrated by example for a group of twelve experts engaged in forecasting the realization of a cognitive information processing computer.

Ishikawa *et al.* (1993) cite limitations in traditional and fuzzy Delphi methods and propose the New Fuzzy Delphi Method (NFDM). The NFDM has the following advantages: (i) fuzziness is inescapably incorporated in the Delphi findings; (ii) the number of rounds in the Delphi is reduced; (iii) the semantic structure of forecast items is refined; and (iv) the individual attributes of the expert are clarified. The NFDM consists of two methodologies: the Min-Max Delphi Method, and the Fuzzy Delphi Method via Fuzzy Integration. The Max-Min Delphi Method clarifies data of each forecaster by expertise, pursues the accuracy of the forecast from the standpoint of an interval representing possibility and impossibility, and identifies the cross point as the most attainable period. The Fuzzy Delphi Method via Fuzzy Integration employs the expertise of each expert as a fuzzy measure and identifies a point estimate as the most attainable period by the fuzzy integration of each membership function. The two methodologies are illustrated by way of a technological forecasting example using members of the Japan Society for Fuzzy Theory and Systems as experts.

Song and Chissom (1993a) provide a theoretic framework for fuzzy time series modeling. A fuzzy time series is applicable when a process is dynamic and has historical data that are fuzzy sets or linguistic values. Fuzzy relational equations are employed to develop fuzzy relations among observations occurring at different time periods. Two classes of fuzzy time series models are defined: time-variant and time-invariant. A seven-step procedure is outlined for conducting a forecast using the time-invariant fuzzy time series model. Song and Chissom (1993b) apply a first order, time-invariant time series model to forecast enrollments of the University of Alabama based on twenty years of historical data. The data was fuzzified and seven fuzzy sets were defined to describe "enrollments". The corresponding linguistic values ranged from "not many" to "too many". The seven-step procedure outlined in Song and Chissom (1993a) is used to fuzzify the data, develop the time series model, and calculate and interpret the output. The errors in the forecasted enrollments ranged from 0.1% to 11%, with an average error of 3.9%. The error resulting from the fuzzy time series model is claimed to be on par with error rates cited in the literature on enrollment forecasting.

Cummins and Derrig (1993) present a fuzzy-decision making procedure for selecting the best forecast subject to a set of vague or fuzzy criteria. Selecting the best forecast, as opposed to selecting the best forecasting model, most efficiently utilizes the historical information available to the forecaster. Fuzzy set theory is used to rank candidate forecasting methods in terms of their membership values in the fuzzy set of "good" forecasts. The membership values are then used to conclude the best forecast. The methodology is demonstrated in developing a forecast of a trend component for an insurance loss cost problem. A set of 72 benchmark forecasting methods are consolidated using fuzzy set theory. On the basis of three fuzzy objectives, a fuzzy set of "good" trend factors results. The authors conclude that fuzzy set theory provides an effective method for combining statistical and judgmental criteria in actuarial decision making. Song and Chissom (1994) apply a first order, time-variant forecasting model to the

Song and Chissom(1993b) university enrollment data set. The time variant-model relaxes the invariant-model assumption that at any time t, the possible values of the fuzzy time series are the same. A 3-layer backpropagation neural network was found to be the most effective method for defuzzifying the forecast fuzzy set. The neural network defuzzification method yielded the smallest average forecasting error over a range of model structures. Sullivan and Woodall (1994) provide a detailed review of the time-invariant and time-variant time series models set forth by Song and Chissom (1993a, 1994). The authors present a Markov model for forecasting enrollments. The Markov approach utilizes linguistic labels and uses probability distributions, as opposed to membership functions, to reflect ambiguity. Using the enrollment data set of Song and Chissom, a Markov model is described and the parameter estimation procedure is compared with that of the fuzzy time series method. The Markov approach resulted in slightly more accurate forecasts than the fuzzy time series approach. The Markov and fuzzy time series models are compared to three conventional time-invariant time series models, a first-order autoregressive model, and two second-order auto- regressive models. The conventional models which use the actual crisp enrollment data, outperformed the fuzzy time series or Markov model and are most effective when predicting values outside the range of the historical data. Song *et al.* (1995) note that the fuzzy time series models defined in Song and Chissom (1993a), and those used to forecast university enrollments in Song and Chissom (1993b, 1994), require fuzzification of crisp historical data. The authors present a new model based on the premise that the historical data are fuzzy numbers. The new model is based on a theorem derived to relate the current value of a fuzzy time series with its past. The theorem expresses the value of a homogeneous fuzzy time series as a linguistic summation of previous values and the linguistic differences of different orders. A second form of the model relates the value of the fuzzy time series as the linguistic summation of various order linguistic backward differences. The authors note that the findings of this study are limited only to homogeneous fuzzy time series with fuzzy number observations. Chen (1996) presents a modified version Song and Chissom's time-invariant fuzzy time series model. The modified model is claimed to be "obviously" more efficient than Song and Chissom's (1993b) model. The claim is based on the proposed model's use of simplified arithmetic operations rather than the max-min composition operators found in Song and Chissom's model. Forecasts based on the modified model and Song and Chissom's model are compared to the historical enrollment data. The forecasted results of the proposed model deviate slightly from Song and Chissom. The statistical significance of the comparison is not addressed.

Tanaka *et al.* (1982) introduce fuzzy linear regression as a means to model casual relationships in systems when ambiguity or human judgment inhibits a crisp measure of the dependent variable. Unlike conventional regression analysis, where deviations between observed and predicted values reflect measurement error, deviations in fuzzy regression reflect the vagueness of the system structure expressed by the fuzzy parameters of the regression model. The fuzzy parameters of the model are considered to be possibility distributions which correspond to the fuzziness of the system. The fuzzy parameters are determined by a linear programming procedure which minimizes the fuzzy deviations subject to constraints of the degree of membership fit. The fuzzy regression forecasting model is demonstrated by a multiple regression example in which the prices of prefabricated houses is determined based on quality of material, floor space, and number of rooms. Heshmaty and Kandel (1985) utilize fuzzy regression analysis to build forecasting models for predicting sales of computers and peripheral equipment. The independent variables are: user population expansion, microcomputer sales, minicomputer sales, and the price of microcomputers. The forecasts for the sales of computers and peripheral equipment are given as fuzzy sets with triangular membership functions. The decision-maker then selects the forecast figure from within the interval of the fuzzy set. Fuzzy set theory has been used to develop quantitative forecasting models such as time series analysis and regression analysis, and in qualitative models such as the Delphi method. In these applications, fuzzy set theory provides a language by which indefinite and imprecise demand factors can be captured. The structure of fuzzy forecasting models are often simpler yet more realistic than non-fuzzy models which tend to add layers of complexity when attempting to formulate an imprecise underlying demand structure. When demand is definable only in linguistic terms, fuzzy forecasting models must be used.

Chang (1997) presents a fuzzy forecasting technique for seasonality in the time-series data using the following procedure. First, with the fuzzy regression analysis the fuzzy trend of a time-series is analyzed. Then the fuzzy seasonality is defined by realizing the membership grades of the seasons to the fuzzy regression model. Both making fuzzy forecast and crisp forecast are investigated. Seasonal fuzziness and trends are analyzed.

Chen and Wang (1999) apply fuzzy concepts in forecasting product price and sales in the semiconductor industry which is often conceived as a highly dynamic environment. First, two fuzzy forecasting methods including fuzzy interpolation and fuzzy linear regression are developed and discussed. Forecasts generated by these methods are fuzzy-valued. Next, the subjective beliefs about whether the industry is booming or slumping and the speed at which this change in prosperity takes place during a given period are also considered. Two subjective functions are defined and used to adjust fuzzy forecasts. Practically, fuzzy forecasts are incorporated with fuzzy programming like fuzzy linear programming or fuzzy nonlinear programming for mid-term or long-term planning.

In modeling a fuzzy system with fuzzy linear functions, the vagueness of the fuzzy output data may be caused by both the indefiniteness of model parameters and the vagueness of the input data. This situation occurs as the input data are envisaged as facts or events of an observation which are uncontrollable or uninfluenced by the observer rather than as the controllable levels of factors in an experiment. Lee and Chen (2001) concentrate on such a situation and refer to it as a generalized fuzzy linear function. Using this generalized fuzzy linear function, a generalized fuzzy regression model is formulated. A nonlinear programming model is proposed to identify the fuzzy parameters and their vagueness for the generalized regression model. A manpower forecasting problem is used to demonstrate the use of the proposed model.

Kuo (2001) utilizes a fuzzy neural network with initial weights generated by genetic algorithm (GFNN) for the sake of learning fuzzy IF-THEN rules for promotion obtained from marketing experts. The result from GFNN is further integrated with an ANN forecast using the time series data and the promotion length from another ANN. Model evaluation results for a convenience store (CVS) company indicate that the proposed system can perform more accurately than the conventional statistical method and a single ANN.

Sales forecasting plays a very prominent role in business strategy. Numerous investigations addressing this problem have generally employed statistical methods, such as regression or autoregressive and moving average (ARMA). However, sales forecasting is very complicated owing to influence by internal and external environments. Recently, ANNs have also been applied in sales forecasting since their promising performances in the areas of control and pattern recognition. However, further improvement is still necessary since unique circumstances, e.g. promotion, cause a sudden change in the sales pattern. Kuo et al. (2002) utilize a proposed fuzzy neural network (FNN), which is able to eliminate the unimportant weights, for the sake of learning fuzzy IF-THEN rules obtained from the marketing experts with respect to promotion. The result from FNN is further integrated with the time series data through an ANN. Both the simulated and real-world problem results show that FNN with weight elimination can have lower training error compared with the regular FNN. Besides, real-world problem results also indicate that the proposed estimation system outperforms the conventional statistical method and single ANN in accuracy.

Reliable prediction of sales can improve the quality of business strategy. Chang and Wang (2005) integrate fuzzy logic and artificial neural network into the fuzzy back-propagation network (FBPN) for sales forecasting in printed circuit board industry. The fuzzy back propagation network is constructed to incorporate production-control expert judgments in enhancing the model's performance. Parameters chosen as inputs to the FBPN are no longer considered as of equal importance, but some sales managers and production control experts are requested to express their opinions about the importance of each input parameter in predicting the sales with linguistic terms, which can be converted into pre-specified fuzzy numbers.

4 Conclusions

This chapter has discussed an extensive literature review and survey of fuzzy set theory in industrial engineering. Throughout the course of this study, it has been observed that (1) fuzzy set theory has been applied to most traditional areas of industrial engineering, and (2) research on fuzzy set theory in industrial engineering has grown in recent years. Fuzzy research in quality management, forecasting, and job shop scheduling have experienced tremendous growth in recent years. The appropriateness and contribution of fuzzy set theory to problem solving in industrial engineering may be seen by paralleling its use in operations research. Zimmerman (1983) identifies that fuzzy set theory can be used in operations research as a language to model problems which contain fuzzy phenomena or relationships, as a tool to analyze such models in order to gain better insight into the problem and as an algorithmic tool to make solution procedures more stable or faster. We hope this chapter should give industrial engineering researchers new tools and ideas on how to approach industrial engineering problems using fuzzy set theory. It also provides a basis for fuzzy set researchers to expand on the toolset for industrial engineering problems.

References

Adenso-Díaz, B., González, I., Tuya, J. (2004): Incorporating fuzzy approaches for production planning in complex industrial environments: the roll shop case. Engineering Applications of Artificial Intelligence, Vol. 17, No. 1, 73-81

Aliev, R.A., Fazlollahi, B., Aliev, R.R. (2004): Soft Computing and its Applications in Business and Economics, Vol. 157, Springer-Verlag

Araz, C., Selim H., Ozkarahan, I. (2005): A fuzzy multi-objective covering-based vehicle location model for emergency services. Computers & Operations Research, In Press

Barelli, L., Bidini, G. (2005): Design of the measurements validation procedure and the expert system architecture for a cogeneration internal combustion engine. Applied Thermal Engineering, Vol. 25, No. 17-18, 2698-2714

Bector, C.R., Chandra, S., (2005): Fuzzy Mathematical Programming and Fuzzy Matrix Games, Vol. 169, Springer-Verlag

Bell, P.M., Crumpton, L. (1997): A fuzzy linguistic model for the prediction of carpal tunnel syndrome risks in an occupational environment. Ergonomics, Vol. 40, No. 8, 790-799

Ben-Arieh, D., Kumar R.R., Tiwari, M. K. (2004): Analysis of assembly operations' difficulty using enhanced expert high-level colored fuzzy Petri net model. Robotics and Computer-Integrated Manufacturing, Vol. 20, No. 5, 385-403

Bhattacharya, U., Rao, J.R., Tiwari, R.N. (1992): Fuzzy multi-criteria facility location problem. Fuzzy Sets and Systems, Vol. 51, No. 3, 277-287

Bhattacharya, U., Rao, J.R., Tiwari, R.N. (1993): Bi-criteria multi facility location problem in fuzzy environment, Fuzzy Sets and Systems, Vol. 56, No. 2, 145-153

Bradshaw, C.W. (1983): A fuzzy set theoretic interpretation of economic control limits. European Journal of Operational Research, Vol. 13, No. 4, 403-408

Buckley, J. J. (1989): Fuzzy PERT, in Applications of Fuzzy Set Methodologies in Industrial Engineering, Evans, G.W., Karwowski, W., Wilhelm, M.R. (eds.), Elsevier Science Publishers B. V., Amsterdam, 103-114

Buckley, J.J., Eslami, E., Feuring, T. (2002): Fuzzy Mathematics in Economics and Engineering, Vol. 91, Springer-Verlag

Carlsson, C., Fuller, R., (2002): Fuzzy Reasoning in Decision Making and Optimization, Vol. 82, Springer-Verlag

Castillo, O., Melin, P., (2003) Soft Computing and Fractal Theory for Intelligent Manufacturing, Vol. 117, Springer-Verlag

Chakraborty, M., Chandra, M.K. (2005): Multicriteria decision making for optimal blending for beneficiation of coal: A fuzzy programming approach, Omega, Vol. 33, No. 5, 413-418

Chakraborty, T.K. (1988): A single sampling attribute plan of given strength based on fuzzy goal programming. Opsearch, Vol. 25, No. 4, 259-271

Chakraborty, T.K. (1992): A class of single sampling plans based on fuzzy optimization. Opsearch, Vol. 29, No. 1, 11-20

Chakraborty, T.K. (1994a): Possibilistic parameter single sampling inspection plans. Opsearch, Vol. 31, No. 2, 108-126

Chakraborty, T.K. (1994b): A class of single sampling inspection plans based on possibilistic programming problem. Fuzzy Sets and Systems, Vol. 63, No.1, 35-43

Chanas, S., Kamburowski, J. (1981): The use of fuzzy variables in PERT. Fuzzy Sets and Systems, Vol. 5, No.1, 11-19

Chanas, S., Kasperski, A. (2004): Possible and necessary optimality of solutions in the single machine scheduling problem with fuzzy parameters. Fuzzy Sets and Systems, Vol. 142, No. 3, Pages 359-371

Chang P.-C., Wang, Y.-W. (2005): Fuzzy Delphi and back-propagation model for sales forecasting in PCB industry, Expert Systems with Applications, In Press

Chang, P.C., Liao, T.W. (2005): Combining SOM and fuzzy rule base for flow time prediction in semiconductor manufacturing factory, Applied Soft Computing, In Press

Chang, P.-T. (1997): Fuzzy seasonality forecasting. Fuzzy Sets and Systems, Vol. 90, No. 1, Pages 1-10

Chang, P.-T. (2005): Fuzzy strategic replacement analysis. European Journal of Operational Research, Vol. 160, No. 2, 532-559

Chang, P.-T., Hung, K.-C. (2005): Applying the fuzzy-weighted-average approach to evaluate network security systems. Computers & Mathematics with Applications, Vol. 49, No. 11-12, 1797-1814

Chang, P.-T., Yao, M.-J., Huang, S.-F., Chen, C.-T. (2005): A genetic algorithm for solving a fuzzy economic lot-size scheduling problem. International Journal of Production Economics, In Press

Chang, S., Tsujimura, Y., Gen, M., Tozawa, T. (1995): An efficient approach for large scale project planning based on fuzzy Delphi method. Fuzzy Sets and Systems, Vol. 76, No. 2, 277-288

Chen, S.-M. (1996): Forecasting enrollments based on fuzzy time series. Fuzzy Sets and Systems, Vol. 81, No. 3, 311-319

Chen, S., Hwang, C., (1992): Fuzzy Multiple Attribute Decision Making: Methods and Applications, Lecture Notes in Economics and Mathematical Systems, Springer-Verlag, Germany

Chen, T., Wang, M.-J.J. (1999): Forecasting methods using fuzzy concepts. Fuzzy Sets and Systems, Vol. 105, No. 3, 339-352

Chen, Y.-J., Huang, T.-C., Hwang, R.-C. (2004): An effective learning of neural network by using RFBP learning algorithm, Information Sciences, Vol. 167, No. 1-4, 77-86

Chen, Y.-M., Wang, S.-C. (2005): Framework of agent-based intelligence system with two-stage decision-making process for distributed dynamic scheduling. Applied Soft Computing, In Press

Chen, Y.-W., Larbani, M. (2006): Two-person zero-sum game approach for fuzzy multiple attribute decision making problems. Fuzzy Sets and Systems, Vol. 157, No. 1, 34-51

Chen, Z.C., Dong, Z., Vickers, G.W. (2003): Automated surface subdivision and tool path generation for -axis CNC machining of sculptured parts. Computers in Industry, Vol. 50, No. 3, 319-331

Cheng, C.-B. (2005): Fuzzy process control: construction of control charts with fuzzy numbers. Fuzzy Sets and Systems, Vol. 154, No. 2, 287-303

Chien C.-J., Tsai, H.-H. (2000): Using fuzzy numbers to evaluate perceived service quality. Fuzzy Sets and Systems, Vol. 116, No. 2, 289-300

Cho, C., Kim, S., Lee, J., Lee, D.W. (2006): A tandem clustering process for multimodal datasets. European Journal of Operational Research, Vol. 168, No. 3, 998-1008

Chu, T.-C. (2002): Facility location selection using fuzzy TOPSIS under group decisions. International Journal of Uncertainty, Fuzziness and Knowledge-Based Systems, Vol. 10, No. 6, 687-702

Chung, K., Tcha, D. (1992): A fuzzy set-theoretic method for public facility location. European Journal of Operational Research, Vol. 58, No. 1, 90-98

Cochran J.K., Chen, H.-N. (2005): Fuzzy multi-criteria selection of object-oriented simulation software for production system analysis. Computers & Operations Research, Vol. 32, No. 1, 153-168

Cummins, J.D., Derrig, R.A. (1993): Fuzzy trends in property-liability insurance claim costs. Journal of Risk and Insurance, Vol. 60, No. 3, 429-465

Darzentas, J. (1987): A discrete location model with fuzzy accessibility measures. Fuzzy Sets and Systems, Vol. 23, No.1, 149-154

Deba, S.K., Bhattacharyyab, B. (2005): Fuzzy decision support system for manufacturing facilities layout planning. Decision Support Systems, Vol. 40, 305–314

Demirli K., and Türksen, I.B. (2000): Sonar based mobile robot localization by using fuzzy triangulation. Robotics and Autonomous Systems, Vol. 33, No. 2-3, 109-123

DePorter, E.L., Ellis, K.P. (1990): Optimization of project networks with goal programming and fuzzy linear programming. Computers and Industrial Engineering, Vol. 19, No.1-4, 500-504

Dompere, K.K. (2004): Cost-Benefit Analysis and the Theory of Fuzzy Decisions, Vol. 160, Springer-Verlag

Dubois, D., Prade, H. (1985): Possibility Theory: An Approach to Computerized Processing of Uncertainty, Plenum Press: New York

Dweiri, F. (1999): Fuzzy development of crisp activity relationship charts for facilities layout Computers & Industrial Engineering, Vol. 36, 1-16

Dweiri, F., Meier, F.A. (1996): Application of fuzzy decision-making in facilities layout planning. International Journal of Production Research, Vol. 34, No.11, 3207-3225

Economakos, E. (1979): Application of fuzzy concepts to power demand forecasting. IEEE Transactions on Systems, Man and Cybernetics, Vol. 9, No.10, 651-657

Ertay, T., Ruan, D., Rıfat Tuzkaya, U.R. (2006): Integrating data envelopment analysis and analytic hierarchy for the facility layout design in manufacturing systems. Information Sciences, Vol. 176, No. 3, 237-262

Evans, G.W., Karwowski, W., Wilhelm, M.R. (1989): Applications of Fuzzy Set Methodologies in Industrial Engineering, Elsevier

Evans, G.W., Wilhelm, M.R., Karwowski, W. (1987): A layout design heuristic employing the theory of fuzzy sets. International Journal of Production Research, Vol. 25, No.10, 1431-1450

Fang, S.-C., Nuttle, H.W.L., Wang, D. (2004): Fuzzy formulation of auctions and optimal sequencing for multiple auctions. Fuzzy Sets and Systems, Vol. 142, No. 3, 421-441

Fodor, J., De Baets, B., Perny, P. (Eds) (2000): Preferences and Decisions under Incomplete Knowledge Vol. 51, Springer-Verlag

Fonseca D.J., Knapp, G.M. (2001): A fuzzy scheme for failure mode screening. Fuzzy Sets and Systems, Vol. 121, No. 3, 453-457

Gaines, B.R., Kohout, L.J. (1977): The fuzzy decade: a bibliography of fuzzy systems and closely related topics. International Journal of Man-Machine Studies, Vol. 9, No.1, 1-68

Gen, M., Syarif, A. (2005): Hybrid genetic algorithm for multi-time period production/distribution planning. Computers & Industrial Engineering, Vol. 48, 799–809

Gen, M., Tsujimura, Y., Ida, K. (1992): Method for solving multiobjective aggregate production planning problem with fuzzy parameters. Computers and Industrial Engineering, Vol. 23, No. 1-4, 117-120

Gholamian, M.R., Ghomi, S.M.T.F. (2005): Meta knowledge of intelligent manufacturing: An overview of state-of-the-art, Applied Soft Computing, In Press

Gholamian, M.R., Ghomi, S.M.T.F., Ghazanfari, M. (2005): A hybrid systematic design for multiobjective market problems: a case study in crude oil markets. Engineering Applications of Artificial Intelligence, Vol. 18, No. 4, 495-509

Glushkovsky, E.A., Florescu, R.A. (1996): Fuzzy sets approach to quality improvement. Quality and Reliability Engineering International, Vol. 12, No. 1, 27-37

Grobelny, J. (1987a): On one possible 'fuzzy' approach to facilities layout problems. International Journal of Production Research, Vol. 25, No. 8, 1123-1141

Grobelny, J. (1987b): The fuzzy approach to facilities layout problems. Fuzzy Sets and Systems, Vol. 23, No. 2, 175-190

Guiffrida, A.L., Nagi, R. (1998): Fuzzy set theory applications in production management research: A literature survey. Journal of Intelligent Manufacturing, Vol. 9, 39-56

Gülbay, M., Kahraman, C., Ruan Da (2004): α-Cut fuzzy control charts for linguistic data. International Journal of Intelligent Systems, Vol. 19, 1173–1195

Gutierrez, I., Carmona, S. (1995): Ambiguity in multicriteria quality decisions, International Journal of Production Economics, Vol. 38 No.2/3, 215-224

Hapke, M., Jaszkiewicz, A., Slowinski, R. (1994): Fuzzy project scheduling system for software development. Fuzzy Sets and Systems, Vol. 67, No. 1, 101-117

Hapke, M., Slowinski, R., (1996): Fuzzy priority heuristics for project scheduling, Fuzzy Sets and Systems, Vol. 83, No. 3, 291-299

Heshmaty, B., Kandel, A. (1985): Fuzzy linear regression and its applications to forecasting in uncertain environment. Fuzzy Sets and Systems, Vol. 15, 159-191

Hop, N.V. (2006): A heuristic solution for fuzzy mixed-model line balancing problem, European Journal of Operational Research, Vol. 168, No. 3, 798-810

Inuiguchi, M., Sakawa, M., Kume, Y. (1994): The usefulness of possibilistic programming in production planning problems. International Journal of Production Economics, Vol. 33, No.1-3, 45-52

Ishikawa, A., Amagasa, M., Tomizawa, G., Tatsuta, R., Mieno, H. (1993): The max-min Delphi method and fuzzy Delphi method via fuzzy integration. Fuzzy Sets and Systems, Vol. 55, No.3, 241-253

Kahraman, C., Beskese A., Ruan, D. (2004): Measuring flexibility of computer integrated manufacturing systems using fuzzy cash flow analysis. Information Sciences, Vol. 168, No. 1-4, 77-94

Kahraman, C., Ruan D., Tolga, E. (2002): Capital budgeting techniques using discounted fuzzy versus probabilistic cash flows. Information Sciences, Vol. 142, No. 1-4, 57-76

Kahraman, C., Ruan, D., Dogan, I. (2003): Fuzzy group decision-making for facility location selection, Information Sciences, Vol. 157, 135–153

Kahraman, C., Tolga E., Ulukan, Z. (2000): Justification of manufacturing technologies using fuzzy benefit/cost ratio analysis. International Journal of Production Economics, Vol. 66, No. 1, 45-52

Kanagawa, A., Ohta, H. (1990): A design for single sampling attribute plan based on fuzzy sets theory. Fuzzy Sets and Systems, Vol. 37, No. 2, 173-181

Kanagawa, A., Tamaki, F., Ohta, H. (1993): Control charts for process average and variability based on linguistic data. International Journal of Production Research, Vol. 31, No. 4, 913-922

Kandel, A. (1986): Fuzzy Mathematical Techniques with Applications, Addison-Wesley: Reading, MA

Kandel, A., Yager, R. (1979): A 1979 bibliography on fuzzy sets, their applications, and related topics, in Advances in Fuzzy Set Theory and Applications, Gupta, M.M., Ragade, R. K. and Yager, R. R. (eds.), North-Holland: Amsterdam, 621-744

Karwowski, W., Evans, G. W. (1986): Fuzzy concepts in production management research: A review. International Journal of Production Research, Vol. 24, No. 1, 129-147

Kasperski, A. (2005): A possibilistic approach to sequencing problems with fuzzy parameters. Fuzzy Sets and Systems, Vol. 150, No. 1, 77-86

Kaufmann, A., Gupta, M.M. (1988): Fuzzy Mathematical Models in Engineering and Management Science, North-Holland: Amsterdam

Kaya, M.D., Hasiloglu, A.S., Bayramoglu, M., Yesilyurt, H., Ozok, A.F. (2003): A new approach to estimate anthropometric measurements by adaptive neuro-fuzzy inference system. International Journal of Industrial Ergonomics, Vol. 32, No. 2, 105-114

Khoo, L.P., Ho, N.C. (1996): Framework of a fuzzy quality deployment system. International Journal of Production Research, Vol. 34, No. 2, 299-311

Kim, K.-J., Moskowitz, H., Dhingra, A., Evans, G. (2000): Fuzzy multicriteria models for quality function deployment. European Journal of Operational Research, Vol. 121, No. 3, 504-518

Kim, K.W., Gen, M., Yamazaki, G. (2003): Hybrid genetic algorithm with fuzzy logic for resource-constrained project scheduling. Applied Soft Computing, Vol. 2, No. 3, 174–188

Klir, G.J., Yuan, B., (1995): Fuzzy Sets and Fuzzy Logic: Theory and Applications, Englewood CliOEs: Prentice-Hall

Kuchta, D. (2000): Fuzzy capital budgeting. Fuzzy Sets and Systems, Vol. 111, No. 3, 367-385

Kulak, O., Durmusoglu, M.D., Kahraman, C. (2005): Fuzzy multi-attribute equipment selection based on information axiom. Journal of Materials Processing Technology, Vol. 169, No. 3, 337-345

Kuo, R.J. (2001): A sales forecasting system based on fuzzy neural network with initial weights generated by genetic algorithm. European Journal of Operational Research, Vol. 129, No. 3, 496-517

Kuo, R.J., Kuo, Y.P., Chen, K.-Y. (2005): Developing a diagnostic system through integration of fuzzy case-based reasoning and fuzzy ant colony system, Expert Systems with Applications, Vol. 28, No. 4, 783-797

Kuo, R.J., Wu, P., Wang, C.P. (2002): An intelligent sales forecasting system through integration of artificial neural networks and fuzzy neural networks with fuzzy weight elimination. Neural Networks: The Official Journal Of The International Neural Network Society, Vol. 15, No. 7, 909-925

Lai, Y.-J., Hwang, C.-L. (1994): Fuzzy Multiple Objective Decision Making Methods and Applications, Springer-Verlag: Berlin

Lee, H.T., Chen, S.H. (2001); Fuzzy regression model with fuzzy input and output data for manpower forecasting. Fuzzy Sets and Systems, Vol. 119, No. 2, 205-213

Lee, Y.Y., Kramer, B.A., Hwang, C.L. (1991): A comparative study of three lot-sizing methods for the case of fuzzy demand. International Journal of Operations and Production Management, Vol. 11, No. 7, 72-80

Leiviskä, K. (Ed.) (2001): Industrial Applications of Soft Computing: Paper, Mineral and Metal Processing Industries, Vol. 71, Springer-Verlag

Li, D.-C., Wu, C.-S., Tsai T.-I., Chang, F.M. (2006): Using mega-fuzzification and data trend estimation in small data set learning for early FMS scheduling knowledge. Computers & Operations Research, Vol. 33, No. 6, 1857-1869

Lootsma, F.A. (1989): Stochastic and Fuzzy PERT. European Journal of Operational Research, Vol. 43, No. 2, 174-183

Lorterapong, P. (1994): A fuzzy heuristic method for resource-constrained project scheduling. Project Management Journal, Vol. 25, No. 4, 12-18

Maiers, J., Sherif, Y.S. (1985): Applications of fuzzy set theory. IEEE Transactions on Systems, Man and Cybernetics, Vol. 15, No. 1, 175-189

Maimon, O., Kandel, A., Last, M. (2001): Information-theoretic fuzzy approach to data reliability and data mining. Fuzzy Sets and Systems, Vol. 117, No. 2, 183-194

McCahon, C.S., (1993): Using PERT as an approximation of fuzzy project-network analysis. IEEE Transactions on Engineering Management, Vol. 40, No. 2, 146-153

McCahon, C.S., Lee, E.S. (1988): Project network analysis with fuzzy activity times. Computers and Mathematics with Applications, Vol. 15, No. 10, 829-838

Mital A., Karwowski, W. (1986): Towards the development of human work-performance standards in futuristic man-machine systems: A fuzzy modeling approach. Fuzzy Sets and Systems, Vol. 19, No. 2, 133-147

Mital, A., Karwowski, W. (1989): A framework of the fuzzy linguistic approach to facilities location problem, in Applications of Fuzzy Set Methodologies in Industrial Engineering, Evans, G.W., Karwowski, W. and Wilhelm, M. R. (eds.), Elsevier Science Publishers B.V., Amsterdam, 323-330

Mital, A., Kromodihardjo, S., Metha, M., Karwowski, W. (1988): Facilities location: quantifying subjective criteria using fuzzy linguistic approach, in Recent Developments in Production Research, Mital, A. (ed.), Elsevier Science Publishers B. V., Amsterdam, 307-314

Modarres, M., Nasrabadi, E., Nasrabadi, M.M. (2005): Fuzzy linear regression models with least square errors. Applied Mathematics and Computation, Vol. 163, No. 2, 977-989

Murray, T. J., Pipino, L.L., vanGigch, J.P. (1985): A pilot study of fuzzy set modification of Delphi, Human Systems Management, Vol. 5, No. 1, 76-80

Narasimhan, R. (1979): A fuzzy subset characterization of a site-selection problem. Decision Sciences, Vol. 10, No. 4, 618-628

Nasution, S.H. (1994): Fuzzy critical path. IEEE Transactions on Systems, Man and Cybernetics, Vol. 24, No. 1, 48-57

Niskanen, V.A., (2004): Soft Computing Methods in Human Sciences, Vol. 134, Springer-Verlag

Ohta, H., Ichihashi, H. (1988): Determination of single- sampling-attribute plans based on membership functions. International Journal of Production Research, Vol. 26, No. 9, 1477-1485

Olson D.L., Wu, D. (2005): Simulation of fuzzy multiattribute models for grey relationships. European Journal of Operational Research, In Press

Onwubolu, G.C., Babu, B.V., (2004): New Optimization Techniques in Engineering, Vol. 141, Springer-Verlag

Pai, P.-F. (2003): Capacitated Lot size problems with fuzzy capacity. Mathematical and Computer Modelling, Vol. 38, No. 5-6, 661-669

Palmero, S.G.I., Santamaria, J.J., de la Torre E.J.M., González, J.R.P. (2005): Fault detection and fuzzy rule extraction in AC motors by a neuro-fuzzy ART-based system. Engineering Applications of Artificial Intelligence, Vol. 18, No. 7, 867-874

Park, J., Han, S.H. (2004): A fuzzy rule-based approach to modeling affective user satisfaction towards office chair design. International Journal of Industrial Ergonomics, Vol. 34, No. 1, 31-47

Prade, H. (1979): Using fuzzy set theory in a scheduling problem: A case study. Fuzzy Sets and Systems, Vol. 2, No. 2, 153-165

Raoot, A., Rakshit, A. (1991): A 'fuzzy' approach to facilities lay-out planning. International Journal of Production Research, Vol. 29, No.4, 835-857

Raoot, A., Rakshit, A. (1993):A 'linguistic pattern' approach for multiple criteria facility layout problems. International Journal of Production Research, Vol. 31, No.1, 203-222

Raoot, A., Rakshit, A. (1994): A 'fuzzy' heuristic for the quadratic assignment formulation to the facility layout problem. International Journal of Production Research, Vol. 32, No. 3, 563-581

Raz, T., Wang, J. (1990): Probabilistic and membership approaches in the construction of control charts for linguistic data. Production Planning and Control, Vol. 1, No. 3, 147-157

Rinks, D.B. (1981): A heuristic approach to aggregate production scheduling using linguistic variables, in Applied Systems and Cybernetics - Vol. VI, Lasker, G. E. (ed.), Pergamon Press: New York, 2877-2883

Rinks, D.B. (1982a): The performance of fuzzy algorithm models for aggregate planning under differing cost structures, in Fuzzy Information and Decision Processes, Gupta, M.M. and Sanchez, E. (eds.), North-Holland Publishing: Amsterdam, 267-278

Rinks, D.B. (1982b) A heuristic approach to aggregate planning production scheduling using linguistic variables: Methodology and application, in Fuzzy Set and Possibility Theory, Yager, R. R. (ed.), Pergamon Press: New York, 562-581

Samanta B., Al-Araimi, S.A., (2001): An inventory control model using fuzzy logic. International Journal of Production Economics, Vol. 73, No. 3, 217-226

Shipley, M.F., De Korvin, A., Omer, K. (1996): A fuzzy logic approach for determining expected values: A project management application, Journal of the Operational Research Society, Vol. 47, No. 4, 562-569

Singh, N., Mohanty, B.K. (1991): A fuzzy approach to multi-objective routing problem with applications to process planning in manufacturing systems. International Journal of Production Research, Vol. 29, No. 6, 1161-1170

Song, Q., Chissom, B.S. (1993a): Fuzzy time series and its models. Fuzzy Sets and Systems, Vol. 54, No. 3, 269-277

Song, Q., Chissom, B.S. (1993b): Forecasting enrollments with fuzzy time series - part I. Fuzzy Sets and Systems, Vol. 54, No. 1, 1-9

Song, Q., Chissom, B.S. (1994): Forecasting enrollments with fuzzy time series - part II. Fuzzy Sets and Systems, Vol. 62, No. 1, 1-8

Song, Q., Leland, R.P., Chissom, B.S. (1995): A new fuzzy time-series model of fuzzy number observations. Fuzzy Sets and Systems, Vol. 73, No. 3, 341-348

Srithongchai, S., Intaranont, K. (1996): A study of impact of shift work on fatigue level of workers in a sanitary-ware factory using a fuzzy set model. Journal of Human Ergology, Vol. 25, No. 1, 93-99

Sullivan, J., Woodall, W.H. (1994): A comparison of fuzzy forecasting and Markov modeling. Fuzzy Sets and Systems, Vol. 64, No. 3, 279-293

Tanaka, H., Ichihashi, H., Asai, K. (1986): A value of information in FLP problems via sensitivity analysis. Fuzzy Sets and Systems, Vol. 18, No. 2, 119-129

Tanaka, H., Uejima, S., Asai, K. (1982): Linear regression analysis with fuzzy model. IEEE Transactions on Systems, Man and Cybernetics, Vol. 12, No. 6, 903-907

Tang, J., Wang, D., Fung, R.Y.K. (2000): Fuzzy formulation for multi-product aggregate production planning, Production Planning & Control, Vol. 11, No. 7, 670-676

Tolga, E., Demircan M.L., Kahraman, C. (2005): Operating system selection using fuzzy replacement analysis and analytic hierarchy process. International Journal of Production Economics, Vol. 97, No. 1, 89-117

Tsai, H.-H., Lu, I.-Y. (2006): The evaluation of service quality using generalized Choquet integral. Information Sciences, Vol. 176, No. 6, 640-663

Tsaur, R.-C., Yang J.-C. O, Wang, H.F. (2005): Fuzzy relation analysis in fuzzy time series model, Computers & Mathematics with Applications, Vol. 49, No. 4, 539-548

Turksen, I.B. (1988a): Approximate reasoning for production planning. Fuzzy Sets and Systems, Vol. 26, No. 1, 23-37

Turksen, I.B. (1988b): An approximate reasoning framework for aggregate production planning, in Computer Integrated Manufacturing, Turksen, I. B. (ed.), NATO ASI Series, Vol. 49, Springer-Verlag: Berlin, 243-266

Verdegay, J.-L. (2003): Fuzzy Sets Based Heuristics for Optimization, Vol. 126, Springer-Verlag

Wang, H.-F., Kuo, C.-Y. (2004): Factor analysis in data mining. Computers & Mathematics with Applications, Vol. 48, No. 10-11, 1765-1778

Wang, J. (2002): A fuzzy project scheduling approach to minimize schedule risk for product development. Fuzzy Sets and Systems, Vol. 127, 99–116

Wang, J., Dai, C.-H. (2004): A fuzzy constraint satisfaction approach for electronic shopping assistance. Expert Systems with Applications, Vol. 27, No. 4, 593-607

Wang, J.-H., Raz, T. (1990): On the construction of control charts using linguistic variables, International Journal of Production Research, Vol. 28, No. 3, 477-487

Wang, R.-C., Chen, C.-H. (1995): Economic statistical np-control chart designs based on fuzzy optimization. International Journal of Quality and Reliability Management, Vol. 12, No.1, 82-92

Wang, R.-C., Liang, T.-F. (2004): Application of fuzzy multi-objective linear programming to aggregate production planning. Computers & Industrial Engineering, Vol. 46, 17–41

Wang, R-C., Fang, H.-H. (2001): Aggregate production planning with multiple objectives in a fuzzy environment. European Journal of Operational Research, Vol. 133, 521-536

Wang, Z., Leung, K.-S., Klir, G.J. (2005): Applying fuzzy measures and nonlinear integrals in data mining. Fuzzy Sets and Systems, Vol. 156, No. 3, 371-380

Ward, T.L., Ralston, P.A.S., Davis, J.A. (1992): Fuzzy logic control of aggregate production planning. Computers and Industrial Engineering, Vol. 23, No.1-4, 137-140

Yen, J., Langari, R., Zadeh, L.A. (1995): Industrial Applications of Fuzzy Logic and Intelligent Systems, IEEE Press, New York

Yongting, C. (1996): Fuzzy quality and analysis on fuzzy probability. Fuzzy Sets and Systems, 83(2), 283-290

Yoshida, Y. (Ed.), (2001): Dynamical Aspects in Fuzzy Decision Making, Vol. 73, Springer-Verlag

Yu, X., Kacprzyk, J. (Eds.) (2003): Applied Decision Support with Soft Computing, Vol. 124, Springer-Verlag
Zadeh, L.A. (1965): Fuzzy Sets, Information and Control, Vol. 8, 338-353
Zhang, H.-C., Huang, S.H. (1994): A fuzzy approach to process plan selection. International Journal of Production Research, Vol. 32, No. 6, 1265-1279
Zimmerman, H.-J. (1983): Using fuzzy sets in operational research. European Journal of Operational Research, Vol. 13, No. 3, 201-216
Zimmermann, H.-J. (1996): Fuzzy Set Theory and Its Applications, Kluwer, Massachusetts

Part I

Fuzzy Control Techniques and Reliability in Industry

Design of Fuzzy Process Control Charts for Linguistic and Imprecise Data

Murat Gülbay and Cengiz Kahraman[1]

Istanbul Technical University, Faculty of Management,
Industrial Engineering Department, 34367 Maçka, İstanbul, Turkey
{gulbaym,kahramanc}@itu.edu.tr

Summary. Even the first control chart was proposed during the 1920's by Shewhart, today they are still subject to new application areas that deserve further attention. If the quality-related characteristics cannot be represented in numerical form, such as characteristics for appearance, softness, color, etc., then control charts for attributes are used. Except for the special cases, fuzzy control charts are used for attributes control charts such as *p* or *c* charts. The theory of classical control charts requires all the data to be exactly known. The major contribution of fuzzy set theory is its capability of representing vague data. Fuzzy logic offers a systematic base in dealing with situations, which are ambiguous or not well defined. Fuzzy control charts based on the fuzzy transformation methods are reviewed and a design for the control charts in the case of vague data using fuzzy sets as real valued interpretations of uncertainty and vagueness is proposed.

Key words. Fuzzy control charts; linguistic data; fuzzy process control; fuzzy p-charts; fuzzy c-charts.

1 Introduction

A control chart is a device for describing in a precise manner what is meant by statistical control and is widely used tool for monitoring and examining production processes. The power of control charts lies in their ability to detect process shifts and to identify abnormal conditions in a production process. The theory of classical control charts requires all the data to be exactly known. In this chapter, we suggest a design for the control charts in the case of vague data using fuzzy sets as real valued

[1] Corresponding author, Phone: +90 212 2931300 (2035 Ext.), Fax: +90 212 2407260

M. Gülbay and C. Kahraman: *Design of Fuzzy Process Control Charts for Linguistic and Imprecise Data*, StudFuzz **201**, 59–88 (2006)
www.springerlink.com

interpretations of uncertainty and vagueness. The major contribution of fuzzy set theory is its capability of representing vague data. Fuzzy logic offers a systematic base in dealing with situations, which are ambiguous or not well defined.

Even the first control chart was proposed during the 1920's by Shewhart, who published his work in 1931, today they are still subject to new application areas that deserve further attention. Shewhart control charts introduced by Shewhart (1931) were designated to monitor processes for shifts in the mean or variance of a *single* quality characteristic, and so called *univariate control charts*. Further developments are focused on the usage of the probability and fuzzy set theory integrated with the control charts. In the literature, there exist some fuzzy control charts developed for linguistic data, which are mainly based on membership and probabilistic approaches (Raz and Wang, 1990; Wang and Raz, 1990; Kanagawa *et al.*, 1993; Höppner, 1994; Höppner and Wolff, 1995; Grzegorzewski, 1997; Grzegorzewski and Hryniewicz, 2000; Grzegorzewski, 2002; Gülbay *et al.*, 2004). Different procedures are proposed to monitor multinomial processes when products are classified into mutually exclusive linguistic categories. Bradshaw (1983) used fuzzy set theory as a basis for interpreting the representation of a graded degree of product conformance with quality standard. Bradshaw (1983) stressed that fuzzy economic control chart limits would be advantageous over traditional acceptance charts in that fuzzy economic control charts provide information on severity as well as the frequency of product nonconformance. Raz and Wang (1990) proposed an approach based on fuzzy set theory by assigning fuzzy sets to each linguistic term, and then combining for each sample using rules of fuzzy arithmetic. The result is a single fuzzy set. A measure of centrality of this aggregate fuzzy set is then plotted on a Shewhart-type control chart. Wang and Raz (1990) developed two approaches called *fuzzy probabilistic approach* and *membership approach*. Probabilistic approach is based on the fuzzy theory and probability theory and so called fuzzy probabilistic approach. In fuzzy probabilistic approach, fuzzy subsets associated with the linguistic terms are transformed into their respective representative values with one of the transformation methods. Center line (*CL*) corresponds to the arithmetic means of representative values of the samples initially available. Then, *upper control limit* (*UCL*) and *lower control limit* (*LCL*) are determined from the formulae for control charts for variables. *Membership approach* is based on the fuzzy theory to combine all observations in only one fuzzy subset using fuzzy arithmetic. Membership control limits are based on membership functions. In membership approach, center line is located as the value of the representative value of the aggregate fuzzy subset. In the control charts proposed by Raz and Wang (1990), and Wang and Raz (1990), *CL, LCL,* and *UCL* are crisp values since samples denoted by linguistic variables are transformed into crisp values by the use of fuzzy transformation, and control limits are then calculated on the base of these transformed crisp values.

Apart from fuzzy probabilistic and fuzzy membership approach, Kanagawa *et al.* (1993) introduced modifications to the construction of control charts given by Wang and Raz (1988, 1990). Their study aimed at directly controlling the underlying probability distributions of the linguistic data, which were not considered by Wang and Raz (1990). Kanagawa *et al.* (1993) proposed control charts for linguistic data from a standpoint different to that of Wang and Raz in order not only to control the process average but also to control the process variability. They presented new linguistic control charts for process average and process variability based on the estimation of probability distribution existing behind the linguistic data. They defined center line as the average mean of the sample cumulants and then calculated the control limits using Gram-Charlier series and the probability method. The main difficulty of this approach is that unknown probability distribution function cannot be determined easily. These procedures are reviewed by Woodall *at al.* (1997) and discussed by Laviolette *et al.* (1995) and Asai (1995). Wang and Chen (1995) presented a fuzzy mathematical programming model and heuristic solution for the economic design of statistical control charts. They argued that under the assumptions of the economic statistical model, the fuzzy set theory procedure presented an improved economic design of control charts by allowing more flexibility in the modeling of the imprecision that exists when satisfying type I and Type II error constraints. Kahraman *et al.* (1995) used triangular fuzzy numbers in the tests of control charts for unnatural patterns. Chang and Aw (1996) proposed a neural fuzzy control chart for identifying process mean shifts. A supervised multi-layer backpropagation neural network is trained off-line to detect various mean shifts in a production process. In identifying mean shifts in real-time usage, the neural network's outputs are classified into various decision regions using a fuzzy set scheme. The approach offers better performance and additional advantages over conventional control charts. Woodall *et al.* (1997) gave a review of statistical and fuzzy control charts based on categorical data. An investigation into the use of fuzzy logic to modify statistical process control (SPC) rules, with the aim of reducing the generation of false alarms and also improving the detection and detection-speed of real faults is studied by El-Shal and Morris (2000). Rowlands and Wang (2000) explored the integration of fuzzy logic and control charts to create and design a fuzzy-SPC evaluation and control method based on the application of fuzzy logic to the SPC zone rules. Hsu and Chen (2001) described a new diagnosis system based on fuzzy reasoning to monitor the performance of a discrete manufacturing process and to justify the possible causes. The diagnosis system consists chiefly of a knowledge bank and a reasoning mechanism. Taleb and Limam (2002) discussed different procedures of constructing control charts for linguistic data, based on fuzzy and probability theory. A comparison between fuzzy and probability approaches, based on the Average Run Length and samples under control, is conducted for real data. Contrary to the conclusions of Raz and Wang (1990) the choice of degree of fuzziness affected the sensitivity of control charts. Grzegorzewski and Hryniewicz (2000) proposed a new fuzzy control chart based on the necessity index. This work has a significant worth in the literature of the fuzzy control charts since this index has natural interpretation and effectiveness in solving real problems, but they gave up the control lines in traditional form.

In this chapter, the construction of fuzzy control charts in the case of uncertain or vague data has been studied. In Section 2, fuzzy control charts based on the fuzzy to scalar transformations are explained. A fuzzy control chart methodology that does not use any transformation method is proposed in Section 3. Finally, conclusions are given in Section 4.

2 Fuzzy Control Charts Based on The Fuzzy to Scalar Transformations

If the quality-related characteristics cannot be represented in numerical form, such as characteristics for appearance, softness, color, etc., then control charts for attributes are used. Except for the special cases, fuzzy control charts are used for attributes control charts such as *p* or *c* charts. In the literature, the data for fuzzy control charts are represented by either linguistic terms such as 'perfect,' 'good,' 'medium,' 'poor,' and 'bad,' (Wang and Raz, 1988; Wang and Raz, 1990) or uncertain (or imprecise) values such as "approximately 5" or "between 2 and 5," (Gülbay et al., 2004) or 'standard,' 'second choice,' 'third choice,' and 'chipped' (Taleb and Limam, 2002). Then fuzzy control charts have been developed by converting the fuzzy sets associated with linguistic or uncertain values into scalars referred to as representative values (Wang and Raz, 1990; Taleb and Limam, 2002, Gülbay et al., 2004). This conversion, which facilitates the plotting of observations on the chart, may be performed in a number of ways as long as the result is intuitively representative of the range of the base variable included in the fuzzy set. Some ways, which are similar in principle to the measures of central tendency used in descriptive statistics, are fuzzy mode, α-level fuzzy midrange, and fuzzy median. It should be pointed out that there is no theoretical basis supporting any one specifically and the selection between them should be mainly based on the ease of computation or preference of the user (Wang and Raz, 1990).

The fuzzy mode of a fuzzy set f shown in Figure 1 is the value of the base variable, X, where the membership function equals to 1. This can be stated as:

$$f_{\text{mod}} = \{x \in X \mid \mu_f(x) = 1\} \tag{1}$$

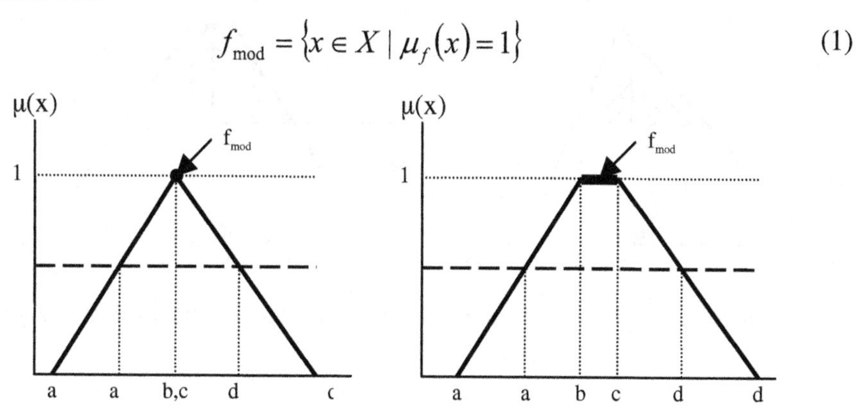

Fig. 1. Representation of fuzzy mode for triangular and trapezoidal fuzzy numbers

The α-level fuzzy midrange, f_{mr}^{α}, is defined as the midpoint of the ends of the α-cut. If a^{α} and d^{α} (See Figure 2) are the end points of α-cut, then

$$f_{mr}^{\alpha} = \frac{1}{2}\left(a^{\alpha} + d^{\alpha}\right) \tag{2}$$

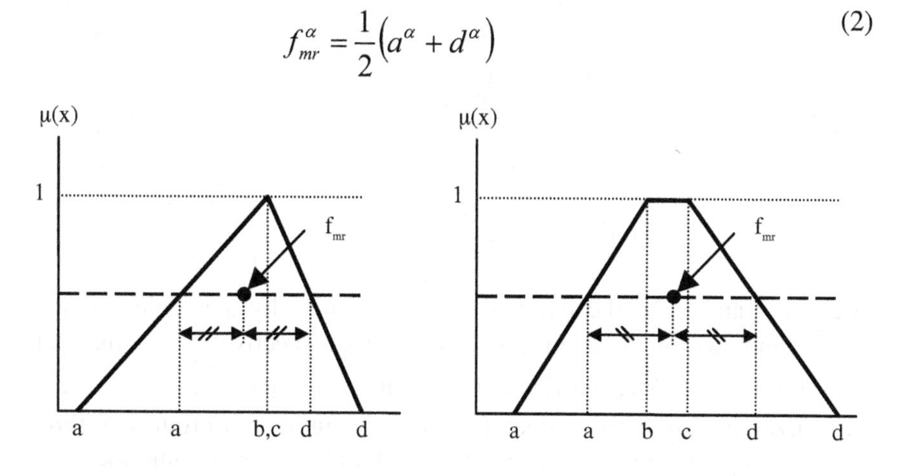

Fig. 2. Representation of fuzzy midrange for triangular and trapezoidal fuzzy numbers

In fact, the fuzzy mode is a special case of α-level fuzzy midrange when α=1.

The α-level fuzzy median, f_{med}^{α}, is the point which partitions the membership function of a fuzzy set into two equal regions at α-level.

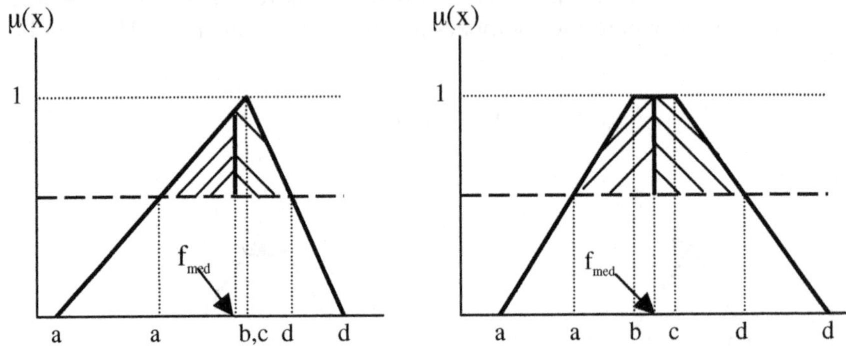

Fig. 3. Representation of fuzzy midrange for triangular and trapezoidal fuzzy numbers

2.1 Fuzzy p Control Charts

In the crisp case, the control limits for fraction rejected (*p* charts) are calculated by the following equations.

$$CL = \overline{p},$$
$$LCL = \overline{p} - 3\sqrt{\frac{\overline{pq}}{n}},$$
$$and$$
$$UCL = \overline{p} + 3\sqrt{\frac{\overline{pq}}{n}}. \tag{3}$$

Let m, j, and n_j be the number of samples initially available, sample number, and size of the sample $j = 1,2,\ldots,m$, respectively. Assume that each sample group is categorized by the linguistic categories $i = 1,2,\ldots,t$ with a fuzzy representative value of r_i. Then the number of products falling to each linguistic category can be represented as k_{ij}. The sample mean of the j_{th} sample, M_j, is the average sample mean and can be expressed by Eq. 4.

$$M_j = \frac{\sum_{j=1}^{m} k_{ij} r_i}{n_j} \quad i=1,2,\ldots,t \; ; j=1,2,\ldots,m \; ; \; 0 \le \overline{M}_j \le 1 \tag{4}$$

M_j can be represented by a triangular fuzzy number as shown in Figure 4 to reflect the human judgment of r_i and vagueness of k_{ij}.

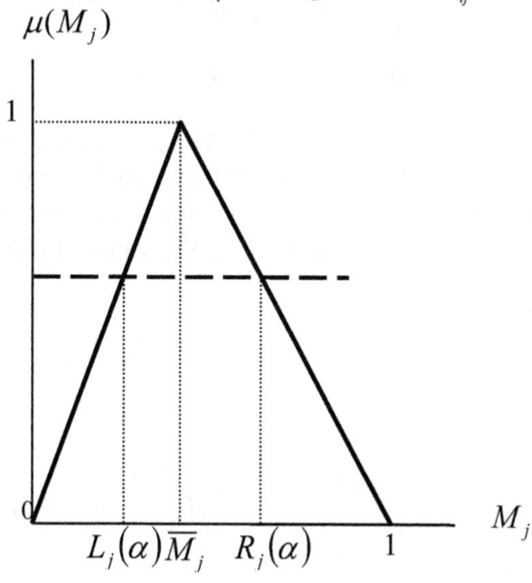

Fig. 4. Representation of M_j by a triangular fuzzy number

For each sample mean, M_j, $L_j(\alpha)$ and $R_j(\alpha)$ can be calculated using Eq. 5.

$$L_j(\alpha) = M_j \, \alpha, \tag{5}$$
$$R_j(\alpha) = 1 - [(1 - M_j) \, \alpha].$$

The center line (CL) is then calculated as the grand average of samples' M_j as given in Eq. 6.

$$CL = \overline{M_j} = \frac{\sum_{j=1}^{m} M_j}{m} \tag{6}$$

The CL is also a fuzzy triangular number. The membership function of the $\overline{M}$, or CL can be written as:

$$\mu_{M_j}(x) = \begin{cases} 0, & if\ x \leq 0, \\ \dfrac{x}{\overline{M}}, & if\ 0 \leq x \leq \overline{M}, \\ \dfrac{1-x}{1-\overline{M}}, & if\ \overline{M} \leq x \leq 1, \\ 0, & if\ x \geq 1. \end{cases} \tag{7}$$

The control limits for α-cut can also be represented by TFNs. Since the membership function of CL is divided into two components, then, each component will have its own CL, LCL, and UCL. The membership function of the control limits depending upon the value of α is given below.

$$Control\ Limits(\alpha) = \tag{8}$$

$$\begin{cases} \left\{ \begin{array}{l} CL^L = \overline{M}\alpha \\ LCL^L = \max\left\{ CL^L - 3\sqrt{\dfrac{(CL^L)(1-CL^L)}{\overline{n}}}, 0 \right\} \\ UCL^L = \min\left\{ CL^L + 3\sqrt{\dfrac{(CL^L)(1-CL^L)}{\overline{n}}}, 1 \right\} \end{array} \right\}, & if\ 0 \leq M_j \leq \overline{M}, \\ \left\{ \begin{array}{l} CL^R = 1 - \left[\left(1 - \overline{M}\alpha \right) \alpha \right] \\ LCL^R = \max\left\{ CL^R - 3\sqrt{\dfrac{(CL^R)(1-CL^R)}{\overline{n}}}, 0 \right\} \\ UCL^R = \min\left\{ CL^R + 3\sqrt{\dfrac{(CL^R)(1-CL^R)}{\overline{n}}}, 1 \right\} \end{array} \right\}, & if\ \overline{M} \leq M_j \leq 1. \end{cases}$$

where $\overline{n}$ is the average sample size (*ASS*). When the *ASS* is used, the control limits do not change with the sample size. Hence, the control limits for all samples are the same. A general illustration of these control limits is shown in Figure 5.

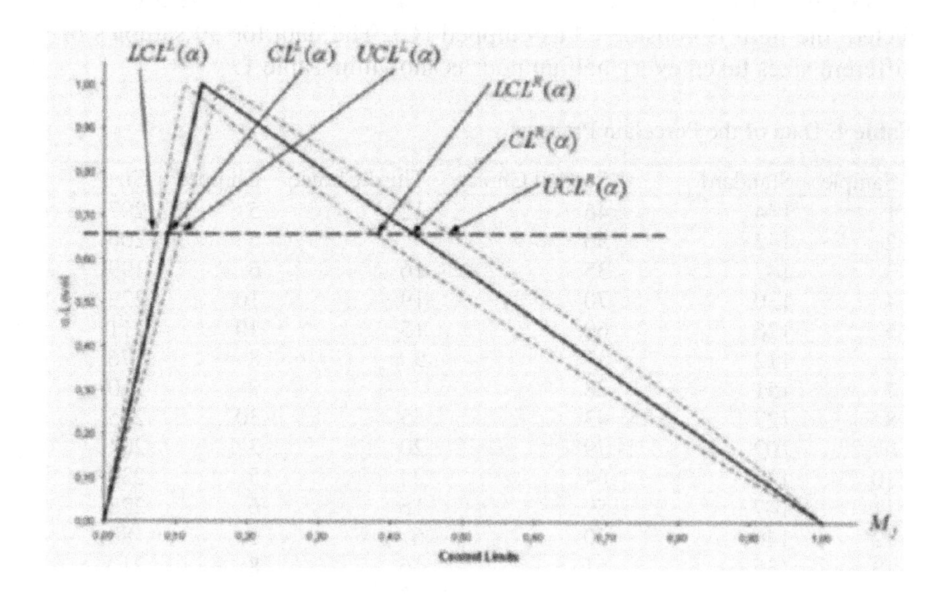

Fig. 5. Illustration of control limits for α-cut

For the variable sample size (*VSS*), $\overline{n}$ should be replaced by the size of the j th sample, n_j. Hence, the control limits change for each sample depending upon the size of the sample. Therefore, each sample has its own control limits.

The decision that whether process is in control (1) or out of control (0) for both *ASS* and *VSS* is as follows:

Process control =

$$\begin{cases} 1, & \text{if } LCL^L(\alpha) \le L_j(\alpha) \le UCL^L(\alpha) \\ & \wedge\, LCL^R(\alpha) \le R_j(\alpha) \le UCL^R(\alpha), \\ 0, & \text{otherwise.} \end{cases} \tag{9}$$

In order to clarify this approach, a numerical example of Tunisie Porcelaine problem stated by Wang and Raz (1990) and Taleb and Limam (2002) will be handled. In the example presented, Taleb and Limam (2002) classified porcelain products into four categories with respect to the quality. When a product represents no default, or an invisible minor default, it is classified as a standard product (*S*). If it presents a visible minor default that does not affect the use of the product, then it is classified as second choice (*SC*). When there is a visible major default that does not affect the product use, it is called as third choice (*TC*). Finally, when the use is affected, the item is considered as chipped (*C*). The data for 30 samples of different sizes taken every half an hour is shown in Table 1.

Table 1. Data of the Porcelain Process

Sample	Standard	Second Choice	Third Choice	Chipped	Size
1	144	46	12	5	207
2	142	50	9	5	206
3	142	35	16	6	199
4	130	70	19	10	229
5	126	60	15	10	211
6	112	47	9	8	176
7	151	28	22	9	210
8	127	43	45	30	245
9	102	79	20	3	204
10	137	64	24	5	230
11	147	59	16	6	228
12	146	30	6	6	188
13	135	51	16	8	210
14	186	82	23	7	298
15	183	53	11	9	256
16	137	65	26	4	232
17	140	70	10	3	223
18	135	48	15	9	207
19	122	52	23	10	207
20	109	42	28	9	188
21	140	31	9	4	184
22	130	22	3	8	163
23	126	29	11	8	174
24	90	23	16	2	131
25	80	29	19	8	136
26	138	55	12	12	217
27	121	35	18	10	184
28	140	35	15	6	196
29	110	15	9	1	135
30	112	37	28	11	188

For each sample, the membership function of the fuzzy subset corresponding to the sample observations is determined. The membership function for the porcelain process is as follows:

$$\mu_S(x)=\begin{cases}0, & if \quad x\le 0,\\ -x+1, & if \quad 0\le x\le 1,\\ 0, & if \quad x\ge 1.\end{cases} \tag{10-1}$$

$$\mu_{SC}(x)=\begin{cases}0, & if \quad x\le 0,\\ 4x, & if \quad 0\le x\le \frac{1}{4},\\ -\frac{4}{3}x+\frac{4}{3}, & if \quad \frac{1}{4}\le x\le 1,\\ 0, & if \quad x\ge 1.\end{cases} \tag{10-2}$$

$$\mu_{TC}(x)=\begin{cases}0, & if \quad x\le 0,\\ 2x, & if \quad 0\le x\le \frac{1}{2},\\ 2-2x, & if \quad \frac{1}{2}\le x\le 1,\\ 0, & if \quad x\ge 1.\end{cases} \tag{10-3}$$

$$\mu_C(x)=\begin{cases}0, & if \quad x\le 0,\\ x, & if \quad 0\le x\le 1,\\ 0, & if \quad x\ge 1.\end{cases} \tag{10-4}$$

By the use of the fuzzy mode transformation, the representative values for fuzzy subsets shown in Table 4 are determined.

Table 2. Representative values of linguistic terms

Linguistic Term	Representative Value
S	0
SC	0.25
TC	0.5
C	1

The membership functions for the porcelain data are also illustrated in Figure 6.

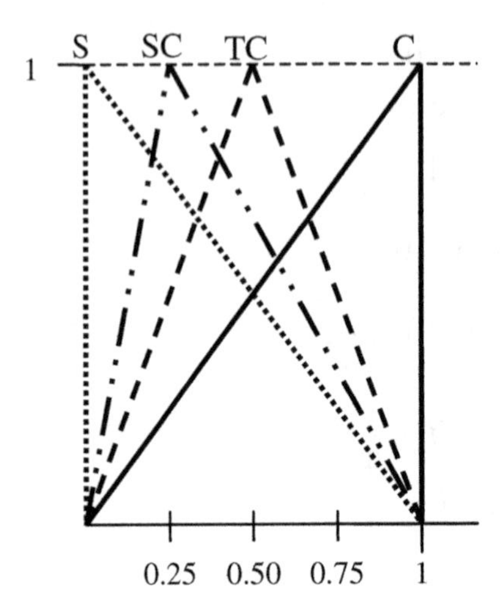

Fig. 6. Membership functions for the porcelain data.

By applying membership approach to the porcelain data, the fuzzy membership control chart is obtained as in Figure 7. As can be seen from Figure 7, only samples 8 and 29 are an out of control state. Note that fuzzy control limits, here, are calculated as follows (Taleb and Limam, 2002).

$$LCL = \max\{0,[CL - k\sigma]\},$$
$$UCL = \min\{1,[CL + k\sigma]\}, \tag{11}$$

where

$$\sigma = \frac{1}{m}\sum_{j=1}^{m} SD_j, \tag{12}$$

and

$$SD_j = \sqrt{\frac{1}{n-1}\sum_{i=1}^{t} k_{ij}(r_j - M_j)^2}. \tag{13}$$

The value of k is calculated by the use of *Monte Carlo* simulation so that a pre-specified type I error probability yields. In this example, the value of k, used here, is approximately 0.2795, CL is 0.136, and σ is 0.229. Then UCL and LCL are calculated as 0.200 and 0.072, respectively.

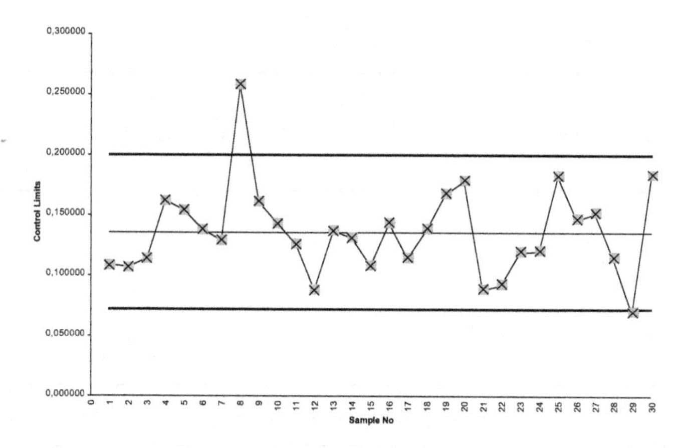

Fig. 7. Fuzzy membership control chart with fuzzy mode transformation.

Assume that the quality control expert decides for reduced inspection, say α = 0.30. If we apply the ASS approach, then the center lines and control limits for α=0.30 can be determined using Eq. 8, as:

$CL^L(\alpha = 0.30) = 0.040800$ $\quad$ $CL^R(\alpha = 0.30) = 0.740800$

$LCL^L(\alpha = 0.30) = 0.0$ $\quad$ $LCL^R(\alpha = 0.30) = 0.648321$

$UCL^L(\alpha = 0.30) = 0.082550$ $\quad$ $UCL^R(\alpha = 0.30) = 0.833279$

As can be seen from Figure 8, corresponding control chart for α=0.30, all the samples are in control.

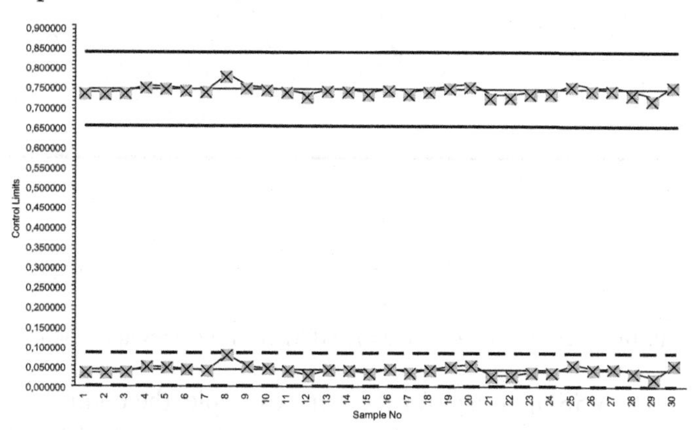

Fig. 8. α-cut fuzzy control chart for α=0.30 (ASS approach)

For a tighter inspection with α=0.50, control chart is obtained as shown in Figure 9. Note that sample 8 begins to be out of control while α is chosen as 0.39 or greater.

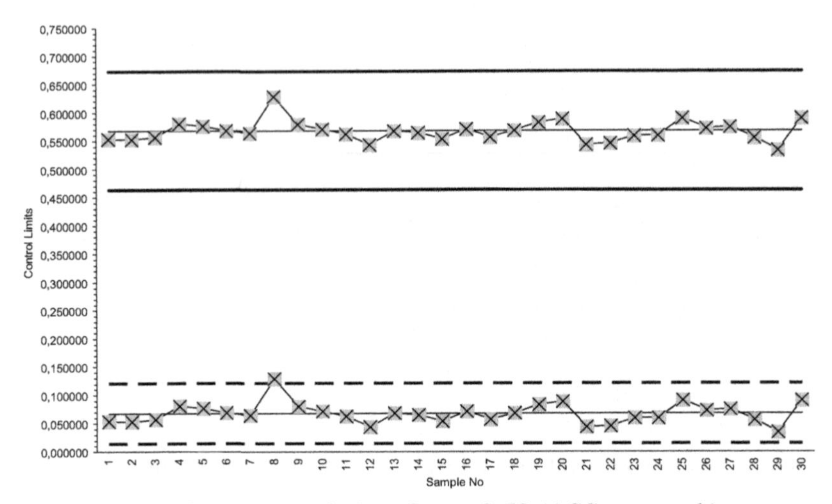

Fig. 9. α-cut fuzzy control chart for α=0.50 (ASS approach)

When α = 1.0, the crisp control limits are obtained as in Figure 10.

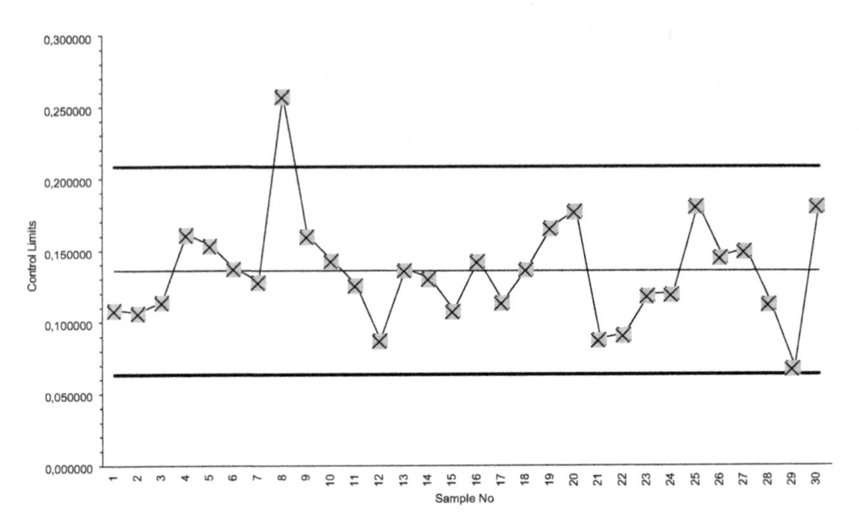

Fig. 10. α-Cut fuzzy control chart for α=1.0 (Crisp Case, ASS approach)

If we use the VSS approach for the same example, the control charts for α=0.30, α=0.50, and α=1.0 are obtained as in Figures 11-13. While increasing α-cut, namely tightening the inspection, sample 8 starts to be out of control while $\alpha \geq 0.33$.

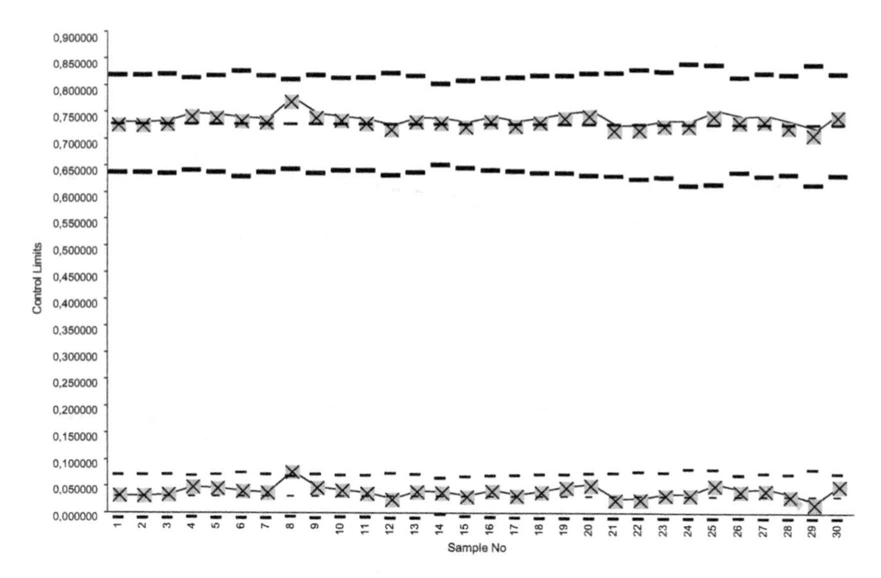

Fig. 11. α-cut fuzzy control chart for α=0.30 (VSS approach)

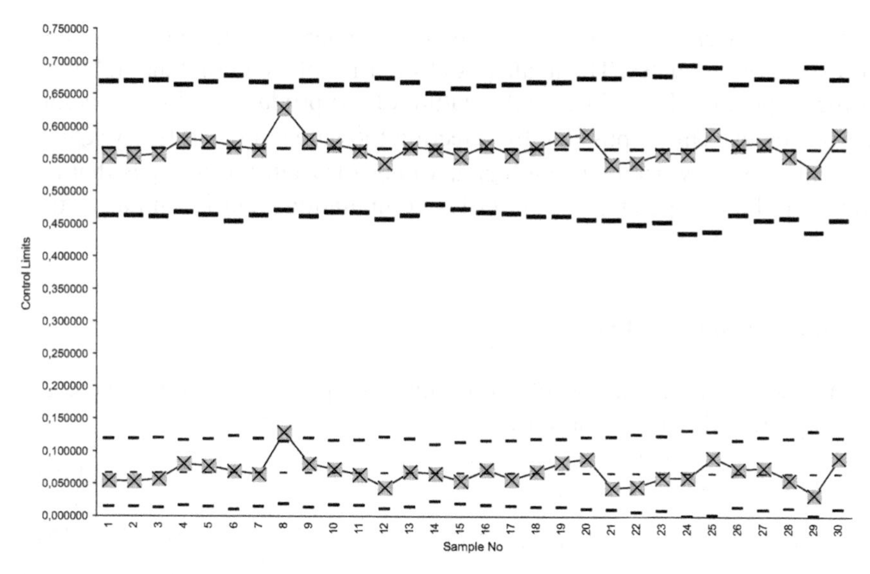

Fig. 12. α-Cut fuzzy control chart for α=0.50 (VSS approach)

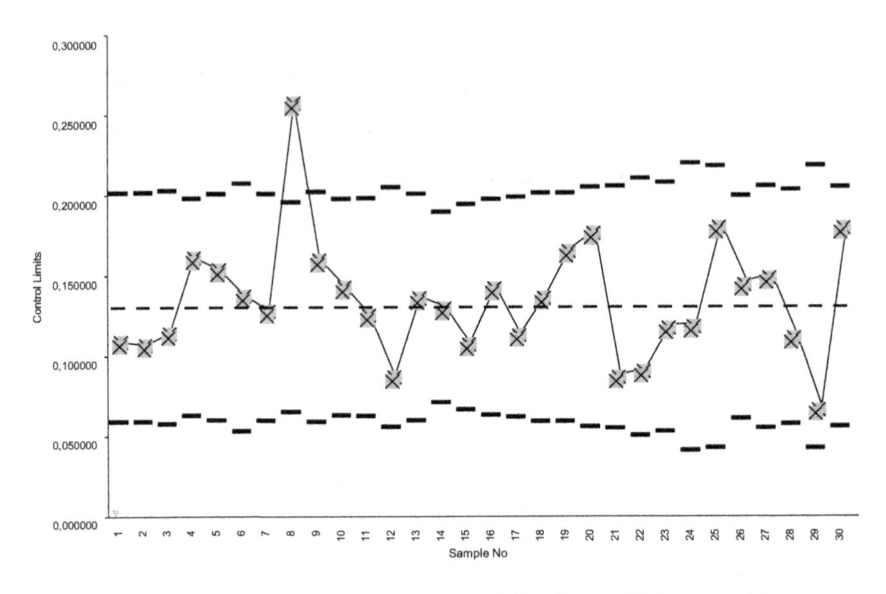

Fig. 13. α-Cut fuzzy control chart for α=1.0 (Crisp Case, VSS approach)

This approach differs from the previous studies from the point of view of inspection tightness. The quality controller is able to define the tightness of the inspection depending on the nature of the products and manufacturing processes. This is possible by selecting the value of α-cut freely. Quality controller may decide using higher values of α-cut for products that require a tighter inspection. It simplifies computations and is similar to the crisp control charts.

2.2 Fuzzy c control charts

In the crisp case, control limits for number of nonconformities are calculated by the following equations.

$$CL = \overline{c}, \tag{14}$$

$$LCL = \overline{c} - 3\sqrt{\overline{c}}, \tag{15}$$

and

$$UCL = \overline{c} + 3\sqrt{\overline{c}}, \tag{16}$$

where $\overline{c}$ is the mean of the nonconformities. In the fuzzy case, each sample, or subgroup, is represented by a trapezoidal fuzzy number (a, b, c, d) or a triangular fuzzy number (a, b, d), or (a, c, d) as shown in Figure 1.

Note that a trapezoidal fuzzy number becomes triangular when b=c. For the ease of representation and calculation, a triangular fuzzy number is also represented as trapezoidal by (*a*, *b*, *b*, *d*) or (*a*, *c*, *c*, *d*).

The center line, $\tilde{CL}$, given in Eq. 4, is the mean of fuzzy samples, and it is shown as $\left(\bar{a},\bar{b},\bar{c},\bar{d}\right)$ where $\bar{a}$, $\bar{b}$, $\bar{c}$, and $\bar{d}$ are the arithmetic means of the *a*, *b*, *c*, and *d*, respectively.

$$\tilde{CL} = \left(\frac{\sum_{j=1}^{m} a_j}{m}, \frac{\sum_{j=1}^{m} b_j}{m}, \frac{\sum_{j=1}^{m} c_j}{m}, \frac{\sum_{j=1}^{m} d_j}{m} \right) = \left(\bar{a},\bar{b},\bar{c},\bar{d}\right), \quad (17)$$

where *m* is the number of fuzzy samples.

Since the $\tilde{CL}$ is a fuzzy set, it can be represented by a fuzzy number whose fuzzy mode (multimodal) is the closed interval of $\left[\bar{b},\bar{c}\right]$. $\tilde{CL}$, $\tilde{LCL}$, and $\tilde{UCL}$ are calculated as shown in the following:

$$\tilde{CL} = \left(\bar{a},\bar{b},\bar{c},\bar{d}\right) = \left(CL_1, CL_2, CL_3, CL_4\right), \quad (18)$$

$$\begin{aligned} \tilde{LCL} &= \tilde{CL} - 3\sqrt{\tilde{CL}} = \left(\bar{a},\bar{b},\bar{c},\bar{d}\right) - 3\sqrt{\left(\bar{a},\bar{b},\bar{c},\bar{d}\right)} \\ &= \left(\bar{a} - 3\sqrt{\bar{d}}, \bar{b} - 3\sqrt{\bar{c}}, \bar{c} - 3\sqrt{\bar{b}}, \bar{d} - 3\sqrt{\bar{a}}\right) = \left(LCL_1, LCL_2, LCL_3, LCL_4\right), \end{aligned} \quad (19)$$

and

$$\begin{aligned} \tilde{UCL} &= \tilde{CL} + 3\sqrt{\tilde{CL}} = (\bar{a},\bar{b},\bar{c},\bar{d}) + 3\sqrt{(\bar{a},\bar{b},\bar{c},\bar{d})} \\ &= \left(\bar{a} + 3\sqrt{\bar{a}}, \bar{b} + 3\sqrt{\bar{b}}, \bar{c} + 3\sqrt{\bar{c}}, \bar{d} + 3\sqrt{\bar{d}}\right) \\ &= (UCL_1, UCL_2, UCL_3, UCL_4). \end{aligned} \quad (20)$$

An α-cut is a nonfuzzy set which comprises all elements whose membership is greater than or equal to α. Applying -cuts of fuzzy sets (Figure 1), values of a^{α} and d^{α} are determined as follows:

$$a^{\alpha} = a + \alpha(b - a), \quad (21)$$

$$d^{\alpha} = d - \alpha(d - c). \tag{22}$$

Similarly the α-cut fuzzy control limits can be stated as follows:

$$\tilde{CL}^{\alpha} = (\bar{a}^{\alpha}, \bar{b}, \bar{c}, \bar{d}^{\alpha}) = (CL_1^{\alpha}, CL_2, CL_3, CL_4^{\alpha}), \tag{23}$$

$$\begin{aligned} \tilde{LCL}^{\alpha} &= \tilde{CL}^{\alpha} - 3\sqrt{\tilde{CL}^{\alpha}} = (\bar{a}^{\alpha}, \bar{b}, \bar{c}, \bar{d}^{\alpha}) - 3\sqrt{(\bar{a}^{\alpha}, \bar{b}, \bar{c}, \bar{d}^{\alpha})} \\ &= \left(\bar{a}^{\alpha} - 3\sqrt{\bar{d}^{\alpha}}, \bar{b} - 3\sqrt{\bar{c}}, \bar{c} - 3\sqrt{\bar{b}}, \bar{d}^{\alpha} - 3\sqrt{\bar{a}^{\alpha}}\right) \\ &= (LCL_1^{\alpha}, LCL_2, LCL_3, LCL_4^{\alpha}), \end{aligned} \tag{24}$$

and

$$\begin{aligned} \tilde{UCL}^{\alpha} &= \tilde{CL}^{\alpha} + 3\sqrt{\tilde{CL}^{\alpha}} = (\bar{a}^{\alpha}, \bar{b}, \bar{c}, \bar{d}^{\alpha}) + 3\sqrt{(\bar{a}^{\alpha}, \bar{b}, \bar{c}, \bar{d}^{\alpha})} \\ &= \left(\bar{a}^{\alpha} + 3\sqrt{\bar{a}^{\alpha}}, \bar{b} + 3\sqrt{\bar{b}}, \bar{c} + 3\sqrt{\bar{c}}, \bar{d}^{\alpha} + 3\sqrt{\bar{d}^{\alpha}}\right) \\ &= (UCL_1^{\alpha}, UCL_2, UCL_3, UCL_4^{\alpha}). \end{aligned} \tag{25}$$

The results of these equations can be illustrated as in Figure 14.

Fuzzy Mode

If the membership function is unimodal, the fuzzy mode is unique. The fuzzy numbers of nonconformities of initially available and incoming samples are transformed to crisp numbers via fuzzy mode transformation. Since the trapezoidal membership function is multimodal, the fuzzy mode is the set of points between b and c, $[b,c]$. $\tilde{UCL}$, $\tilde{CL}$, and $\tilde{LCL}$ given in Eqs. 18-20 are transformed to their representative values using fuzzy mode to determine control limits for these charts. Since $\tilde{UCL}$, $\tilde{CL}$, and $\tilde{LCL}$ are multimodal, their fuzzy modes are closed intervals whose membership degrees are 1.

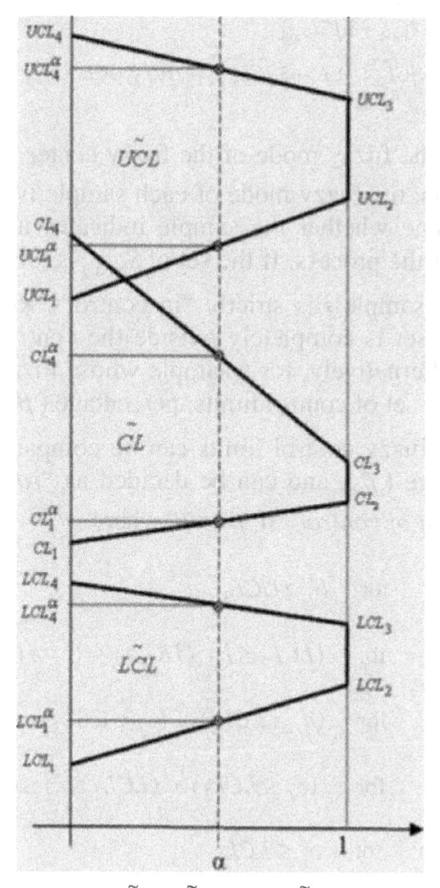

Fig. 14. Representations of $\tilde{UCL}$, $\tilde{CL}$, and $\tilde{LCL}$

Referring to the representation in Figure 14, the fuzzy mode of sample j, $S_{\text{mod},j}$, and corresponding closed intervals of CL, LCL, and UCL are then determined by the following equations:

$$S_{\text{mod},j} = \left[b_j, c_j\right], \tag{26}$$

$$CL_{\text{mod}} = f_{\text{mod}}(\tilde{CL}) = \left[CL_2, CL_3\right], \tag{27}$$

$$\begin{aligned} LCL_{\text{mod}} &= CL_{\text{mod}} - 3\sqrt{CL_{\text{mod}}} \\ &= \left[\left(CL_2 - 3\sqrt{CL_2}\right), \left(CL_3 - 3\sqrt{CL_3}\right)\right] = \left[LCL_2, LCL_3\right], \end{aligned} \tag{28}$$

and

$$UCL_{\text{mod}} = CL_{\text{mod}} + 3\sqrt{CL_{\text{mod}}}$$
$$= \left[\left(CL_2 + 3\sqrt{CL_2}\right), \left(CL_3 + 3\sqrt{CL_3}\right)\right] = \left[UCL_2, UCL_3\right], \tag{29}$$

where $f_{\text{mod}}(\tilde{CL})$ is the fuzzy mode of the fuzzy center line. After calculation of control limits, the fuzzy mode of each sample is compared to these intervals to determine whether the sample indicates in-control or out of control situation for the process. If the set of $S_{\text{mod},j}$ is totally covered by the control limits, then sample j is strictly "in control". Reversely, a sample whose fuzzy mode set is completely outside the control limits is strictly "out of control". Alternatively, for a sample whose fuzzy mode set is partially included in the set of control limits, percentage (β_j) of the set falling into the interval of fuzzy control limits can be compared to a predefined acceptable percentage (β), and can be decided as *"rather in control"* if $\beta_j \geq \beta$ or *"rather out of control"* if $\beta_j < \beta$ where

$$\beta_j = \begin{cases} 0\ , & \text{for} \quad b_j \geq UCL_3, \\ \dfrac{UCL_3 - b_j}{c_j - b_j}\ , & \text{for} \quad (LCL_2 \leq b_j \leq UCL_3) \wedge (c_j \geq UCL_3), \\ 1\ , & \text{for} \quad (b_j \geq LCL_2) \wedge (c_j \leq UCL_3), \\ \dfrac{LCL_2 - b_j}{c_j - b_j}\ , & \text{for} \quad (b_j \leq LCL_2) \wedge (LCL_2 \leq c_j \leq UCL_3), \\ 0\ , & \text{for} \quad c_j \leq LCL_2. \end{cases} \tag{30}$$

These decisions are also illustrated in Figure 15.

The process control conditions can be stated as below.

$$\text{Process Control} = \begin{cases} \text{in-control, for } \beta = 1\ (b_j \geq LCL_2 \ \wedge\ c_j \leq UCL_3), \\ \text{out-of-control, for } \beta{=}0\ (b_j \geq UCL_3 \ \vee\ c_j \leq LCL_2), \\ \left.\begin{array}{l} \text{rather in-control, for } \beta_j \geq \beta \\ \text{rather out of control, for } \beta_j < \beta \end{array}\right\} \text{otherwise.} \end{cases} \tag{31}$$

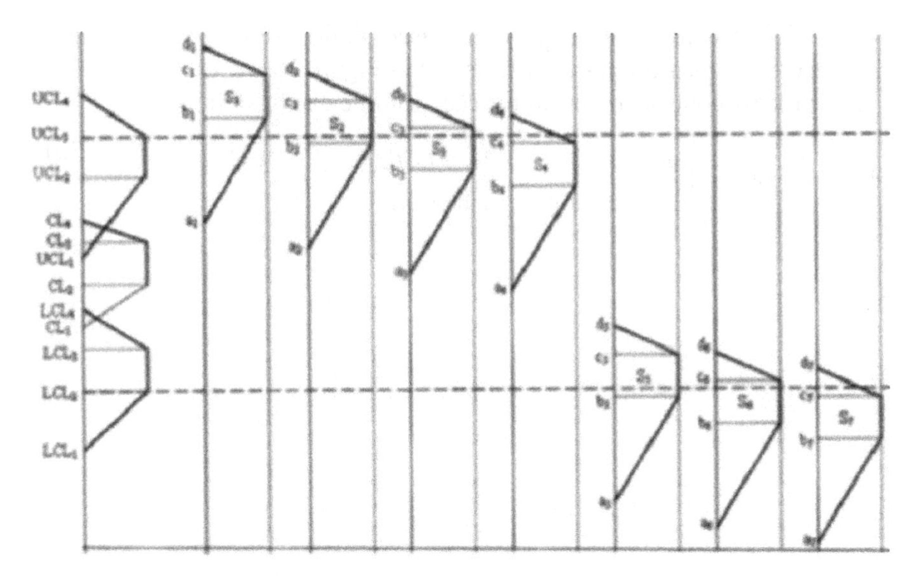

Fig. 15. Example samples (S_j, $j=1,2,\ldots,7$) resulting with four types of different decisions for $\beta = 0.50$:

S_1, S_7: "out of control" (mode set of S_1 and S_7 are completely outside the fuzzy control limits, $\beta_1 = \beta_7 = 0$),

S_2, S_6: "rather out of control", ($\beta = 0.50$, $\beta_2, \beta_6 \le 0.50$),

S_3, S_5: "rather in control" ($\beta = 0.50$, $\beta_3, \beta_5 \ge 0.50$),

S_4: "in control" (mode set of S_4 is completely inside the fuzzy control limits, $\beta_4 = 1$)

α-level Fuzzy Midrange

α-level fuzzy midrange is used as the fuzzy transformation method when calculating control limits (Eqs. 18-20).

$$CL_{mr}^{\alpha} = f_{mr}^{\alpha}(\tilde{CL}) = \frac{CL_1^{\alpha} + CL_4^{\alpha}}{2} = \frac{CL_1 + CL_4 + \alpha\left[(CL_2 - CL_1) - (CL_3 - CL_4)\right]}{2}, \tag{32}$$

$$LCL_{mr}^{\alpha} = CL_{mr}^{\alpha} - 3\sqrt{CL_{mr}^{\alpha}}, \tag{33}$$

and

$$UCL_{mr}^{\alpha} = CL_{mr}^{\alpha} + 3\sqrt{CL_{mr}^{\alpha}}. \tag{34}$$

The possible process control decisions based on the α-level (α=0.70) fuzzy midrange for some samples are illustrated in Figure 16.

The process control condition for each sample can be defined as:

$$\text{Process Control} = \begin{cases} \text{in-control} & , \text{ for } \; LCL^{\alpha}_{mr} \leq S^{\alpha}_{mr,j} \leq UCL^{\alpha}_{mr}, \\ \text{out of control} & , \text{ otherwise.} \end{cases} \tag{35}$$

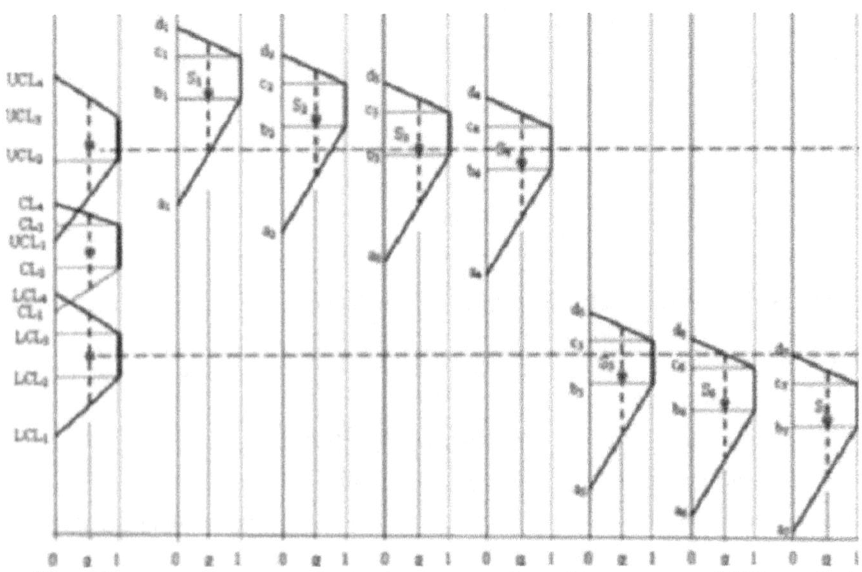

Fig. 16. Example samples (Sj, j=1,2,…,7) resulting with two types of different decisions for:
S1, S2, S5, S6 , S7 : "out of control" (α-level fuzzy midrange (•) of S1, S2, S5 , S6 , and S7 outside the control limits),
S3, S4 : "in control" (α-level midrange of S3 and S4 are inside the control limits)

3 Fuzzy Control Charts in Fuzzy Spaces

In this approach, the linguistic data are not transformed into representative values using fuzzy transformation in order to save representative properties of the fuzzy samples: Both the samples and control limits are represented and operated as fuzzy numbers. The fuzzy control limits, $\tilde{UCL}$, $\tilde{CL}$, and $\tilde{LCL}$ are as given in Eq. 18-20. α-level fuzzy control limits,

$\tilde{UCL}^{\alpha}$, $\tilde{CL}^{\alpha}$, and, $\tilde{LCL}^{\alpha}$, can be determined by fuzzy arithmetic as shown in Eqs. 36-38.

$$\tilde{CL}^{\alpha} = \left(CL_1^{\alpha}, CL_2, CL_3, CL_4^{\alpha}\right) \tag{36}$$

$$\tilde{LCL}^{\alpha} = \tilde{CL}^{\alpha} - 3\sqrt{\tilde{CL}^{\alpha}} = \left(CL_1^{\alpha}, CL_2, CL_3, CL_4^{\alpha}\right) - 3\sqrt{\left(CL_1^{\alpha}, CL_2, CL_3, CL_4^{\alpha}\right)} \tag{37}$$

$$\tilde{LCL}^{\alpha} = \left(CL_1^{\alpha} - 3\sqrt{CL_4^{\alpha}}, CL_2 - 3\sqrt{CL_3}, CL_3 - 3\sqrt{CL_2}, CL_4^{\alpha} - 3\sqrt{CL_1^{\alpha}}\right) = \left(LCL_1^{\alpha}, LCL_2, LCL_3, LCL_4^{\alpha}\right)$$

$$\tilde{UCL}^{\alpha} = \tilde{CL}^{\alpha} + 3\sqrt{\tilde{CL}^{\alpha}} = \left(CL_1^{\alpha}, CL_2, CL_3, CL_4^{\alpha}\right) + 3\sqrt{\left(CL_1^{\alpha}, CL_2, CL_3, CL_4^{\alpha}\right)} \tag{38}$$

$$\tilde{UCL}^{\alpha} = \left(CL_1^{\alpha} + 3\sqrt{CL_1^{\alpha}}, CL_2 + 3\sqrt{CL_2}, CL_3 + 3\sqrt{CL_3}, CL_4^{\alpha} + 3\sqrt{CL_4^{\alpha}}\right) =$$

where,

$$CL_1^{\alpha} = CL_1 + \alpha\left(CL_2 - CL_1\right) \tag{39}$$

$$CL_4^{\alpha} = CL_4 - \alpha\left(CL_4 - CL_3\right)$$

The decision about whether the process is in control can be made according to the percentage area of the sample which remains inside the $\tilde{UCL}$ and/or $\tilde{LCL}$ defined as fuzzy sets. When the fuzzy sample is completely involved by the fuzzy control limits, the process is said to be "*in-control*". If a fuzzy sample is totally excluded by the fuzzy control limits, the process is said to be "*out of control*". Otherwise, a sample is partially included by the fuzzy control limits. In this case, if the percentage area which remains inside the fuzzy control limits (β_j) is equal or greater than a predefined acceptable percentage (β), then the process can be accepted as "*rather in-control*"; otherwise it can be stated as "*rather out of control*". Possible decisions resulting from DFA are illustrated in Figure 17. Parameters for determination of the sample area outside the control limits for á-level fuzzy cut are LCL_1, LCL_2, UCL_3, UCL_4, a, b, c, d, and á. The shape of the control limits and fuzzy sample are formed by the lines of $\overline{LCL_1LCL_2}$, $\overline{UCL_3LCL_4}$, $\overline{ab}$, and $\overline{cd}$. A flowchart to calculate area of the fuzzy sample outside the control limits is given in Figure 18.

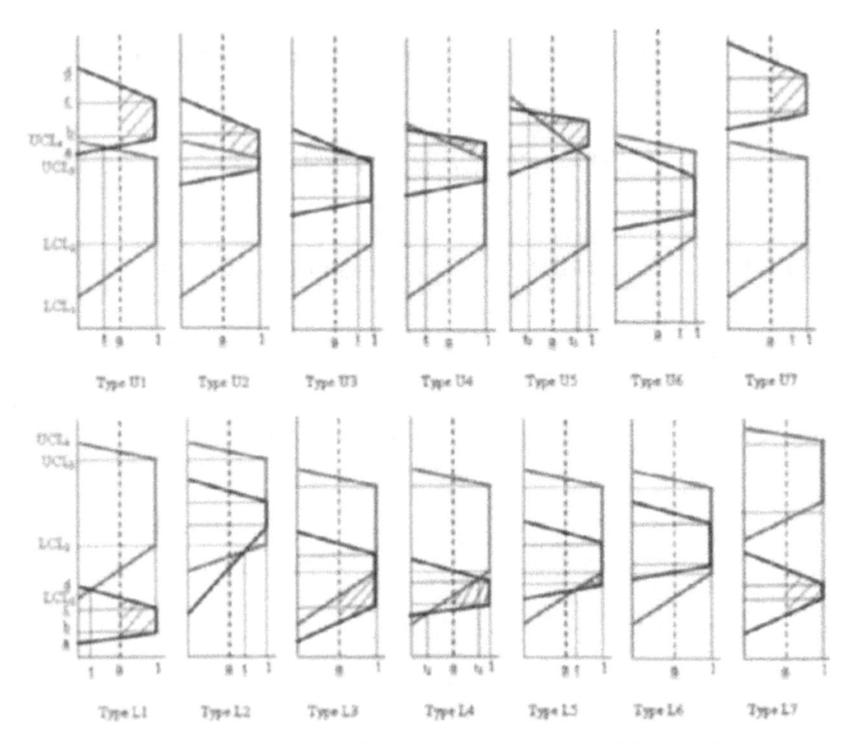

Fig. 17. Illustration of all possible sample areas outside the fuzzy control limits at α-level cut.

Sample area above the upper control limits, A_{out}^{U}, and sample area falling below the lower control limits, A_{out}^{L} ,can be calculated according to the flowchart given in Figure 18. Equations to compute A_{out}^{U} and A_{out}^{L} are given in Appendix A. Then, total sample area outside the fuzzy control limits, A_{out}, is the sum of the areas below fuzzy lower control limit and above fuzzy upper control limit. Percentage sample area within the control limits is calculated as given in Eq. 40.

$$\beta_j^{\alpha} = \frac{S_j^{\alpha} - A_{out,j}^{\alpha}}{S_j^{\alpha}} \tag{40}$$

where S_j^{α} is the sample area at á-level cut.

In contrast to the methods, which use fuzzy transformations, this approach is very flexible and would be more accurate since both the linguistic data and control limits are not transformed into representative values to prevent the loss of properties of the samples.

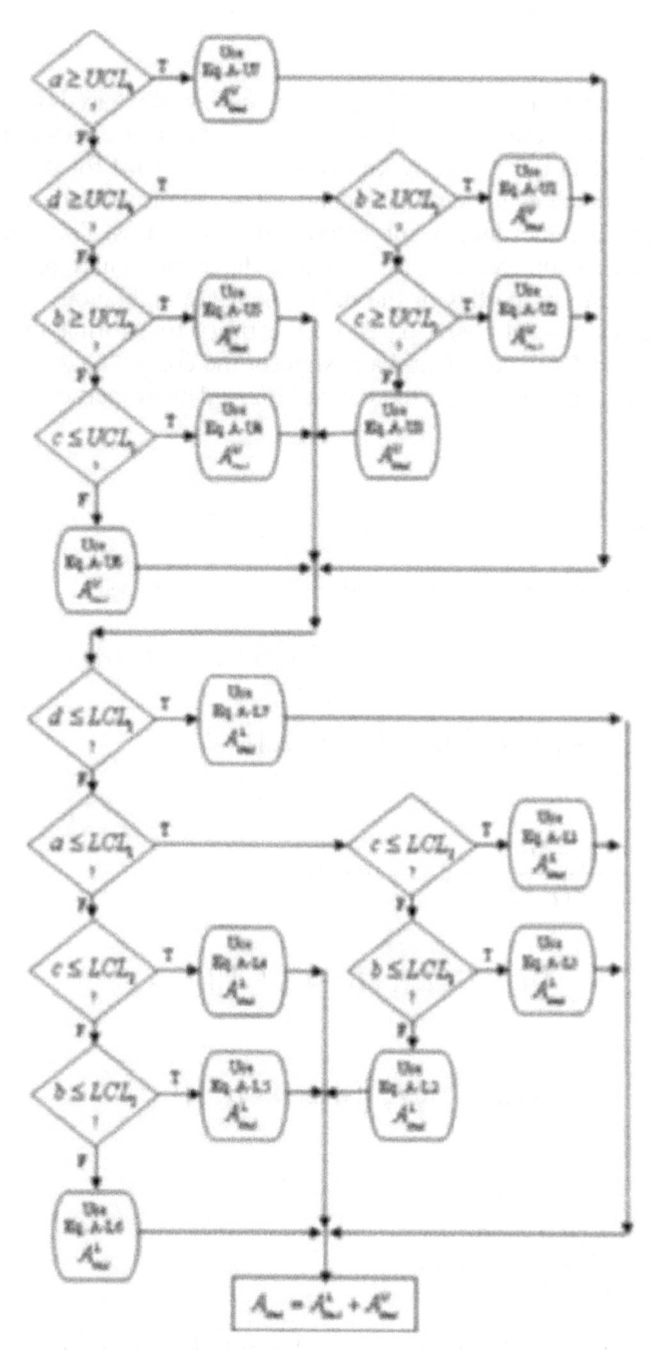

Fig. 18. Flowchart to compute the area of a fuzzy sample (a, b, c, d) falling outside the fuzzy control limits. (See appendix for the equations)

4 Conclusion

The fuzzy control charts in the literature are commonly based on the fuzzy transformations to crisp cases. The theory of classical control charts requires all the data to be exactly known that is very hard for the attributes control charts such as p-charts or c-charts. The data to construct attributes control charts includes human subjectivity and vagueness. Transformations to scalar values simplify the construction of the control chart, but results in the loss of many properties of the data which are ambiguous or not well defined. Fuzzy control charts based on the fuzzy transformation methods are reviewed and developed, and a design for the control charts in the case of vague data using fuzzy sets to keep the fuzzy data properties is handled.

APPENDIX

Equations to compute sample area outside the control the limits.

$$A_{out}^{U}=\frac{1}{2}\left[\left(d^{\alpha}-UCL_{4}^{\alpha}\right)+\left(d^{t}-UCL_{4}^{t}\right)\right]\left(\max\left(t-\alpha,0\right)\right)+ \frac{1}{2}\left[\left(d^{z}-a^{z}\right)+\left(c-b\right)\right]\left(\min\left(1-t,1-\alpha\right)\right) \tag{A-U1}$$

where,

$$t=\frac{UCL_{4}-a}{\left(b-a\right)+\left(c-b\right)} \text{ and } z=\max\left(t,\alpha\right)$$

$$A_{out}^{U}=\frac{1}{2}\left[\left(d^{\alpha}-UCL_{4}^{\alpha}\right)+\left(c-UCL_{3}\right)\right]\left(1-\alpha\right) \tag{A-U2}$$

$$A_{out}^{U}=\frac{1}{2}\left(d^{\alpha}-UCL_{4}^{\alpha}\right)\left(\max\left(t-\alpha,0\right)\right) \tag{A-U3}$$

where

$$t=\frac{UCL_{4}-d}{\left(UCL_{4}-UCL_{3}\right)-\left(d-c\right)}$$

$$A_{out}^{U}=\frac{1}{2}\left[\left(c-UCL_{3}\right)+\left(d^{z}-UCL_{4}^{z}\right)\right]\left(\min\left(1-t,1-\alpha\right)\right) \tag{A-U4}$$

where

$$t=\frac{UCL_4-d}{(UCL_4-UCL_3)-(d-c)} \text{ and } z=\max(t,\alpha)$$

$$A_{out}^U=\frac{1}{2}\left[\left(d^{z_2}-UCL_4^{z_2}\right)+\left(d^{t_1}-UCL_4^{t_1}\right)\right]\left(\min\left(\max\left(t_1-\alpha,0\right),t_1-t_2\right)\right) \quad \text{(A-U5)}$$

$$\frac{1}{2}\left[\left(d^{z_1}-a^{z_1}\right)+(c-b)\right]\left(\min\left(1-t_1,1-\alpha\right)\right)$$

where

$$t_1=\frac{UCL_4-a}{(b-a)+(UCL_4-UCL_3)},$$

$$t_2=\frac{UCL_4-d}{(UCL_4-UCL_3)-(d-c)},$$

$$z_1=\max(\alpha,t_1) \text{, and } z_2=\max(\alpha,t_2)$$

$$A_{out}^U=0 \quad \text{(A-U6)}$$

$$A_{out}^U=\frac{1}{2}\left[\left(d^{\alpha}-a^{\alpha}\right)+(c-b)\right](1-\alpha) \quad \text{(A-U7)}$$

$$A_{out}^L=\frac{1}{2}\left[\left(LCL_1^{\alpha}-a^{\alpha}\right)+\left(LCL_1^{t}-a^{t}\right)\right]\left(\max\left(t-\alpha,0\right)\right)+ \quad \text{(A-L1)}$$

$$\frac{1}{2}\left[\left(d^{z}-a^{z}\right)+(c-b)\right]\left(\min\left(1-t,1-\alpha\right)\right)$$

where

$$t=\frac{d-LCL_1}{(LCL_2-LCL_1)+(d-c)} \text{ and } z=\max(\alpha,t)$$

$$A_{out}^L=\frac{1}{2}\left[\left(d^{\alpha}-a^{\alpha}\right)+(c-b)\right](1-\alpha) \quad \text{(A-L2)}$$

$$A_{out}^{L}=\frac{1}{2}\left[\left(LCL_1^{\alpha}-a^{\alpha}\right)+\left(LCL_2-b\right)\right]\left(1-\alpha\right) \tag{A-L3}$$

$$A_{out}^{L}=\frac{1}{2}\left[\left(LCL_1^{z_2}-a^{z_2}\right)+\left(LCL_1^{t_1}-a^{t_1}\right)\right]\left(\min\left(\max\left(t_1-\alpha,\right.\right.\right. \tag{A-L4}$$

$$\frac{1}{2}\left[\left(d^{z_1}-a^{z_1}\right)+\left(c-b\right)\right]\left(\min\left(1-t,1-\alpha\right)\right)$$

where

$$t_1=\frac{d-LCL_1}{\left(LCL_2-LCL_1\right)+\left(d-c\right)},$$

$$t_2=\frac{a-LCL_1}{\left(LCL_2-LCL_1\right)-\left(b-a\right)}$$

$z_1=\max\left(\alpha,t_1\right)$, and $z_2=\max\left(\alpha,t_2\right)$

$$A_{out}^{L}=\frac{1}{2}\left[\left(LCL_1^{z}-a^{z}\right)+\left(LCL_2-b\right)\right]\left(\min\left(1-t,1-\alpha\right)\right) \tag{A-L5}$$

where

$$t=\frac{a-LCL_1}{\left(LCL_2-LCL_1\right)-\left(b-a\right)}$$, and $z=\max\left(\alpha,t\right)$

$$A_{out}^{L}=0 \tag{A-L6}$$

$$A_{out}^{L}=\frac{1}{2}\left[\left(d^{\alpha}-a^{\alpha}\right)+\left(c-b\right)\right]\left(1-\alpha\right) \tag{A,L7}$$

References

Asai, K. (1995), Fuzzy Systems for Management, *IOS Press*, Amsterdam.

Bradshaw, C. W. (1983), A fuzzy set theoretic interpretation of economic control limits, *European Journal of Operational Research*, 13, 403-408.

Chang S.I., Aw, C.A. (1996), A neural fuzzy control chart for detecting and classifying process mean shifts, *International Journal of Production Research*, 34, 2265-2278.

El-Shal, S.M., Morris, A.S. (2000), A fuzzy rule-based algorithm to improve the performance of statistical process control in quality systems, *Journal of Intelligent & Fuzzy Systems*, 9, 207-223.

Grzegorzewski P., Control Charts for Fuzzy Data, In: Proceedings of the 5th European Congress on Intelligent Techniques and Soft Computing EUFIT'97, Aachen, 1997, ss. 1326-1330.

Grzegorzewski P., Hryniewicz O., Soft Methods in Statistical Quality Control, Control and Cybernetics, 29 (2000), 119-140.

Grzegorzewski P., Testing Fuzzy Hypotheses with Vague Data, In: Statistical Modeling, Analysis and Management of Fuzzy Data, Bertoluzza C., Gil M.A., Ralescu D. (Eds.), Springer - Physica Verlag, Heidelberg, 2002, pp. 213-225.

Gülbay, M., Kahraman, C., Ruan, D. (2004), α-cut fuzzy control charts for linguistic data, *International Journal of Intelligent Systems*, 19, 1173-1196.

Höppner J., Statistiche Proceßkoontrolle mit Fuzzy-Daten, Ph.D. Dissertation, Ulm University, 1994.

Höppner J., Wolff H., The Design of a Fuzzy-Shewart Control Chart, Research Report, Würzburg University, 1995.

Kahraman, C., Tolga, E., Ulukan, Z. (1995), Using triangular fuzzy numbers in the tests of control charts for unnatural patterns, in Proceedings of *INRIA / IEEE Conference on Emerging Technologies and Factory Automation*, October, 10-13, Paris-France, 3, 291- 298.

Kanagawa, A., Tamaki, F., Ohta, H. (1993), Control charts for process average and variability based on linguistic data, *International Journal of Production Research*, 2, 913-922.

Laviolette, M., Seaman, J. W., Barrett, J. D., Woodall, W. H. (1995), A probabilistic and statistical view of fuzzy methods, with discussion., *Technometrics*, 37, 249-292.

Raz, T., Wang, J-H. (1990), Probabilistic and membership approaches in the construction of control charts for linguistic data, *Production Planning and Control*, 1, 147-157.

Rowlands, H., Wang, L.R. (2000), An approach of fuzzy logic evaluation and control in SPC, *Quality and Reliability Engineering International*, 16, 91-98.

Shewhart, W. A. (1931), Economic control of quality of manufactured product, *D. Van Nostrand, In.c,* Princeton, N. J.

Taleb, H., Limam, M. (2002), On fuzzy and probabilistic control charts, *International Journal of Production Research*, 40, 2849-2863.

Wang, J-H., Chen, C-H. (1995), Economic statistical *np*-control chart designs based on fuzzy optimization, *International Journal of Quality and Reliability Management*, 12, 88-92.

Wang, J-H., Raz, T. (1988), Applying fuzzy set theory in the development of quality control charts, *International Industrial Engineering Conference proceedings,* Orlando, FL, 30-35.

Wang, J-H., Raz, T. (1990), On the construction of control charts using linguistic variables, *International Journal of Production Research*, 28, 477-487.

Woodall, W., Tsui, K-L., Tucker, G.L. (1997), A review of statistical and fuzzy control charts based on categorical data, *Frontiers in Statistical Quality Control 5 (Heidelberg, Germany: Physica-Verlag)*, 83-89.

Fuzzy Rule Reduction and Tuning of Fuzzy Logic Controllers for a HVAC System*

R. Alcalá, J. Alcalá-Fdez, M.J. Gacto, and F. Herrera

Dept. of Computer Science and Artificial Intelligence
University of Granada, E-18071 – Granada, Spain
{alcala,jalcala,herrera}@decsai.ugr.es, mjgacto@ugr.es

Summary. Heating, Ventilating and Air Conditioning (HVAC) Systems are equipments usually implemented for maintaining satisfactory comfort conditions in buildings. The design of Fuzzy Logic Controllers (FLCs) for HVAC Systems is usually based on the operator's experience. However, an initial rule set drawn from the expert's experience sometimes fail to obtain satisfactory results, since inefficient or redundant rules are usually found in the final Rule Base. Moreover, in our case, the system being controlled is too complex and an optimal controller behavior is required.

Rule selection methods directly obtain a subset of rules from a given fuzzy rule set, removing inefficient and redundant rules and, thereby, enhancing the controller interpretability, robustness, flexibility and control capability. On the other hand, different parameter optimization techniques could be applied to improve the system accuracy by inducing a better cooperation among the rules composing the final Rule Base.

In this chapter, we present a study of how several tuning approaches can be applied and combined with a rule selection method to obtain more compact and accurate FLCs concerning energy performance and indoor comfort requirements of a HVAC system. This study has been performed considering a physical modelization of a real test environment.

Keywords: HVAC systems; Fuzzy logic controller; tuning approaches ; rule selection methods; parameter optimization techniques.

1 Introduction

HVAC Systems are equipments usually implemented for maintaining satisfactory comfort conditions in buildings. The energy consumption as well as indoor comfort aspects of ventilated and air conditioned buildings are highly dependent on the design, performance and control of their HVAC systems and

* Supported by the Spanish CICYT Project TIC-2002-04036-C05-01 (KEEL).

R. Alcalá et al.: *Fuzzy Rule Reduction and Tuning of Fuzzy Logic Controllers for a HVAC System*, StudFuzz **201**, 89–117 (2006)
www.springerlink.com

equipments. Therefore, the use of appropriate automatic control strategies, as FLCs [18, 38, 39], for HVAC systems control could result in important energy savings when compared to manual control, specially when they explicitly try to minimize the energy consumption [1, 5, 31, 43].

In current systems [5, 7, 22, 31, 37, 43, 44, 51, 52], several criteria are individually considered, thermal regulation, energy consumption or comfort improvement (the next section includes a deeper explanation of these works). However, different criteria must be considered jointly in order to reduce the energy consumption maintaining a desired comfort level. In our case, five criteria will be optimized and 17 variables are considered by the FLC. Furthermore, control systems in buildings are often designed using rules of thumb not always compatible with the controlled equipment requirements and energy performance. Therefore, the different involved criteria should be optimized for a good performance of the HVAC system.

A way to solve these problems is removing rules that degrade the system behavior [32, 33, 36] (*rule selection* methods). Other technique that improves the FLC performance is the tuning of parameters. Two different tuning approaches could be considered:

- *The classical tuning*: This approach consists of a tuning of the parameters that define the linguistic labels [24, 28, 34, 35, 40]. In this way, considering triangular-shaped membership functions three parameters are optimized.
- *The lateral tuning*: This technique was presented in [4], to reduce the size of the search space in complex problems, since the 3 parameters considered per label are reduced to only one symbolic translation parameter.

The smart combination of rule selection with tuning techniques can improve even more the system behavior [3, 23]. In this work, we present a study of how these tuning approaches can be applied and combined with a rule selection method to obtain more compact and accurate FLCs concerning energy performance and indoor comfort requirements of a HVAC system.

This contribution is arranged as follows. In the next section, the basics of the HVAC systems control problem are presented, studying how FLCs can be applied to it. In Section 3, the proposed real test site and the control objectives are introduced, establishing the concrete problem that will be solved. Section 4 introduces the rule selection, the classical and the lateral tuning. Section 5 describes the different evolutionary post-processing algorithms. Experimental results are shown in Section 6. Finally, Section 7 points out some conclusions.

2 Heating, Ventilating, and Air Conditioning Systems

A HVAC system is comprised by all the components of the appliance used to condition the indoor air of a building. The HVAC system is needed to provide the occupants with a comfortable and productive working environment which satisfies their physiological needs. Therefore, in a quiet and energy-efficient way at low life-cycle cost, a HVAC system should achieve two main tasks:

- To dilute and remove emission from people, equipment and activities and to supply clean air (Indoor Air Quality).
- To maintain a good thermal quality (Thermal Climate).

There are no statistical data collected on types and sizes of HVAC systems delivered to each type of building in different European countries. Therefore, to provide a HVAC system compatible with the ambiance is a task of the Building Energy Management System (BEMS) designer depending on its own experience. In Figure 1, a typical office building HVAC system is presented. This system consists of a set of components to be able to raise and lower the temperature and relative humidity of the supply air.

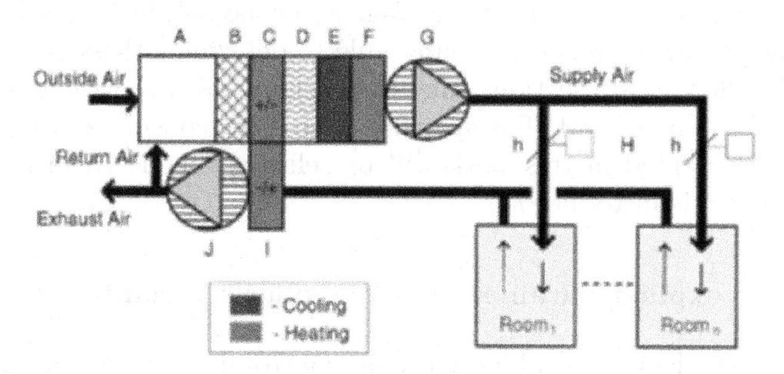

A - This module mixes the return and the outside air to provide supply air, and also closes outside air damper and opens return air damper when fan stops. **B** - It is a filter to reduce the outside air emissions to supply air. **C** - The preheater/heat recovery unit preheats the supply air and recovers energy from the exhaust air. **D** - A humidifier raising the relative humidity in winter. **E** - This is a cooler to reduce the supply air temperature and/or humidity. **F** - An after-heater unit to raise the supply air temperature after humidifier or to raise the supply air temperature after latent cooling (dehumidifier). **G** - The supply air fan. **H** - The dampers to demand controlled supply air flow to rooms. **I** - It is a heat recovery unit for energy recovery from exhaust air. **J** - The exhaust air fan.

Fig. 1. Generic structure of an office building HVAC system

2.1 The HVAC System Control Problem

Temperature and relative humidity are essential factors in meeting physiological requirements. When temperature is above or below the comfort range, the environment disrupts person's metabolic processes and disturbs his activities.

Therefore, a HVAC system is essential to a building in order to keep occupants comfortable. A well-designed operated, and maintained HVAC system is essential for a habitable and functional building environment. Outdated, inappropriate, or misapplied systems result in comfort complaints, indoor air quality issues, control problems, and exorbitant utility costs. Moreover, many

HVAC systems do not maintain an uniform temperature throughout the structure because those systems employ unsophisticated control algorithms. In a modern intelligent building, a sophisticated control system should provide excellent environmental control [5].

Within this framework (building automation), the objective of a global controller is to maintain the indoor environment within the desired (or stipulated) limits. In our case, to maintain environmental conditions within the comfort zone and to control the indoor air quality. Furthermore, other important objectives are usually required, e.g, energy savings (our main objective), system stability, etc. In any case, numerous factors have to be considered in order to achieve these objectives. It makes the system being controlled very complex and present a strong non linearity.

To obtain an optimal controller, control and controlled parameters [2] have to be chosen regarding the control strategy being implemented, the technical feasibility of the measurements as well as economic considerations. Fortunately, the BEMS designer is usually able to determine these parameters.

In the following subsections, the most usually used control and controlled parameters are presented. The specific parameters considered in the test site (building) presented in this work will be selected among them in Section 3, where this site is introduced.

Control or explicit parameters: Controller's variables

To identify the FLC's variables, various (control or explicit) parameters may be considered depending on the HVAC system, sensors and actuators. Usually, these parameters are selected among the following ones:

- *Predicted Mean Vote (PMV) index for thermal comfort*: Instead of only using air temperature as a thermal comfort index, we could consider the more global PMV index selected by international standard organization ISO 7730 (http://www.iso.org/iso/en/ISOOnline.frontpage), incorporating relative humidity and mean radiant temperature.
- *Difference between supply and room temperatures*: Possible disturbances can be related to the difference between supply and mean air temperature. When ventilation systems are used for air conditioning, such a criterion can be important.
- *CO_2 concentration*: Indoor air quality was found to be critical. As CO_2 concentration is a reliable index of the pollution emitted by occupants, it can be selected as indoor air quality index. It is therefore supposed that

[2] Control or explicit parameters are variables which may be used as inputs or outputs for a control strategy (controller's variables), whilst controlled or implicit parameters are variables which are affected by the action of a controlled device, and may be considered in order to evaluate the performance of such controller (problem's objectives).

both the building and the HVAC system have been properly designed and that occupants actually are the main source of pollution.

- *Outdoor temperature*: Outdoor temperature also needs to be accounted for, since during mid-season periods (or even mornings in summer periods) its cooling (or heating) potential through ventilation can be important and can reduce the necessity of applying mechanical cooling (or heating).
- *HVAC system actuators*: They directly depends on the concrete HVAC system, e.g., valve positions, operating modes, fan speeds, etc.

Controlled or implicit parameters: Problem's objectives

To identify global indices for assessment of the indoor building environment, various (controlled or implicit) parameters may be measured depending on the objectives of the control strategy. In these kinds of problems, these parameters could be selected among:

- Thermal comfort parameters: Indoor climate control is one of the most important goals of intelligent buildings. Among indoor climate characteristics, thermal comfort is of major importance. This might include both global and local comfort parameters.
- Indoor air quality parameters: Indoor air quality is also of major concern in modern buildings. It is controlled either at the design stage by reducing possible pollutants in the room and during operation thanks to the ventilation system. As our work is dedicated to HVAC systems, indoor air quality is also an important parameter to account for.
- Energy consumption: If appropriate indoor air quality and thermal comfort levels have to be guaranteed in offices, this has to be achieved at a minimum energy cost. Therefore, energy consumption parameters would need to be incorporated.
- HVAC system status: A stable operation of the controlled equipments is necessary in order to increase life cycle and thus reduce the maintenance cost. Information of the status of the equipments at the decision time step or on a longer period must thus be considered.
- Outdoor climate parameters: Indoor conditions are influenced by outdoor conditions (air temperature, solar radiation, wind). Moreover, in an air distribution HVAC system, the power required to raise or lower the supply temperature is a function of outdoor temperature and humidity. Some of these parameters would thus need to be selected.

2.2 Fuzzy Control of HVAC Systems

Nowadays, there is a lot of real-world applications of FLCs like intelligent suspension systems, mobile robot navigation, wind energy converter control, air conditioning controllers, video and photograph camera autofocus and imaging stabilizer, anti-sway control for cranes, and many industrial automation applications [30].

In these kinds of problems (HVAC system controller design), various criteria are considered independently, thermal regulation, maintaining a temperature setpoint or range, which only considers implicit energy savings [5, 22, 31, 44, 51, 52]. In [7], the more global PMV is used to control thermal comfort (incorporating relative humidity and mean radiant temperature), but again it does not explicitly optimize the energy consumption, the system stability or the indoor air quality (CO_2 concentration). In [37], an adaptive neuro-fuzzy inference system (ANFIS) is employed to optimization of the system energy consuption by the control in-building section of HVAC system (indoor air loop and chilled water loop). In [43], a FLC involving 7 variables (5 inputs and 2 outputs) is optimized by means of an evolutionary algorithm to decrement the energy consumption and to maintain a temperature setpoint, which also set aside some important criteria.

However, in this work, various different criteria must be considered in order to reduce the energy consumption maintaining a desired comfort level. Therefore, many variables have to be considered from the controlled system, which makes the problem very complex. In our case, five criteria will be optimized and 17 variables are considered by the FLC.

In current systems, the Knowledge Base (KB) is usually constructed based on the operator's experience. However, FLCs sometimes fail to obtain satisfactory results with the initial rule set drawn from the expert's experience [31]. Moreover, in our case the system being controlled is too complex and optimal FLCs are required. Therefore, this approach needs of a modification of the initial KB to obtain an optimal controller with an improved performance.

Many different possibilities to improve Linguistic Fuzzy Modeling have been considered in the specialized literature [8]. They can also be applied to the framework of fuzzy control (e.g., a tuning on the semantics of a FLC previously obtained from human experience could be performed by modification of the Data Base components [1, 2]). All of these approaches share the common idea of improving the way in which the linguistic fuzzy model/controller performs the interpolative reasoning by inducing a better cooperation between the rules in the KB.

There are two of these approaches presenting complementary characteristics, the parameter tuning and the rule selection. In this work, we combines the tuning methods (classical and lateral tuning) with rule selection, which present a positive synergy, reducing the search space, easing the system readability and even improving the system accuracy.

On the other hand, to evaluate the FLC performance a physical modelization of the controlled buildings and equipments is usually needed. These models have been developed by BEMS designers using building simulation tools, and they are able to account for all the parameters considered in the control process. Thus, we will have the chance to evaluate the FLCs designed in the simulated system with the desired environmental conditions. In the same way, these system models can be used by the experts to validate the initial KB before the automatic optimization process. Besides, it is of major importance to assess the fitness function in this process.

3 The GENESYS Test Cell

Within the framework of the JOULE-THERMIE programme under the GENESYS [3] project, a real test site (building) provided by a French private enterprise —whose name must remain anonymous— was available for experimentation. From now on, this site will be called the GENESYS test site.

Located in France, this test environment consists of seven single zone test cells. Around the walls of these cells, an artificial climate can be created at any time (winter conditions can be simulated in summer and *viceversa*). The cells considered are medium weight constructions. Figure 2 illustrates this environment and presents its main characteristics. Two adjacent twin cells were available for our experiments, the cells number four and five. Both test cells were equipped with all sensors required according to the selected control and controlled parameters. The HVAC system tested was a fan coil unit supplied by a reverse-cycle heat pump, and a variable fan speed mechanical extract for ventilation.

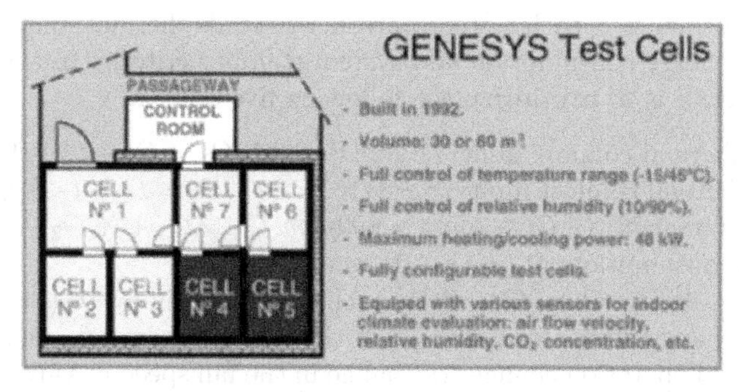

Fig. 2. Representation and main characteristics of the GENESYS test cells

The first task was to develop the thermal model of this test site. The main achievement was the development of a full monozone building model. This model was built from scratch within the Matlab-Simulink environment, being developed as a general purpose model which could be used for any other conditions, projects or applications in the future. However, in order to improve its performance, it was later customized to suit the GENESYS test site. The thermal simulation was based on finite-differences methods for the conduction

[3] GENESYS Project: Fuzzy controllers and smart tuning techniques for energy efficiency and overall performance of HVAC systems in buildings, European Commission, Directorate-General XII for Energy (contract JOE-CT98-0090).

model. The maximum value for the time-step of the simulation was calculated using the stability condition according to the discretization scheme. Simulation time step could be reduced to 60 seconds. Due to the relatively small thickness and large thermal conductivity of windows, the heat conduction model for the windows was considered constant. Convective heat exchanges were based on constant heat convection coefficients. Radiant temperature was calculated as a function of surface temperature, weighted by their relative area. The HVAC system model was based on manufacturers data and modules developed in the frame of *IEA (International Energy Agency) task 22* provided by the Royal Technical Institute of Stockholm.

Data were available and used for model calibration. The main problems in the calibration concerned the modelization of the HVAC equipment as well as solar radiation effects on internal heat gains. *The experimentation of this work has been performed considering the calibrated and validated GENESYS test cell simulation model.* Concretely, the GENESYS summer-season model.

3.1 Objectives and Fitness Function

As said, our main optimization objective was the energy performance but maintaining the required indoor comfort levels. Therefore, we should consider the development of a fitness function aiming at characterizing the performance of each tested controller towards thermal comfort, indoor air quality, energy consumption and system stability criteria. In this way, the global objective is to **minimize** the following five criteria:

O_1 Upper thermal comfort limit: $if\ PMV > 0.5, O_1 = O_1 + (PMV - 0.5)$.
O_2 Lower thermal comfort limit: $if\ PMV < -0.5, O_2 = O_2 + (-PMV - 0.5)$.
O_3 IAQ requirement: $if\ CO_2\ conc. > 800ppm, O_3 = O_3 + (CO_2 - 800)$.
O_4 Energy consumption: $O_4 = O_4 +$ Power at time t.
O_5 System stability: $O_5 = O_5 +$ System change from time t to $(t-1)$, where system change states for a change in the system operation, i.e., it counts the system operation changes (a change in the fan speed or valve position).

In our case, these criteria are combined into one overall objective function by means of a vector of weights. This technique (objective weighting) has much sensitivity and dependency toward weights. However, when trustworthy weights are available, this approach reduces the size of the search space providing the adequate direction into the solution space and its use is highly recommended. Since trustworthy weights were obtained from experts, we followed this approach.

Hence, an important outcome was to assign appropriate weights to each criterion of the fitness function. The basic idea in this weight definition was to find financial equivalents for all of them. Such equivalences are difficult to define and there is a lack of confident data on this topic. Whereas energy consumption cost is easy to set, comfort criteria are more difficult. Recent studies have shown that a 18% improvement in people's satisfaction about indoor climate corresponds to a 3% productivity improvement for office workers. Based on typical salaries and due to the fact that PMV and CO_2 concentrations are related to people's satisfaction, such equivalences can be defined. The same strategy can be applied to the systems stability criterion, life-cycle of various systems being related to number of operations. Based on this, weights can be obtained for each specific building (test site). Thus, trusted weights for the GENESYS test cell objective weighting fitness function were obtained by the experts with the following values: $w_1^O = 0.0083022$, $w_2^O = 0.0083022$, $w_3^O = 0.00000456662$, $w_4^O = 0.0000017832$ and $w_5^O = 0.000761667$. Finally, the fitness function to be minimized was computed as:

$$F = \sum_{i=1}^{n} w_i^O \cdot O_i \ .$$

3.2 FLC Variables and Architecture

A hierarchical FLC architecture considering the PMV, CO_2 concentration, previous HVAC system status and outdoor temperature was proposed by the BEMS designer for this site. The GENESYS summer-season FLC architecture, variables and initial Rule Base are presented in Figure 3 and Figure 4.

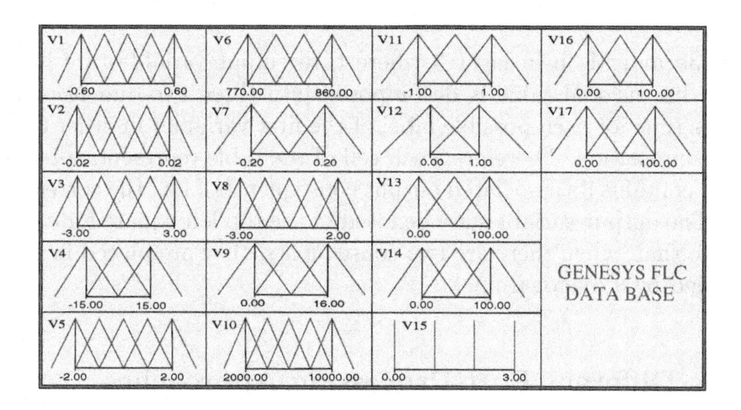

Fig. 3. Data Base of the GENESYS FLC

As Data Base, we considered symmetrical fuzzy partitions of triangular-shaped membership functions for each variable. These membership functions were labeled from $L1$ to Ll_i, with l_i being the number of membership functions of the i-th variable. Figure 3 depicts the Data Base. Both, the initial Rule Base and the Data Base, were provided by the BEMS designer.

Notice that, Figure 4 represents the decision tables of each module of the hierarchical FLC considered in terms of these labels. When the Rule Base considers more than two input variables (as in the case of modules M-2 in layer

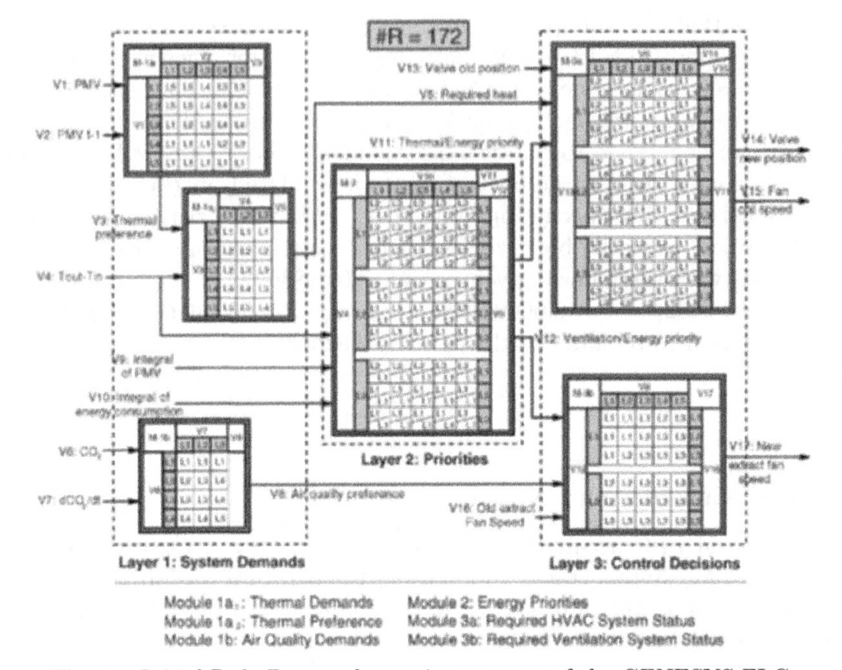

Fig. 4. Initial Rule Base and generic structure of the GENESYS FLC

2 and M-3a and M-3b in layer 3 where three input variables are involved), the three-dimensional table is decomposed into three two-dimensional decision tables (one for each possible label of the first variable) in order to clearly show its composition. Therefore, each cell of the table represents a fuzzy subspace and contains its associated output consequent(s), i.e., the corresponding label(s). The output variables are denoted in the top left square for each module. Notice that, when there are two consequents, they are placed in the same cell (divided by a diagonal line).

4 Three Different Post-Processing Approaches

This section introduces the three different post-processing approaches considered in this work: rule selection, classical tuning and lateral tuning.

4.1 Rule Selection

In complex multidimensional problems with highly nonlinear input-output relations many redundant, inconsistent and conflicting rules are usually found in the obtained Rule Base (especially in the case when they are generated by only considering the expert's knowledge). On the other hand, in high-dimensional problems, the number of rules in the Rule Base grows exponentially as more inputs are added. A large rule set might contain many redundant, inconsistent and conflicting rules. These kinds of rules are detrimental to the FLC performance and interpretability.

Rule Selection methods directly aggregate multiple rules and/or select a subset of rules from a given fuzzy rule set in order to minimize the number of rules while at the same time maintaining (or even improving) the system performance [12, 13, 25, 47, 48, 49, 53]. Inconsistent and conflicting rules that degrade the performance are eliminated thus obtaining a fuzzy rule set with better cooperation. Using Genetic Algorithms (GAs) to search for an optimized subset of rules is motivated in the following situations:

- the integration of an expert rule set and a set of fuzzy rules extracted by means of automated learning methods [27],
- the selection of a cooperative set of rules from a candidate fuzzy rule set [14, 15, 16, 32, 33, 36],
- the selection of rules from a given KB together with the selection of the appropriate labels for the consequent variables [11],
- the selection of rules together with the tuning of membership functions by coding all of them (rules and parameters) in a chromosome [23], and
- the derivation of compact fuzzy models through complexity reduction combining fuzzy clustering, rule reduction by orthogonal techniques, similarity driving simplification and genetic optimization [45].

Two of them are of particular interest in our case, the second and the fourth. In this work, we propose the selection of a cooperative set of rules from a candidate fuzzy rule set together with the tuning of parameters coding all in a chromosome. This pursues the following aims:

- To improve the FLC accuracy selecting the set of rules best cooperating while a tuning of membership functions is performed.
- To obtain easily understandable FLCs by removing unnecessary rules.

4.2 Classical Tuning of Membership Functions

This approach, usually called data base tuning, involves refining the membership function shapes from a previous definition once the remaining FRBS components have been obtained [14, 24, 28, 34, 35, 40].

The classical way to refine the membership functions is to change their definition parameters. For example, if the following triangular-shape membership function is considered:

$$\mu(x) = \begin{cases} \frac{x-a}{b-a}, & \text{if } a \le x < b \\ \frac{c-x}{c-b}, & \text{if } b \le x \le c \\ 0, & \text{otherwise} \end{cases},$$

changing the basic parameters — a, b, and c — will vary the shape of the fuzzy set associated to the membership function, thus influencing the FRBS performance (See Figure 5). The same yields for other shapes of membership functions (trapezoidal, gaussian, sigmoid, etc.).

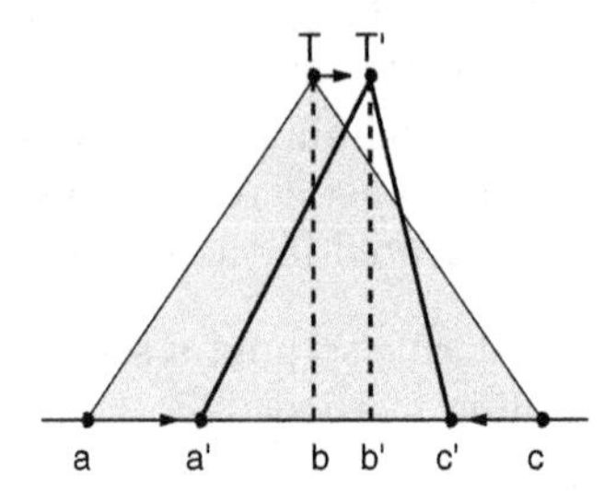

Fig. 5. Tuning by changing the basic membership function parameters

Tuning membership functions involves fitting the characterization of the membership functions associated to the primary linguistic terms considered in the system. Thus, the meaning of the linguistic terms is changed from a previous definition (an initial data base). In order to ensure a good interpretability through the membership functions optimization process [9, 41, 42], some researchers have proposed several properties. Considering one or more of these properties several constraints can be applied in the design process in order to obtain a BD maintaining the linguistic model comprehensibility to the higher possible level [6, 14, 10, 21].

An example of evolutionary tuning can be seen in Figure 6, where each membership function is encoded by means of three gene values representing its definition points.

4.3 The Lateral Tuning of Membership Functions

The lateral tuning is a new model of tuning considering the linguistic 2-tuples representation to laterally tune the support of a label, which maintains the interpretability associated to the FLC. A new model for rule representation based on the linguistic 2-tuples is introduced. This concept is presented in [29] and allow a lateral displacement of the labels named symbolic translation. The symbolic translation of a linguistic term is a number within the interval [-0.5, 0.5) that expresses the domain of a label when it is moving between its two lateral labels. Formally, we have the couple,

$$(s_i, \alpha_i),\ \ s_i \in S,\ \ \alpha_i \in [0.5, -0.5).$$

Figure 7 shows the lateral displacement of the label M. The new label "y_2" is located between B and M, being enough smaller than M but closer to M.

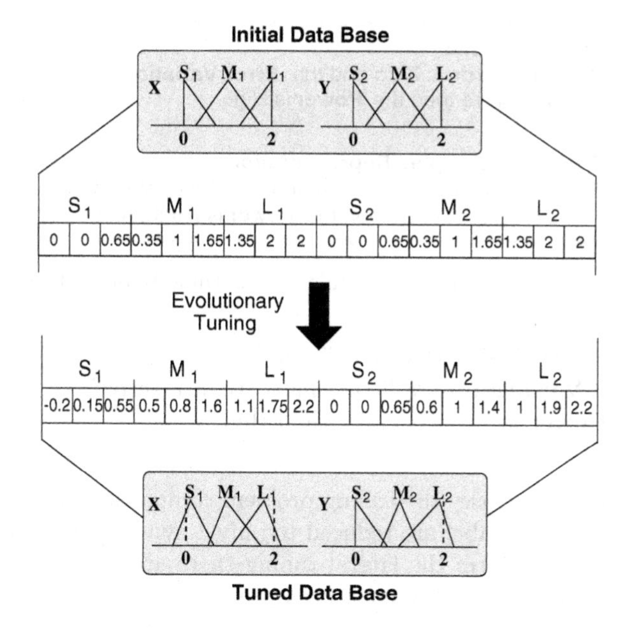

Fig. 6. Example of evolutionary tuning

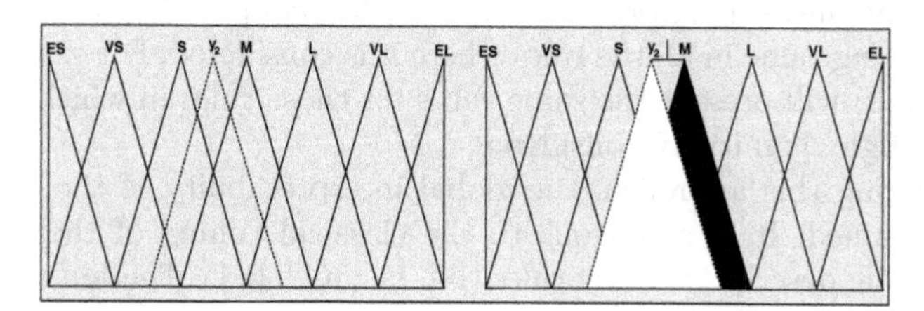

Fig. 7. Lateral Displacement of the Linguistic Label M

In [29], both the linguistic 2-tuples representation model and the needed elements for linguistic information comparison and aggregation are presented and applied to the Decision Making framework. In the context of Fuzzy Modeling and control, we are going to see its use in the linguistic rule representation. In the next we present this approach considering a simple control problem.

Let us consider a control problem with two input variables, one output variable and a Data Base defined from experts with the following labels:

$$Error \rightarrow \{N, Z, P\},\ \triangledown Error \rightarrow \{N, Z, P\},\ Power \rightarrow \{L, M, H\}\ .$$

Figure 8 shows the concept of classical rule and linguistic 2-tuples represented rule. Analized from the rule interpretability point of view, we could interpret the tuned rule as:

Classical Rule:
R1: If the **Error** is Zero and the **Error Variation** is Positive then the **Power** is High

Rules with 2-tuples Representation:
R1: If the **Error** is (Zero, 0.3) and the **Error Variation** is (Positive, -0.2) then the **Power** is (High, -0.1)

Fig. 8. Classical Rule and Rule with 2-Tuple Representation

If the **Error** is "higher than Zero" and
the **Error Variation** is "a little smaller than Positive"
then the **Power** is "a bit smaller than High".

This proposal decreases the tuning problem complexity, since the 3 parameters considered per label are reduced to only 1 symbolic translation parameter. As to how perform the lateral tuning there are two possibilities, the most interpretable one and the most accurate one:

- Global Tuning of the Semantics. In this case, the tuning is applied to the level of linguistic partition. In this way, the couple (X_i, label) takes the same tuning value in all the rules where it is considered. For example, X_i is (High, 0.3) will present the same value for those rules in which the couple "X_i is High" is initially considered.
  Considering this approach, the global interpretability of the final model is maintained. It is analogous to the classical tuning of the Data Base considering descriptive fuzzy rules [14], i.e., a global collection of fuzzy sets is considered by all the fuzzy rules. Therefore, this approach obtains more interpretable but less accurate linguistic models than the local approach.
- Local Tuning of the Rules. In this case, the tuning is applied to the level of rule. The couple (X_i, label) is tuned in a different way in each rule, based on the quality measures associated to the tuning method (usually the system error).

  Rule k: X_i is (High, 0.3) (more than high)
  Rule j: X_i is (High, -0.2) (a little lower than high)

  In this case, the global interpretability is lost to some degree and, the obtained model should be interpreted from a local point of view. This approach is analogous to the classical tuning of approximate fuzzy rules [14], i.e., each fuzzy rule has associated its own local fuzzy sets. However, in our case, the tuned labels are still related to the initial ones, preserving the global interpretability to some degree. Anyway, this approach presents more accuracy but less interpretability than the global approach.

5 Five Different Optimization Methods

Once the rule selection and two different tuning approaches have been presented, the genetic optimization algorithms developed for rule selection [32, 46], classical tuning [24, 28], lateral tuning [4] and, the combined action of rule selection with both tuning approaches (rule selection and classical tuning, rule selection and lateral tuning) are proposed in this section. However, only four of them will be presented, the second one, the third one and the last ones, since the algorithm for rule selection can be obtained as a part of this last, the C_S part.

In the following, the common parts of the said algorithms are introduced to later present its application (coding scheme and operators) to the classical and lateral tuning methods and to the combined action of rule selection with both tuning approaches.

5.1 Common Aspects of the Algorithms

It consists of a GA based on the well-known steady-state approach and considering an objective weighting-based fitness function. The steady-state approach [50] consists of selecting two of the best individuals in the population and combining them to obtain two offspring. These two new individuals are included in the population replacing the two worst individuals if the former are better adapted than the latter. An advantage of this technique is that good solutions are used as soon as they are available. Therefore, the convergence is accelerated while the number of evaluations needed is decreased (in our case it is very important since the model evaluation takes several minutes).

In order to make the method robust and more independent from the weight selection for the fitness function, the use of fuzzy goals for dynamically adapting the search direction in the space of solutions will be considered. The selection scheme is based on the Baker's stochastic universal sampling together with an elitist selection.

Evaluating the chromosome:

The fitness function (see Section 3.1) has been modified in order to consider the use of fuzzy goals that decrement the importance of each individual fitness value whenever it comes to its respective goal or penalize each objective whenever its value worse with respect to the initial solution. To do so, a function modifier parameter is considered, $\delta_i(x)$ (taking values over 1.0). A penalization rate, p_i, has been included in $\delta_i(x)$, allowing the user to set up priorities in the objectives (0 less priority and 1 more priority). Therefore, the global fitness is evaluated as:

$$F' = \sum_{i=1}^{5} w_i^O \cdot \delta_i(O_i) \cdot O_i \ ,$$

Two situations can be presented according to the value of the goal g_i, and the value of the initial solution i_i. Depending on these values, two different δ functions will be applied:

- When the value of g_i is lesser than the value of i_i, the objective is not considered if the goal is met and penalized if the initial results are worsened (see Figure 9).

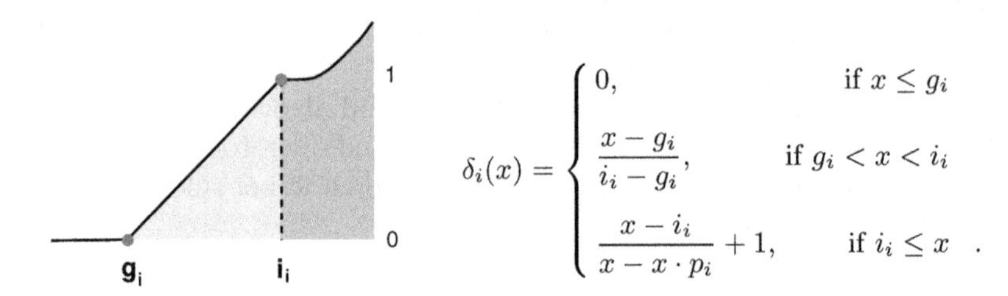

$$\delta_i(x) = \begin{cases} 0, & \text{if } x \leq g_i \\ \dfrac{x - g_i}{i_i - g_i}, & \text{if } g_i < x < i_i \\ \dfrac{x - i_i}{x - x \cdot p_i} + 1, & \text{if } i_i \leq x \end{cases} .$$

Fig. 9. $\delta_i(x)$ when $g_i \leq i_i$

- When the value of i_i is lesser than the value of g_i, the initial results can be worsened while the goal is met and, it is penalized otherwise (see Figure 10).

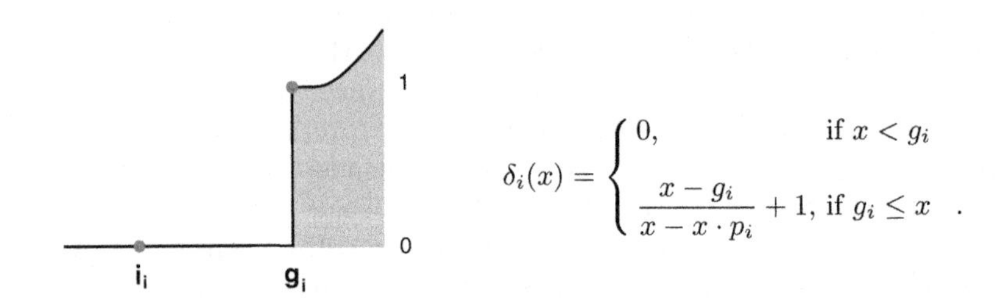

$$\delta_i(x) = \begin{cases} 0, & \text{if } x < g_i \\ \dfrac{x - g_i}{x - x \cdot p_i} + 1, & \text{if } g_i \leq x \end{cases} .$$

Fig. 10. $\delta_i(x)$ when $g_i > i_i$

Restarting approach:

Finally, to get away from local optima, this algorithm uses a restart approach [19]. Thus, when the population of solutions converges to very similar results (practically the same fitness value in all the population), the entire population but the best individual is randomly generated within the corresponding variation intervals. It allows the algorithm to perform a better exploration of the search space and to avoid getting stuck at local optima.

5.2 Evolutionary Algorithm for Classical Tuning

In this subsection, the coding scheme and genetic operators of the algorithm proposed for classical tuning are explained. To do so, the WMC-SSGA algorithm presented in [2] for classical tuning will be briefly described.

The coding scheme represents a solution by joining the representation of the m^i labels of each one of the n variables composing the Data Base:

$$C_i = (a_1^i, b_1^i, c_1^i, \ldots, a_{m^i}^i, b_{m^i}^i, c_{m^i}^i), \; i = 1, \ldots, n \; ,$$
$$C_T = C_1 C_2 \ldots C_n \; .$$

To make use of the existing knowledge, the Data Base previously obtained from experts is included in the population as an initial solution. The remaining individuals are randomly generated maintaining their genes within their respective variation intervals. These intervals are computed from the initial Data Base, having the same interval the group composed by the vertex of a label and the nearest points of the next and the previous labels. From these groups, the interval extremes are obtained computing the middle point between the nearest points of the corresponding consecutive groups [2]. Finally, these intervals are dynamically adapted from the best individual for each generation.

Since a real coding scheme is considered, the crossover and mutation operators have been selected according to this aspect: the Max-Min-Arithmetical crossover and Michale-wicz's non-uniform mutation (more complete information on these operators can be found in [2, 17]). Once the mutation operator is applied on the four offspring generated by the crossover operator, the two best are selected as the final descendents.

5.3 Evolutionary Algorithm for Lateral Tuning

This subsection presents the coding scheme and genetic operators of the lateral tuning algorithm.

Coding scheme and initial gene pool:

Taking into account that two different types of tuning have been proposed (global tuning of the semantics and local tuning of the rules), there are two different kinds of coding schemes. In both cases, a real coding is considered, i.e., the real parameters are the GA representation units (genes).

In the following both schemes are presented:

- Global tuning of the semantics: Joint of the parameters of the fuzzy partitions. Let us consider the following number of labels per variable: $(m^1, m^2, \ldots, m^n)$, with n being the number of system variables. Then, a chromosome has the following form (where each gene is associated to the tuning value of the corresponding label),

$$C_T = (c_{11}, \ldots, c_{1m^1}, c_{21}, \ldots, c_{2m^2}, \ldots, c_{n1}, \ldots, c_{nm^n}).$$

See the C_T part of Figure 11 (in the next section) for an example of coding scheme considering this approach.

- Local tuning of the rules: Joint of the rule parameters. Let us consider that the FRBS has M rules: $(R1, R2, \ldots, RM)$, with n system variables. Then, the chromosome structure is,

$$C_T = (c_{11}, \ldots, c_{1n}, c_{21}, \ldots, c_{2n}, \ldots, c_{M1}, \ldots, c_{Mn}).$$

To make use of the available information, the initial FRBS obtained from automatic fuzzy rule learning methods or from expert's knowledge is included in the population as an initial solution. To do so, the initial pool is obtained with the first individual having all genes with value '0.0', and the remaining individuals generated at random.

Genetic operators:

The genetic operator considered is crossover. No mutation is considered in this case in order to improve the algorithm convergence. A description of the crossover operator is presented in the following.

The BLX-α crossover [20] and a hybrid between a BLX-α and an arithmetical crossover [26] are considered. In this way, if two parents, $C_T^v = (c_{T1}^v, \ldots, c_{Tk}^v, \ldots, c_{Tm}^v)$ and $C_T^w = (c_{T1}^w, \ldots, c_{Tk}^w, \ldots, c_{Tm}^w)$, are going to be crossed, two different crossovers are considered:

1. Using the BLX-α crossover [20] (with α being a constant parameter chosen by the GA designer), one descendent $C_T^h = (c_{T1}^h, \ldots, c_{Tk}^h, \ldots, c_{Tm}^h)$ is obtained, with c_{Tk}^h being randomly generated within the interval $[I_{L_k}, I_{R_k}] = [c_{min} - I \cdot \alpha, c_{max} + I \cdot \alpha]$, $c_{min} = min(c_{Tk}^v, c_{Tk}^w)$, $c_{max} = max(c_{Tk}^v, c_{Tk}^w)$ and $I = c_{max} - c_{min}$.
2. The application of the arithmetical crossover [26] in the wider interval considered by the BLX-α, $[I_{L_k}, I_{R_k}]$, results in the next descendent:

$$C_T^h \text{ with } c_{Tk}^h = aI_{L_k} + (1 - a)I_{R_k},$$

with $a \in [0, 1]$ being a random parameter generated each time this crossover operator is applied. In this way, this operator performs the same gradual adaptation in each gene, which is an interest characteristic.

5.4 Evolutionary Algorithm for Rule Selection + Tuning

In this subsection, the coding scheme and genetic operators of the algorithms combining rule selection with both tuning approaches are presented.

Coding scheme and initial gene pool:

A double coding scheme ($C = C_S + C_T$) for both *rule selection* and *tuning* is used:

- For the C_S part, the coding scheme generates binary-coded strings of length m (with m being the number of fuzzy rules in the existing FLC, obtained from expert knowledge). Depending on whether a rule is selected or not, the alleles '1' or '0' will be respectively assigned to the corresponding gene. Thus, the corresponding part C_S^p for the p-th chromosome will be a binary vector representing the subset of rules finally obtained.

$$C_S^p = (c_{S1}^p, \dots, c_{Sm}^p) \mid c_{Si}^p \in \{0, 1\}$$

- The C_T part represent the coding scheme previously explained for the classical or lateral tuning algorithm.

Finally, a chromosome C^p is coded in the following way:

$$C^p = C_S^p C_T^p$$

An example of the coding scheme considering the global lateral tuning with rule selection can be seen in Figure 11.

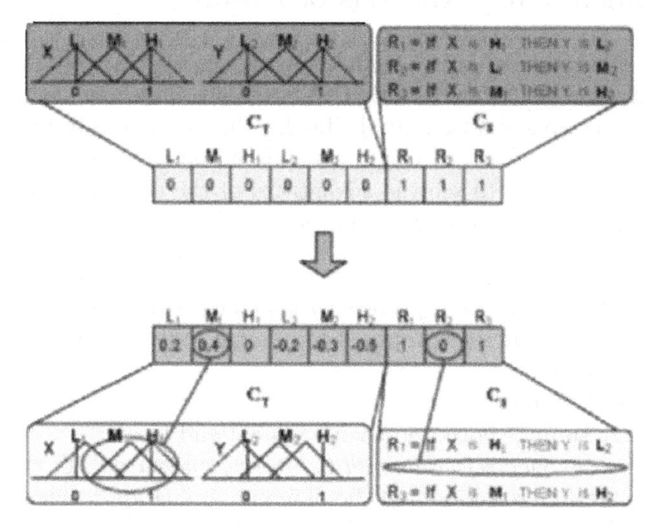

Fig. 11. Example of Coding Scheme Considering the Global Lateral Tuning and Rule Selection

To make use of the available information, the FLC previously obtained from expert knowledge is included in the population as an initial solution. To do so, the initial pool is obtained with an individual having all genes with value '1' in the C_S part and the initialization previously explained for the classical or lateral tuning algorithms in the C_T part. The remaining individuals are generated at random.

Genetic operators:

The crossover operator will depend on the chromosome part where it is applied: in the C_S part, the standard two-point crossover is used, whilst in the C_T part for the classical tuning is applied the Max-Min-Arithmetical operator and for the lateral tuning is applied a hybrid between a BLX-α and an arithmetical crossover. The two-point crossover involves exchanging the fragments of the parents contained between two points selected at random (resulting two different descendents). Finally, eight/four offspring are generated by combining the two ones from the C_S part with the four/two ones from the C_T part (classical/lateral tuning).

As regards the mutation operator, it flips the gene value in the C_S part. In the C_T part, for classical tuning the Michale-wicz's non-uniform mutation is used and for lateral tuning no mutation is applied. In this way, once the mutation operator is applied over the offspring obtained from the crossover operator, the resulting descendents are the two best individuals.

6 Experiments and Analysis of Results

To evaluate the goodness of the studied techniques, several experiments have been carried out considering the GENESYS test site. The main characteristics, the control objectives and the initial FLC for this site have been presented in Section 3. In this section, the experiments performed with the said algorithms are presented. In order to see the advantages of the combined action of the rule selection and the tuning techniques, three different studies have been performed:

1. *Considering the said post-processing approaches separately.* In this case, we consider the different proposed techniques individually:
   - Rule Selection.
   - Classical Tuning.
   - Lateral Tuning (both approaches, global and local).
2. *Combining the rule selection with the tuning approaches.* In this case, we consider the rule selection and the different tuning approaches jointly:
   - Rule Selection and Classical Tuning.
   - Rule Selection and Lateral Tuning (both approaches, global and local).
3. *Analysis of the different algorithms.* A comparison will be performed pointing out the good performance obtained when both, rule selection and tuning, are combined.

To assess the proposed techniques for fitness computation, accurate models of this controlled building (as well as the corresponding initial FLC) were provided by experts. The proposed optimization strategy was assessed with simulations of 10 days with the corresponding climatic conditions.

The FLCs obtained from the proposed technique will be compared to the performance of a classic On-Off controller and to the performance of the initial FLC. *The goals and improvements will be computed with respect to this classical controller as done in the GENESYS [3] project.* The intention from experts was to try to have 10% energy saving (O_4) together with a global improvement of the system behavior compared to On-Off control. Comfort parameters could be slightly increased if necessary (no more than 1.0 for criteria O_1 and O_2).

Table 1. Initial results and fitness function (F') parameters

		Fitness		PMV		CO_2	Energy		Stability	
MODEL	#R	F	%	O_1	O_2	O_3	O_4	%	O_5	%
ON-OFF	–	6.58	–	0.0	0	0	3206400	–	1136	–
FLC	172	6.32	4	0.0	0	0	2901686	9.50	1505	-32.48
Goals (g_i)	–	–	–	1.0	1	7	2000000	–	1000	–
Rates (p_i)	–	–	–	1	1	1	0.9	–	0.97	–

Table 1 presents the results obtained with the On-Off and the initial FLC controllers together with the parameters considered to compute the fitness function in the GA (F'), fuzzy goals and penalization rates (the objective weights can be seen in Section 3.1). Notice that, the goals imposed to the algorithm are higher than the ones initially required by the experts since we are trying to obtain even better results. No improvement percentages have been considered in the table for $O_1 \ldots O_3$, since these objectives always met the experts requirements and the On-Off controller presents zero values for these objectives.

Finally, the values of the parameters used in all of these experiments are presented as follows: 31 individuals, 0.2 as mutation probability per chromosome (except for the lateral tuning which has no mutation), 0.3 for the factor α in the hybrid crossover operator and 0.35 as factor a in the max-min-arithmetical crossover. The termination condition will be the development of 2000 evaluations, in order to perform a fair comparative study. In order to evaluate the GA good convergence, three different runs have been performed considering three different seeds for the random number generator.

6.1 Results Separately Considering the Said Post-Processing Approaches

The methods considered in this study are shown in Table 2. The models presented in Table 3, where % stands for the improvement rate with respect to the On-Off controller for each criterion and #R for the number of fuzzy rules, correspond to the best individuals from the last population considering the three runs performed. The time required for each model evaluation is 215 seconds approximately. Therefore, the run times are approximately four days (evaluations × evaluation time).

Table 2. Methods Considered for Comparison

Method	Description
S	Rule Selection
C	Classical Tuning
GL	Global Lateral Tuning
LL	Local Lateral Tuning

Table 3. Results obtained with rule selection and tuning approaches

		PMV		CO_2	Energy		Stability	
MODEL	#R	O_1	O_2	O_3	O_4	%	O_5	%
ON-OFF	–	0.0	0	0	3206400	–	1136	–
FLC	172	0.0	0	0	2901686	9.50	1505	-32.48
				Rule Selection				
S1	147	0.2	0	0	2867692	10.56	991	12.76
S2	162	0.0	0	0	2889889	9.87	1441	-26.85
S3	172	0.0	0	0	2901686	9.50	1505	-32.48
				Classical Tuning				
C1	172	0.0	0	0	2575949	19.66	1115	1.85
C2	172	0.0	0	0	2587326	19.31	1077	5.19
C3	172	0.0	0	0	2596875	19.01	1051	7.48
				Global Lateral Tuning				
GL1	172	0.7	0	0	2378784	25.81	1069	5.90
GL2	172	1.0	0	0	2327806	27.40	1066	6.16
GL3	172	0.9	0	0	2268689	29.25	1080	4.93
				Local Lateral Tuning				
LL1	172	0.9	0	0	2386033	25.59	896	21.13
LL2	172	0.8	0	0	2343409	26.92	943	16.99
LL3	172	0.3	0	0	2377596	25.85	938	17.43

From the obtained results, the tuning approaches present better results in energy and stability than the rule selection, On-Off controller and the initial

FLC controller. However, the rule selection technique minimizes the number of rules presenting significant improvements respect to the On-Off controller.

Regarding the tuning approaches, the lateral tuning presents a good trade-off between energy and stability, since this approach reduces the size of the search space of this complex problem. The lateral tuning techniques are robust and perform a better exploration of the search space, avoiding getting stuck at local optima. Note that, the local lateral tuning obtains more accurate results than the global approach, since this technique presents more freedom degrees and locally tunes each parameter. The local tuning presents improvement rates of about a 26% in energy and about a 18% in stability.

6.2 Results Combining the Rule Selection with the Tuning Approaches

The methods considered in this study are shown in Table 4. The models presented in Table 5 correspond to the best individuals from the last population considering the three proposed seeds (once again % stands for the improvement rate with respect to the On-Off controller and #R for the number of fuzzy rules). Again, the run times are approximately four days.

Table 4. Methods Considered for Comparison

Method	Description
C-S	Classical Tuning and Rule Selection
GL-S	Global Lateral Tuning and Rule Selection
LL-S	Local Lateral Tuning and Rule Selection

In view of the obtained results, we can point out that all the controllers derived by the studied methods achieve significant improvements over both, the On-Off controller and the initial FLC controller. In this case, all the goals required by experts were met, amply exceeding the expected results.

A good trade-off between energy and stability was achieved for all the obtained models, maintaining the remaining criteria within the optimal values. GL-S presents improvement rates of about a 28.6% in energy and about a 29.6% in stability, with the remaining criteria for comfort and air quality within the requested levels. Moreover, the proposed algorithm presents a good convergence and seems to be independent of random factors.

Figure 12 represents the initial and the final data base of the FLC obtained by GL-S1 in Table 5. It shows that small variations in the membership functions cause large improvements in the FLC performance. Figure 13 represents the decision tables of the FLC obtained from GL-S1 (see Section 3.2). In this case, a large number of rules have been removed from the initial FLC, obtaining much simpler models (more or less 59 rules were eliminated in each

Table 5. Results obtained combining rule selection with the tuning approaches

MODEL	#R	PMV O_1	O_2	CO_2 O_3	Energy O_4	%	Stability O_5	%
ON-OFF	–	0.0	0	0	3206400	–	1136	–
FLC	172	0.0	0	0	2901686	9.50	1505	-32.48
Selection with Classical Tuning								
C-S1	94	0.0	0	0	2540065	20.78	1294	-13.91
C-S2	109	0.1	0	0	2492462	22.27	989	12.94
C-S3	100	0.1	0	0	2578019	19.60	887	21.92
Selection with Global Lateral Tuning								
GL-S1	105	1.0	0	0	2218598	30.81	710	37.50
GL-S2	115	0.4	0	0	2358405	26.45	818	27.99
GL-S3	118	0.8	0	0	2286976	28.68	872	23.24
Selection with Local Lateral Tuning								
LL-S1	133	0.5	0	0	2311986	27.90	788	30.63
LL-S2	104	0.6	0	0	2388470	25.51	595	47.62
LL-S3	93	0.5	0	0	2277807	28.96	1028	9.51

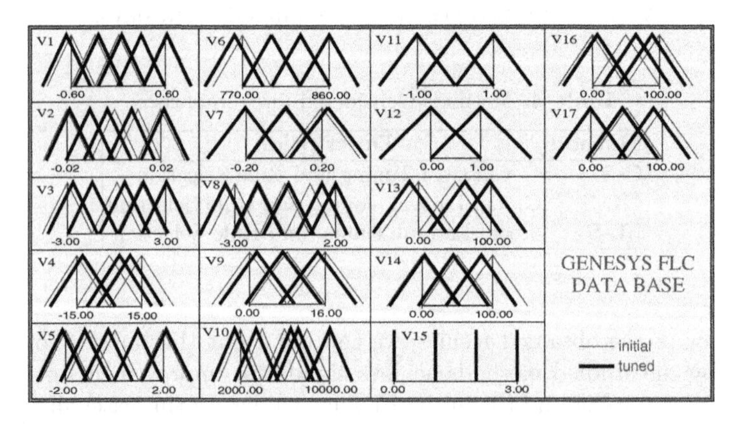

Fig. 12. Initial and Tuned Data Base of a Model Obtained with GL-S (Seed 1)

run). This fact improves the system readability, and allows us to obtain simple and accurate FLCs.

6.3 Analyzing Both Approaches

In order to see how the consideration of the rule selection affects to the tuning approaches, Table 6 presents a comparison. The averaged results obtained from the three different runs performed in the previous subsections are shown in the table.

The methods combining the rule selection with the tuning approaches has yielded much better results than the different post-processing approaches

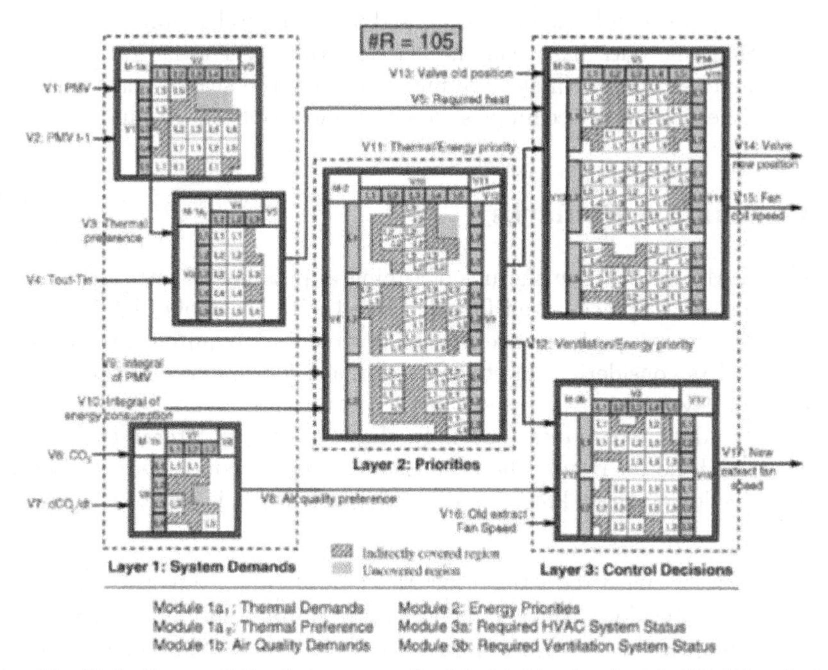

Fig. 13. Rule Base and final structure of a Model Obtained with GL-S (seed 1)

Table 6. Comparison among the different methods

MODEL	#R	PMV O_1	O_2	CO_2 O_3	Energy O_4	%	Stability O_5	%
ON-OFF	–	0.0	0	0	3206400	–	1136	–
FLC	172	0.0	0	0	2901686	9.50	1505	-32.48
Averaged Results								
$\overline{S}$	160	0.1	0	0	2886422	9.98	1312	-15.52
$\overline{C}$	172	0.0	0	0	2586717	19.33	1081	4.84
$\overline{GL}$	172	0.9	0	0	2325093	27.49	1072	5.66
$\overline{LL}$	172	0.7	0	0	2369013	26.12	926	18.52
$\overline{C-S}$	109	0.1	0	0	2536849	20.88	1057	6.98
$\overline{GL-S}$	113	0.7	0	0	2287993	**28.64**	800	**29.58**
$\overline{LL-S}$	110	0.5	0	0	2326088	27.46	804	29.26

separately, specially in the case of the global lateral tuning together with the rule selection. Moreover, in the case of GL-S, the interpretability level obtained is very near to the original one, since the initial rules and membership function shapes remain fixed. It is notorious the fact that, in general, the simplified FLCs only maintain a 64% of the initial rules. Furthermore, considering rule selection helps to reduce the search space and favors the ability of the tuning techniques to obtain good solutions.

On the other hand, the methods based on the lateral tuning present better results than the ones based on the classical tuning. The lateral tuning is a particular case of the classical tuning, however, the search space reduction helps to these kinds of techniques to obtain more optimal results.

7 Concluding Remarks

In this work, we propose the use of tuning approaches together with a rule selection to develop accurate FLCs dedicated to the control of HVAC systems concerning energy performance and indoor comfort requirements. To do so, different GAs considering an efficient approach for tuning and rule selection have been developed.

The studied techniques, specially those based on lateral tuning, have yielded much better results than the classical On-Off controller showing their good behavior on these kinds of complex problems. It is due to the following reasons:

- The search space reduction that the lateral tuning involves in complex problems. It allows to these techniques to obtain more optimal FLCs.
- The complementary characteristics that the use of the tuning approaches and the rule selection approach present. The ability of the rule selection to reduce the number of rules by only selecting the rules presenting a good cooperation is combined with the tuning accuracy improvement, obtaining accurate and compact FLCs.

References

1. R. Alcalá, J. Casillas, J.L. Castro, A. González, F. Herrera, A multicriteria genetic tuning for fuzzy logic controllers, Mathware and Soft Computing 8:2 (2001) 179–201.
2. R. Alcalá, J.M. Benítez, J. Casillas, O. Cordón, R. Pérez, Fuzzy control of HVAC systems optimized by genetic algorithms, Applied Intelligence 18 (2003) 155–177.
3. R. Alcalá, J. Alcalá-Fdez, J. Casillas, O. Cordón, F. Herrera, Hybrid Learning Models to Get the Interpretability-Accuracy Trade-Off in Fuzzy Modeling, International Journal of Soft Computing (2004) in press.
4. R. Alcalá, F. Herrera, Genetic tuning on fuzzy systems based on the linguistic 2-tuples representation, in Proc. of the 2004 IEEE International Conference on Fuzzy Systems 1 (Budapest, Hungary, 2004) 233–238.
5. M. Arima, E.H. Hara, J.D. Katzberg, A fuzzy logic and rough sets controller for HVAC systems, Proceedings of the IEEE WESCANEX'95 1 (NY, 1995) 133–138.
6. U. Bodenhofer and P. Bauer, A formal model of interpretability of linguistic variables, in Interpretability issues in fuzzy modeling, J. Casillas, O. Cordón, F. Herrera, L. Magdalena (Eds.), Springer-Verlag (2003) 524–545.

7. F. Calvino, M.L. Gennusa, G. Rizzo, G. Scaccianoce, The control of indoor thermal comfort conditions: introducing a fuzzy adaptive controller, Energy and Buildings 36 (2004) 97–102.
8. J. Casillas, O. Cordón, F. Herrera, L. Magdalena (Eds.), Accuracy improvements in linguistic fuzzy modeling, Studies in Fuzziness and Soft Computing 129 (Springer-Verlag, Heidelberg, Germany, 2002).
9. J. Casillas, O. Cordón, F. Herrera, L. Magdalena (Eds.), Interpretability issues in fuzzy modeling, Springer-Verlag (2003).
10. F. Cheong and R. Lai, Constraining the optimization of a fuzzy logic controller using an enhanced genetic algorithm, IEEE Transactions on Systems, Man, and Cybernetics—Part B: Cybernetics 30:1 (2000) 31–46.
11. T. C. Chin, X. M. Qi, Genetic algorithms for learning the rule base of fuzzy logic controller, Fuzzy Sets and Systems 97:1 (1998) 1–7.
12. S. Chiu, Fuzzy model identification based on cluster estimation, Journal of Intelligent and Fuzzy Systems 2 (1994) 267–278.
13. W. E. Combs, J. E. Andrews, Combinatorial rule explosion eliminated by a fuzzy rule configuration, IEEE Transactions on Fuzzy Systems 6:1 (1998) 1–11.
14. O. Cordón, F. Herrera, A three-stage evolutionary process for learning descriptive and approximative fuzzy logic controller knowledge bases from examples, International Journal of Approximate Reasoning 17:4 (1997) 369–407.
15. O. Cordón, M. J. del Jesús, F. Herrera, Genetic learning of fuzzy rule-based classification systems cooperating with fuzzy reasoning methods, International Journal of Intelligent Systems 13:10-11 (1998) 1025–1053.
16. O. Cordón, F. Herrera, A proposal for improving the accuracy of linguistic modeling, IEEE Transaction on Fuzzy Systems 8:3 (2000) 335–344.
17. O. Cordón, F. Herrera, F. Hoffmann, and L. Magdalena, Genetic fuzzy systems—evolutionary tuning and learning of fuzzy knowledge bases, World Scientific (2001).
18. D. Driankov, H. Hellendoorn, M. Reinfrank, An introduction to fuzzy control (Springer-Verlag, 1993).
19. L.J. Eshelman, The CHC adaptive search algorithm: How to have safe search when engaging in nontraditional genetic recombination, in: G.J.E. Rawlins (Ed.), Foundations of Genetic Algorithms (Morgan Kauffman, San Mateo, CA, 1991) 265–283.
20. L.J. Eshelman, J.D. Schaffer, Real-coded genetic algorithms and interval-schemata, in: Foundations of Genetic Algorithms 2 (Morgan Kauffman, San Mateo, CA, 1993) 187–202.
21. J. Espinosa and J. Vandewalle, Constructing fuzzy models with linguistic integrity from numerical data-afreli algorithm, IEEE Transactions on Fuzzy Systems—Part B: Cybernetics 8:5 (2000) 591–600.
22. P.Y. Glorennec, Application of fuzzy control for building energy management, in: Building Simulation: International Building Performance Simulation Association 1 (Sophia Antipolis, France, 1991) 197–201.
23. A. F. Gómez-Skarmeta, F. Jiménez, Fuzzy modeling with hybrid systems, Fuzzy Sets and Systems 104 (1999) 199–208.
24. H. B. Gürocak, A genetic-algorithm-based method for tuning fuzzy logic controllers, Fuzzy Sets and Systems, 108:1 (1999) 39–47.
25. S. Halgamuge, M. Glesner, Neural networks in designing fuzzy systems for real world applications, Fuzzy Sets and Systems 65:1 (1994) 1–12.

26. F. Herrera, M. Lozano, J.L. Verdegay, Fuzzy connectives based crossover operators to model genetic algorithms population diversity, Fuzzy Sets and Systems 92:1 (1997) 21–30.
27. F. Herrera, M. Lozano, J.L. Verdegay, A learning process for fuzzy control rules using genetic algorithms, Fuzzy Sets and Systems 100 (1998) 143–158.
28. F. Herrera, M. Lozano, and J. L. Verdegay, Tuning fuzzy controllers by genetic algorithms, Int. J. of Approximate Reasoning 12 (1995) 299–315.
29. F. Herrera and L. Martínez, A 2-tuple fuzzy linguistic representation model for computing with words, IEEE Transactions on Fuzzy Systems 8 (2000) 746–752.
30. K. Hirota (Ed.), Industrial applications of fuzzy technology (Springer-Verlag, 1993).
31. S. Huang, R.M. Nelson, Rule development and adjustment strategies of a fuzzy logic controller for an HVAC system - Parts I and II (analysis and experiment), ASHRAE Transactions 100:1 (1994) 841–850, 851–856.
32. H. Ishibuchi, T. Murata, I. B. Türksen, Single-objective and two-objective genetic algorithms for selecting linguistic rules for pattern classification problems, Fuzzy Sets and Systems 89 (1997) 135–150.
33. H. Ishibuchi, K. Nozaki, N. Yamamoto, H. Tanaka, Selecting fuzzy if-then rules for classification problems using genetic algorithms, IEEE Transactions on Fuzzy Systems 9:3 (1995) 260–270.
34. Y. Jin, W. von Seelen, and B. Sendhoff, On generating FLC^3 fuzzy rule systems from data using evolution strategies, IEEE Transactions on Systems, Man, and Cybernetics 29:4 (1999) 829–845.
35. C.L. Karr, Genetic algorithms for fuzzy controllers, AI Expert 6:2 (1991) 26–33.
36. A. Krone, H. Krause, T. Slawinski, A new rule reduction method for finding interpretable and small rule bases in high dimensional search spaces, Proceedings of the 9th IEEE International Conference on Fuzzy Systems, San Antonio, TX, USA, 2000, 693–699.
37. L. Lu, W. Cai, L. Xie, S. Li, Y.C. Soh, HVAC system optimization in building section, Energy and Buildings 37 (2005) 11–22.
38. E.H. Mamdani, Applications of fuzzy algorithms for control a simple dynamic plant, Proceedings of the IEEE 121:12 (1974) 1585–1588.
39. E.H. Mamdani, S.Assilian, An experiment in linguistic synthesis with a fuzzy logic controller, International Journal of Man-Machine Studies 7 (1975) 1–13.
40. D. Nauck and R. Kruse, Neuro-fuzzy systems for function approximaton, Fuzzy Sets and Systems 101:2 (1999) 261–271.
41. J. V. de Oliveira, Semantic constraints for membership function optimization, IEEE Transactions on Systems, Man, and Cybernetics—Part A: Systems and Humans 29:1 (1999) 128–138.
42. J. V. de Oliveira, Towards neuro-linguistic modeling: constraints for optimization of membership functions, Fuzzy Sets and Systems 106:3 (1999) 357–380.
43. J. Pargfrieder, H. JÖRGL, An integrated control system for optimizing the energy consumption and user comfort in buildings, Proceedings of the 12th IEEE International Symposium on Computer Aided Control System Design (Glasgow, Scotland, 2002) 127–132.
44. A. Rahmati, F. Rashidi, M. Rashidi, A hybrid fuzzy logic and PID controller for control of nonlinear HVAC systems, Proceedings of the IEEE International Conference on Systems, Man and Cybernetics 3 (Washington, D.C., USA, 2003) 2249–2254.

45. H. Roubos, M. Setnes, Compact fuzzy models through complexity reduction and evolutionary optimization, Proceedings of the 9th IEEE International Conference on Fuzzy Systems 2 (San Antonio, Texas, USA, 2000) 762–767.
46. H. Roubos and M. Setnes, Compact and transparent fuzzy models through iterative complexity reduction, IEEE Transactions on Fuzzy Systems 9:4 (2001) 515–524.
47. R. Rovatti, R. Guerrieri, G. Baccarani, Fuzzy rules optimization and logic synthesis, Proceedings of the 2nd IEEE International Conference on Fuzzy Systems 2 (San Francisco, USA, 1993) 1247–1252.
48. M. Setnes, R. Babuska, U. Kaymak, H. R. van Nauta-Lemke, Similarity measures in fuzzy rule base simplification, IEEE Transactions on Systems, Man, and Cybernetics—Part B: Cybernetics 28 (1998) 376–386.
49. M. Setnes, H. Hellendoorn, Orthogonal transforms of ordering and reduction of fuzzy rules, Proceedings of the 9th IEEE International Conference on Fuzzy Systems 2 (San Antonio, Texas, USA, 2000) 700–705.
50. D. Whitley, J. Kauth, GENITOR: A different genetic algorithm, Proceedings of the Rocky Mountain Conference on Artificial Intelligence, Denver (1988) 118–130.
51. J. Wu, W. Cai, Development of an adaptive neuro-fuzzy method for supply air pressure control in HVAC system, Proceedings of the IEEE International Conference on Systems, Man and Cybernetics 5 (Nashville, Tennessee, USA, 2000) 3806–3809.
52. I.H. Yang, M.S. Yeo, K.W. Kim, Application of artificial neural network to predict the optimal start time for heating system in building, Energy Conversion and Management 44 (2003) 2791–2809.
53. J. Yen, L. Wang, Simplifying fuzzy rule-based models using orthogonal transformation methods, IEEE Transactions on Systems, Man, and Cybernetics—Part B: Cybernetics 29 (1999) 13–24.

A Study of 3-gene Regulation Networks Using NK-Boolean Network Model and Fuzzy Logic Networking

Trina Kok and Paul Wang

Pratt School of Engineering, Duke University

1 Introduction

Boolean network theory, proposed by Stuart A. Kauffman about 3 decades ago, is more general than the cellular automata theory of von Neumann. This theory has many potential applications, and one especially significant application is in the modeling of genetic networking behavior. In order to understand the genomic regulations of a living cell, one must investigate the chaotic phenomena of some simple Boolean networks.

This chapter studies a very basic and simple 3-genes regulation network. Different combinations of the three basic logic elements: AND, OR and COMPLEMENT results in different logic functions. We study the influence of these logic functions on steady states behavior of the attractors and limit cycle patterns of cells.

In evaluating the degrees of gene expression using Boolean network theory, it is necessary to quantize the expression levels to "1" and "0". "1" indicates that the gene is expressed and a protein is formed; "0" indicates that the gene is not expressed at all. However, gene expression occurs in many stages, and it is not uncommon for the expression of a gene to cease in one of the intermediate steps. Thus, there is a need for the development of a model to represent the varying degrees of gene expression. We used Fuzzy Logic Networking to circumvent the information loss associated with quantization.

Hopefully, a complete dictionary of the classification or taxonomy, of all possible chaotic patterns can be established, as it is useful in the sense that more complex chaotic behavior resulted from gene regulation can be derived from the basic patterns in it. It is highly possible that the "reverse engineering" problem can be completely solved theoretically for the 3-gene networks.

T. Kok and P. Wang: *A Study of 3-gene Regulation Networks Using NK-Boolean Network Model and Fuzzy Logic Networking*, StudFuzz **201**, 119–151 (2006)
www.springerlink.com

2 Biological Background

Deoxyribonucleic acid (DNA) was discovered in 1869 by a Swiss biochemist, Johann Freidrich Miescher when he prepared a pure sample of nucleic acid from salmon sperm [2]. As various biologists such as Frederick Griffith [3], Hershey and Chase [4], and Oswald Avery et al [5] confirmed that DNA composes the genetic material in all cells (eukaryotic and prokaryotic), DNA has become the focus of many biological studies. Of these studies, Watson and Crick's elucidation of the DNA structure [6] remains the most important and famous. Today, DNA transcription, translation and mutation, and gene expression and regulation are enigma no more in the scientific community.

The human genome is a term used to describe the total genetic information (DNA content) in human cells, and is composed of the nuclear genome and mitochondrial genome. The nuclear genome accounts for 99.9995% of the total genetic information, the bulk of which codes for protein synthesis on cytoplasmic ribosomes, while mitochondrial genome accounts for the remaining 0.0005%. [7] The starting product of genome expression is the transcriptome, which is a term used to represent the agglomeration of ribonucleic acid (RNA) molecules synthesized from protein-coding genes [2]. The end product of genome expression is the proteome, which is the collection of proteins that subsequently contribute to the functioning of the cell [2]. In order to understand gene expression and regulation, knowledge of the processes from genome to proteome is requisite.

The gene is a part of the genome that codes for a particular protein, and composes of a sequence of nucleotides. The flow of genetic information from gene to protein is largely one-way: from DNA to RNA to protein. [7] However, the intricate details of protein synthesis are far from simple. An outline of the important reactions follows.

An enzyme, RNA polymerase, carries out transcription of RNA from DNA. In cells, DNA is packaged tightly into chromatin, which is in turn attached to various proteins that must be displaced so that RNA polymerase can contact the genes. Before transcription, the unnecessary proteins are removed and chromatin is unwound to expose the DNA. RNA polymerase and its various accessory proteins then assemble to form the transcription initiation complex. This complex binds to promoter elements on the exposed part of the DNA to signal that RNA synthesis is about to begin. As transcription begins, RNA polymerase dissociates from the transcription initiation complex and begins to catalyze the synthesis of RNA [2].

In most eukaryotic cells, the RNA transcript undergoes a series of processing reactions. This largely involves splicing and capping. During splicing, non-coding regions (introns) of RNA are removed and coding regions (exons) are ligated to create a continuous sequence of information. During capping, a nucleotide linkage is added to the 5' end of the RNA, and adenylate (AMP) residues are sequentially added to the 3' end to form a poly (A) tail. The cap and poly (A) tail serve to facilitate movement of the RNA molecule from the nucleus to the cytoplasm [7].

In the cytoplasm, information encoded on the RNA molecule is translated into proteins via the ribosomes. Ribosomes are RNA-protein complexes that help thread various amino acids in the order defined by the RNA sequence. Proteins formed from the ribosome then undergo post-translational modification where specific chemical groups are added or removed. These chemical groups tag the different proteins for different functions [7].

Gene regulation occurs at every stage of the cascade leading from genome to proteome. Proteins that make up the transcription initiation complex bind to promoter regions of the DNA to activate transcription. Depending on the nature of the binding, different amounts of RNA transcripts are produced. The proteins would thus be known as transcriptional activators. Transcriptional repressors are also found in the cell, and as the name suggests, they suppress the production of RNA transcripts. Together with post-transcriptional and post-translational regulators, transcriptional regulators coordinate the production of active protein in response to cell cycle changes and environmental stimulants.

Most of the above-mentioned regulators are proteins, products from the expression of other genes. Thus it is conceivable that the expression of one gene influences another. With the estimated 30,000 to 35,000 genes in human cells [8], the interrelation between genes sets up a convoluted network. Using location analysis and expression data, Simon et al [9] found that transcriptional activators responsible for one stage of the yeast cell cycle regulate transcriptional activators responsible for the next stage, setting up a complex regulatory network.

3 Regulatory Networks

The advent of DNA microarray technology and oligonucleotide chips [10-14] has presented much data regarding gene expression and activity profiles. The employment of gene expression data has enabled the classification of breast cancer [15], leukemias [16], and blue-cell cancers [17]. Recently, many studies have applied this data to reverse engineering. In reverse engineering, researchers strive to evaluate a single set of regulatory interactions from samples of expression data by Schmulevich et al [18]. The ability to evaluate regulatory networks from expression data is projected to facilitate the identification of drug targets. Various models and algorithms were studied in an attempt to elucidate genetic regulatory networks, some of which include synchronous and asynchronous Boolean models [18-20], probabilistic Boolean models [18], cellular automata [21,22], Bayesian networks [23,24], Artificial Neural networks [25], (quasi) linear [26] and linear [27] networks, Petri Nets [28], Mjolsness models [29], ordinary differential equations [30], genetic programming [31], fuzzy logic [32], qualitative reasoning [33], S-systems [33], clustering [34] and yet more other approaches.

Clustering is often coupled with other algorithms and models to provide an integrative regulatory genetic network. Genes with similar expression profiles are likely to be regulated by the same processes. Clustering allows for identification of such groups of genes and further elucidation of their individual regulation. However, they fail to provide a holistic topography of the genetic regulatory network, so that various other algorithms and models have to fill the void [34].

Of all the models, Boolean models pioneered by Kauffman [35] – [36] are still the most studied. The model considers each gene symbolically as either ON/1 (expressed) or OFF/0 (not expressed), so that the continuous data obtained from microarray technology has to be quantized to these two levels. States of genes in time ‘t’ regulate the states of genes in time ‘t+1’ via logic functions consisting of AND, OR and COMPLEMENT connectors. As regulation proceeds among genes in a parallel manner, a synchronized genetic regulatory network evolves. An NK-Boolean network is set up by the evolution of N genes with K connectivity, where N refers to the total number of genes in the network, and K refers to the maximum number of genes that regulate some single gene. The number of possible states for such a network and the amount of data necessary for its elucidation is 2^N (Please refer to Table I). Assuming a network with maximum connectivity (K=N) like those studied by Wang et al [37-40], there are

$$\left(2^{2^N} - 2\right)^N \tag{1}$$

possible logic functions to realize a typical gene regulatory network.

Thus it may be concluded that the Boolean network model results in an uninformative discrete representation of gene expression and activity profiles, and lead to an intractable solution.

In addressing the above concerns, it was found that real regulatory networks typically have low connectivity, which translates into low K values [41]. Only a small fraction of the 2N possible gene expression states are fulfilled where unfulfilled states represent unstable states. A low K connectivity results in a tractable solution with a smaller number of possible logic functions. Working from the hypothesis that if the quantized expression and activity profiles do not provide sufficient information to separate classes of tumors, the Boolean network model is an unrealistic representation of genetic networks, Shmulevich and Zhang set out to evaluate the credibility of the Boolean network model [42]. In that study, Shmulevich and Zhang found that the Boolean network model was able to provide a clear separation between different classes of sarcomas and different subclasses of gliomas, indicating that the Boolean network model retains sufficient biological information to realistically model genetic regulatory networks [42]. Intuitively, the Boolean network model is a suitable representation of genetic networks because genetic manipulation often involves either over-expression or deletion of a gene [20].

Table 1. from [34]

Model	Data needed
Boolean, fully connected	2^N
Boolean, connectivity K	2^K (K + log(N))
Boolean, connectivity K, linearly separable	K log(N/K)
Continuous, fully connected, additive	N+1
Continuous, connectivity K, additive	K log (N/K) (*)
Pairwise correlation comparisons (clustering)	log(N)

Fully connected is where each gene can receive regulatory inputs from all other genes. Connectivity K: at most K regulatory inputs per gene. Additive, linearly separable: regulation can be modeled using a weighted sum. Pairwise correlation: significance level for pairwise comparisons based on correlation must decrease inversely proportional to number of variables. (*) conjecture.

As a Boolean network evolves in time, a sequence of states results and converges to limit cycles or attractors eventually. Information from the initial states are no longer as important and only a small number of all the possible configurations actually occur [43], composing the limit cycle or attractor.

1. *Attractor*: An attractor is a set of states, invariant under the dynamically progrssion, towards which the neighboring states in a given basin of attraction asymptotically approach in the course of dynamic evolution. An attractor is defined as the smallest unit which cannot be itself decomposed into two or more attractors with distinct basins of attraction. [44]
2. *Basin of attraction:* The set of points in the state vector space of system state variables such that initial conditions chosen in this set dynamically evolve to a particular attractor. [45]
3. *Limit cycle:* An attracting set of state vectors to which orbits or trajectories converge and upon which trajectories are periodic. [46]
4. *Length of a limit cycle:* In the above sense, the length of a limit cycle represents its fundamental period and is equal to the number of states contained within the cycle.
5. *Basin number:* The basin number is the number of reachable states to a limit cycle or attractor.

Attractors and limit cycles of the Boolean network model can be interpreted in two ways. First, they can be seen to represent stable phenotypes of differentiated cells- muscle vs. nerve cells, or healthy vs. sick cells [35] [47]. In a non-chaotic network, Kauffman indicates that the number of attractors and limit cycles corresponds to the number of biological cell types [35]. Second, attractors and limit cycles can be regarded as cellular states-differentiation, apoptosis and cell cycle [20]. Both interpretations capture the concept of homeostasis perfectly. Homeostasis occurs when cells maintain their state despite minor disturbances in their environmental and internal stimuli. These perturbations can be interpreted as changes in state configurations of the cell, but as long as they reside within the same basin of attraction, the same attractors or limit cycles will be reached. Thus attractor or limit cycle stability increases with the size of the basin of attraction.

Regarding attractors and limit cycles as cellular states, cancer can be represented as a shift from the usually stable “differentiation” state to the “growth” state. Mutations might have reduced the size of the basin of attraction leading to the “differentiation” state, thus rendering it less stable and more susceptible to perturbations. Cancer drugs should then strive to push the cell from its “growth” state back into “differentiation” state. [20]

4 Investigation of 3-gene Boolean Network

The Boolean network model is realistic in its representation of genetic networks, capturing the essence of cell development and leading to a tractable solution. Here in this paper, we consider the Boolean network, which are special cases of NK-networks, where each site takes on binary values of either 0 or 1, and represents gene expression states.

6. *Cellular Automata*: Cellular automata are simple mathematical idealizations of natural systems. They consist of a lattice of discrete sites, each site taking on a finite set of, say, integer values. The values of the sites evolve in discrete time steps according to deterministic rules that specify the value of each site in terms of the values of the neighboring sites. [43]

However, the cellular automata model is not a realistic model for biological natural systems.

In this study, we attempt to evaluate the evolution pattern of a NK-Boolean network whose evolution depends only on its two neighboring sites and itself. In addition, we assume N (total number of genes in the network) =3 and K (connectivity) =3, so that there are $2^N = 8$ possible states and

$$\left(2^{2^N} - 2\right)^N = 16{,}387{,}064 \tag{2}$$

possible combinations of logic function. As we believe that networks with larger N's may be broken down into networks of N = 2 or 3, we studied NK-Boolean network of the 3-gene network. (A study on the 2-gene network can be found in [38]). Of the 16,387,064 possible logic functions, about 150 examples were evaluated by hand, resulting in diagrams similar to Figure 1 and 2. A', B', C' represent genes at time 't+1' and A, B, C, represent genes at time 't'. Figure 1 illustrates two limit cycles, one with length 2 (L2) and the other with length 6 (L6). Both limit cycles have a basin number of 0. Figure 2 illustrates two attractors, one with basin number 0 (B0) and the other with basin number 6 (B6). In the syntax of this study, logic functions that involve no logic connectors are termed "PLAIN". Conversely, logic functions that involve AND, OR, COMPLEMENT connectors are termed "AND", "OR" and "NOT" respectively.

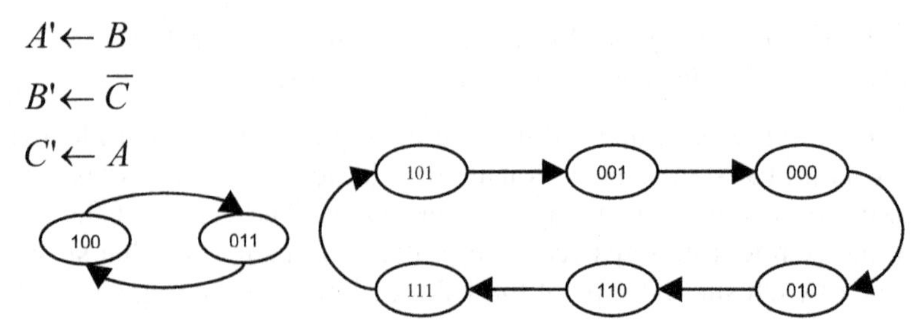

Fig. 1. An example of a 3-gene network with two limit cycles

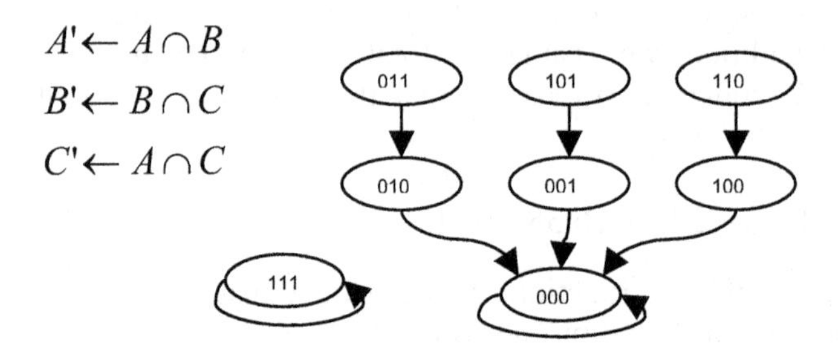

Fig. 2. An example of a 3-gene network with two attractors

5 Evolutionary Patterns

We evaluated and grouped the evolution patterns resulting from different types of logic functions. Several general rules that govern the evolution of NK-Boolean network patterns were discovered. To better summarize and compare these NK-Boolean network patterns, the logic functions and their corresponding evolution patterns are grouped into tables as follow. Figures are attached in a separate appendix for easy reference. A comprehensive dictionary is also appended as a classification of the logic functions and the evolution patterns.

Since "Reverse Engineering" has been identified by biologists as an extremely important problem, we chose to explore the possibility of solving this problem via a dictionary approach. Because it may be possible to develop some effective algorithms via heuristic arguments, having some insight about this issue is very important, even if currently far from solving the real biological problem. Many algorithms have been discovered based on biologically inspired problems.

Table 2. NK-Boolean Network patterns of PLAIN Functions

Logic Function	Logic Description	Network Description	Fig
A' ← A B' ← B C' ← C	3 self regulations	8 B0 attractors	i
A' ← B B' ← C C' ← A	No self regulation No repeated regulation	2 L3, B0 limit cycles 2 B0 attractors	ii
A' ← B B' ← A C' ← C	1 self regulation No repeated regulation	2 L2, B0 limit cycles 4 B0 attractors	iii
A' ← B B' ← A C' ← A	No self regulation 1 repeated regulation	1 L2, B2 limit cycles 2 B1 attractors	iv
A' ← A B' ← A C' ← B	1 self regulation and it is repeated	0 limit cycle 2 B3 attractors	v
A' ← A B' ← B C' ← A	2 self regulation 1 repeated regulation	0 limit cycle 4 B1 attractors	vi
A' ← A B' ← A C' ← A	1 self regulation 2 repeated regulation	0 limit cycle 2 B3 attractors	vii

Logic functions represent one example only.

Observations of PLAIN functions:

1. Attractors always include "000" and "111".

Table 3. NK-Boolean Network Patterns of NOT Functions

Logic Function	Logic Description	Network Description	Fig
A' ← $\overline{B}$ B' ← C C' ← A	1 NOT No self regulation No repeated regulation	1 L6, B0 limit cycle 1 L2, B0 limit cycle 0 attractor	viii
A' ← $\overline{C}$ B' ← B C' ← A	1 NOT 1 self regulation ≠ NOT No repeated regulation	2 L4, B0 limit cycles 0 attractor	ix

A' ← $\overline{A}$ B' ← C C' ← B	1 NOT 1 self regulation = NOT No repeated regulation	4 L2, B0 limit cycles 0 attractor	x
A' ← $\overline{C}$ B' ← $\overline{A}$ C' ← B	2 NOTs No self regulation No repeated regulation	2 L3, B0 limit cycles 2 B0 attractors	ii
A' ← A B' ← $\overline{C}$ C' ← $\overline{B}$	2 NOTs 1 self regulation ≠ NOT No repeated regulation	2 L2, B0 limit cycles 4 B0 attractors	iii
A' ← $\overline{B}$ B' ← A C' ← $\overline{C}$	2 NOTs 1 self regulation = NOT No repeated regulation	2 L4, B0 limit cycles 0 attractor	ix
A' ← $\overline{C}$ B' ← $\overline{A}$ C' ← $\overline{B}$	3 NOTs No self regulation No repeated regulation	1 L6, B0 limit cycle 1 L2, B0 limit cycle 0 attractor	viii
A' ← $\overline{C}$ B' ← $\overline{B}$ C' ← $\overline{A}$	3 NOTs 1 self regulation = NOT No repeated regulation	4 L2, B0 limit cycles 0 attractor	x
A' ← f(A) B' ← f(B) C' ← f(C)	1/2/3 NOTs 3 self regulation No repeated regulation	4 L2, B0 limit cycles 0 attractor	x
A' ← $\overline{A}$ B' ← $\overline{A}$ C' ← $\overline{C}$	1 repeated regulation, regardless self or not	Basin number = Length	eg. iv

Logic functions represent one example only.

Observations of NOT functions:

1. NOT functions always lead to at least 1 limit cycle.
2. If there is repeated regulation, a NK-Boolean network pattern will result where the basin number is equal to the length of the limit cycle.

Table 4. NK-Boolean Network Patterns of AND Function

Logic Function	Logic Description	Network Description	Fig
A' ← A∩B B' ← B∩C C' ← A∩C	3 (AND 2)s No repeated regulation	1 B6 attractor 1 B0 attractor	xi
A' ← A∩C B' ← B∩C C' ← B∩C	3 (AND 2)s 1 or 2 repeated regulation	No. of branches leading to attractor =3 + No. of repeated regulation	eg. xii
A' ←A∩B∩C B' ← A C' ← B	1 (AND 3)s	No general pattern observed	eg. xiv
A' ← A∩B∩C B' ← C C' ← A∩B∩C	2 (AND 3)s un-AND gene is not self regulated	1 B6 attractor 1 B0 attractor	xvi
A' ← A∩B∩C B' ← B C' ← A∩B∩C	2 (AND 3)s un-AND gene is self regulated	1 B3 attractor 1 B2 attractor 1 B0 attractor	xvii
A' ← A∩B∩C B' ← A∩B∩C C' ← A∩B∩C	3 (AND 3)s	1 B6 attractor 1 B0 attractor	xviii

Logic functions represent one example only

Observations of AND functions

1. AND functions always lead to attractors of "000" and "111"
2. "000" is always the dominant attractor with a greater basin number, while "111" is the least dominant attractor with the smallest basin number

Table 5. NK-Boolean Network Patterns of OR Functions

Logic Function	Logic Description	Network Description	Fig
A' ← A∪B B' ← B∪C C' ← A∪C	3 (OR 2)s No repeated regulation	1 B6 attractor 1 B0 attractor	xi
A' ← A∪B B' ← A∪B C' ← A∪C	3 (OR 2)s 1 or 2 repeated regulation	No. of branches leading to attractor = 3+ No. of repeated regulation	eg. xix

A' ← A ∪ B ∪ C B' ← A C' ← B	1 (OR 3)s	No general pattern observed	eg. xx
A' ← A ∪ B ∪ C B' ← C C' ← A ∪ B ∪ C	2 (OR 3)s un-OR gene is not self regulated	1 B6 attractor 1 B0 attractor	xvi
A' ← A ∪ B ∪ C B' ← B C' ← A ∪ B ∪ C	2 (OR 3)s un-OR gene is self regulated	1 B3 attractor 1 B2 attractor 1 B0 attractor	xvii
A' ← A ∪ B ∪ C B' ← A ∪ B ∪ C C' ← A ∪ B ∪ C	3 (OR 3)s	1 B6 attractor 1 B0 attractor	xviii

Logic Functions represent one example only.

Observations of OR functions

1. OR functions always lead to attractors of "000" and "111"
2. "111" is always the dominant attractor with a greater basin number, while "000" is the least dominant attractor with the smallest basin number

More complicated functions that involve combination of logic connectors like those shown in Figure 3 and 4 were also evaluated, but no general evolution pattern was discovered for any of these functions.

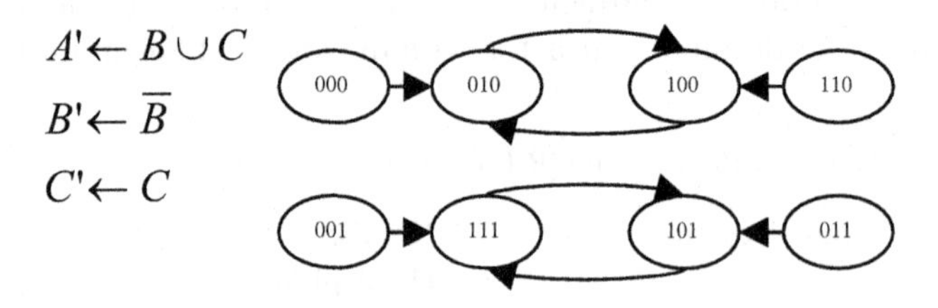

Fig. 3. An example of a complicated 3-gene network

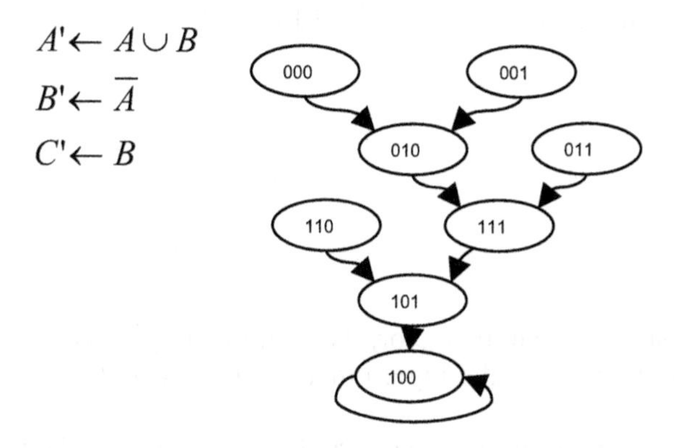

Fig. 4. An example of a complicated 3-gene network

6. Results

The concept of a "dictionary-like" solution to the "reverse-engineering" problem of a (N, K) = (3, 3) Boolean network yields a total possible combinations of logic functions of $(2^{2^N}-2)^N = (2^{2^3}-2)^3 = 16,387,064$ seem hopelessly complex. The theoretical solution, to say the least, is insurmountable complex. On the optimistic side, the situation is much better than one would think. Based upon the results presented in this paper, we have developed many heuristic rules via induction-deduction methods. Furthermore, the objective of a dictionary-like solution is attractive and feasible because the problem itself is tractable in some way if one sets up the problem as a patterns classification problem.

To begin with, one may designate a pattern vector space with 16,387,064 vector elements. In this pattern space, we may define the following features in the feature vectors space.

- $x^a_i = 1,2, \ldots\ldots, 8$, with i = 8 as maximum = number of attractors
- $x^a_i(B(i)) = x^a_1(B(1)), x^a_2(B(2)), \ldots \ldots, x^a_8(B(8))$, with $x^a_8(B(8)) = 0$ as limit = number of attractors basin
- $x^l_j = 1,2,3,4$, with j = 4 as maximum = number of limit cycles
- $x^l_j(B(j)) = x^l_1(B(1)), x^l_2(B(2)), x^l_3(B(3)), x^l_4(B(4))$, with $x^l_4(B(4)) = 0$ as limit
- $x^l_j(L(j)) = x^l_1(L(1)), x^l_2(L(2)), x^l_3(L(3)), x^l_4(L(4))$, with $x^l_4(L(4)) = 2$ as limit = the length of limit cycle

To illustrate an example, say for the case of Figure (xii).

- $x^a_3 = 3$
- $x^a_1(B(1)) = 5$,
- $x^a_2(B(2)) = 0$,
- $x^a_3(B(3)) = 0$,
- $x^1_1 = 0$,
- $x^1_1\ (L(1)) = 0$

With the above specific feature vector, the significant question to be answered now would be, what is the logic functions for this output observations?

It is more interesting to observe that strictly speaking, this dictionary-like solution really is not a pattern recognition problem. Usually, the reverse problem for pattern recognition problem is not one-to-one (or unique inverse mapping). This approach nevertheless is very interesting in that the problem resembles the mathematical linguistic pattern recognition problem.

Close observation of the above tables leads to a number of general rules that govern the evolution of NK-Boolean network patterns. These links between logic functions and NK-Boolean network evolution patterns will aid efforts in "reverse engineering". We list our observations as below

- Rule 1: NOT functions produce at least 1 limit cycle. Conversely, pure AND and OR functions only produce attractors
- Rule 2: For NOT functions, the below function is a special case where 4 limit cycles with length 2 always result.

  A' ← f(A)
  B' ← f(B)
  C' ← f(C)
- Rule 3: For NOT functions, the presence of repeated regulation will result in a NK-Boolean network pattern with length equal to its basin number.
- Rule 4: For AND and OR functions, attractors of "000" and "111" are always present.
- Rule 5: For AND functions, the dominant attractor is "000", and the least dominant attractor is "111".
- Rule 6: For OR functions, the dominant attractor is "111", and the least dominant attractor is "000".
- Rule 7: For 3 (AND 2) functions such as the below, the number of branches leading to the attractor is 3+number of repetitions. Therefore,

the range of branches leading to the attractors is between 3 and 5 (since there can be at most 2 repetitions)

$A' \leftarrow A \cap B$
$B' \leftarrow B \cap C$
$C' \leftarrow A \cap C$

In summary, we view the results of this study fructified, insightful and satisfactory.

7 Problems of the Boolean Network

Whilst the Boolean network provides a suitable and tractable solution in modeling gene regulatory networks, it fails to address several inherent characteristics of real biological genetic networks.

First, the Boolean network is deterministic and real genetic regulatory networks are stochastic [48]. A deterministic network consists of gene expression states with only one output, but a stochastic network consists of gene expression states with two or more outputs. With the same inputs and the same initial gene expression state, a stochastic network may produce a different output at one time and another output at a later time. This problem could be addressed by probabilistic Boolean networks that take into account the probability of each output occurrence [18].

Second, the Boolean network does not accommodate noise introduced by microarray and measurement techniques [18]. However, this property of real genetic regulatory networks is considered in the algorithm by Akutsu et al [33] in their evaluation of Boolean genetic networks with noise.

Third, the absolute binary values of Boolean network does not account for the varying degrees at which regulators affect gene expression. By placing weights on certain regulators, neural networks attempt to model the real genetic regulatory networks more realistically [25].

Fourth, negative feedback with a moderate feedback gain is often used in real genetic regulatory networks to stabilize the system, but negative feedbacks in Boolean networks only serve to destabilize the system [34].

Fifth, the Boolean network operates synchronously as each gene expression state gets updated in time-steps, but real genetic networks operate asynchronously.

8 Investigation of Fuzzy Logic Networking

In addressing the problem of representing varying degrees of gene expression, we study Fuzzy Logic Networking as a tool. Here, ***X*** represents the set of expressed genes and μ_x represents a function that maps elements from the universal set ***U*** to ***X***. ***X*** connotes the idea of "expressed" genes, and μ_x corresponds to the degree of expression of each gene. μ_x is known as the membership function and X(*u*) is degree of membership of *u* in the fuzzy subset of X [49].

Table 6. Five Fuzzy Logics from [50]

		x or y	x and y
Logic 1	CFMQVS	min(1,x+y)	x*y
Logic 2	max/min	max(x, y)	min(x,y)
Logic 3	probabilistic	x+y-x*y	x*y
Logic 4	MV	min(1,x+y)	max(0,x+y-1)
Logic 5	gcd/lcm	gcd(x,y)	lcm(x,y)

In the case of fuzzy subsets, we employ the same operations used in the Boolean representation i.e. AND, OR and COMPLEMENT. Due to the range of membership values, there is no one way of carrying out the AND, OR and COMPLEMENT operations on the fuzzy subsets. Table 6 presents five logics that were discussed by Reiter [50]. As observed by Reiter, each of the logic has different characteristics that are worth noting. Logic 2 is a common operation based on maximum and minimum replacing, and it used in both [49] and [51]. Reiter observed that Logic 1 has a fairly high-valued "OR" and a fairly low-valued "AND"; Logic 3 has the same "AND" as Logic 1 but the "OR" values are sometimes lower; Logic 4 has the same "OR" as Logic 1 but there are more 0 values present in "AND". Based on the operation on two fuzzy subsets X and Y, we generated figures that showed the differences amongst these. The plots compare only Logic 1 to Logic 4 as Logic 5 does not map back to the interval of [0,1]. Figure 5 shows that Logic 2 gives higher "AND" values than Logic 3, and Logic 3 in turn gives higher "AND" values than Logic 1 and 4. Figure 6 shows that Logic 4 gives higher "OR" values than Logic 1 and 3, and Logic 1 and 3 gives higher "OR" values than Logic 2. From the plots, the logic functions are approximately similar.

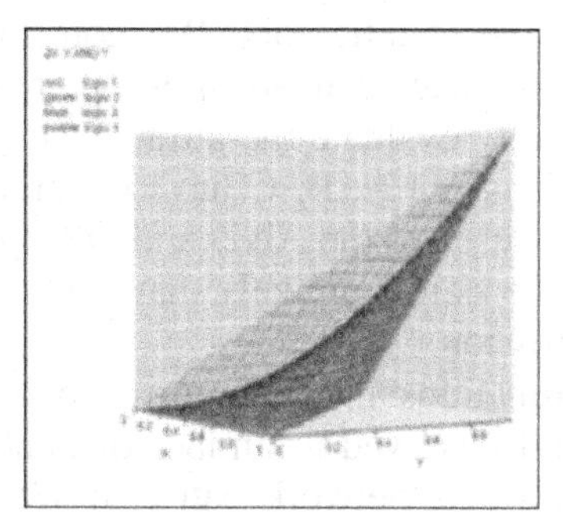

Fig. 5. Comparison of "AND" values. Logic 1 is same as logic 3.

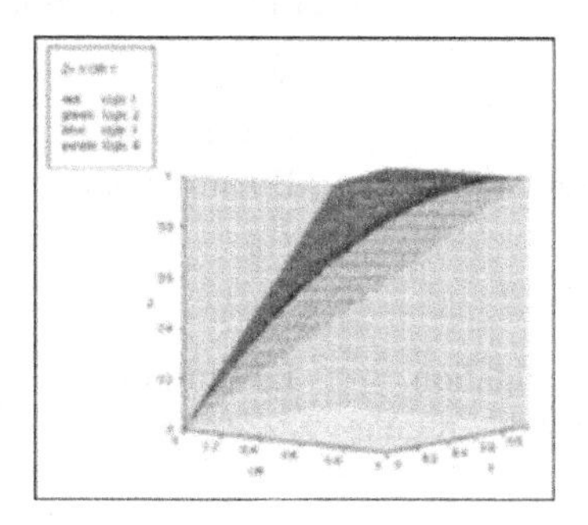

Fig. 6. Comparison of "OR" values. Logic 1 is same as logic 4.

9 Fuzzy Study of Regulation Networks

A preliminary study of 3-gene regulation networks using fuzzy sets was carried out. About 150 examples were evaluated by hand. We observed that different (i) logic operations (Logic1, 2, 3 or 4), (ii) logic functions, and (iii) initial membership values led to different attractor and limit cycles for the 3-gene regulation network. Compared to the study of 3-gene regulation networks using NK-Boolean network, (i) and (iii) are additional parameters in the evolution pattern of the 3-gene network.

Evolution patterns of 3-gene networks governed by PLAIN or NOT functions are unaffected by parameter (i). Assuming different membership values at time 't_0', 3-gene regulation networks governed by PLAIN or NOT functions generally evolve to limit cycles, as long as self-regulation does not take place. 3-gene regulation networks governed by AND or OR functions generally evolve to attractors. We note the similarity between fuzzy analysis and NK-Boolean analysis of 3-gene regulation networks.

In order to meaningfully evaluate the attractor and limit cycle patterns, we made two assumptions. 1.) A, B, C have different initial membership values and 2.) Logic 2 is used to the AND and OR operations. Assumption 1 is intuitively reasonable as it is unlikely that two genes will have exactly the same extent of expression.

We grouped the logic functions and their corresponding evolution patterns into tables as below and observed a number of general rules that govern the evolution of Fuzzy Logic network patterns. We list our observations below after Table 7-10.

Table 7. Fuzzy Logic Network Patterns of PLAIN Functions

Logic Function	Logic Description	Network Description	Reason
A' ← A B' ← B C' ← C	3 self regulations	B0 attractors	
A' ← B B' ← C C' ← A	No self regulation No repeated regulation	L3 limit cycle	Exchange of results- requires 3 steps to repeat
A' ← B B' ← A C' ← C	1 self regulation No repeated regulation	L2 limit cycle	Toggling of values between 2 non-self regulated gene
A' ← B B' ← A C' ← A	No self regulation 1 repeated regulation	L2 limit cycle	Toggling between 2 sets of values
A' ← A B' ← A C' ← B	1 self regulation and it is repeated	B2 attractors to value of repeated gene	Requires 2 steps before all follows repeated gene
A' ← A B' ← B C' ← A	2 self regulation 1 repeated regulation	B1 attractor	Since 2 genes are self-regulated, requires only step to force non-self regulated gene
A' ← A B' ← A C' ← A	1 self regulation 2 repeated regulation	B1 attractor	

Table 8. Fuzzy Logic Network Patterns of NOT functions

Logic Function	Logic Description	Network Description	Reason
$A' \leftarrow \overline{B}$ $B' \leftarrow C$ $C' \leftarrow A$	1 NOT No self regulation No repeated regulation	L6 limit cycle	
$A' \leftarrow \overline{C}$ $B' \leftarrow B$ $C' \leftarrow A$	1 NOT 1 self regulation ≠ NOT No repeated regulation	L4 limit cycle	NOT adds double the steps
$A' \leftarrow \overline{A}$ $B' \leftarrow C$ $C' \leftarrow B$	1 NOT 1 self regulation = NOT No repeated regulation	L2 limit cycle	
$A' \leftarrow \overline{C}$ $B' \leftarrow \overline{A}$ $C' \leftarrow B$	2 NOTs No self regulation No repeated regulation	L3 limit cycle	
$A' \leftarrow A$ $B' \leftarrow \overline{C}$ $C' \leftarrow \overline{B}$	2 NOTs 1 self regulation ≠ NOT No repeated regulation	L2 limit cycle	Two non-self regulated genes are negated, takes two steps to revert back
$A' \leftarrow \overline{B}$ $B' \leftarrow A$ $C' \leftarrow \overline{C}$	2 NOTs 1 self regulation = NOT No repeated regulation	L4 limit cycle	
$A' \leftarrow \overline{C}$ $B' \leftarrow \overline{B}$ $C' \leftarrow \overline{A}$	3 NOTs No self regulation No repeated regulation	L6 limit cycle	NOT adds double the steps
$A' \leftarrow f(A)$ $B' \leftarrow f(B)$ $C' \leftarrow f(C)$	1/2/3 NOTs 3 self regulation No repeated regulation	L2 limit cycle	Self regulation and NOT ensures only a toggle of 2 values

Table 9. Fuzzy Logic Network patterns of AND Functions

Logic Function	Logic Description	Network Description	Reason
A' ← A∩B B' ← B∩C C' ← A∩C	3 (AND 2)s No repeated regulation	B2 attractor	Need two steps to pick the minimum of 3 genes
A' ← C B' ← A∩B C' ← B∩C	un-AND gene is not self regulated	B3 attractor	
A' ← A∩B B' ← B C' ← B∩C	2 (AND 2)s un-AND gene is self regulated	If un-AND gene holds highest value: B0 attractor If un-AND gene holds middle value: B1 attractor If un-AND gene holds middle value: B2 attractor	
A' ← A∩B∩C B' ← C C' ← A∩B∩C	2 (AND 3)s un-AND gene is not self regulated	B2 attractor	Since un-AND gene is not self regulated, require 2 steps to achieve minimum of 3
A' ← A∩B∩C B' ← B C' ← A∩B∩C	2 (AND 3)s un-AND gene is self regulated	B1 attractor	Since un-AND gene is self regulated, require 1 step to achieve minimum of 3
A' ← A∩B∩C B' ← A∩B∩C C' ← A∩B∩C	3 (AND 3)s	B1 attractor	

Table 10. Fuzzy Logic Network Patterns of OR Functions

Logic Function	Logic Description	Network Description	Reason
A' ← A ∪ B B' ← B ∪ C C' ← A ∪ C	3 (OR 2)s No repeated regulation	B2 attractor	Need two steps to pick the maximum of 3 genes
A' ← C B' ← A ∪ B C'← B ∪ C	un-OR gene is not self regulated	B3 attractor	
A' ← A ∪ B B' ← B C' ← B ∪ C	2 (OR 2)s un-OR gene is self regulated	If un-OR gene holds highest value: B2 attractor If un-OR gene holds middle value: B1 attractor If un-OR gene holds middle value: B0 attractor	
A' ← A ∪ B ∪ C B' ← C C' ← A ∪ B ∪ C	2 (OR 3)s un-OR gene is not self regulated	B2 attractor	Since un-OR gene is not self regulated, require 2 steps to achieve maximum of 3
A' ← A ∪ B ∪ C B' ← B C' ← A ∪ B ∪ C	2 (OR 3)s un-OR gene is self regulated	B1 attractor	Since un-OR gene is self regulated, require 1 step to achieve maximum of 3
A' ← A ∪ B ∪ C B' ← A ∪ B ∪ C C' ← A ∪ B ∪ C	3 (OR 3)s	B1 attractor	

We observed the following general rules

- Rule 1: NOT functions produce limit cycles.
- Rule 2:

Table 11. Table for Rule 2

Example	Logic Description	Rules
A' ← A ∩ B B' ← B C' ← B ∩ C	2 (AND 2)s un-AND gene is self regulated	If un-AND gene holds highest value: B0 attractor If un-AND gene holds middle value: B1 attractor If un-AND gene holds middle value: B2 attractor

- Rule 3:

Table 12. Table for Rule 3

Example	Logic Description	Rules
A' ← A ∪ B B' ← B C' ← B ∪ C	2 (OR 2)s un-OR gene is self regulated	If un-OR gene holds highest value: B2 attractor If un-OR gene holds middle value: B1 attractor If un-OR gene holds middle value: B0 attractor

- Rule 4:

Table 13. Table for Rule 4

Example	Logic Description	Rules
A' ← A ∩ B B' ← B ∩ C C' ← A ∩ C	3 (AND 2)s No repeated regulation	B2 attractor results as two steps are needed to pick the minimum of 3 genes

- Rule 5:

Table 14. Table for Rule 5

Example	Logic Description	Rules
A' ← A ∪ B B' ← B ∪ C C' ← A ∪ C	3 (OR 2)s No repeated regulation	B2 attractor as two steps are needed to pick the maximum of 3 genes

10 Conclusion

As the cell ages and undergoes irreversible changes, it either goes into an invariant state or progresses through a periodic cycle. The irreversible evolution of the NK-Boolean network and Fuzzy Logic network mimics such changes very well. Future work would build on the basic connectors studied in this paper, and include computer simulations of more complicated functions involving combinations of logic connectors.

The most significant result obtained from this study is a fundamental understanding of the stochastic behavior of a cell with simple assumption of a simple three-gene network. It is affirmative that a "reverse-engineering" problem can be solved, at least theoretically via induction and reduction approach. In general, the realistic biological behavior is so complicated that it can be seen to be at the edge-of-the-chaotics. Usually, a theoretical and mathematical derivation is very difficult. Nevertheless, a simplified analysis based upon a simple model would definitely provide a much needed visualization of the biological behavior and associated phenomenon. It is only through mathematical analysis would there be a better chance in understanding the complex phenomenon. This is partially true when the simplified model can be viewed as the basic building blocks of a complicated situation. Only then would there be some hope for some progress.

Appendix of Figures

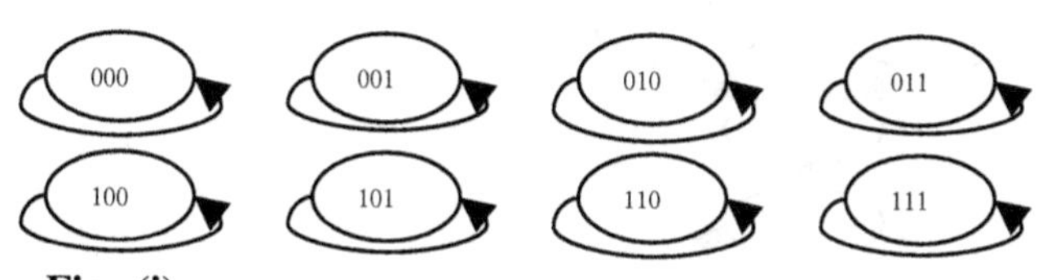

Fig. (i).

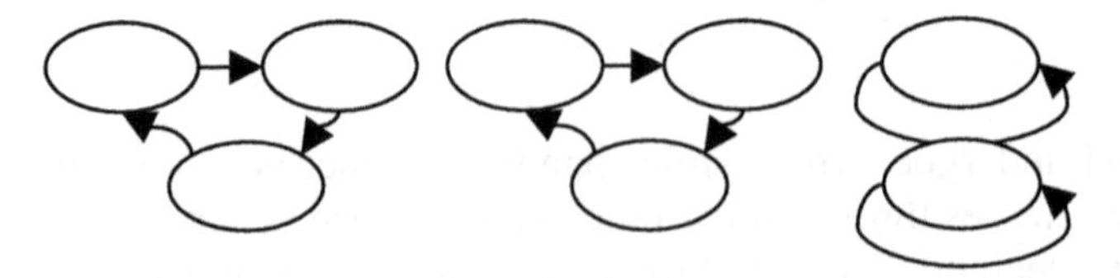

Fig. (ii).

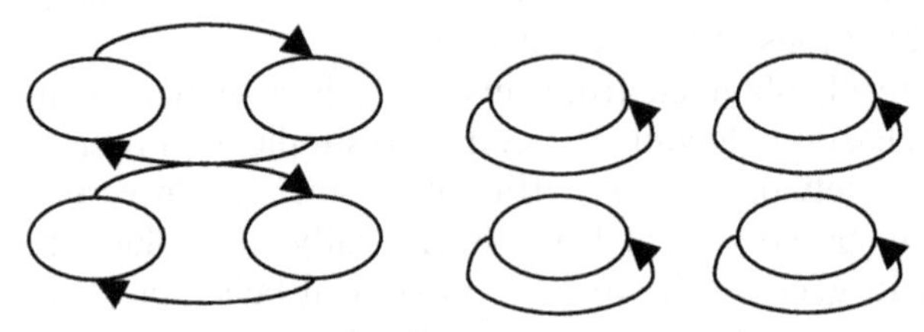

Fig. (iii).

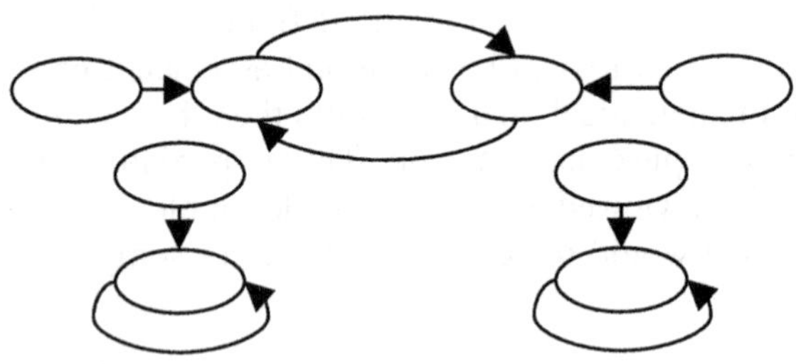

Fig. (iv).

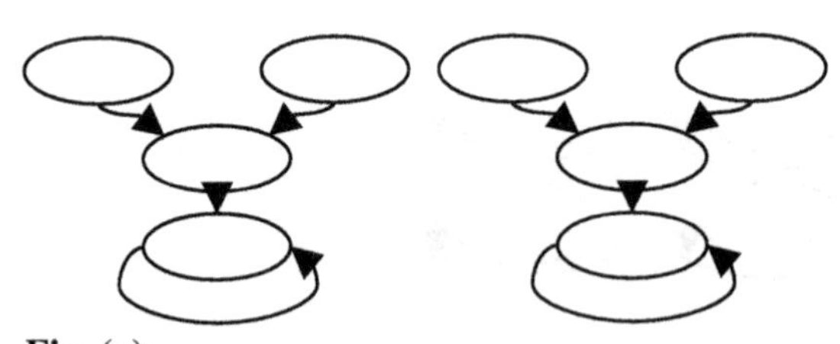

Fig. (v).

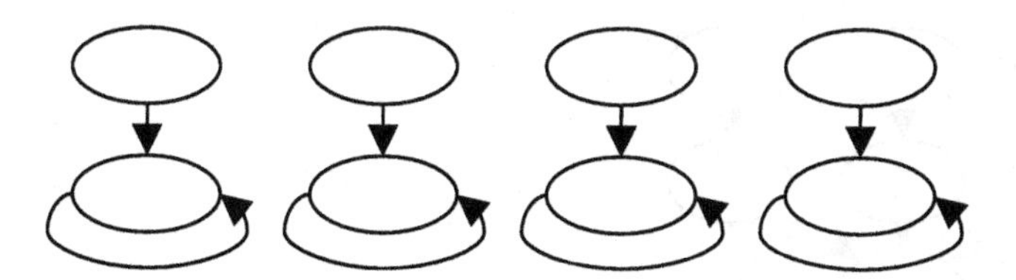

Fig. (vi).

Fig. (vii).

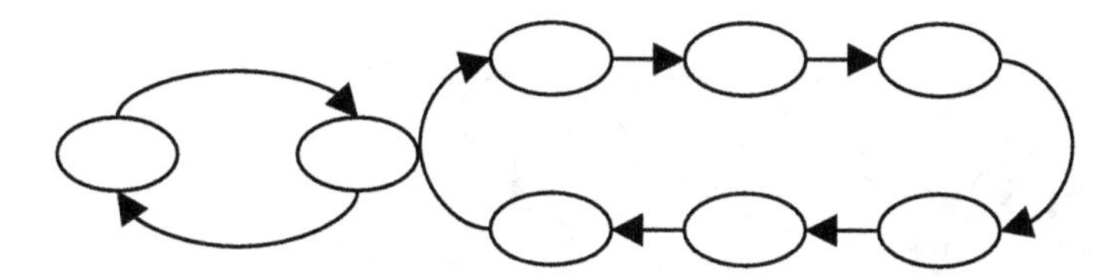

Fig. (viii).

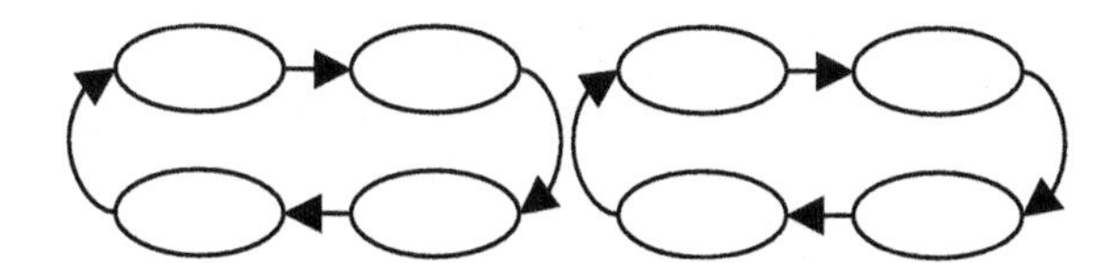

Fig. (ix).

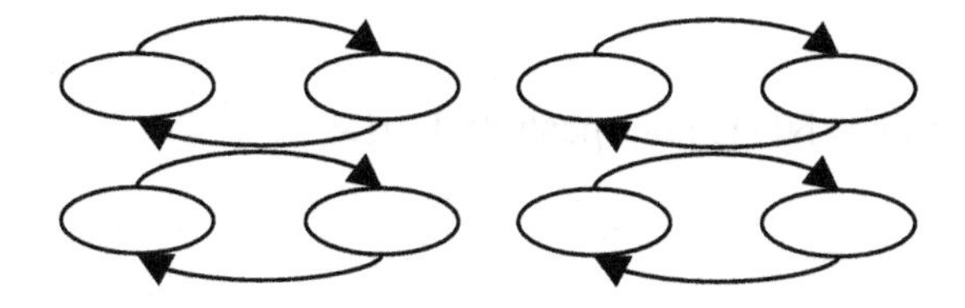

Fig. (x).

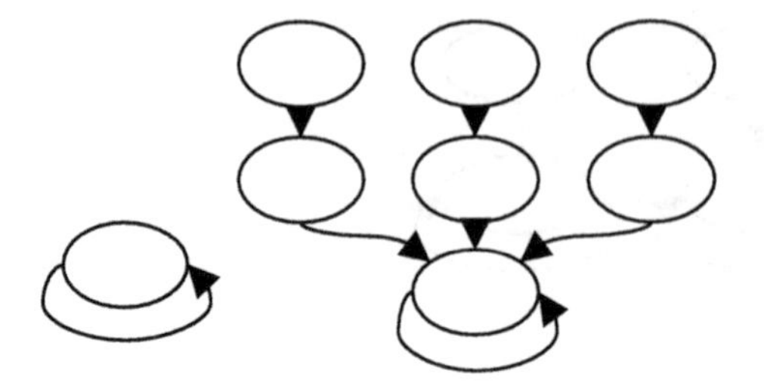

Fig. (xi)

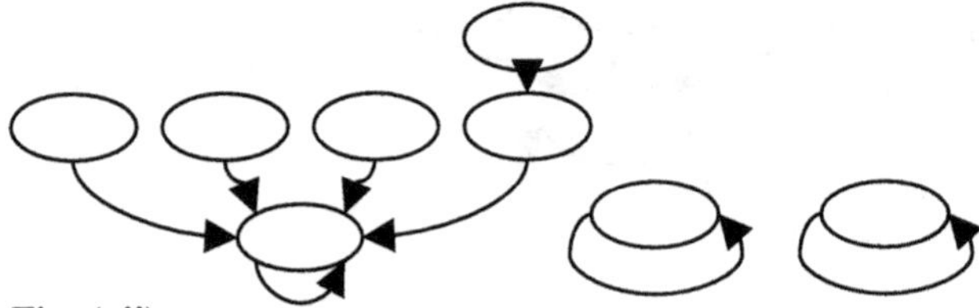

Fig. (xii).

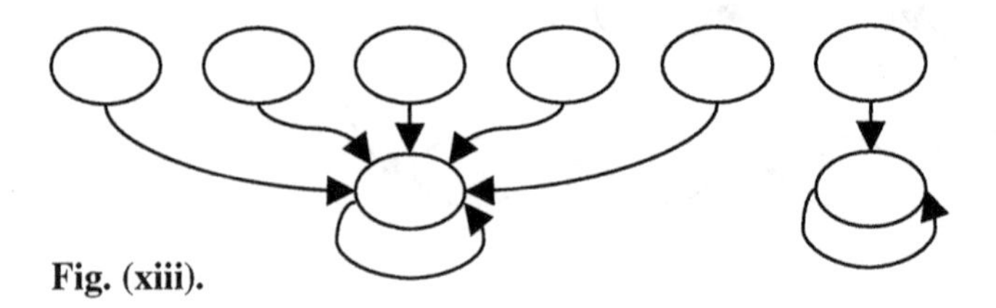

Fig. (xiii).

Appendix of Dictionary

Pure Limit Cycles

- 1 limit cycle:
- 2 limit cycle:
  (a) 1 NOT (1 self regulation ≠ NOT, no repeats) = L4/L4

  eg.

  A' ← $\overline{C}$

  B' ← B

  C' ← A

(b) 1 NOT (no self regulation, no repeats) = L6/L2

eg.

A' ← $\overline{B}$
B' ← C
C' ← A

(c) 2 NOTs (1 self regulation = NOT, no repeats) = L4/L4

eg.

A' ← $\overline{B}$
B' ← A
C' ← $\overline{C}$

(d) 3 NOTS (no self regulation, no repeats) = L6/L2

eg.

A' ← $\overline{C}$
B' ← $\overline{A}$
C' ← $\overline{B}$

- 3 limit cycles:
- 4 limit cycles:

(a) 1 NOT (1 self regulation = NOT, no repeats) = L2/L2/L2/L2

eg.

A' ← $\overline{A}$
B' ← C
C' ← B

(b) 3 NOTs (1 self regulation = NOT, no repeats) = L2/L2/L2/L2

eg.

A' ← f(A)
B' ← f(B)
C' ← f(C)

Pure Attractors

- 1 attractor:
- 2 attractor:

(a) 2(AND 3) (un-AND gene is not self-regulated) = B6/B0

eg.

$A' \leftarrow A \cap B \cap C$
$B' \leftarrow C$
$C' \leftarrow A \cap B \cap C$

(b) 3(AND 3) = B6/B0

eg.

$A' \leftarrow A \cap B \cap C$
$B' \leftarrow A \cap B \cap C$
$C' \leftarrow A \cap B \cap C$

(c) 2(OR 3) (un-OR gene is not self regulated) = B6/B0

eg.

$A' \leftarrow A \cup B \cup C$
$B' \leftarrow C$
$C' \leftarrow A \cup B \cup C$

(d) 3(OR 3) = B6/B0

eg.

$A' \leftarrow A \cup B \cup C$
$B' \leftarrow A \cup B \cup C$
$C' \leftarrow A \cup B \cup C$

– 3 attractor:

(a) 2(AND 3)(un-AND gene is self regulated)= B3/B2/B0

eg.

$A' \leftarrow A \cap B \cap C$
$B' \leftarrow B$
$C' \leftarrow A \cap B \cap C$

(b) 2(OR 3) (un-OR gene is self regulated) = B3/B2/B0

eg.

$A' \leftarrow A \cup B \cup C$
$B' \leftarrow B$
$C' \leftarrow A \cup B \cup C$

– 4 attractor:

(a) Plain (2 self regulation, 1 repeat) = B1/B1/B1/B1

$A' \leftarrow A$		A		A		B		A		C
$B' \leftarrow B$	or	A	or	B	or	B	or	C	or	B
$C' \leftarrow A$		C		B		C		C		C

– 5 attractor:

– 6 attractor:

- 7 attractor:
- 8 attractor:
  (a) Plain(3 self regulations) = B0/B0/B0/B0/B0/B0/B0/B0

      A’ ← A
      B’ ← B
      C’ ← C

Mixture

- 2 limit cycles and 4 attractors

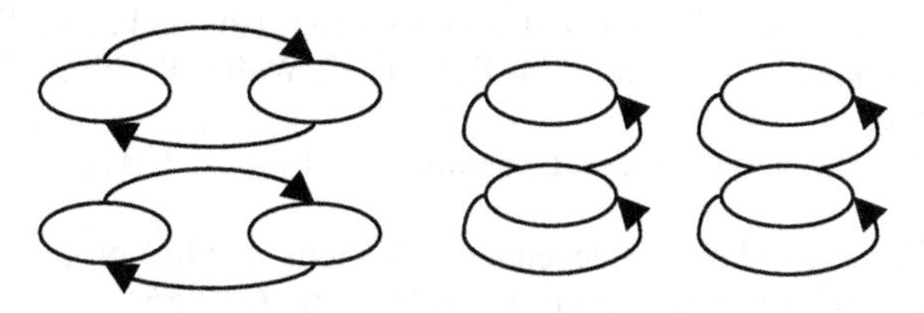

  (a) Plain (1 self regulation, no repeats) = 4 B0 attractors, 2 L2 limit cycles

      A’ ← B   A   C
      B’ ← A or C or B
      C’ ← C   B   A

  (b) 2 NOTs (1 self regulation ≠ NOT, no repeats) = 4 B0 attractors, 2 L2 limit cycles

      eg.
      A’ ← A
      B’ ← $\overline{C}$
      C’ ← $\overline{B}$

- 2 limit cycles and 2 attractors

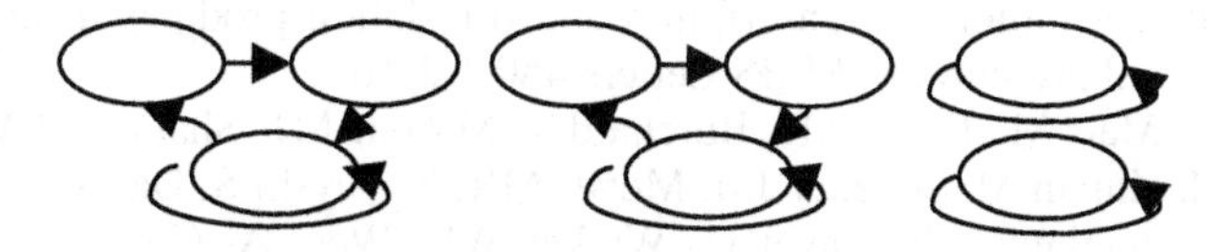

  (a) Plain (no self regulation, no repeats) = 2 B0 attractors, 2 L3 limit cycles

      A’ ← B   C
      B’ ← C or A
      C’ ← A   B

(b) 2 NOTs (no self regulation, no repeats) = 2 B0 attractors, 2 L3 limit cycles

eg.

A’ ← C^c
B’ ← $\overline{A}$
C’ ← B

References

1. Brown TA (2002) Genomes 2. John Wiley & Sons Inc, 1-69 and 219-347
2. Alberts B, Johnson A, Lewis J, Raff M, Roberts K, Walter P (2002) The Cell. Garland Science, 375-394
3. Griffith F (1928) The significance of pneumococcal types. J Hygiene 27:113-159
4. Hershey AD, Chase M (1952) Independent functions of viral protein and nucleic acid in growth of bacteriophage. J Gen Physiol 36:39-56
5. Avery OT, MacLeod CM, McCarty M (1944) Studies on the chemical nature of the substance inducing transformation of pneumococcal types. J Exp Med 79:137-158
6. Watson JD, Crick FHC (1953) Molecular structure of nucleic acids: a structure for deoxyribose nucleic acid. Nature 171:737-738
7. Strachan T, Read A (1999) Human Molecular Genetics 2. Wiley-Liss, Chapter 7
8. Human Genome Management Information System (2003) Quick Facts, http://www.ornl.gov/sci/techresources/Human_Genome/resource/media.shtml
9. Simon I, Barnett J, Hannett N, Harbison CT, Rinaldi NJ, Volkert TL, Wyrick JJ, Zeitliner J, Gifford DK, Jaakola TS, Young RA (2001) Serial regulation of transcriptional regulators in the yeast cell cycle. Cell 106:697-708
10. Schena M, Shalon D, Davis RW, Brown PO (1995) Quantitative monitoring of gene expression patterns with a complementary DNA microarray. Science 270:467-470
11. Celis JE, Kruhoffer M, Gromova I, Frederiksen C, Ostergaard M, Thyjaer T, Gromov P, Yu J, Palsdottir H, Magnusson N, Orntoft TF (2000) Gene expression profiling: Monitoring transcription and translation products using DNA microarrays and proteomics. FEBS Letters 480:1:2-16
12. Hughes TR, Mao M, Jones AR, Burchard J, Marton MJ, Shannon KW, Lefkowitz SM, Ziman M, Schelter JM, Meyer MR, Kobayshi S, Davis C, Dai H, He YD, Stephaniants SB, Cavet G, Walker WL, West A, Coffey E, Shoemaker DD, Stoughton R, Blanchard AP, Friend SH, Linsley PS (2001) Expression profiling using microarrays fabricated by an ink-jet oligonucleotide synthesizer. Nature Biotechnol 19:342-347
13. Lipshutz RJ, Fodor SPA, Gingeras TR, Lockart DJ (1999) High density synthetic oligonucletide arrays. Nature genetics 21:20-49

14. Lockhart DJ, Winzeler, EA (2000) Genomics, gene expression and DNA arrays. Nature 405:827-836
15. Hedenfalk I, Duggan D, Chen Y, Radmacher M, Bittner M, Simon R, Meltzer P, Gusterson B, Estellar M, Raffeld M, Yakhini Z, Ben-Dor A, Dougherty E, Kononen J, Bubendorf L, Fehrle W, Pittaluga S, Gruverger S, Loman N, Johannsson O, Olsson H, Wifond B, Sauter G, Kallioniemi OP, Borg A, Trent J (2001) Gene expression profiles distinguish hereditary breast cancers. New England J Med 34:539-548
16. Golub TR, Slonim DK, Tamayo P, Huard C, Gaasenbeek M, Mesirov JP, Coler H, Loh ML, Downing JR, Caligiuri MA, Bloomfield CD, Lander ES (1999) Molecular classification of cancer: Class discovery and class prediction by gene expression monitoring. Science 286:531-537
17. Khan J, Wei JS, Ringner M, Saal LH, Ladanyi M, Westermann F, Berthold F, Schwab M, Antonescu CR, Peterson C, Meltzer PS (2001) Classification and diagnostic prediction of cancers using gene expression profiling and artificial neural networks. Nature Medicine 7:673-679
18. Shmulevich I, Dougherty ER, Zhang W (2002) From Boolean to probabilistic Boolean networks as models of genetic regulatory networks. Proceedings of the IEEE 90:11:1778-1792
19. Liang S, Fuhrman S, Somogyi R (1998) REVEAL, A general reverse engineering algorithm for inference of genetic network architectures. Pacific Symposium on Biocomputing 3:18-29
20. Huang S (1999) Gene expression profiling, genetic networks, and cellular state: an integrating concept for tumorigenesis and drug discovery. J Mol Med 77:469-480
21. de Sales JA, Martins ML, Stariolo DA (1997) Cellular automata model for genetic networks. Physical Review E 55:3:3262-327
22. Ermentrout GB, Edelstein-Keshet L (1993) Cellular automata approaches to biological modeling. J theor Biol 160:97-133
23. Friedman N, Linial M, Nachman I, Pe'er D (2000) Using Bayesian networks to analyze expression data. J Computation Biology 7:601-620
24. Friedman N, Goldszmidt M, Wyner A (1999) Data analysis with Bayesian networks: A bootstrap approach. Proc Fifteenth Conf On Uncertainty in Artificial Intelligence (UAI)
25. Weaver D, Workman C, Stormo G (1999) Modeling regulatory networks with matrices. Pacific Symposium on Biocomputing 4:112-123
26. D'Haeseleer P, Wen X, Fuhrman S, Somogyi R (1999) Linear modeling of mRNA expression levels during cns development and injury. Pacific Symposium on Biocomputing 4:29-40
27. van Someren EP, Wessels LFA, Reinders MJT (2000) Linear modeling of genetic networks from experimental data. Proceedings of the Eighth International Conference on Intelligent Systems for Molecular Biology 355-366
28. Matsuno H, Doi A, Nagasaki M, Miyano S (2000) Hybrid Petri Net representation of gene regulatory network. Proc Pacific Symposium on Biocomputing 5:341-352

29. Mjolsness EDS, Reinitz J (1991) A connectionist model of development. J of Theoretical Biology 152:429-453
30. Chen T, He H and Church G (1999) Modeling gene expression with differential equations. Pacific Symposium on Biocomputing 4:29-40
31. Koza J, Mydlowec W, Lanza G, Yu J, Keane MA (2001) Reverse engineering of metabolic pathways from observed data using genetic programming. Pacific Symposium on Biocomputing 434-445
32. Thieffry D, Thomas R (1995) Dynamical behavior of biological regulatory networks-II Immunity control in bacteriophage lambda. Bull Math Biol, 57: 277-297
33. Akutsu T, Miyano S, Kuhara S (2000) Algorithms for inferring qualitative models of biological networks. Pacific Symposium on Biocomputing 5:290-301
34. D'Haeseleer P, Liang S, Somogyi R (2000) Genetic network inference: from co-expression clustering to reverse engineering. Bioinformatics 16:8: 707-26
35. Kauffman SA (1993) The Origins of Order, Self-Organization, and Selection in Evolution. Oxford University Press
36. Kauffman SA (1969) Metabolic stability and epigenesis in randomly connected nets. J Theoret Biol 22:437-467
37. Wang PP, Robinson J (2003) What is SORE? The 7th Proceedings of the Joint Conferences on Information Sciences
38. Wang PP, Cao Y, Robinson J, Tokuta A (2003). A study of two-genes network- The simplest special case of SORE. The 7th Proceedings of the Joint Conferences on Information Science
39. Wang PP, Tao H (2003). A novel method of error correcting code generation based on SORE. The 7th Proceedings of the Joint Conferences on Information Sciences
40. Wang PP, Yu J (2003) SORE, Self Organizable & Regulating Engine- A powerful classifier. The 7th Proceedings of the Joint Conferences on Information Sciences
41. Thieffry D, Huerta AM, Perez-Rueda E, Collado Vides J (1998) From specific gene regulation to genomic networks: a global analysis of transcriptional regulation in Escherichia coli. Bioessays 20:433-440
42. Shmulevich I, Zhang W (2002) Binary analysis and optimization-based normalization of gene expression data. Bioinformatics 18:4:555-565
43. Wolfram S (1983) Cellular Automata. Los Alamos Science 9:2-21
44. Weisstein WE "Limit Cycle." MathWorld—A Wolfram Web Resource. http://mathworld.wolfram.com/Attractor.html
45. Weisstein WE "Limit Cycle." MathWorld—A Wolfram Web Resource. http://mathworld.wolfram.com/BasinofAttraction.html
46. Weisstein WE "Limit Cycle." MathWorld—A Wolfram Web Resource. http://mathworld.wolfram.com/Limitcycle.html
47. Somogyi R, Sniegoski C (1996) Modeling the complexity of genetic networks. Complexity 1:6:45-63

48. Szallasi Z (1999) Genetic network analysis in light of massively parallel biological data acquisition. Pacific Symposium on Biocomputing 4:5-16
49. Nguyen HT, Walker EA (2000) A First Course in Fuzzy Logic. Chapman & Hall/CRC 2-12
50. Reiter CA (2002) Fuzzy Automata and Life. Complexity, 7:3:19-29
51. Wang L (1997) A Course In Fuzzy Systems And Control. Prentice Hall PTR 29-32

Optimizing Nuclear Reactor Operation Using Soft Computing Techniques

J.O. Entzinger[1] and D. Ruan[2]

[1] University of Twente
Department of Mechanical Automation
PO-Box 217
NL-7500 AE Enschede, The Netherlands
Jorg@Entzinger.nl

[2] Belgian Nuclear Research Centre (SCK•CEN)
Department of Reactor Physics & Myrrha
Boeretang 200
B-2400 Mol, Belgium
DRuan@SCKCEN.be

Summary. Safety regulations for nuclear reactor control are very strict, which makes it difficult to implement new control techniques. One such technique could be fuzzy logic control (FLC), which can provide very desirable advantages over classical control, like robustness, adaptation and the capability to include human experience into the controller. Simple fuzzy logic controllers have been implemented for a few nuclear research reactors, among which the Massachusetts Institute of Technology (MIT) research reactor [1] in 1988 and the first Belgian reactor (BR1) [2] in 1998, though only on a temporal basis.

The work presented here is a continuation of earlier research on adaptive fuzzy logic controllers for nuclear reactors at the SCK•CEN [2, 3, 4] and [5] (pp 65-82). A series of simulated experiments has been carried out using adaptive FLC, genetic algorithms (GAs) and neural networks (NNs) to find out which strategies are most promising for further research and future application in nuclear reactor control.

Hopefully this contribution will lead to more research on advanced FLC in this domain and finally to an optimised and intrinsically safe control strategy.

Key words: Nuclear Reactors, BR1, Nuclear Plant Operation, Fuzzy Logic Control, Genetic Algorithms, Neural Networks

J.O. Entzinger and D. Ruan: *Optimizing Nuclear Reactor Operation Using Soft Computing Techniques*, StudFuzz **201**, 153–173 (2006)
www.springerlink.com

1 Introduction

In a broad range of applications fuzzy logic (FL) has established itself as a serious alternative. However when it comes to nuclear reactor operation, strict safety regulations prevent engineers from quickly developing and implementing new control methods. This makes the use of artificial intelligent control in this field still a tremendous challenge. Although research is going on and some successful implementations of FL were reported [1, 2], most implementations use very basic controllers with a static rule base. A good overview of FL applications in nuclear reactor operation can be found in [6] and in [7] for more recent developments.

1.1 Nuclear Reactor Control

The problem faced is how to control the power output of a nuclear reactor in an optimally safe and efficient way. The description below is based on the behaviour of the Belgian Reactor 1 (BR1), which is a graphite-moderated and air-cooled system fuelled with natural uranium (Fig. 1).
Nuclear reactors have three key elements:

- radioactive fuel
- a moderator
- control rods

When the fuel is bombarded by free, low-energy neutrons, the atoms split into two major fission fragments and release high-energy neutrons as well as energy in the form of heat. The new, free, but high-energy neutrons are less likely to trigger other atoms to split. Therefore a moderator such as carbon (graphite) or hydrogen (in the form of water) is needed, so the high-energy neutrons lose energy by colliding into and bouncing off the moderator atoms. Now the free neutrons are slowed down they will trigger new fuel atoms to split, causing a chain reaction as shown in Fig. 1(b).

To control the chain reaction, control rods are inserted into or withdrawn from the reactor core. These rods are made of a material that absorbs neutrons, so when inserted further, more free neutrons are taken out of the chain reaction and the process is tempered. Of course for a steady-state process it is important to have one effective free neutron left from each fission. To raise the power output the control rods are temporarily withdrawn a little to keep slightly more free neutrons, increasing the number of fissions and thus the power output. When the power is near the desired level, they are inserted again to have a new steady-state, keeping one free neutron from each fission.

(a) The BR1 reactor.

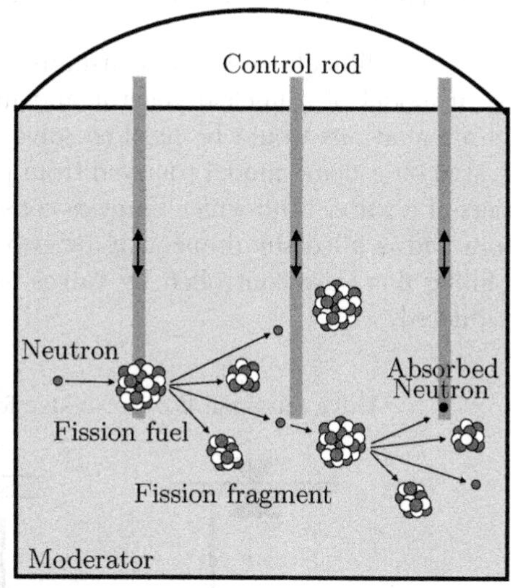

(b) Working principle of a nuclear fission reactor.

Fig. 1. The nuclear reactor.

There are three types of control rods:

- shim rods (also C-rods, for coarse control)
- regulating rods (also A-rods, for fine adjustments)
- safety rods (for very fast shutdown)

For a normal controller only the first two types are interesting, because the safety rods will only be controlled manually or by a supervisory controller. Normally only the regulating rods are active to level out small changes in the power output. Such changes could for instance be caused by a change in reactivity due to a raise of the reactor core temperature. When the regulation rods are almost fully inserted or withdrawn, the shim rods are moved slightly, allowing the regulation rods to move back to a central position. Further the shim rods are only used at startup, shutdown and setpoint changes.

The BR1 nuclear reactor is currently controlled by a conventional simple on/off controller which only controls the regulation rods. The shim rods have to be moved manually, so the controller can only maintain a steady-state power level and is not used for setpoint changes. Of course also these setpoint changes should be carried out by a controller in future implementations. An early study of fuzzy logic control for the BR1 reactor ([2, 3]) resulted in a 1-year operation of a controller which actually did combine the shim rod and regulation rod control, however it was only used in steady-state situations.

1.2 Control of a Demonstration Model

Because it is difficult and time consuming to make a well resembling mathematical model of a nuclear reactor and moreover, the set of nonlinear differential equations would be hard to solve, newly developed controller types are tested on a demo model (derived from [4] and Chap. 4 of [5]). This model consists of a water tank which empties constantly through a small hole in the bottom and is filled simultaneously by two water flows as depicted in Fig. 2. The filling flows are controlled by valves, so the water level in the tank can be regulated.

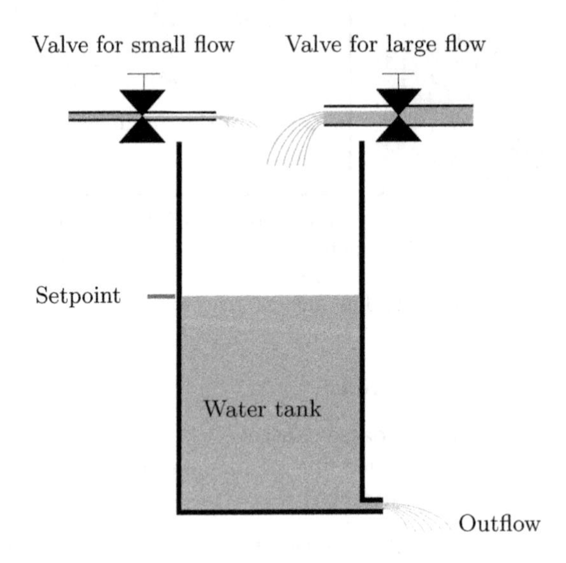

Fig. 2. Water tank demo model.

In this model the water level in the tank resembles the power output of the nuclear reactor. The speed of opening or closing the valves reflects the speed of inserting or withdrawing the control rods. One of the filling flows (VL) is made significantly larger than the other (VS) to copy the difference between shim rods and control rods. The control inputs are the difference between the current water level and the setpoint (D) and the change rate of the water level (DD).

Though this water tank demo model is much simpler than the nuclear reactor, it is still a non-linear system which is difficult to control. Therefore the demo is a good and representative testbed for the controllers investigated.

1.3 Directions of Exploration

For nuclear power plants it is important to quickly respond to a change in power demand. At the same time overshoot should be minimal for both safety and economic reasons. With these problems we find the classical control problem: the controller should be both fast and accurate.

The performance of the examined controllers will be evaluated based on it's response to time to time setpoint changes, which is a more distinct and more interesting criterion for future applications than small disturbances in a steady-state.

Several options are investigated to improve a fuzzy logic controller (FLC) for speed, stability and accuracy. In this investigation soft computing techniques are used, mainly because of their robustness, their learning capabilities and their high level of problem independence. Options investigated are:

- Adaptive rule generation (Sect. 2)
- Input scaling (Sect. 2.4)
- Increasing the number of membership functions (Sect. 2.4)
- Optimization of membership functions by Genetic Algorithms (Sect. 3)
- Use of Neural Networks for plant simulation (Sect. 4.1)
- Fuzzy Neural Network controllers (Sect. 4.4)

A setup with an adaptive fuzzy rule base was already present from earlier research ([4] and Chap. 4 of [5]). Most tests are performed with both a static rule base and an adaptive version.

Only a few possibilities are investigated, but some other options like different rule base adaptation mechanisms, membership function adaptation, plant identification techniques and advanced neuro-fuzzy systems might be interesting as well. Also non-fuzzy control and combinations of fuzzy logic with classical control have not been investigated, but are very well possible. It must be stressed that this is only an explorative investigation and no directly implementable result should be expected.

2 Rule Base Adaptation

A distinction will be made between the 'static' (reference) controller which has a fixed rule base and an 'adaptive' controller which has a rule base that is updated at each evaluation. The adaptive rule base introduces great flexibility, not only because it makes the controller (almost) problem independent, but also for instance because the controller can keep track of moving equilibria. However, in practice it is almost impossible to get an adaptive controller to work on-line in a nuclear reactor, due to safety regulations.

2.1 A Static Controller

All fuzzy variables can take five different values: Positive Big, Positive Small, Zero, Negative Small and Negative Big or PB, PS, ZE, NS and NB for short. For the valves VL and VS positive means opening and negative means closing, for the water level error D positive means too high and negative means too low and for the rate DD positive is raising and negative is dropping.

The membership functions (MFs) are shown in Fig. 3 and the static rule base is given in Table 2.1. The rule base table should be read as following: the value PS in the upper right corner of the rule base for VL means: '**If** $D = PB$ **and** $DD = NB$ **then** $VL = PS$'.

Table 1. Static rule base for VL and VS

VL rule base

DD \ D	NB	NS	ZE	PS	PB
NB	PB	PB	PB	PS	PS
NS	PB	PB	PS	PS	PS
ZE	PB	PS	ZE	NS	NS
PS	PS	ZE	NB	NB	NB
PB	ZE	NB	NB	NB	NB

VS rule base

DD \ D	NB	NS	ZE	PS	PB
NB	ZE	ZE	ZE	ZE	ZE
NS	ZE	ZE	NS	PS	NS
ZE	ZE	PS	ZE	PS	PS
PS	ZE	NS	ZE	NB	NB
PB	ZE	NB	ZE	NB	NB

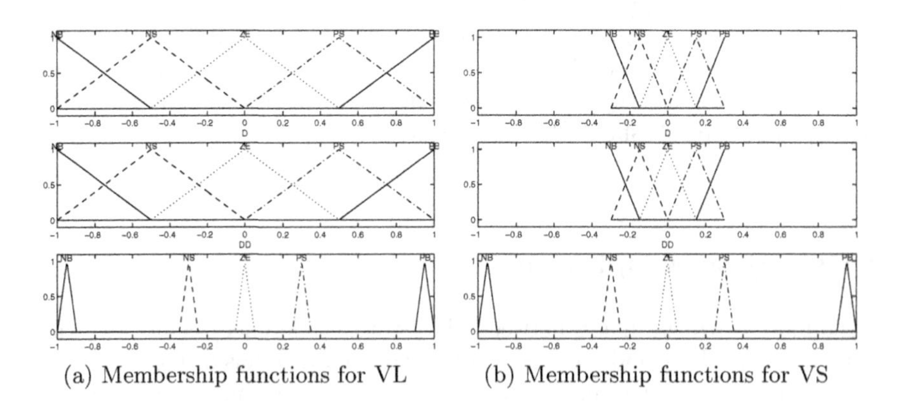

(a) Membership functions for VL (b) Membership functions for VS

Fig. 3. Input and output membership functions

2.2 An Adaptive Controller

The adaptive controller is just like the static controller, only the conclusion part of all rules is initialised to Negative Big (which is intrinsically safe) and it's rule base is adapted while controlling the system according to two simple rules:

- If the water level is too low and still dropping, open the valve faster (or close it slower)
- If the water level is too high, close the valve faster (or open it slower)

or in more technical terms:

- If $D < 0$ and $DD \leq 0$ then raise the conclusion part of the most triggered rule with 1
- If $D > 0$ then lower the conclusion part of the most triggered rule with 1

Raising here means that the next value out of $\{NB, NS, ZE, PS, PB\}$ will be taken, i.e., NB becomes NS, NS becomes ZE etcetera. The conclusion part for $D = ZE$ & $DD = ZE$ is always set to ZE to make sure the system will stabilize.

These rules are chosen in such a way that overshoot is minimized, thus restraining the speed of the controller. Other rules and additional rules have been tried but did not provide a generally better solution: speed could only be gained at the expense of more overshoot.

2.3 Comparison of Simulation Results

The adaptive controller initializes itself very quick and its response is often much faster than that of the static controller (see Fig. 4), which means that requests for more power can be met faster and that there is less waste of energy when the power need decreases. Disadvantages are of course the overshoot (although it is only slight) and the (short) fluctuation when meeting a new setpoint value.

The adapted rule base is presented in Table 2.3, where the changed rules are printed bold and italic. It may seem that not many rules have adapted, but this is quite logical because the four rules in the lower left corner and the centre rule will never change when using the proposed adaptation rules and the rules applying to a too high water level (i.e., the two most right columns) are supposed to make the water level decrease as soon as possible, which means they should have NB as a conclusion part. This means only the eight upper left rules (centre rule excluded) should change, which is exactly what happens. The upper rows do not change because of an improper scaling of the DD input variable.

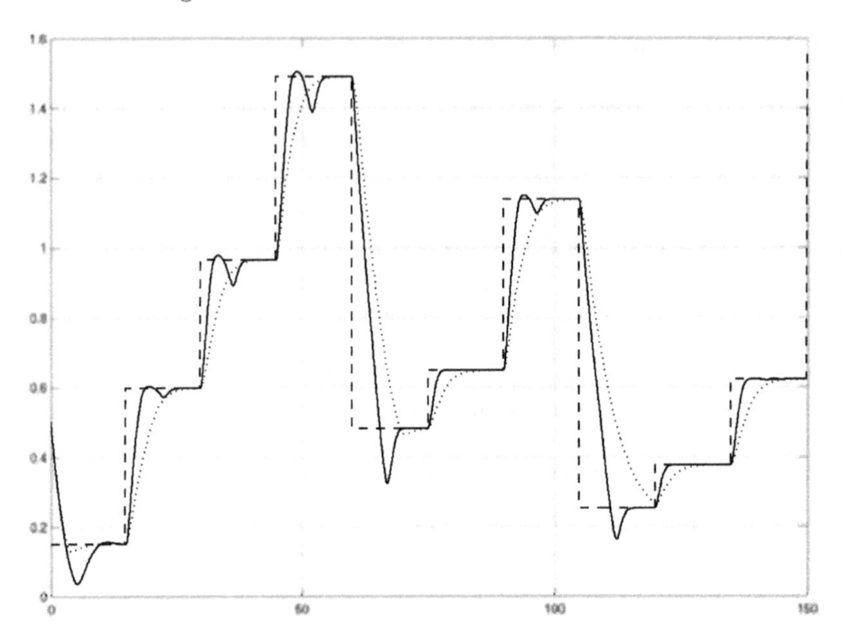

Fig. 4. Comparison of the static (*dotted*) and adaptive (*solid*) controller responses.

Table 2. Rule base for VL and VS after adaptation

VL rule base

DD \ D	NB	NS	ZE	PS	PB
NB	NB	NB	NB	NB	NB
NS	NB	NB	NB	NB	NB
ZE	***PB***	***PB***	ZE	NB	NB
PS	NB	NB	NB	NB	NB
PB	NB	NB	NB	NB	NB

VS rule base

DD \ D	NB	NS	ZE	PS	PB
NB	NB	NB	NB	NB	NB
NS	NB	***PB***	***PB***	NB	NB
ZE	***NS***	***PB***	ZE	NB	NB
PS	NB	NB	NB	NB	NB
PB	NB	NB	NB	NB	NB

2.4 Enhancements

Some enhancements can be made to the adaptive controller to improve its performance. The first thing would be a scaling of the inputs. The level in the water tank can change with a rate roughly between $-0.3\,m/s$ and $0.9\,m/s$ due to the design of the tank. This means a scaling of the DD input could be to multiply the value of DD with a factor 3 if the rate is negative and leave it as is when it is positive. Although the level can vary between 0 and 2 (so D can vary between -2 and 2) the input D will not be scaled because errors larger than 1 rarely occur and even if they do, they are dealt with just as well with the current scaling.

A second enhancement solves the problem that the adaptation algorithm sometimes gets trapped during initialisation. This is due to a limit approach of the equilibrium between in- and outflow such that $DD > 0$ so neither of both adaptation rules will be triggered, even not when a setpoint change occurs. Changing the first adaptation rule from 'If $D < 0$ and $DD \leq 0$ then ...' to 'If $D < 0$ and $DD \leq 10^{-2}$ then ...' solves this problem.

The performance of the adaptive controller with these enhancements is much better, as can be seen in Fig. 5. The only problem with the scaling is that it makes the controller much more problem specific. A better way would be to scale inputs dynamically, or to have an enormous input range and then dynamically set the positions of membership functions within a meaningful range.

A test with seven instead of five membership functions did not show a significantly better response.

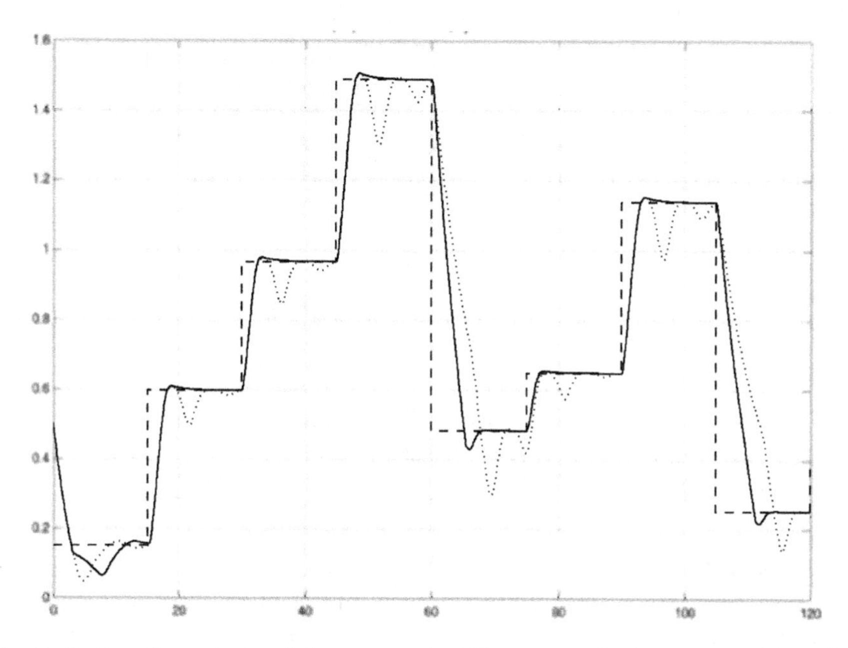

Fig. 5. Scaling the water level change rate (*solid*) removes the fluctuations (*dotted*).

2.5 Performance Comparison with Changing Flow Rates

To test the adaptive controller for robustness, simulations have been carried out with a changing maximum flow rate through the refilling pipes. The rates now increase or decrease with values which change over time and can run up to 25% of the original maximal flows. In the first test this value is chosen randomly every few seconds, in the second test the change is sinusoidal to resemble the slow and continuous changes in reactor physics.

Both the static and the adaptive controller still do their work properly when subjected to the changing flow rates. The adaptive controller still has a much faster response and is still able to maintain a relatively stable water level at the setpoint value.

To measure the influence this distortion has on the controller, the difference between the water level in the original system has been compared to the level in the system with changing flow rates. For a long run (1000 seconds) the mean difference in water level and the mean percentage difference are calculated. The percentage difference is interesting because it compensates for the fact that for high water levels the valves will be opened wider, so a sudden change in flow will have a larger effect.

Table 3. Robustness of the static and adaptive controller to random flow changes.

	Mean difference Random flow changing every			Mean percentage difference Random flow changing every		
	1s	2s	5s	1s	2s	5s
Static	0.0110	0.0102	0.0069	1.47	1.33	1.21
Adaptive	0.0210	0.0102	0.0054	3.44	1.43	0.81

Table 4. Robustness of the static and adaptive controller to sinusoidal flow changes.

	Mean difference Sine frequency			Mean percentage difference Sine frequency		
	0.1rad/s	0.05rad/s	0.01rad/s	0.1rad/s	0.05rad/s	0.01rad/s
Static	0.0068	0.0048	0.0029	1.16	0.87	0.58
Adaptive	0.0060	0.0028	0.0018	1.34	0.44	0.27

Table 3 shows that the adaptive controller needs some time to adapt itself to every flow rate change: when the flow rate changes every 5 seconds the adaptive controller is robuster than the static one, but when the changes come faster after each other the static controller is robuster. To the more gradual flow changes of the sinusoidal distortion the adaptive controller is always robuster, as Table 4 shows.

3 Optimization of Fuzzy Logic Controllers

For a fuzzy logic controller there are many parameters that can be tuned, for instance:

- the rules
- the weights applied to each rule
- the number of membership functions
- the shape of the membership functions
- the positioning of the membership functions in the input and output space

The rules are already tuned by the adaptation algorithm, which appears to work very well. Adapting the weights of the rules could be interesting, but a lot of parameters would be needed. The shape and number of MFs usually have a negligible effect. The positioning of the membership functions within the input space seems an interesting topic because it could solve the scaling problem that we saw in Sect. 2.4 in a much more problem independent way.

3.1 Genetic FLC Optimization

Genetic algorithms (GAs) are a soft computing technique to perform optimisations in a way comparable to evolutionary development [8, 9]. They can handle very complex and coupled systems without the need of derivative information to define a search direction. Also GAs can very well handle search spaces with a lot of local optima, which make them very suitable for many applications [10, 11, 12, 13]. A drawback is that convergence is relatively slow and large numbers of optimisation variables will slow down the process even more. Considering all, the GAs can very well be applied to optimize the performance of an FLC for an arbitrary system by tuning the locations of the MFs.

Genetic Parametrisation

To optimize the membership functions, a set of parameters must be found that satisfyingly defines their positions in the input space. Satisfyingly here means that we want as few parameters as possible and the parameters must still have as much 'physical relevance' as possible, to make the crossover operator a meaningful tool to reach convergence. Also the parametrization must prevent ambiguity (switching the positions of MFs) and practically impossible or unwanted variants.

The parametrization chosen uses the 'remain range', the space left over between the MFs that are already set and the input ranges, to determine the positioning of the tops of the MFs (see Fig. 6). A few constraints were added to the parametrization:

- The central MF always has its top at $x = 0$
- The x position of the top of one MF is one of the base points for the neighbouring MFs (so no gaps between MFs can arise)
- The MFs for VL and VS are the same, they are just scaled down for VS (like in the static controller case).

Some optimizations with independent MFs for VL and VS have been performed, the results however were not better and often even worse than optimizations with the same MFs for VL and VS. This could be due to the fact that the parameter vector is twice as long in this case, which makes it more difficult to for the GA to converge.

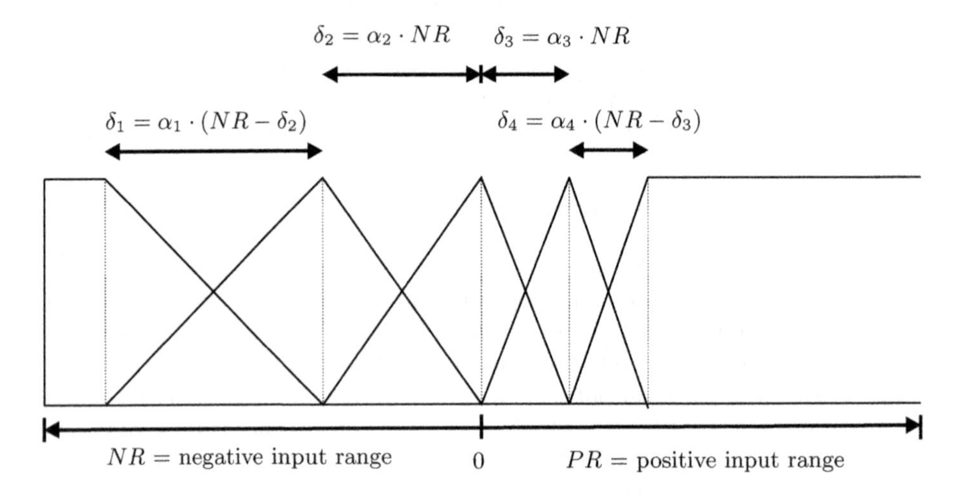

Fig. 6. Derivation of the membership functions from a parameter vector α.

The GA will create different individuals (possible solutions) using this parametrization and apply crossover, mutation and selection to reach an optimum. Actually an individual is only a vector of eight values: the four top-positions for the D input and the four for the DD input.

Fitness Function

The fitness function is the heart of the genetic algorithm, because it defines how well each individual performs. The optimum obtained by the GA is highly dependent on the fitness function, therefore it has to be designed with care. It is important to make an inventory of all characteristics of good (or bad) controller performance, but also a representative control simulation has to be designed.

Designing a representative control simulation is not trivial. The problems the controller will come across during actual operation should all be present in the simulation in the right proportions. At the same time the simulation should be as short as possible to save time. For an actual application some more time should be invested in the simulation design (i.e., the setpoint changes over time), also some checks on robustness might be added.

To see whether a certain simulation or fitness function is general enough to cover the whole operation range of the system, it might be a good idea to design multiple simulations and compare the resulting fitnesses for some different controllers. If both simulations are considered 'meaningful', a better fitness according to the one simulation should also have a better fitness according to the other.

At the moment four different fitness functions have been implemented. These differences can be found in the simulation time (30 or 250 seconds) and in the fitness criteria:

- **The simple fitness functions** only use the error area (difference between water level and setpoint integrated over time) as fitness value, where positive errors (i.e., overshoot) are multiplied by a factor larger than 1 because they are considered to be worse because of their possible dangerous consequences in a nuclear reactor.
- **The extensive fitness functions** also consider the mean settling time, mean overshoot, mean positive overshoot and mean negative overshoot, each with a different weight.

Optimisation Results

The improvement reached can be seen in Fig. 7 where the original (slow) response of the static controller and the response of the controller with optimized MFs are both plotted. In Fig. 8 the original and optimized membership functions are depicted. For this optimization the extensive fitness function was used with 250s simulation runs.

In this response we see the speed of the controller has improved significantly, at the cost of a little overshoot. The overshoot could be suppressed more by applying a higher weight to that in the fitness function. This is of course a decision that should be made according to the wishes of the controller designer and the specifications given.

When we compare the optimised MFs for the level (the upper plot in Fig. 8(b)) to the original ones (Fig. 8(a)) we see that the three central MFs are put closer together. This is not very surprising, because most control actions will be in this range and surely the accuracy in this range will be more important. When looking at the MFs for the rate (the lower ones) the most eye-catching thing is the disappearance of the NB membership function. This can very well be due to the fact that there is not much of a change in the rules between $DD = NS$ and $DD = NB$ (see Table 2.1). Also the fact that rates smaller than -0.3 are physically impossible might play a role here.

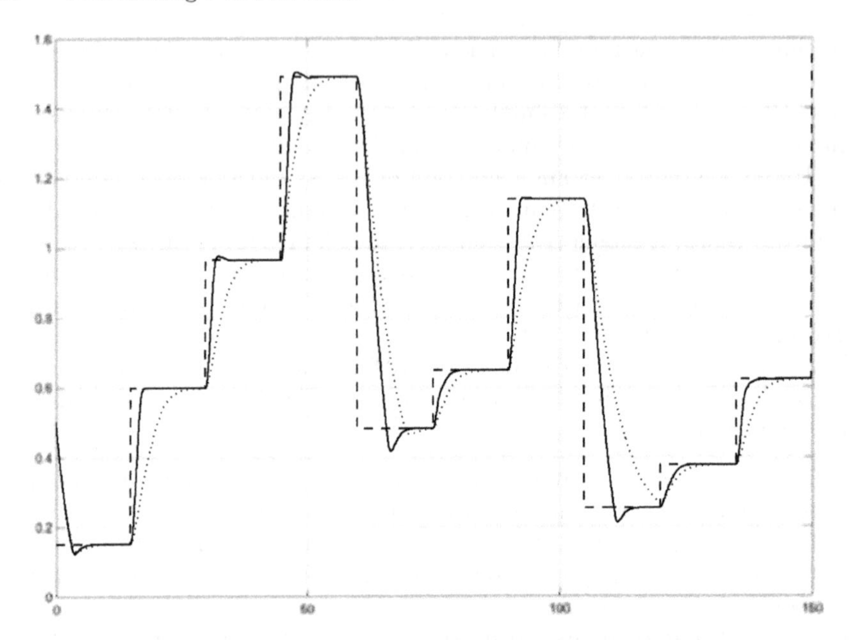

Fig. 7. Response of the original (*dotted*) and the optimized (*solid*) static controller.

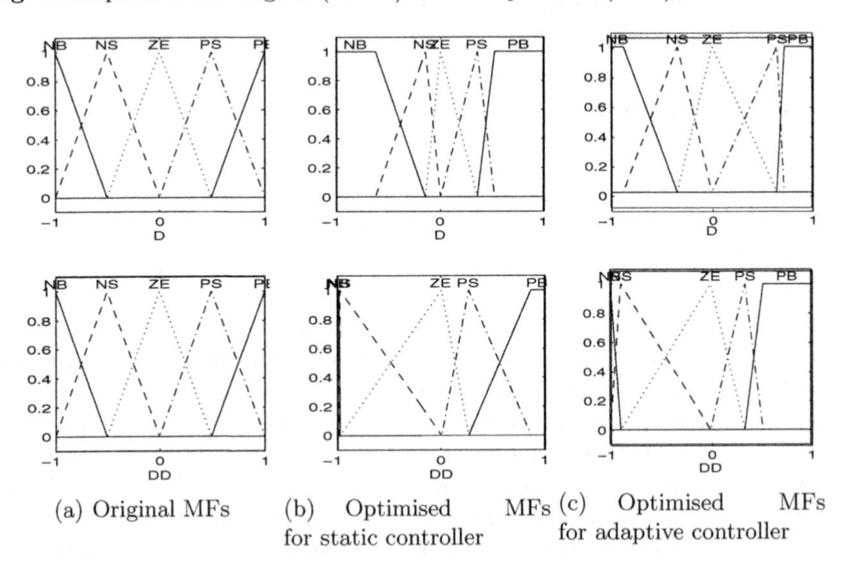

Fig. 8. The original and optimised membership functions.

3.2 Genetic Adaptive FLC Optimization

Now we have seen the potential of membership function for a static FLC we will extend this principle for an adaptive controller. The same parametrization and the extensive fitness function as described in Sect. 3.1 are used. The main difference is that the optimization has now become an iterative process, where the controller first gets some time to adapt (using the adaptation mechanism from Sect. 2.2), after which the membership functions are optimized, followed by a new adaptation run and a new optimization run etcetera.

This process is likely to converge relatively fast, because the rule base and the MFs keep the same mutual relation. Also the demands on convergence can be at a lower priority, because if the GA has not fully converged in the first run, it can carry on in the second run. To make this principle work, the best found solution from the last optimization should be added to the initial population of the new GA run.

Figure 9 shows that the genetic optimization is able to remove the fluctuations in the original adaptive controller without doing much harm to the rest of the performance (there is even no overshoot anymore). This means the optimization algorithm provides a very general way to improve controller performance, because engineering knowledge like applied for the input scaling in Sect. 2.4 is not needed any more. The improved MFs (Fig. 8(c)) have the same characteristics as the optimised static controller MFs (Fig. 8(b)).

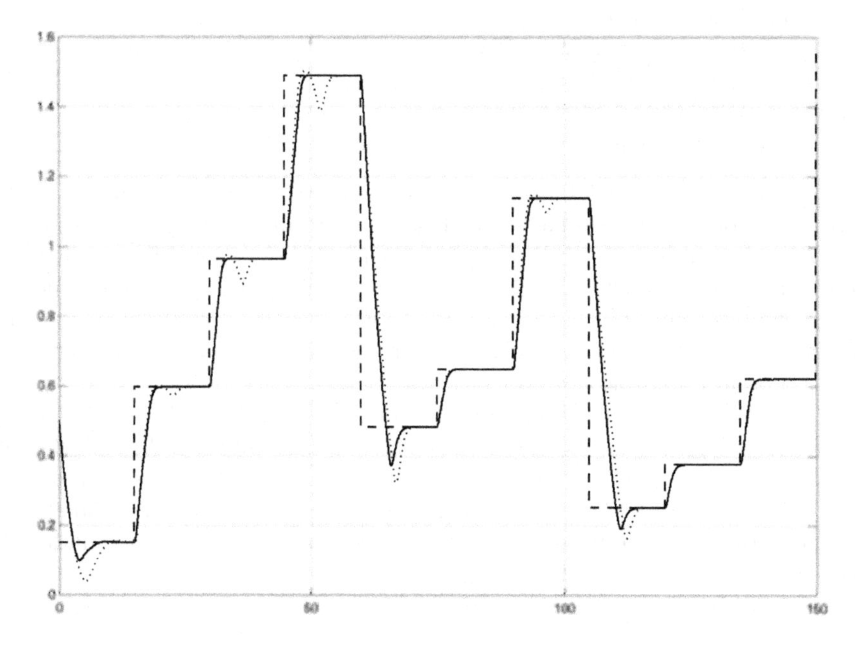

Fig. 9. Response of the original (*dotted*) and the optimized (*solid*) adaptive controller.

4 Application of Neural Networks

A problem with the setup as used in Sects. 2 and 3 is that in the nuclear reactor case safety regulations make it impossible to let a controller adapt itself on-line (i.e., when it is actually controlling the reactor). Therefore it is important to obtain a good model of the system so the controller can be adapted off-line which is fully safe. The adapted controller can now be checked for stability by engineers before it is actually implemented on-line.

It is difficult to obtain a good and fast model of a system (plant), especially when its dynamics are not fully known. This is often the case, because even if a mathematical formulation exists, most of the parameters in that formulation are uncertain or even change during operation. A neural network could be trained to mimic a plant's behaviour [14, 15, 16, 17] without any mathematical knowledge of the process and thus is a very general solution.

4.1 Tested Neural Networks

Several standard network types have been investigated:

1. Back Propagation (BP) networks
   - Feed-Forward back propagation (FF)
   - Cascade-Forward back propagation (CF)
   - Elman back propagation (ELM)
2. Radial Basis networks
   - Standard Radial Basis (RB)
   - Generalized Regression (GR)
3. Neuro-Fuzzy networks
   - Adaptive Neuro-Fuzzy Inference Systems (ANFIS)

All tests were performed using the standard networks provided by MATLAB [18]. For the back propagation networks the values of weights and biasses are set using an iterative training scheme such as Levenberg-Marquardt. The radial basis networks use the distances of an input to all training sets to determine the influence of each neuron. An ANFIS is a true hybrid system where no clear distinction can be made any more between the fuzzy logic controller and the neural network.

The NNs were trained to predict the new D_{t+1} and rate DD_{t+1} based on the control inputs VL_t and VS_t and the last level D_t and rate DD_t. Different settings (such as number of neurons, number of hidden layers and spread factors for RB and GR NNs) and different data set sizes (2500, 5000 and 10000 data sets) have been used. Also an implementation with estimation of the level (D_{t+1}) only has been tested.

4.2 Results of the Back Propagation Networks

If we check the performance of the networks with a test-data set, the CF networks come out best, the FF networks come second and the ELM networks worst. For CF and FF networks it seems that large networks (more layers, more neurons) are better off with large data sets and small networks are better off with small sets. For ELM networks things only get worse when more neurons are added.

If we check the performance using the estimation error in an actual control simulation, things turn out very different. The ELM networks still perform miserable, but now the performance of the CF and FF networks are much more mixed and small data sets seem to be the way to go, preferably in combination with a small network. However, the best performance (by a CF network with 10 neurons, 2500 data sets, see Fig. 10) is still too poor to be really useful.

When only estimating the level things do not get much better. CF still seems the best choice, but now middle-class networks with small training sets perform best. Based on the simulation this is almost the same case. Training a network with level estimation costs about 2/3 of the time needed for training a network estimating both level and rate.

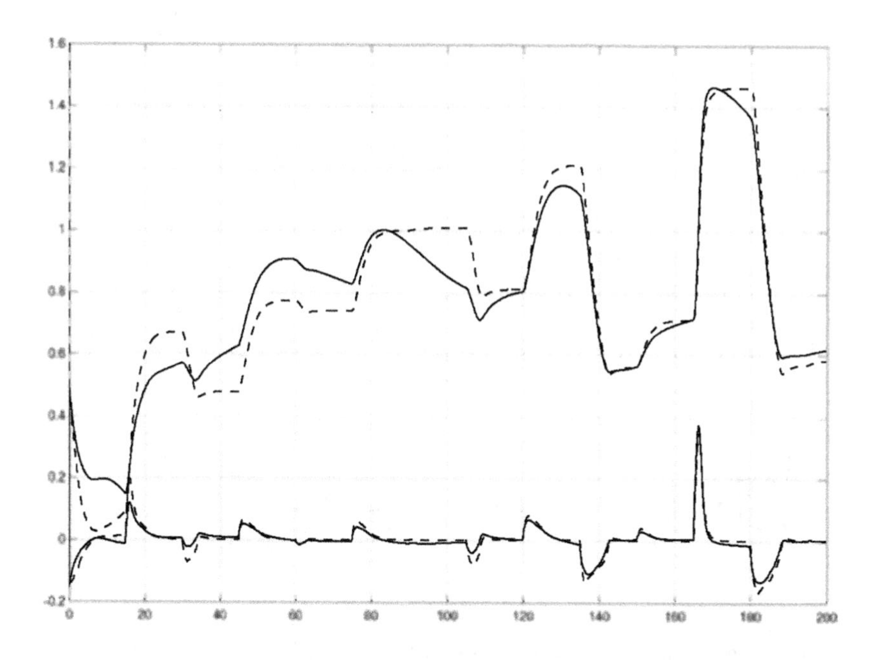

Fig. 10. Simulation using the best tested BP network providing a level and rate estimation (*solid*). The actual plant response (reference) is *dashed*

4.3 Results of the Radial Basis Networks

It is very difficult to get normal RB networks to mimic the plant accurately. When using too many data sets the performance gets worse, probably due to conflicting data (almost the same inputs, but different outputs). It is therefore important to make a well-balanced dataset and choose the appropriate spread factor (first hidden layer bias value). When adaptive RB NNs are used (adding neurons one by one) results do not get any better and training takes a lot longer. Also training the level only comes with some problems: when the flow has stabilized, the NN does not see that the rate is 0 and estimates a slightly positive or negative value, thus changing the level when it should be steady.

The GR networks (Fig. 11, solid lines) seem to give better results than the BP-NNs. The shape is right and the lines are straight. However, at some points (5-15 s, 105-110 s, 135-165 s, 185-200 s) things are going wrong, as can be clearly seen in the rate estimate (the lower line). The problem between 5-15 s could be because the training data contain to few sets with these very low levels. However, if we would eliminate this first error we get the dotted line in Fig. 11, and we see that the estimate would still not be accurate enough to design a controller upon.

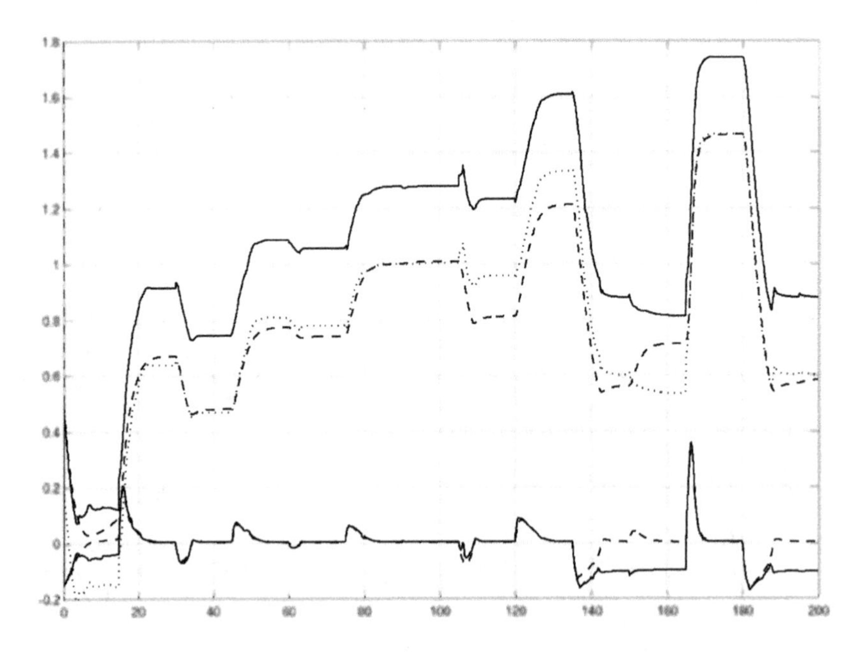

Fig. 11. Simulation using the best tested GR network providing a level and rate estimation (*solid*). The actual plant response (reference) is *dashed*, the estimation with compensation for the initial error is *dotted*

4.4 Using the ANFIS as a Controller

Instead of mimicking a plant, NNs could also be used as a controller. Especially the ANFIS should be suited for such a task when trained with data that reflect 'good control'. This means an ANFIS might be used to convert the expert knowledge of a manual operator into a static controller. Another option would be to try inverse model control. This theory is based on the idea that if we know the transfer function of our plant, we can control it in an exact manner if we make a controller with the exact inverse transfer function. In classical control this is impossible because these inverse transfer functions do not exist in practice, but it should be possible by training an ANFIS.

Training an ANFIS with *inverse* plant data however, is almost the same as training an ANFIS to mimic the plant. Because none of the ANFIS plants trained made any sense in a control simulation, it is quite logic that ANFIS controllers did not work either.

5 Conclusion

When dealing with soft computing techniques it is difficult to make hard statements on performance, due to a lack of in-depth investigations at the current stage. Actually the vagueness which is the strength of these techniques is also their weakness when it comes to drawing conclusions. However some very meaningful conclusions can be drawn from the research done.

As shown by almost any test in this chapter fuzzy logic provides a powerful way to control a strongly non-linear system, even when the system is not fully defined or has changing properties. Although the strength of the applied soft computing methods would be that no detailed (mathematical) description of the plant would be needed, even when such a description is available FLC still has many benefits, such as its robustness and transparency.

A static controller (Sect. 2.1) is the simplest and safest solution in an industrial environment, however designing the rule base by hand costs a lot of time and engineering skills. Moreover, such controllers are very problem specific and are likely not to provide optimal control in real life applications.

Adaptive control can solve these problems because they need very limited knowledge of the system. Actually for the problem investigated only two simple 'common sense' rules are needed to fully control the strongly non-linear system, as shown in Sect. 2.2.

Tuning of the FLC is important and can be done by hand (Sect. 2.4) when some insight in the controller and the system to control is available. An automated tuning method using Genetic Algorithm optimization techniques was proposed and successfully tested in Sect. 3 for both the static and the adaptive controller. In both cases it appears to be very profitable to let the algorithm tune the controllers according to predefined desires such as minimum overshoot and short settling time.

Due to safety regulations and practical problems controllers cannot be adapted and optimised on-line. Therefore an accurate model of the plant (i.e., the controlled system) is needed. To maintain the problem independence and superfluity of a mathematical description of the plant, Neural Networks were proposed in Sect. 4. From all tested networks, the Generalized Regression (GR) networks seem to be most suitable.

Although the simulation results of the GR NNs look quite reasonable, the exact values are by far not accurate enough to be applied in a real system. This may be due to the fact that neural networks generally provide an approximation instead of an exact answer combined with the fact that errors in the simulation accumulate. More balanced training data or other or adjusted types of networks might help to find a workable solution but it might just as well be impossible by principle to make an accurate plant simulation model using NNs. A first step in further research would be to find successful implementations in literature.

A promising solution has been found to make high performance FLCs for application in for instance nuclear reactor control. A subject for further research would be how to make the proposed methods suitable for practical application in an industrial environment.

References

1. J.A. Bernard. Use of a rule-based system for process control. *IEEE Control Systems Magazine*, 8(5):3–13, 1988.
2. D. Ruan. Initial experiments on fuzzy control for nuclear reactor operations at the belgian reactor 1. *Nuclear Technology*, 143:227–240, August 2003.
3. A.J. van der Wal D. Ruan. Controlling the power output of a nuclear reactor with fuzzy logic. *Information Sciences*, 110:151–177, 1998.
4. D. Ruan. Implementation of adaptive fuzzy control for a real-time control demo-model. *Real-Time Systems*, 21:219–239, 2001.
5. D. Ruan, editor. *Fuzzy Systems and Soft Computing in Nuclear Engineering.* Studies in Fuzzyness and Soft Computing. Physica-Verlag, 2000. ISBN 3-7908-1251-X.
6. M. Jamshidi S. Heger, N.K. Alang-Rashid. Application of fuzzy logic in nuclear reactor control, part i: An assessment of state-of-the-art. *Nuclear Safety*, 36(1): 109, 1996.
7. P.F. Fantoni D. Ruan, editor. *Power Plant Surveillance and Diagnostics.* Studies in Fuzzyness and Soft Computing. Springer, 2002. ISBN 3-540-43247-7.
8. Z. Michalewicz. *Genetic Algorithms + Data Structures = Evolution Programs.* Springer Verlag, 1997. ISBN 3-540-60676-9.
9. D. Beasley J. Heitktter. The hitch-hiker's guide to evolutionary computation. http://www.cs.bham.ac.uk/Mirrors/ftp.de.uu.net/EC/clife/www/, 2000.

10. S.E. Haupt R.L. Haupt. *Practical Genetic Algorithms.* Wiley-Interscience, 1997. ISBN 0-471-18873-5.
11. D. Goldberg. *Genetic Algorithms in Search, Optimization, and Machine Learning.* Addison-Wesley, 1989. ISBN 0-471-18873-5.
12. A.S. Wu & H. Yu D.C. Marinescu, H.J.Siegel. A genetic approach to planning in heterogeneous computing environments. In *IPDPS*, page 97, 2003.
13. W. Ruijter J.O. Entzinger, R. Spallino. Multilevel distributed structure optimization. *Proceedings of the 24th International Congress of the Aeronautical Sciences (ICAS), Yokohama*, 2004.
14. S.F. de Azevedo A. Andršik, A. Mszros. On-line tuning of a neural pid controller based on plant hybrid modeling. *Computers and Chemical Engineering*, 28 (2004):1499–1509, 2003.
15. S. Mukhopadhyay J.F. Briceno, H. El-Mounayri. Selecting an artificial neural network for efficient modeling and accurate simulation of the milling process. *International Journal of Machine Tools & Manufacture*, 42(2002):663–674, 2002.
16. R. Babuška H.B. Verbruggen J.A. Roubos, S. Mollov. Fuzzy model-based predictive control using takagi-sugeno models. *International Journal of Approximate Reasoning*, 22(1999):3–30, 1999.
17. A. Dourado H. Duarte-Ramos P. Gil, J. Henriques. Fuzzy model-based predictive control using takagi-sugeno models. *Proceedings of ESIT'99 European Symposium on Intelligent Techniques, Crete, Greece*, (1999), june 1999.
18. Inc The MathWorks. Matlab 6.5, the language of technical computing. Software with online help (www.mathworks.com), 1984-2002.

Part II

Fuzzy Engineering Economics and Investment Analysis Techniques

Applications of Fuzzy Capital Budgeting Techniques

Cengiz Kahraman[1], Murat Gülbay[1], and Ziya Ulukan[2]

[1]Istanbul Technical University, Department of Industrial Engineering, Maçka 34367, Istanbul, Turkey
[2]Galatasaray University, Faculty of Engineering and Technology, Ortaköy 34357, Istanbul, Turkey

Summary. In an uncertain economic decision environment, an expert's knowledge about dicounting cash flows consists of a lot of vagueness instead of randomness. Cash amounts and interest rates are usually estimated by using educated guesses based on expected values or other statistical techniques to obtain them. Fuzzy numbers can capture the difficulties in estimating these parameters. In this chapter, the formulas for the analyses of fuzzy present value, fuzzy equivalent uniform annual value, fuzzy future value, fuzzy benefit-cost ratio, and fuzzy payback period are developed and some numeric examples are given. Then the examined cash flows are expanded to geometric and trigonometric cash flows and using these cash flows fuzzy present value, fuzzy future value, and fuzzy annual value formulas are developed for both discrete compounding and continuous compounding. Finally, a fuzzy versus stochastic investment analysis is examined by using the probability of a fuzzy event.

Key words: fuzzy number, capital budgeting, cash flow, ranking method

1 Introduction

The purpose of this chapter is to develop the fuzzy capital budgeting techniques. The analyses of fuzzy future value, fuzzy present value, fuzzy rate of return, fuzzy benefit/cost ratio, fuzzy payback period, fuzzy equivalent uniform annual value are examined for the case of discrete compounding.

To deal with vagueness of human thought, Zadeh [1] first introduced the fuzzy set theory, which was based on the rationality of uncertainty due to imprecision or vagueness. A major contribution of fuzzy set theory is its capability of representing vague knowledge. The theory also allows mathematical operators and programming to apply to the fuzzy domain.

C. Kahraman et al.: *Applications of Fuzzy Capital Budgeting Techniques*, StudFuzz **201**, 177–203 (2006)
www.springerlink.com

A fuzzy number is a normal and convex fuzzy set with membership function $\mu_A(x)$ which both satisfies normality: $\mu_A(x)=1$, for at least one $x \in R$ and convexity: $\mu_A(x') \geq \mu_A(x_1) \Lambda \mu_A(x_2)$, where $\mu_A(x) \in [0,1]$ and $\forall x' \in [x_1, x_2]$. '$\Lambda$' stands for the minimization operator.

Quite often in finance future cash amounts and interest rates are estimated. One usually employs educated guesses, based on expected values or other statistical techniques, to obtain future cash flows and interest rates. Statements like *approximately between \$ 12,000 and \$ 16,000* or *approximately between 10% and 15%* must be translated into an exact amount, such as *\$ 14,000* or *12.5%* respectively. Appropriate fuzzy numbers can be used to capture the vagueness of those statements.

A tilde will be placed above a symbol if the symbol represents a fuzzy set. Therefore, $\tilde{P}, \tilde{F}, \tilde{G}, \tilde{A}, \tilde{i}, \tilde{r}$ are all fuzzy sets. The membership functions for these fuzzy sets will be denoted by $\mu(x|\tilde{P}), \mu(x|\tilde{F}), \mu(x|\tilde{G})$, etc. A fuzzy number is a special fuzzy subset of the real numbers. A triangular fuzzy number (TFN) is shown in Figure 1. The membership function of a TFN ($\tilde{M}$) defined by

$$\mu(x|\tilde{M}) = (m_1, f_1(y|\tilde{M}) / m_2, m_2 / f_2(y|\tilde{M}), m_3) \tag{1}$$

where $m_1 \prec m_2 \prec m_3$, $f_1(y|\tilde{M})$ is a continuous monotone increasing function of y for $0 \leq y \leq 1$ with $f_1(0|\tilde{M}) = m_1$ and $f_1(1|\tilde{M}) = m_2$ and $f_2(y|\tilde{M})$ is a continuous monotone decreasing function of y for $0 \leq y \leq 1$ with $f_2(1|\tilde{M}) = m_2$ and $f_2(0|\tilde{M}) = m_3$. $\mu(x|\tilde{M})$ is denoted simply as $(m_1 / m_2, m_2 / m_3)$.

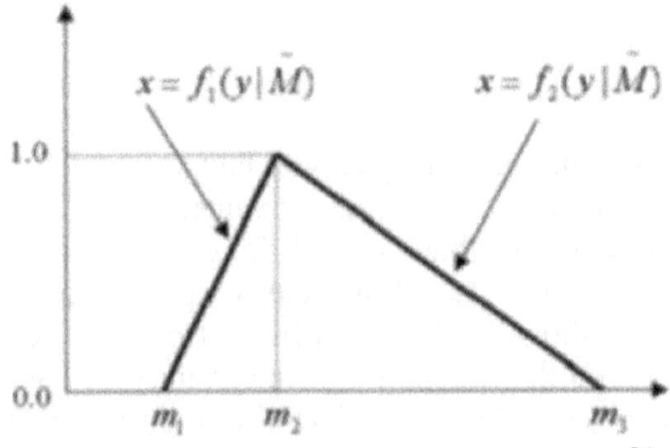

Fig. 1. A Triangular Fuzzy Number, $\tilde{M}$

A flat fuzzy number (FFN) is shown in Figure 2. The membership function of a FFN, $\widetilde{V}$ is defined by

$$\mu(x|\widetilde{V}) = (m_1, f_1(y|\widetilde{V}) / m_2, m_3 / f_2(y|\widetilde{V}), m_4) \tag{2}$$

where $m_1 \prec m_2 \prec m_3 \prec m_4$, $f_1(y|\widetilde{V})$ is a continuous monotone increasing function of y for $0 \le y \le 1$ with $f_1(0|\widetilde{V}) = m_1$ and $f_1(1|\widetilde{V}) = m_2$ and $f_2(y|\widetilde{V})$ is a continuous monotone decreasing function of y for $0 \le y \le 1$ with $f_2(1|\widetilde{V}) = m_3$ and $f_2(0|\widetilde{V}) = m_4$. $\mu(y|\widetilde{V})$ is denoted simply as $(m_1 / m_2, m_3 / m_4)$.

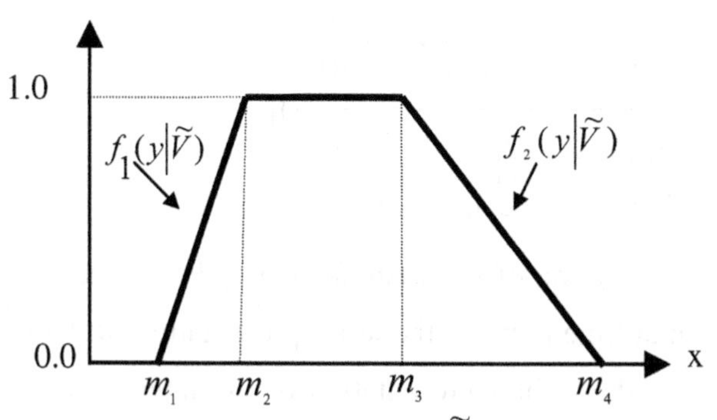

Fig. 2. A Flat Fuzzy Number, $\widetilde{V}$

The fuzzy sets $\widetilde{P}, \widetilde{F}, \widetilde{G}, \widetilde{A}, \widetilde{i}, \widetilde{r}$ are usually fuzzy numbers but n will be discrete positive fuzzy subset of the real numbers [2]. The membership function $\mu(x|\widetilde{n})$ is defined by a collection of positive integers n_i, $1 \le i \le K$, where

$$\mu(x|\widetilde{n}) = \begin{cases} \mu(n|\widetilde{n}) = \lambda, & 0 \le \lambda \le 1 \\ 0, & otherwise \end{cases} \tag{3}$$

2 Fuzzy Present Value (PV) Method

The present-value method of alternative evaluation is very popular because future expenditures or receipts are transformed into equivalent dollars now. That is, all of the future cash flows associated with an alternative are converted into present dollars. If the alternatives have different lives, the alternatives must be compared over the same number of years.

Chiu and Park [3] propose a present value formulation of a fuzzy cash flow. The result of the present value is also a fuzzy number with nonlinear membership function. The present value can be approximated by a TFN. Chiu and Park [3]'s formulation is

$$P\tilde{V} = [\sum_{t=0}^{n} (\frac{\max(P_t^{l(y)},0)}{\prod_{t'=0}^{t}(1+r_{t'}^{r(y)})} + \frac{\min(P_t^{l(y)},0)}{\prod_{t'=0}^{t}(1+r_{t'}^{l(y)})}), \tag{4}$$

$$\sum_{t=0}^{n} (\frac{\max(P_t^{r(y)},0)}{\prod_{t'=0}^{t}(1+r_{t'}^{l(y)})} + \frac{\min(P_t^{r(y)},0)}{\prod_{t'=0}^{t}(1+r_{t'}^{r(y)})})]$$

where $P_t^{l(y)}$: the left representation of the cash at time t, $P_t^{r(y)}$: the right representation of the cash at time t, $r_t^{l(y)}$: the left representation of the interest rate at time t, $r_t^{r(y)}$: the right representation of the interest rate at time t.

Buckley's [2] membership function for $\tilde{P}_n$,

$$\mu(x|\tilde{P}_n) = (p_{n1}, f_{n1}(y|\tilde{P}_n)/p_{n2}, p_{n2}/f_{n2}(y|\tilde{P}_n), p_{n3}) \tag{5}$$

is determined by

$$f_{ni}(y|\tilde{P}_n) = f_i(y|\tilde{F})(1+f_k(y|\tilde{r})^{-n} \tag{6}$$

for $i = 1,2$ where $k=i$ for negative $\tilde{F}$ and $k=3-i$ for positive $\tilde{F}$.

Ward [4] gives the fuzzy present value function as

$$P\tilde{V} = (1+r)^{-n}(a,b,c,d) \tag{7}$$

where (a, b, c, d) is a flat fuzzy filter function (4F) number.

Example 1. A $ (-14,000, -12,000, -10,000) investment will return annual benefits $(2,650, 2,775, 2,900) for six years with no salvage value at the end of six years. Compute the fuzzy present worth of the cash flow using an interest of (7.12%, 10.25%, 13.42%) per year.

$$f_{6,1}(y|\widetilde{P}) = \sum_{j=0}^{6} f_{j,1}(y|\widetilde{F}_j)[1 + f_{k(j)}(y|\widetilde{r}_f)]^{-j} \tag{8}$$

$$f_{6,2}(y|\widetilde{P}) = \sum_{j=0}^{6} f_{j,2}(y|\widetilde{F}_j)[1 + f_{k(j)}(y|\widetilde{r}_f)]^{-j} \tag{8}$$

i=1, 2 where $k(j)=i$ for negative $\widetilde{F}_j$ and $k(j) = 3 - i$ for positive $\widetilde{F}_j$.

For $y = 0$, $f_{6,1}(y|\widetilde{P}) = \$ - 3{,}525.57$

For $y = 1$, $f_{6,1}(y|\widetilde{P}) = f_{6,2}(y|\widetilde{P}) = \$ - 24.47$

For $y = 0$, $f_{6,2}(y|\widetilde{P}) = \$ + 3{,}786.34$

The possibility of $NPV = 0$ for this triangular fuzzy number can be calculated using a linear enterpolation:
$x = -3{,}810.81y + 3{,}786.34$

For $x = 0$, $Poss(NPV = 0) = 0.9936$.

3 Fuzzy Capitalized Value Method

A specialized type of cash flow series is a perpetuity, a uniform series of cash flows which continues indefinetely. An infinite cash flow series may be appropriate for such very long-term investment projects as bridges, highways, forest harvesting, or the establishment of endowment funds where the estimated life is 50 years or more.

In the nonfuzzy case, if a present value P is deposited into a fund at interest rate r per period so that a payment of size A may be withdrawn each and every period forever, then the following relation holds between P, A, and r:

$$P = \frac{A}{r} \tag{10}$$

In the fuzzy case, lets assume all the parameters as triangular fuzzy numbers: $\widetilde{P}=(p_1,p_2,p_3)$ or $\widetilde{P}=((p_2-p_1)y+p_1),(p_2-p_3)y+p_3)$ and $\widetilde{A}=(a_1,a_2,a_3)$ or $\widetilde{A}=((a_2-a_1)y+a_1),(a_2-a_3)y+a_3)$ and $\widetilde{r}=(r_1,r_2,r_3)$ or $\widetilde{r}=((r_2-r_1)y+r_1,(r_2-r_3)y+r_3)$, where y is the membership degree of a certain point of A and r axis. If $\widetilde{A}$ and $\widetilde{r}$ are both positive,

$$\widetilde{P}=\widetilde{A}\div\widetilde{r}=(a_1/r_3,a_2/r_2,a_3/r_1) \tag{11}$$

or

$$\widetilde{P}=(((a_2-a_1)y+a_1)/((r_2-r_3)y+r_3),((a_2-a_3)y+a_3)/((r_2-r_1)y+r_1)) \tag{12}$$

If $\widetilde{A}$ is negative and $\widetilde{r}$ is positive,

$$\widetilde{P}=\widetilde{A}\div\widetilde{r}=(a_1/r_1,a_2/r_2,a_3/r_3) \tag{13}$$

or

$$\widetilde{P}=(((a_2-a_1)y+a_1)/((r_2-r_1)y+r_1),((a_2-a_3)y+a_3)/((r_2-r_3)y+r_3)) \tag{14}$$

Now, let $\widetilde{A}$ be an expense every nth period forever, with the first expense occurring at n. For example, an expense of (\$5,000,\$7,000,\$9,000) every third year forever, with the first expense occurring at t=3. In this case, the fuzzy efffective rate $\widetilde{e}$ may be used as in the following:

$$f_i(y|\widetilde{e})=(1+(1/m)f_i(y|\widetilde{r}'))^m-1 \tag{15}$$

where i=1,2; $f_1(y|\widetilde{e})$:a continuous monotone increasing function of y; $f_2(y|\widetilde{e})$:a continuous monotone decreasing function of y; m: the number of compoundings per period; $\widetilde{r}'$: the fuzzy nominal interest rate per period. The membership function of $\widetilde{e}$ may be given as

$$\mu(x|\widetilde{e})=(e_1,f_1(y|\widetilde{e})/e_2,e_2/f_2(y|\widetilde{e}),e_3) \tag{16}$$

If $\widetilde{A}$ and $f_i(y|\widetilde{e})$ are both positive,

$$\widetilde{P} = \widetilde{A} \oslash \widetilde{e} = [((a_2 - a_1)y + a_1) / f_2(y|\widetilde{e}), ((a_2 - a_3)y + a_3) / f_1(y|\widetilde{e})] \tag{17}$$

If $\widetilde{A}$ is negative and $f_i(y|\widetilde{e})$ is positive,

$$\widetilde{P} = \widetilde{A} \oslash \widetilde{e} = [((a_2 - a_1)y + a_1) / f_1(y|\widetilde{e}), ((a_2 - a_3)y + a_3 / f_2(y|\widetilde{e})] \tag{18}$$

$(a_2 - a_1)y + a_1$ and $(a_2 - a_3)y + a_3$ can be symbolized as $f_1(y|\widetilde{a})$ and $f_2(y|\widetilde{a})$ respectively.

Example 2: Project ABC consists of the following requirements. Find the capitalized worth of the project if $\widetilde{r} = (\%12, \%15, \%18)$ annually.

1. A (\$40,000, \$50,000, \$60,000) first cost $(F\widetilde{C})$ at t=0,
2. A (\$4,000, \$5,000, \$6,000) expense $(\widetilde{A}_1)$ every year,
3. A (\$20,000, \$25,000, \$30,000) expense $(\widetilde{A}_2)$ every third year forever, with the first expense occurring at t=3.

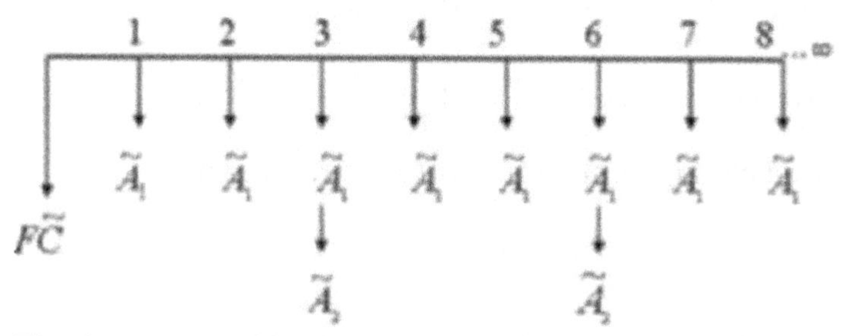

Fig. 3. The Cash Flow Diagram for Example 2

$$\widetilde{P} = F\widetilde{C} \oplus (\widetilde{A}_1 \oslash \widetilde{r}) + (\widetilde{A}_2 \oslash \widetilde{e}) \tag{19}$$

$\widetilde{P} = ((10{,}000y + 40{,}000), (-10{,}000y + 60{,}000)) +$

$((1{,}000y + 4{,}000) / (-0.03y + 0.18), (-1{,}000y + 6{,}000) / (0.03y + 0.12)) +$

$((5{,}000y + 20{,}000) / ((1.18 - 0.03y)^3 - 1),$

$(-5{,}000y + 30{,}000) / ((1.12 + 0.03y)^3 - 1))$

For y=0, $f_{\infty,1}(y|\widetilde{P}) = \$93{,}326.42$

For y=1, $f_{\infty,1}(y|\widetilde{P}) = f_{\infty,2}(y|\widetilde{P}) = \$131{,}317.97$

For y=0, $f_{\infty,2}(y|\widetilde{P}) = \$184{,}074.07$

4 Fuzzy Future Value Method

The future value (*FV*) of an investment alternative can be determined using the relationship

$$FV(r) = \sum_{t=0}^{n} P_t (1+i)^{n-t} \tag{20}$$

where *FV(r)* is defined as the future value of the investment using a minimum attractive rate of return *(MARR)* of $r\%$. The future value method is equivalent to the present value method and the annual value method.

Chiu and Park's [3] formulation for the fuzzy future value has the same logic of fuzzy present value formulation:

$$\{\sum_{t=0}^{n-1}[\max(P_t^{l(y)},0)\prod_{t'=t+1}^{n}(1+r_{t'}^{l(y)})+\min(P_t^{l(y)},0)\prod_{t'=t+1}^{n}(1+r_{t'}^{r(y)})]+P_n^{l(y)}, \tag{21}$$

$$\sum_{t=0}^{n-1}[\max(P_t^{r(y)},0)\prod_{t'=t+1}^{n}(1+r_{t'}^{r(y)})+\min(P_t^{r(y)},0)\prod_{t'=t+1}^{n}(1+r_{t'}^{l(y)})]+P_n^{r(y)}\}$$

Buckley's [2] membership function $\mu(x|\widetilde{F})$ is determined by

$$f_i(y|\widetilde{F}_n) = f_i(y|\widetilde{P})(1+f_i(y|\widetilde{r}))^n \tag{22}$$

For the uniform cash flow series, $\mu(x|\widetilde{F})$ is determined by

$$f_{ni}(y|\widetilde{F}) = f_i(y|\widetilde{A})\beta(n, f_i(y|\widetilde{r})) \tag{23}$$

where i=1,2 and $\beta(n,r) = (((1+r)^n - 1)/r)$ and $\widetilde{A} \succ 0$ and $\widetilde{r} \succ 0$.

5 Fuzzy Benefit/Cost Ratio Method

The benefit/cost ratio (*BCR*) is often used to assess the value of a municipal project in relation to its cost; it is defined as

$$BCR = \frac{B - D}{C} \tag{24}$$

where B represents the equivalent value of the benefits associated with the project, D represents the equivalent value of the disbenefits, and C represents the project's net cost. A BCR greater than 1.0 indicates that the project evaluated is economically advantageous. In BCR analyses, costs are not preceded by a minus sign.

When only one alternative must be selected from two or more mutually exclusive (stand-alone) alternatives, a multiple alternative evaluation is required. In this case, it is necessary to conduct an analysis on the incremental benefits and costs. Suppose that there are two mutually exclusive alterantives. In this case, for the incremental BCR analysis ignoring disbenefits the following ratios must be used:

$$\Delta B_{2-1} / \Delta C_{2-1} = \Delta PV B_{2-1} / \Delta PVC_{2-1} \tag{25}$$

where PVB: present value of benefits, PVC: present value of costs. If $\Delta B_{2-1} / \Delta C_{2-1} \geq 1.0$, the alternative 2 is preferred.

In the case of fuzziness, first, it will be assumed that the largest possible value of Alternative 1 for the cash in year t is less than the least possible value of Alternative 2 for the cash in year t. The fuzzy incremental BCR is

$$\Delta\tilde{B}/\Delta\tilde{C} = \left(\frac{\sum_{t=0}^{n}(B_{2t}^{l(y)} - B_{1t}^{r(y)})(1 + r^{r(y)})^{-t}}{\sum_{t=0}^{n}(C_{2t}^{r(y)} - C_{1t}^{l(y)})(1 + r^{l(y)})^{-t}}, \frac{\sum_{t=0}^{n}(B_{2t}^{r(y)} - B_{1t}^{l(y)})(1 + r^{l(y)})^{-t}}{\sum_{t=0}^{n}(C_{2t}^{l(y)} - C_{1t}^{r(y)})(1 + r^{r(y)})^{-t}}\right) \tag{26}$$

If $\Delta \tilde{B}/\Delta\tilde{C}$ is equal or greater than (1, 1, 1), Alternative 2 is preferred.

In the case of a regular annuity, the fuzzy $\tilde{B}/\tilde{C}$ ratio of a single investment alternative is

$$\tilde{B} / \tilde{C} = \left(\frac{A^{l(y)}\gamma(n, r^{r(y)})}{C^{r(y)}}, \frac{A^{r(y)}\gamma(n, r^{l(y)})}{C^{l(y)}}\right) \tag{27}$$

where $\tilde{C}$ is the first cost and $\tilde{A}$ is the net annual benefit, and $\gamma(n,r) = ((1+r)^n - 1)/(1+r)^n r)$.

The $\Delta\tilde{B} / \Delta\tilde{C}$ ratio in the case of a regular annuity is

$$\Delta\tilde{B} / \Delta\tilde{C} = \left(\frac{(A_2^{l(y)} - A_1^{r(y)})\gamma(n, r^{r(y)})}{C_2^{r(y)} - C_1^{l(y)}}, \frac{(A_2^{r(y)} - A_1^{l(y)})\gamma(n, r^{l(y)})}{C_2^{l(y)} - C_1^{r(y)}}\right) \tag{28}$$

6 Fuzzy Equivalent Uniform Annual Value (EUAV) Method

The *EUAV* means that all incomes and disbursements (irregular and uniform) must be converted into an equivalent uniform annual amount, which is the same each period. The major advantage of this method over all the other methods is that it does not require making the comparison over the least common multiple of years when the alternatives have different lives [5]. The general equation for this method is

$$EUAV = A = NPV\gamma^{-1}(n,r) = NPV[\frac{(1+r)^n r}{(1+r)^n - 1}] \tag{29}$$

where *NPV* is en the fuzzy $EU\widetilde{A}V$ ($\widetilde{A}_n$) will be found. The membership function the net present value. In the case of fuzziness, $N\widetilde{P}V$ will be calculated and th $\mu(x|\widetilde{A}_n)$ for $\widetilde{A}_n$ is determined by

$$f_{ni}(y|\widetilde{A}_n) = f_i(y|N\widetilde{P}V)\gamma^{-1}(n, f_i(y|\widetilde{r})) \tag{30}$$

and *TFN(y)* for fuzzy *EUAV* is

$$\widetilde{A}_n(y) = (\frac{NPV^{l(y)}}{\gamma(n, r^{l(y)})}, \frac{NPV^{r(y)}}{\gamma(n, r^{r(y)})}) \tag{31}$$

Example 3. Assume that $N\widetilde{P}V = (-\$3{,}525.57, -\$24.47, +\$3{,}786.34)$ and $\widetilde{r} = (3\%, 5\%, 7\%)$. Calculate the fuzzy *EUAV*.

$$f_{6,1}(y|\widetilde{A}_6) = (3{,}501.1y - 3{,}525.57)[\frac{(1.03+0.02y)^6(0.02y+0.03)}{(1.03+0.02y)^6 - 1}]$$

$$f_{6,2}(y|\widetilde{A}_6) = (-3{,},810.81y + 3{,}786.34)[\frac{(1.07-0.02y)^6(0.07-0.02y)}{(1.07-0.02y)^6 - 1}]$$

For y=0, $f_{6,1}(y|\widetilde{A}_6) = -\$650.96$

For y=1, $f_{6,1}(y|\widetilde{A}_6) = f_{6,2}(y|\widetilde{A}_6) = -\$4.82$

For y=0, $f_{6,2}(y|\widetilde{A}_6) = +\$795.13$

7 Fuzzy Payback Period (FPP) Method

The payback period method involves the determination of the length of time required to recover the initial cost of investment based on a zero interest rate ignoring the time value of money or a certain interest rate recognizing the time value of money. Let C_{j0} denote the initial cost of investment alternative *j*, and R_{jt} denote the net revenue received from investment *j* during period *t*. Assuming no other negative net cash flows occur, the smallest value of m_j ignoring the time value of money such that

$$\sum_{t=1}^{m_j} R_{jt} \geq C_{j0} \tag{32}$$

or the smallest value of m_j recognizing the time value of money such that

$$\sum_{t=1}^{m_j} R_{jt}(1+r)^{-t} \geq C_{j0} \tag{33}$$

defines the payback period for the investment *j*. The investment alternative having the smallest payback period is the preferred alternative. In the case of fuzziness, the smallest value of m_j ignoring the time value of money such that

$$(\sum_{t=1}^{m_j} r_{1jt}, \sum_{t=1}^{m_j} r_{2jt}, \sum_{t=1}^{m_j} r_{3jt}) \geq (C_{1j0}, C_{2jo}, C_{3j0}) \tag{34}$$

and the smallest value of m_j recognizing the time value of money such that

$$\begin{aligned}&(\sum_{t=1}^{m_j} (R_{jt}^{l(y)} /(1+r^{r(y)})^t), \sum_{t=1}^{m_j} (R_{jt}^{r(y)} /(1+r^{l(y)})^t) \geq \\ &((C_{2j0} - C_{1j0})y + C_{1j0}, (C_{2j0} - C_{3j0})y + C_{3j0})\end{aligned} \tag{35}$$

defines the payback period for investment *j*, where r_{kjt} : the *k*th parameter of a triangular fuzzy R_{jt}; C_{kj0} : the *k*th parameter of a triangular fuzzy C_{j0}; $R_{jt}^{l(y)}$: the left representation of a triangular fuzzy R_{jt}; $R_{jt}^{r(y)}$: the right representation of a triangular fuzzy R_{jt}. If it is assumed that the discount rate changes from one period to another, $(1+r^{r(y)})^t$ and $(1+r^{l(y)})^t$ will be $\prod_{t'=1}^{t}(1+r_{t'}^{r(y)})$ and $\prod_{t'=1}^{t}(1+r_{t'}^{l(y)})$ respectively.

It is now necessary to use a ranking method to rank the triangular fuzzy numbers such as Chiu and Park's [3], Chang's [6] method , Dubois and Prade's [7] method, Jain's [8] method, Kaufmann and Gupta's [9] method, Yager's [10] method. These methods may give different ranking results and most methods are tedious in graphic manipulation requiring complex mathematical calculation. In the following, two of the methods which does not require graphical representations are given. Chiu and Park's (1994) weighted method for ranking TFNs with parameters (*a, b, c*) is formulated as

$$\left((a+b+c)/3\right)+wb \tag{36}$$

where w is a value determined by the nature and the magnitude of the most promising value. The preference of projects is determined by the magnitude of this sum.

Kaufmann and Gupta (1988) suggest three criteria for ranking TFNs with parameters (a,b,c). The dominance sequence is determined according to priority of:

1. Comparing the ordinary number (a+2b+c)/4
2. Comparing the mode, (the corresponding most promise value), b, of each TFN.
3. Comparing the range, c-a, of each TFN.

The preference of projects is determined by the amount of their ordinary numbers. The project with the larger ordinary number is preferred. If the ordinary numbers are equal, the project with the larger corresponding most promising value is preferred. If projects have the same ordinary number and most promising value, the project with the larger range is preferred.

Example 4. Assume that there are two alternative machines that are under consideration to replace an aging production machine. The associated cash flows are given in the following table. Determine the best alternative by using the payback period method recognizing the time value of money. The fuzzy interest rate is (12%, %15, %18) annually.

Table 1. Fuzzy Cash Flow (x$1,000)

End of year	0	1	2	3	4
Alt. A	(-7, -5, -3)	(2, 3, 4)	(4, 4.5, 5)	(1, 1.5, 2)	(3.5, 4, 4.5)
Alt. B	(-12, -10, -8)	(3, 4, 5)	(4.5, 5, 5.5)	(3, 3.5, 4)	(4, 4.5, 5)

If Chiu and Park 's [3] method (CP) is used for ranking TFNs, it is obtained (w=0.3):

For Alternative A,

$$CP_0 = (\frac{C_{1jt} + C_{2jt} + C_{3jt}}{3}) + wC_{2jt} = -6{,}500$$

$$TFN_1 = (\frac{1{,}000y + 2{,}000}{1.18 - 0.03y}, \frac{-1{,}000y + 4{,}000}{1.12 + 0.03y}) = (1695, 2608.7, 3571.4)$$

$$CP_1 = 3{,}407.6$$

$$TFN_2 = (\frac{500y + 4{,}000}{(1.18 - 0.03y)^2}, \frac{-500y + 5{,}000}{(1.12 + 0.03y)^2}) = (2872.7, 3402..6, 3986)$$

$$CP_2 = 4{,}441.2$$

$$\sum_{i=1}^{2} CP_i \succ CP_0 \rightarrow PP_A = 1.656 years .$$

For Alternative B,

$$CP_0 = -13{,}000$$

$$TFN_1 = (2542.4, 3478.3, 4464.3)$$

$$CP_1 = 4538.5$$

$$TFN_2 = (3231.8, 3780.7, 4384.6)$$

$$CP_2 = 4{,}993.2$$

$$TFN_3 = (1825.9, 2301.3, 2847.1)$$

$$CP_3 = 3{,}015.2$$

$$TFN_4 = (2063.2, 2572.9, 3177.6)$$

$$CP_4 = 3{,}376.4$$

$$\sum_{i=1}^{4} CP_i \succ CP_0 \rightarrow PP_B = 3.3 years.$$

Alternative A is the preferred one.

8 Fuzzy Internal Rate of Return (IRR) Method

The IRR method is referred to in the economic analysis literature as the discounted cash flow rate of return, internal rate of return, and the true rate of reurn. The internal rate of return on an investment is defined as the rate of interest earned on the unrecovered balance of an investment. Letting r^* denote the rate of return, the equation for obtaining r^* i

$$\sum_{t=1}^{n} P_t (1+r^*)^{-t} - FC = 0 \tag{37}$$

where P_t is the net cash flow at the end of period t.

Assume the cash flow $\widetilde{F} = \widetilde{F}_0, \widetilde{F}_1, ..., \widetilde{F}_N$ is fuzzy. $\widetilde{F}_n$ is a negative fuzzy number and the other $\widetilde{F}_i$ may be positive or negative fuzzy numbers. The fuzzy $IRR(\widetilde{F}, n)$ is a fuzzy interest rate $\widetilde{r}$ that makes the present value of all future cash amounts equal to the initial cash outlay. Therefore, the fuzzy number $\widetilde{r}$ satisfies

$$\sum_{i=1}^{n} PV_{k(i)}(\widetilde{F}_i, i) = -\widetilde{F}_0 \tag{38}$$

where $\sum$ is fuzzy addition, $k(i)$=1 if $\widetilde{F}_i$ is negative and $k(i)$=2 if $\widetilde{F}_i$ is postive.

Buckley [2] shows that such simple fuzzy cash flows may not have a fuzzy *IRR* and concludes that the *IRR* technique does not extend to fuzzy cash flows. Ward [4] considers Eq. (37) and explains that such a procedure can not be applied for the fuzzy case because the right hand side of Eq. (37) is fuzzy, 0 is crisp, and an equality is impossible.

9 An Expansion to Geometric and Trigonometric Cash Flows

When the value of a given cash flow differs from the value of the previous cash flow by a constant percentage, $j\%$, then the series is referred to as a *geometric series*. If the value of a given cash flow differs from the value of the previous cash flow by a sinusoidal wave or a cosinusoidal wave, then the series is referred to as a *trigonometric series*

9.1 Geometric Series - Fuzzy Cash Flows in Discrete Compounding

The present value of a crisp geometric series is given by

$$P = \sum_{n=1}^{N} F_1 (1+g)^{n-1} (1+i)^{-n} = \frac{F_1}{1+g} \sum_{n=1}^{N} \left(\frac{1+g}{1+i}\right)^n \tag{39}$$

where F_1 is the first cash at the end of the first year. When this sum is made, the following present value equation is obtained:

$$P = \begin{cases} F_1\left[\dfrac{1-(1+g)^N(1+i)^{-N}}{i-g}\right], i \neq g \\ \dfrac{NF}{1+i}, i = g \end{cases} \tag{40}$$

and the future value

$$F = \begin{cases} F_1\left[\dfrac{(1+i)^N-(1+g)^N}{i-g}\right], i \neq g \\ NF_1(1+i)^{N-1}, i = g \end{cases} \tag{41}$$

In the case of fuzziness, the parameters used in Eq.(4) will be assumed to be fuzzy numbers, except project life. Let $\gamma(i,g,N) = [\frac{1-(1+g)^N(1+i)^{-N}}{i-g}], i \neq g$. As it is in Figure 1 and Figure 2, when k=1, the left side representation will be depicted and when k=2, the right side representation will be depicted. In this case, for $i \neq g$

$$f_{Nk}(y|\widetilde{P}_N) = f_k(y|\widetilde{F}_1)\gamma(f_{3-k}(y|\widetilde{i}), f_{3-k}(y|\widetilde{g}), N) \tag{42}$$

In Eq. (42), the least possible value is calculated for $k = 1$ and $y = 0$; the largest possible value is calculated for $k = 2$ and $y = 0$; the most promising value is calculated for $k = 1$ or $k = 2$ and $y = 1$.

To calculate the future value of a fuzzy geometric cash flow, let $\varsigma(i,g,N) = [\frac{(1+i)^N-(1+g)^N}{i-g}], i \neq g$. Then the fuzzy future value is

$$f_{Nk}(y|\widetilde{F}_N) = f_k(y|\widetilde{F}_1)\varsigma(f_k(y|\widetilde{i}), f_k(y|\widetilde{g}), N) \tag{43}$$

In Eq. (43), the least possible value is calculated for $k = 1$ and $y = 0$; the largest possible value is calculated for $k = 2$ and $y = 0$; the most promising value is calculated for $k = 1$ or $k = 2$ and $y = 1$. This is also valid for the formulas developed at the rest of the paper.

The fuzzy uniform equivalent annual value can be calculated by using Eq. (44):

$$f_{Nk}(y|\widetilde{A}) = f_k(y|\widetilde{P}_N)\vartheta(f_k(y|\widetilde{i}), N) \tag{44}$$

where $\vartheta(i,N) = [\frac{(1+i)^N i}{(1+i)^N - 1}]$ and $f(y|\widetilde{P}_N)$ is the fuzzy present value of the fuzzy geometric cash flows.

9.2 Geometric Series - Fuzzy Cash Flows in Continuous Compounding

In the case of crisp sets, the present and future values of discrete payments are given by Eq. (45) and Eq. (46) respectively:

$$P=\begin{cases} F_1\left[\dfrac{1-e^{(g-r)N}}{e^r-e^g}\right], r\neq g \\ \dfrac{NF_1}{e^r}, g=e^r-1 \end{cases} \tag{45}$$

$$F=\begin{cases} F_1\left[\dfrac{e^{rN}-e^{gN}}{e^r-e^g}\right], r\neq g \\ NF_1e^{r(N-1)}, g=e^r-1 \end{cases} \tag{46}$$

and the present and future values of continuous payments are given by Eq. (47) and Eq. (48) respectively:

$$P=\begin{cases} F_1\left[\dfrac{1-e^{(g-r)N}}{r-g}\right], r\neq g \\ \dfrac{NF_1}{1+r}, r=g \end{cases} \tag{47}$$

$$F=\begin{cases} F_1\left[\dfrac{e^{rN}-e^{gN}}{r-g}\right], r\neq g \\ \dfrac{NF_1e^{rN}}{1+r}, r=g \end{cases} \tag{48}$$

The fuzzy present and future values of the fuzzy geometric discrete cash flows in continuous compounding can be given as in Eq. (49) and Eq. (50) respectively:

$$f_{Nk}(y|\widetilde{P}_N)=f_k(y|\widetilde{F}_1)\beta(f_{3-k}(y|\widetilde{r}),f_{3-k}(y|\widetilde{g}),N) \tag{49}$$

$$f_{Nk}(y|\widetilde{F})=f_k(y|\widetilde{F}_1)\tau(f_k(y|\widetilde{r}),f_k(y|\widetilde{g}),N) \tag{50}$$

where $\beta(r,g,N)=F_1[\frac{1-e^{(g-r)N}}{e^r-e^g}], r\neq g$ for present value and $\tau(r,g,N)=F_1[\frac{e^{rN}-e^{gN}}{e^r-e^g}], r\neq g$ for future value.

The fuzzy present and future values of the fuzzy geometric continuous cash flows in continuous compounding can be given as in Eq. (51) and Eq. (52) respectively:

$$f_{Nk}(y|\widetilde{P}_N)=f_k(y|\widetilde{F}_1)\eta(f_{3-k}(y|\widetilde{r}),f_{3-k}(y|\widetilde{g}),N) \quad (51)$$

$$f_{Nk}(y|\widetilde{F}_N)=f_k(y|\widetilde{F}_1)\upsilon(f_k(y|\widetilde{r}),f_k(y|\widetilde{g}),N) \quad (52)$$

where $\eta(r,g,N)=F_1[\frac{1-e^{(g-r)N}}{r-g}]$, $\upsilon(r,g,N)=F_1[\frac{e^{rN}-e^{gN}}{r-g}], r\neq g$

9.3 Trigonometric Series - Fuzzy Continuous Cash Flows

In Figure 4, the function of the semi-sinusodial wave cash flows is depicted. This function, *h(t),* is given by Eq. (53) in the crisp case:

$$h(t)=\begin{cases} D\sin\pi t, 0\leq t\leq 1 \\ 0, \textit{otherwise} \end{cases} \quad (53)$$

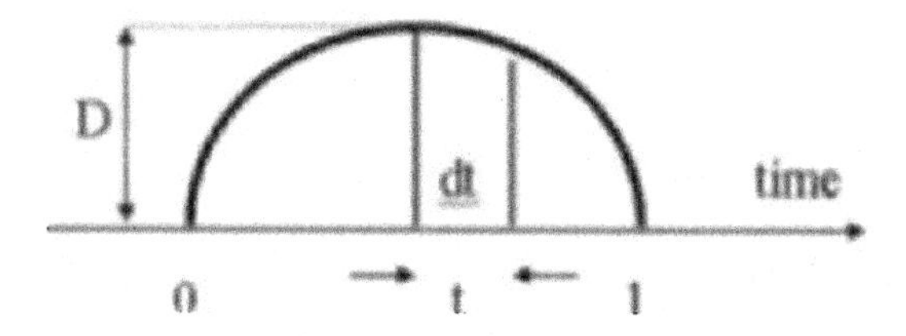

Fig. 4. Semi-Sinusoidal Wave Cash Flow Function

The future value of a semi-sinusoidal cash flow for *T*=1 and *g* is defined by Eq. (54) :

$$V(g,1)=D\int_0^1 e^{r(1-t)}\sin\pi.t.tdt = D[\frac{\pi(2+g)}{r^2+\pi^2}] \quad (54)$$

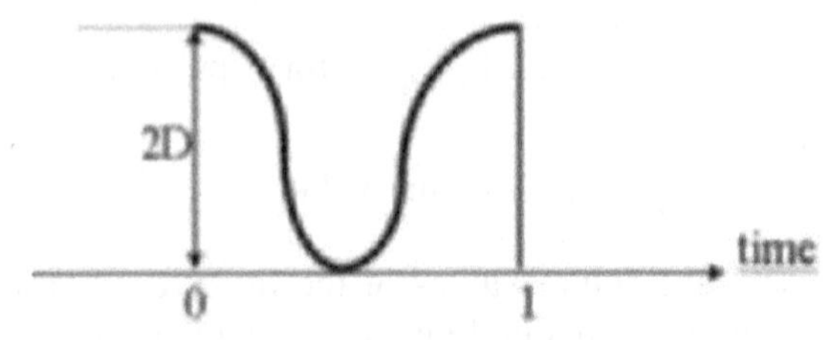

Fig. 5. Cosinusoidal Wave Cash Flow Function

Figure 5 shows the function of a cosinusoidal wave cash flow. This function, $h(t)$, is given by Eq. (55):

$$h(t) = \begin{cases} D(Cos2\pi t + 1), \ 0 \le t \le 1 \\ 0, \ otherwise \end{cases} \tag{55}$$

The future value of a cosinusoidal cash flow for T=1 and g is defined as

$$V(g,1) = D\int_0^1 e^{r(t-1)}(\cos 2\pi.t + 1)dt = D[\frac{gr}{r^2+4\pi^2} + \frac{g}{r}] \tag{56}$$

Let the parameters in Eq. (54), r and g, be fuzzy numbers. The future value of the semi-sinusoidal cash flows as in Figure 6 is given by

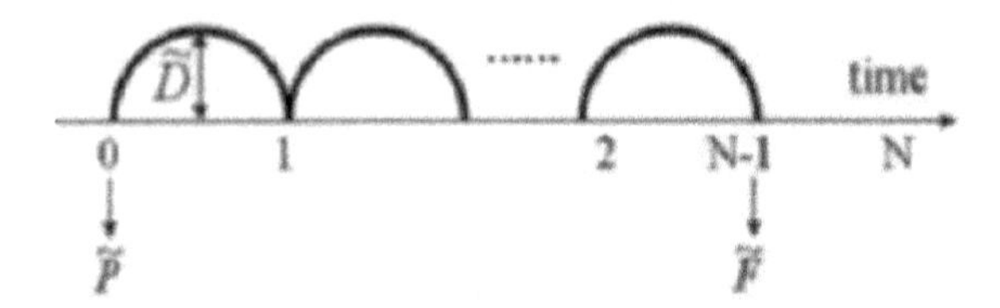

Fig. 6. Fuzzy Sinusoidal Cash Flow Diagram

$$f_{Nk}(y|\widetilde{F}_N) = f_k(y|\widetilde{D})\phi(f_{3-k}(y|\widetilde{r}), f_k(y|\widetilde{g}))\varphi(f_k(y|\widetilde{r}), N) \tag{57}$$

where $\phi(r,g) = \pi(2+g)/(r^2+\pi^2)$, $\varphi(r,N) = (e^{rN}-1)/(e^r-1)$.

The present value of the semi-sinusoidal cash flows is given by Eq. (58):

$$f_{Nk}(y|\widetilde{P}_N) = f_k(y|\widetilde{D})\phi(f_{3-k}(y|\widetilde{r}), f_k(y|\widetilde{g}))\psi(f_{3-k}(y|\widetilde{r}), N) \tag{58}$$

where $\psi(r,N) = (e^{rN}-1)/((e^r-1)e^{rN})$.

The present and future values of the fuzzy cosinusoidal cash flows can be given by Eq. (59) and Eq. (60) respectively:

$$f_{Nk}(y|\widetilde{P}_N) = f_k(y|\widetilde{D})\xi(f_{3-k}(y|\widetilde{r}), f_k(y|\widetilde{g}))\psi(f_{3-k}(y|\widetilde{r}), N) \quad (59)$$

where $\xi(r, g) = [\frac{gr}{r^2 + 4\pi^2} + \frac{g}{r}]$ and the fuzzy future value is

$$f_{Nk}(y|\widetilde{F}_N) = f_k(y|\widetilde{D})\xi(f_{3-k}(y|\widetilde{r}), f_k(y|\widetilde{g}))\varphi(f_k(y|\widetilde{r}), N) \quad (60)$$

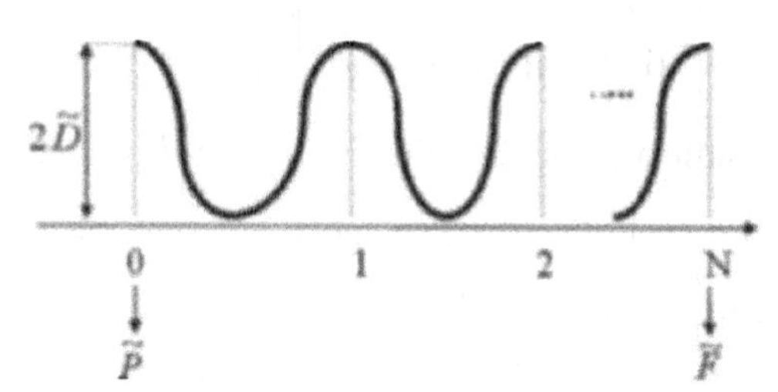

Fig. 7. Fuzzy Cosinusoidal Wave Cash Flow Diagram

Numerical Example-1

The continuous profit function of a firm producing ice cream during a year is similar to semi-sinusoidal wave cash flows whose g is around 4%. The maximum level in \$ of the ice-cream sales is between the end of June and the beginning of July. The profit amount obtained on this point is around \$120,000. The firm manager uses a minimum attractive rate of return of around 10%, compounded continuously and he wants to know the present worth of the 10-year profit and the possibility of having a present worth of \$1,500,000.

'Around \$120,000' can be represented by a TFN, (\$100,000; \$120,000;\$130,000). 'Around 10%' can be represented by a TFN, (9%;10%;12%). 'Around 4%' can be represented by a TFN, (3%;4%;6%).

$$f_1(y|\widetilde{r}) = 0.09 + 0.01y \qquad f_1(y|\widetilde{D}) = 100{,}000 + 20{,}000y$$

$$f_2(y|\widetilde{r}) = 0.12 - 0.02y \qquad f_2(y|\widetilde{D}) = 130{,}000 - 10{,}000y$$

$$f_{10,1}(y|\widetilde{P}_{10}) = f_1(y|\widetilde{D})\Phi(f_2(y|\widetilde{r}), f_1(y|\widetilde{g}))\Psi(f_2(y|\widetilde{r}), 10)$$

$$f_{10,2}(y|\widetilde{P}_{10}) = f_2(y|\widetilde{D})\Phi(f_1(y|\widetilde{r}), f_2(y|\widetilde{g}))\Psi(f_1(y|\widetilde{r}), 10)$$

$$f_1(y|\widetilde{g}) = 0.03 + 0.01y \qquad f_2(y|\widetilde{g}) = 0.06 - 0.02y$$

$$f_{10,1}\left(y\middle|\widetilde{P}_{10}\right)=\left(100{,}000+20{,}000y\right)\times$$

$$\left[\frac{\pi(2.03+0.01y)}{(0.12-0.02y)^2+\pi^2}\right]\left[\frac{e^{(0.12-0.02y)10}-1}{e^{(0.12-0.02y)}-1}\right]\frac{1}{e^{(0.12-0.02y)10}}$$

$$f_{10,2}\left(y\middle|\widetilde{P}_{10}\right)=\left(130{,}000-10{,}000y\right)\times$$

$$\left[\frac{\pi(2.06-0.02y)}{(0.09+0.01y)^2+\pi^2}\right]\left[\frac{e^{(0.09+0.01y)10}-1}{e^{(0.09+0.01y)}-1}\right]\frac{1}{e^{(0.09+0.01y)10}}$$

For y = 0, the smallest possible value is $f_{10,1}\left(y\middle|\widetilde{P}_{10}\right)=\$353{,}647.1$

For y = 1, the most possible value is $f_{10,1}\left(y\middle|\widetilde{P}_{10}\right)=f_{10,2}\left(y\middle|\widetilde{P}_{10}\right)=\$467{,}870.9$

For y = 0, the largest possible value is $f_{10,2}\left(y\middle|\widetilde{P}_{10}\right)=\$536{,}712.8$

It seems to be impossible to have a present worth of $1,500,000.

Numerical Example-2

The continuous cash flows of a firm is similar to cosinusoidal cash flows. The maximum level of the cash flows during a year is around $780,000. The fuzzy nominal cost of capital is around 8% per year. The fuzzy geometric growth rate of the cash flows is around 4% per year. Let us compute the future worth of a 10 year working period.

Let us define

$$\widetilde{D}=(\$300{,}000;\$390{,}000;\$420{,}000)$$

$$f_2\left(y\middle|\widetilde{D}\right)=420{,}000-30{,}000y$$

$$f_1\left(y\middle|\widetilde{D}\right)=350{,}000+40{,}000y$$

$$\widetilde{r}=(6\%,8\%,10\%)$$

$$f_1\left(y\middle|\widetilde{r}\right)=0.06+0.02y$$

$$f_2\left(y\middle|\widetilde{r}\right)=0.10-0.02y$$

$$\widetilde{g}=(3\%,4\%,5\%)$$

$$f_1\left(y\middle|\widetilde{g}\right)=0.03+0.01y$$

$$f_2\left(y\middle|\widetilde{g}\right)=0.05-0.01y$$

$$f_{10,1}\left(y\middle|\widetilde{F}_{10}\right)=\left(350{,}000+40{,}000y\right)\times$$

$$\left[\frac{(0.03+0.01y)(0.10-0.02y)}{(0.19-0.02y)^2+4\pi^2}+\frac{0.03+0.01y}{0.10-0.02y}\right]\left[\frac{e^{(0.06+0.02y)10}-1}{e^{0.06+0.02y}-1}\right]$$

$$f_{10,2}\left(y\middle|\widetilde{F}_{10}\right)=\left(420{,}000-30{,}000y\right)\times$$

$$\left[\frac{(0.05-0.01y)(0.06+0.02y)}{(0.06+0.02y)^2+4\pi^2}+\frac{0.05-0.01y}{0.06+0.02y}\right]\left[\frac{e^{(0.10-0.02y)10}-1}{e^{0.10-0.02y}-1}\right]$$

For $y = 0$, the smallest possible value is $f_{10,1}\left(y\middle|\widetilde{F}_{10}\right)=\$1{,}396{,}331.5$

For $y = 1$, the most possible value is $f_{10,1}\left(y\middle|\widetilde{F}_{10}\right)=f_{10,2}\left(y\middle|\widetilde{F}_{10}\right)=\$2{,}869{,}823.5$

For $y = 0$, the largest possible value is $f_{10,2}\left(y\middle|\widetilde{F}_{10}\right)=\$5{,}718{,}818.9$

10 Investment Analysis under Fuzziness Using Possibilities of Probabilities

A typical investment may involve several factors such as the initial cost, expected life of the investment, the market share, the operating cost and so on. Values for factors are projected at the time the investment project is first proposed and are subject to deviations from their expected values. Such variations in the outcomes of future events, often termed risk, have been of primary concern to most decision makers in evaluating investment alternatives.

The term risk analysis has different interpretations among various agencies and units. However, there is a growing acceptance that risk analysis involves the development of the probability distribution for the measure of effectiveness. Furthermore, the risk associated with an investment alternative is either given as the probability of an unfavorable value for the measure of effectiveness or measured by the variance of the measure of effectiveness.

Probability of a Fuzzy Event

The formula for calculating the probability of a fuzzy event A is a generalization of the probability theory:

$$P(A)=\begin{cases}\int \mu_A(x)P_X(x)dx, & \text{if X is continuous}\\ \sum_i \mu_A(x_i)P_X(x_i), & \text{if X is discrete}\end{cases} \tag{61}$$

where P_x denotes the probability distribution function of X.

Fuzzy verses Stochastic Investment Analyses

Typical parameters for which conditions of risk can reasonably be expected to exist include the initial investment, yearly operating and maintenance expenses, salvage values, the life of an investment, the planning horizon, and the minimum attractive rate of return. The parameters can be statistically independent, correlated with time and/or correlated with each other.

In order to determine analytically the probability distribution for the measure of effectiveness, a number of simplifying assumptions are normally made. The simplest situation is one involving a known number of random and statistically independent cash flows. As an example, suppose the random variable A_j denotes the net cash flow occurring at the end of period j, $j=0,1,..,N$. Hence, the present worth (PW) is given by

$$PW=\sum_{j=0}^{N} A_j\left(1+i\right)^{-j} \tag{62}$$

Since the expected value, $E[.]$, of a sum of random variables equals the sum of the expected values of the random variables, then the expected present worth is given by:

$$E[PW]=\sum_{j=0}^{N} E\left[A_j\right]\left(1+i\right)^{-j} \tag{63}$$

Furthermore, since the A_j's are statistically independent, then the variance, V(.), of present worth is given by:

$$V(PW)=\sum_{j=0}^{N} V\left(A_j\right)\left(1+i\right)^{-2j} \tag{64}$$

The central limit theorem, from probability theory, establishes that the sum of independently distributed random variables tends to be normally distributed as the number of terms in the summation increases. Hence, as N increases, PW tends to be normally distributed with a mean value of E[PW] and a variance of V(PW).

An illustrative example:

A new cost reduction proposal is expected to have annual expenses of \$20,000 with a standard deviation of \$3,000, and it will likely save \$24,000 per year with a standard deviation of \$4,000. The proposed operation will be in effect for 3 years, and a rate of return of 20 percent before taxes is required. Determine the probability that implementation of the proposal will actually result in an overall loss and the probability that the PW of the net savings will exceed \$10,000.

The expected value of the present worth of savings and cost is

$$E[PW] = (\$24{,}000 - \$20{,}000)(P/A, 20\%, 3) = \$8{,}426$$

The variance is calculated from the relation

$$\sigma^2_{savings-costs} = \sigma^2_{savings} + \sigma^2_{costs} \tag{65}$$

to obtain

$$\begin{aligned} Var[PW] &= (\$3{,}000)^2 (P/F,20,2) + \\ &(\$3{,}000)^2 (P/F,20,4) + (\$3{,}000)^2 (P/F,20,6) \\ &+ (\$4{,}000)^2 (P/F,20,2) + (\$4{,}000)^2 (P/F,20,2) \\ &+ (\$4{,}000)^2 (P/F,20,2) = 37{,}790{,}000 \end{aligned}$$

from which

$$\sigma_{PW} = \sqrt{Var[PW]} = \$6{,}147$$

Assuming that the PW is normally distributed, we find that

$$P(loss) = P\left(Z < \frac{0 - \$8{,}426}{\$6{,}147}\right)$$

$$= P(Z < 1.37) = 0.0853$$

and

$$P(PW > \$10{,}000) = P\left(Z > \frac{\$10{,}000 - \$8{,}426}{\$6{,}147}\right)$$

$$= P(Z > 0.256) = 0.40$$

Under fuzziness, the fuzzy expected net present value in a triangular fuzzy number form is calculated as in the following way. The expected fuzzy annual expenses are around \$20,000 with a fuzzy standard deviation of around \$3,000, and it will possibly save around \$24,000 per year with a fuzzy standard deviation of around \$4,000. The proposed operation will be in effect for 3 years, and a rate of return of around 20 percent before taxes is required. Determine the possibility that implementation of the proposal will actually result in an overall loss and the possibility that the PW of the net savings will exceed around $10,000. The fuzzy annual expenses are

$E_e\left[\tilde{X}\right]=(\$19,000;\$20,000;\$24,000)$

$\sigma_e\left(\tilde{X}\right)=(\$2,500;\$3,000;\$3,500)$

The fuzzy annual savings are

$E_s\left[\tilde{X}\right]=(\$23,000;\$24,000;\$25,000)$

$\sigma_s\left(\tilde{X}\right)=(\$3,500;\$4,000;\$4,500)$

The required fuzzy rate of return is

$\tilde{i}_{annual}=(18\%,20\%,22\%)$.

The fuzzy variance of the cash flows is calculated using

$$\tilde{\sigma}^2_{s-e}=\tilde{\sigma}^2_s+\tilde{\sigma}^2_e \tag{66}$$

and it is equal to $\tilde{\sigma}^2_{s-e}=(18,500,000;25,000,000;32,500,000)$ with the left side representation $f_l\left[y|\tilde{\sigma}_{s-e}\right]=\sqrt{(18,500,000+6,500,000y)}$ and the right side representation $f_r\left[y|\tilde{\sigma}_{s-e}\right]=\sqrt{(32,500,000-7,500,000y)}$ where y shows the degree of membership.

The fuzzy present worth is calculated by using the Formula

$$P\tilde{W}=\left[E_s\left[\tilde{X}\right]-E_e\left[\tilde{X}\right]\right]\left(P/A,\tilde{i}_{annual},n\right) \tag{67}$$

The left side representation of the difference between savings and expenses is $f_l\left(y|\tilde{X}_{s-e}\right)=(-1,000+5,000y)$ and the right side representation is $f_r\left(y|\tilde{X}_{s-e}\right)=(6,000-2,000y)$ where y shows the degree of membership. Similarly, the left side representation of the fuzzy interest rate is $f_l\left(y|\tilde{i}_{annual}\right)=(0.18+0.02y)$ and the right side representation is $f_r\left(y|\tilde{i}_{annual}\right)=(022-0.02y)$. Using these representations we find the left side representation of the $P\tilde{W}$ as

$$f_l\left(y|P\tilde{W}\right)=(-1,000+5,000y)\times\left[\frac{(1.22-0.02y)^3-1}{(1.22-0.02y)^3(0.22-0.02y)}\right]$$

and the right side representation as

$$f_r\left(y|P\tilde{W}\right)=(6,000-2,000y)\times\left[\frac{(1.18+0.02y)^3-1}{(1.18+0.02y)^3(0.18+0.02y)}\right]$$

When these two functions are combined on x-y axes, we obtain the graph of fuzzy PW.

Now, to calculate the possibility of loss, we can use the following equalities

$$P_1(loss) = P_1\left(\tilde{Z} \prec \frac{\tilde{0} - f_r\left(y \middle| P\tilde{W}\right)}{f_r\left(y \middle| \tilde{\sigma}_{s-e}\right)} \right) \tag{68}$$

$$P_r(loss) = P_r\left(\tilde{Z} \prec \frac{\tilde{0} - f_l\left(y \middle| P\tilde{W}\right)}{f_l\left(y \middle| \tilde{\sigma}_{s-e}\right)} \right) \tag{69}$$

The graph of these two functions is illustrated in Figure 8.

To calculate the possibility that the PW of the net savings will exceed around \$10,000, the following equations can be used.

$$P_1\left(P\tilde{W} > \$10{,}000\right) = P_1\left(\tilde{Z} \prec \frac{10{,}\tilde{0}00 - f_r\left(y \middle| P\tilde{W}\right)}{f_r\left(y \middle| \tilde{\sigma}_{s-e}\right)} \right)$$

and

$$P_r\left(P\tilde{W} > \$10{,}000\right) = P_r\left(\tilde{Z} \prec \frac{10{,}\tilde{0}00 - f_l\left(y \middle| P\tilde{W}\right)}{f_l\left(y \middle| \tilde{\sigma}_{s-e}\right)} \right)$$

where $\$10{,}00\tilde{0}$ is accepted as $(\$9{,}000;\$10{,}000;\$11{,}000)$ with the left side representation $(\$9{,}000 + \$1{,}000y)$ and the right side representation $(\$11{,}000 - \$1{,}000y)$. The graph of these two functions is illustrated in Figure 8.

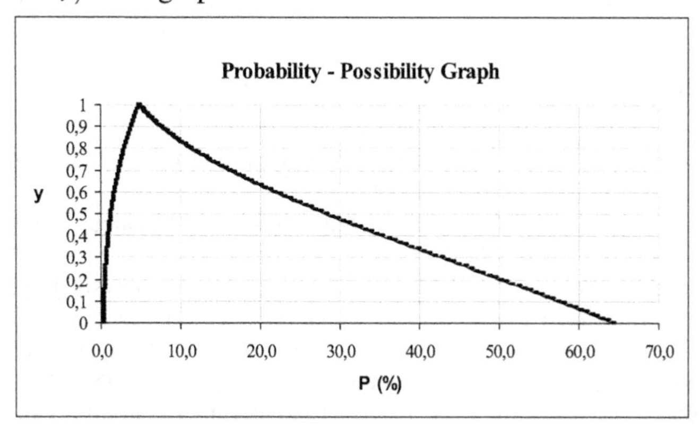

Fig. 8. The possibilities of probabilities P(Loss)

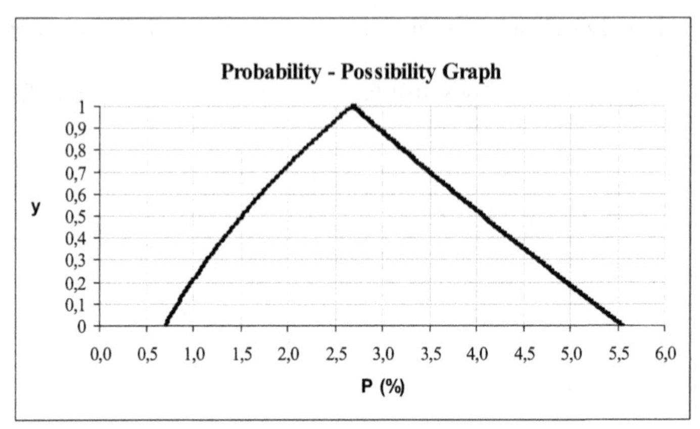

Fig. 9. The possibilities of probabilities P(Profit>10,000)

11 Conclusion

In this chapter, capital budgeting techniques in the case of fuzziness and discrete compounding have been studied. The cash flow profile of some investments projects may be geometric or trigonometric. For these kind of projects, the fuzzy present, future, and annual value formulas have been also developed under discrete and continuous compounding in this chapter. Fuzzy set theory is a powerful tool in the area of management when sufficient objective data has not been obtained. Appropriate fuzzy numbers can capture the vagueness of knowledge. The other financial subjects such as replacement analysis, income tax considerations, continuous compounding in the case of fuzziness can be also applied [11], [12]. Comparing projects with unequal lives has not been considered in this paper. This will also be a new area for further study.

References

[1] Zadeh, L. A. (1965) Fuzzy Sets, Information and Control, Vol. 8, pp. 338-353.

[2] Buckley, J. U. (1987) The Fuzzy Mathematics of Finance, Fuzzy Sets and Systems, Vol. 21, pp. 257-273.

[3] Chiu, C.Y., Park, C.S. (1994) Fuzzy Cash Flow Analysis Using Present Worth Criterion, The Engineering Economist, Vol. 39, No. 2, pp. 113-138.

[4] Ward, T.L. (1985) Discounted Fuzzy Cash Flow Analysis, in 1985 Fall Industrial Engineering Conference Proceedings, pp. 476-481.

[5] Blank, L. T., Tarquin, J. A. (1987) Engineering Economy, Third Edition, McGraw-Hill, Inc.

[6] Chang, W.(1981) Ranking of Fuzzy Utilities with Triangular Membership Functions, Proc. Int. Conf. of Policy Anal. and Inf. Systems, pp. 263-272.

[7] Dubois, D., Prade, H. (1983) Ranking Fuzzy Numbers in the Setting of Possibility Theory, Information Sciences, Vol. 30, pp. 183-224.

[8] Jain, R. (1976) Decision Making in the Presence of Fuzzy Variables, IEEE Trans. on Systems Man Cybernet, Vol. 6, pp. 698-703.

[9] Kaufmann, A., Gupta, M. M. (1988) Fuzzy Mathematical Models in Engineering and Management Science, Elsevier Science Publishers B. V.

[10] Yager, R. R. (1980) On Choosing Between Fuzzy Subsets, Kybernetes, Vol. 9, pp. 151-154.

[11] Kahraman, C., Ulukan, Z. (1997) Continuous Compounding in Capital Budgeting Using Fuzzy Concept, in the Proceedings of 6th IEEE International Conference on Fuzzy Systems (*FUZZ-IEEE'97)*, Bellaterra-Spain, pp. 1451-1455.

[12] Kahraman,C., Ulukan, Z. (1997) Fuzzy Cash Flows Under Inflation, in the Proceedings of Seventh International Fuzzy Systems Association World Congress (IFSA'97), University of Economics, Prague, Czech Republic, Vol. IV, pp. 104-108.

[13] Zimmermann, H. -J. (1994) Fuzzy Set Theory and Its Applications, Kluwer Academic Publishers.

[6] [illegible] (1981) [illegible] Proc. Int. Conf. of Policy Anal. and Inf. Systems, pp. 263-272
[7] [illegible] Systems, Vol. [illegible], pp. [illegible]
[8] [illegible]
[9] [illegible] ing and Management [illegible]
[10] [illegible]
[11] [illegible] Capital Budgeting [illegible]
[12] [illegible]
[13] [illegible]

Fuzzy Capital Budgeting: Investment Project Evaluation and Optimization

Pavel Sevastjanov[1], Ludmila Dimova[1], and Dmitry Sevastianov[2]

[1] Institute of Comp.& Information Sci., Technical University of Czestochowa,Dabrowskiego 73, 42-200 Czestochowa, Poland
sevast@icis.pcz.czest.pl

[2] Reuters Research, Inc.
3 Times Square, 17 floor, location 17-W15
New York, NY 10036, USA
Tel.: +1-646-223-8182; fax: +1-646-223-4123.
dmitry.sevastyanov@reuters.com
The statements made and the views expressed are solely those of the author and should not be attributed to his employer.

Summary. Capital budgeting is based on the analysis of some financial parameters of considered investment projects. It is clear that estimation of investment efficiency, as well as any forecasting, is rather an uncertain problem. In a case of stock investment one can to some extent predict future profits using stock history and statistical methods, but only in a short time horizon. In the capital investment one usually deals with a business-plan which takes a long time — as a rule, some years — for its realization. In such cases, a description of uncertainty within a framework of traditional probability methods usually is impossible due to the absence of objective information about probabilities of future events. This is a reason for the growing for the last two decades interest in applications of interval and fuzzy methods in budgeting. In this paper a technique for fuzzy-interval evaluation of financial parameters is presented. The results of technique application in a form of fuzzy-interval and weighted non-fuzzy values for main financial parameters NPV and IRR as well as the quantitative estimation of risk of an investment are presented.Another problem is that one usually must consider a set of different local criteria based on financial parameters of investments. As its possible solution, a numerical method for optimization of future cash-flows based on the generalized project's quality criterion in a form of compromise between local criteria of profit maximisation and financial risk minimisation is proposed.

Keywords: Capital budgeting; investment project optimisation, fuzzy-interval evaluation; risk minimisation, profit maximisation.

P. Sevastjanov et al.: *Fuzzy Capital Budgeting: Investment Project Evaluation and Optimization*, StudFuzz **201**, 205–228 (2006)
www.springerlink.com

1 Introduction

Consider common non-fuzzy approaches to a capital budgeting problem. There are a lot of financial parameters proposed in literature [1, 2, 3, 4] for budgeting.The main are: Net Present Value (NPV), Internal Rate of Return (IRR), Payback Period (PB), Profitability Index (PI). These parameters are usually used for a project quality estimation, but in practice they have different importance. It is earnestly shown in [5] that the most important parameters are NPV and IRR.

Therefore, further consideration will be based only on the analysis of the NPV and IRR. Good review of other useful financial parameters can be found in [6]. Net Present Value is usually calculated as follows:

$$NPV = \sum_{t=t_n}^{T} \frac{P_t}{(1+d)^t} - \sum_{t=0}^{t_c} \frac{KV_t}{(1+d)^t}, \tag{1}$$

where d - discount rate, t_n - first year of production, t_c - last year of investments, KV_t - capital investment in year t, P_t - income in year t, T - duration of an investment project in years. Usually, the discount rate is taken equal to an average bank interest rate in a country of investment or other value corresponding to a profit rate of alternate capital investments. An economic nature of the Internal Rate of Return (IRR) can be explained as follows. As an alternative to analyzed project, the deposit under some bank interest distributed in time the same way as analyzed investments is considered. All earned profits are also deposited with the same interest rate. If the discount rate is equal IRR, an investment in the project will give the same total income as in a case of the deposit. Thus, both alternatives are economically equivalent. If the actual bank discount rate is less then IRR, the investment into the project is more preferable. Therefore IRR is a threshold discount rate dividing effective and ineffective investment projects. The value of IRR is a solution of a non-linear equation with respect to d:

$$\sum_{t=t_n}^{T} \frac{P_t}{(1+d)^t} - \sum_{t=0}^{t_c} \frac{KV_t}{(1+d)^t} = 0. \tag{2}$$

An estimation of IRR is frequently used as a first step of the financial analysis. Only projects with IRR not below of some accepted threshold value, e.g., 15–20%, can be chosen for further consideration.

There are two conjoint discussable points in the budgeting realm. The first is the multiple roots of Eq. (2), i.e., so called multiple IRR problem. The second is the negative NPV problem. The problem of multiple roots of Eq. (2) rises when the negative cash flows take place after starting investment. In practice, an appearance of some negative cash flow after initial investment is usually treated as a local "force majeur" or even a total project's failure. That is why, on the stage of planning, investors try to avoid situations when

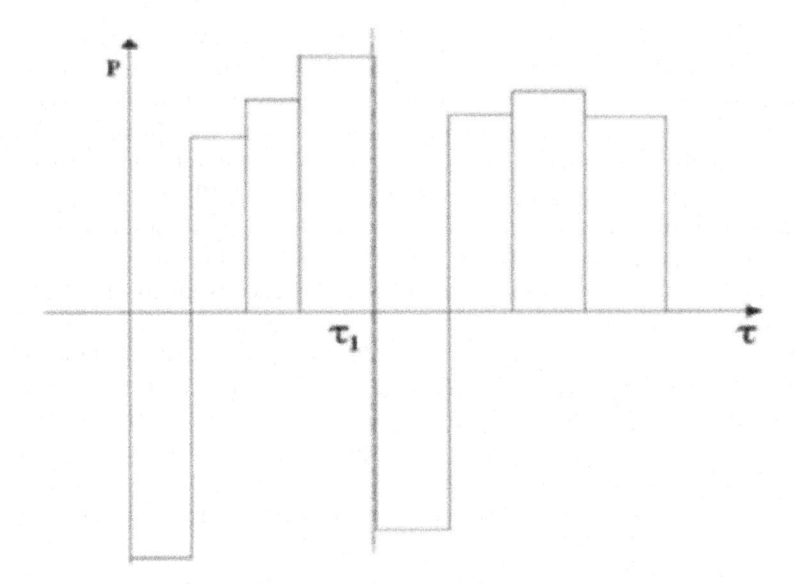

Fig. 1. Two stage investment project.

such negative cash flows are possible, except the cases when they are dealing with long-term projects consist of some phases. Let us see to the Fig. 1. This is a typical two-phase project: after initial investment the project brings considerable profits and at the time τ_1 a part of accumulated earnings and, perhaps, an additional banking credit are invested once again. Factually, an investor buys new production equipment and buildings (in fact creating the new enterprise) and from his/her point of view a quite new project is started. It is easy to see that investor's creditors which are interested in repayment of a credit always analyze phases $\tau < \tau_1$ and $\tau > \tau_1$ separately. It worth noting that what we describe is only an investment planning routine, not some theoretical considerations we can find in finansial books. On the other hand, a separate assessment of different projects' phases reflects economic sense of capital investment better. Indeed, if we consider a two phase project as a whole, we often get the IRRs performed by two roots so different that it is impossible to make any decision. For example, we can obtain IRR_1 =4% and IRR_2 =120%. It is clear that average IRR= (4+120)/2=62% seems as rather fantastic estimation, whereas when considering the two phases of project separately we usually get quite acceptable values, e.g., for the first phase IRR_1 =20% and for the second phase IRR_2 =25%. So we can say that the problem of "multiple IRR values" exists only in some finansial textbooks, not in the practice of capital investment. Therefore, only the case when Eq. (2) has a single root will be analyzed in the current paper. Similarly, the negative NPV

problem seems as a rather artificial one. Obviously, any investment project with negative NPV should be rejected at the planning stage. On the other hand, all possible undesirable events leading to the financial losses or even to the failure of the projects should be taken into account too. In the framework of probabilistic approach, e.g., when using the Monte-Carlo method, there may be local results of calculations with negative NPV and the problem of their interpretation in terms of risk management or in other context arises. The different situation we meet when future cash flows are presented by fuzzy numbers. It is clear the full body of uncertainty is involved in such a description. So if the decision maker find some negative part in predicted cash flow he/she consider such a case as a source of risk and try to improve the project to avoid this risk. As the result in a fuzzy budgeting the negative cash flows and especially NPV, seem rather as the exotics. Nevertheless, the probabilistic approach to interval and fuzzy value comparison we describe in Section 2, makes it possible to deal with such situation as well, i.e., to compare NPV comprising negative part with some real or fuzzy number representing acceptable risk associated with future NPV .

The focus of current paper is that nowadays traditional approach to the evaluation of NPV, IRR and other financial parameters is subjected to quite deserved criticism, since the future incomes P_t, capital investments KV_t and rates d are rather uncertain parameters. Uncertainties which one meets in capital budgeting differ from those in a case of share prices forecasting and cannot be adequately described in terms of the probability theory. In a capital investment one usually deals with a business-plan that takes a long time — as a rule, some years — for its realization. In such cases, the description of uncertainty within a framework of traditional probability methods usually is impossible due to the absence of objective information about probabilities of future events. Thus, what really is available in such cases are some expert estimates. In real-world situations, investors or experts involved are able to predict confidently only intervals of possible values P_t, KV_t and d and sometimes the most expected values inside these intervals. Therefore, during last two decades the growing interest in applications of interval arithmetic [7] and fuzzy sets theory methods [8] in budgeting was observing.

After pioneer works of T.L.Ward [9] and J.U. Buckley [10], some other authors contributed to the development of the fuzzy capital budgeting theory [11, 12, 13, 15, 16, 17, 18, 19, 20, 21, 22, 23, 24, 25]. It is safe to say that almost all problems of the fuzzy NPV estimation are solved now, but an interesting and important problem of project risk assessment using fuzzy NPV gets higher priority.

An unsolved problem is a fuzzy estimation of the IRR. Ward [9] considers Eq. (2) and states that such an expression cannot be applied to fuzzy case because the left side of Eq. (2) is fuzzy, 0 is crisp and an equality is impossible. Hence, the Eq. (2) is senseless from fuzzy viewpoint.

In [23], a method for the fuzzy IRR estimation is proposed where α-cut representation of fuzzy numbers [26] is used. The method is based on an assumption (see [23, p. 380]) that a set of equations for IRR determination on each α-level may be presented as (in our notation)

$$(CF_0^\alpha)_1 + \sum_{i=1}^{n} \frac{(CF_i^\alpha)_1}{(1+IRR_1^\alpha)^i} = 0, \quad (CF_0^\alpha)_2 + \sum_{i=1}^{n} \frac{(CF_i^\alpha)_2}{(1+IRR_2^\alpha)^i} = 0, \tag{3}$$

where $CF_i^\alpha = [(CF_i^\alpha)_1, (CF_i^\alpha)_2]$, $i = 0$ to n, are crisp interval representations of fuzzy cash flows on α-levels. Of course, from Eqs. (3) all crisp intervals $IRR^\alpha = [IRR_1^\alpha, IRR_2^\alpha]$ expressing the fuzzy valued IRR may be obtained. Regrettable, there is a little mistake in (3). Taking into account the conventional interval arithmetic rules, the right crisp interval representation of Eq. (2) on α-levels must be written as

$$(CF_0^\alpha)_1 + \sum_{i=1}^{n} \frac{(CF_i^\alpha)_1}{(1+IRR_2^\alpha)^i} = 0, \quad (CF_0^\alpha)_2 + \sum_{i=1}^{n} \frac{(CF_i^\alpha)_2}{(1+IRR_1^\alpha)^i} = 0. \tag{4}$$

There is no way to get intervals IRR^α from (4), but the crisp ones may be obtained (see Section 3, below). Another problem not presented in literature is an optimization of cash flows. The rest of the paper is set out as follows. In Section 2, a method for a fuzzy estimation of NPV is presented and possible approaches to the risk estimation are considered. In Section 3, a method for crisp solving of Eq. (2) for a case of fuzzy cash flows is described. As an outcome of the method a set of useful crisp parameters is proposed and analyzed. In Section 4, a numerical method of an optimization of cash flows as a compromise between local criteria of a profit maximisation and financial risk minimisation is proposed.

2 Fuzzy *NPV* and Risk Assessment

The technique is based on the fuzzy extension principle [8]. According to it, the values of uncertain parameters P_t, KV_t and d are substituted for corresponding fuzzy intervals. In practice it means that an expert sets lower — P_{t1} (pessimistic value) and upper — P_{t4} (optimistic value) boundaries of the intervals and internal intervals of the most expected values $[P_{t2}, P_{t3}]$ for analyzed parameters (see Fig. 2). The function $\mu(P_t)$ is usually interpreted as a membership function, i.e., a degree to which values of a parameter belong to an interval (in this case $[P_{t1}, P_{t4}]$). A membership function changes continuously from 0 (an area out of the interval) up to maximum value 1 in an area of the most possible values. It is obvious that a membership function is a generalization of a characteristic function of usual set, which equals 1 for all values inside a set and 0 in all other cases.

The linear character of the function is not obligatory, but such a mode is most used and it allows to represent the fuzzy intervals in a convenient form of a quadruple $P_t = \{P_{t1}, P_{t2}, P_{t3}, P_{t4}\}$. Then all necessary calculations are

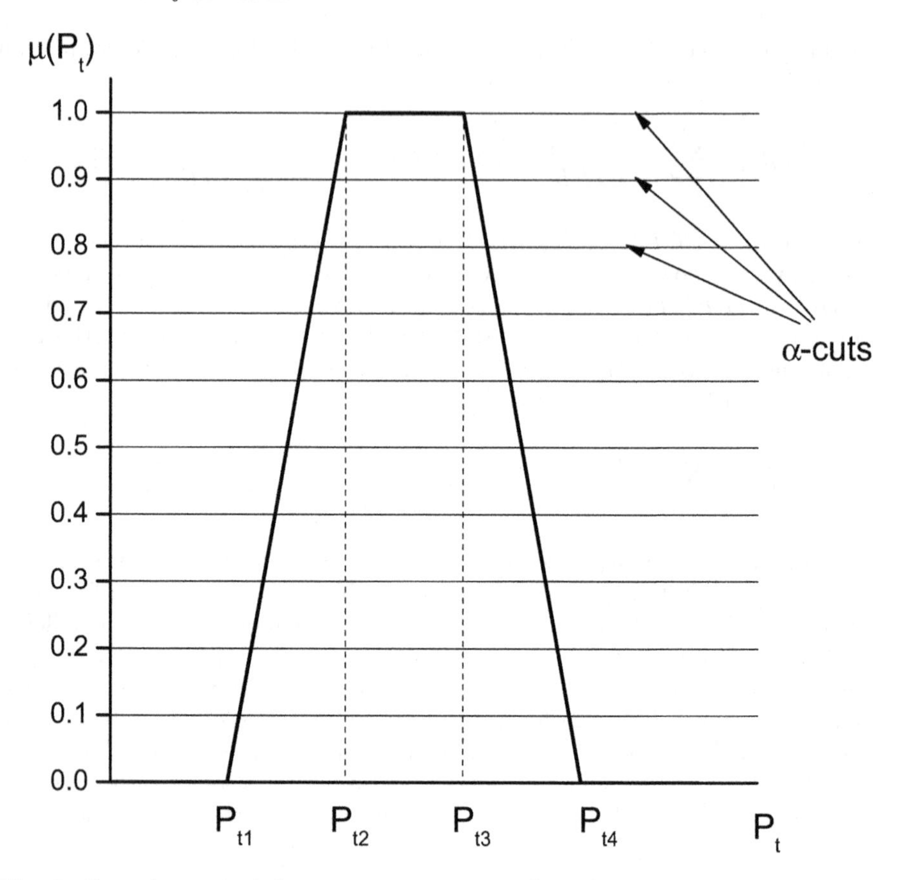

Fig. 2. Fuzzy interval of the uncertain parameter P_t and its membership function $\mu(P_t)$

carried out using special fuzzy-interval arithmetic rules. Consider some basic principles of the fuzzy arithmetic [26]. In general, for an arbitrary form of a membership function the technique of fuzzy-interval calculations is based on representation of initial fuzzy intervals in a form of so-called α-cuts (Fig. 2) which are, in fact, crisp intervals associated with corresponding degrees of the membership. All further calculations are made with those α-cuts according to well known crisp interval-arithmetic rules and resulting fuzzy intervals are obtained as disjunction of corresponding final α-cuts.

Thus, if A is a fuzzy number then $A = \bigcup_{\alpha} \alpha A_\alpha$, where A_α is a crisp interval $\{x : \mu_A(x) \geq \alpha\}$, αA_α is a fuzzy interval $\{(x, \alpha) : x \in A_\alpha\}$. So if A, B, Z are fuzzy numbers (intervals) and @ is an operation from $\{+, -, *, /\}$ then

$$Z = A@B = \bigcup_{\alpha} (A@B)_\alpha = \bigcup_{\alpha} A_\alpha @ B_\alpha. \tag{5}$$

Since in a case of α-cut representation the fuzzy arithmetic is based on crisp interval arithmetic rules, basic definitions of applied interval analysis also must be presented. There are several definitions of interval arithmetic (see [7, 27]), but in practical applications so-called "naive" form proved to be the best one. According to it, if $A = [a_1, a_2]$ and $B = [b_1, b_2]$ are crisp intervals, then

$$Z = A@B = \{z = x@y, \forall x \in A, \forall y \in B\}. \tag{6}$$

As a direct outcome of the basic definition (6) following expressions were obtained:

$$\begin{aligned} A + B &= [a_1 + b_1, b_2 + b_2], \\ A - B &= [a_1 - b_2, a_2 - b_1], \\ A * B &= [\min(a_1 b_1, a_2 b_2, a_1 b_2, a_2 b_1), \max(a_1 b_1, a_2 b_2, a_1 b_2, a_2 b_1)], \\ A/B &= [a_1, a_2] * [1/b_2, 1/b_1] \end{aligned}$$

Of course, there are many internal problems within applied interval analysis, for example, a division by zero-containing interval, but in general, it can be considered as a good mathematical tool for modelling under conditions of uncertainty.

To illustrate, consider an investment project, in which building phase proceeds two years with investments KV_0 and KV_1 accordingly. Profits are expected only after the end of the building phase and will be obtained during two years (P_2 and P_3). It is suggested that the fuzzy interval for the discount d remains stable during the time of project realisation. The sample trapezoidal initial fuzzy intervals are presented in Table 1.

Table 1. Parameters of sample project

KV_0	$\{2, 2.8, 3.5, 4\}$	P_0	$\{0, 0, 0, 0\}$
KV_1	$\{0, 0.88, 1.50, 2\}$	P_1	$\{0, 0, 0, 0\}$
KV_2	$\{0, 0, 0, 0\}$	P_2	$\{6.5, 7.5, 8.0, 8.5\}$
KV_3	$\{0, 0, 0, 0\}$	P_3	$\{5.5, 6.5, 7.0, 7.5\}$

It was assumed that $d = \{0.08, 0.13, 0.22, 0.35\}$. Resulting fuzzy interval NPV calculated using fuzzy extension of Eq. (1) is presented in Fig. 3.

Obtained fuzzy interval allows to estimate the boundaries of possible values of predicted NPV, the interval of the most expected values, and also — that is very important — to evaluate a degree of financial risk of investment. There may be different ways to define the measure of financial risk in the framework of fuzzy sets based methodology. Therefore we consider here only the three, in our opinion, most interesting and scientifically grounded approaches.

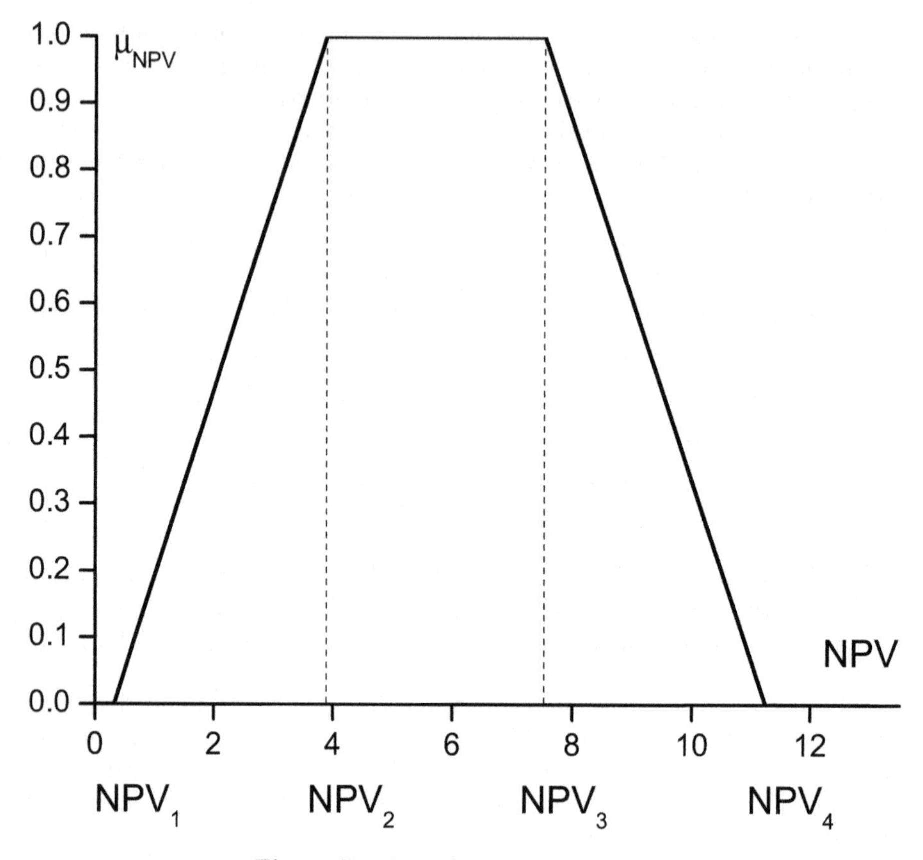

Fig. 3. Resulting fuzzy interval NPV

1. To estimate the financial risk, the following inherent property of fuzzy sets was taken into account. Let A be some fuzzy subset of X, being described by a membership function $\mu(A)$. Then complementary fuzzy subset $\bar{A}$ has a membership function $\mu(\bar{A}) = 1 - \mu(A)$. The principal difference between fuzzy subset and usual precise one is that an intersection of fuzzy A and $\bar{A}$ is not empty, that is $A \cap \bar{A} = B$, where B is also not an empty fuzzy subset. It is clear that the closer A to $\bar{A}$, the more power of set B and more A differs from ordinary sets.

   Using this circumstance R. Yager [30] proposed a set of grades of non-fuzziness of fuzzy subsets

$$D_p(A, \bar{A}) = \frac{1}{n} \left| \sum_{i=1}^{n} |\mu_A(x_i) - \mu_{\bar{A}}(x_i)|^p \right|^{\frac{1}{p}}, p = 1, 2, \ldots, \infty. \tag{7}$$

   Hence, the grade of fuzziness may be defined as

$$dd_p(A, \bar{A}) = 1 - D_p(A, \bar{A}). \tag{8}$$

The definition (8) is in compliance with obvious requests to a grade of fuzziness. If A is a fuzzy subset on X, $\mu(A)$ is its membership function and dd is a corresponding grade of fuzziness, then following properties should be observed:

a) $dd(A) = 0$, if A is a crisp subset.
b) $dd(A)$ has a maximum value if $\mu(A) = 1/2$ for $x \in X$.
c) $dd(A_1) > dd(A)$ if $\mu(x) < \mu(y)(x \in A_1, y \in A)$.

It is proved that introduced measure is similar to the Shannon entropy measure [30].

In the most useful case ($p = 1$), expression (8) is transformed to

$$dd = 1 - \frac{1}{n} \sum_{i=1}^{n} |2\mu_A(x_i) - 1| . \tag{9}$$

It is clear (see Eq. (9)) that the grade of fuzziness is rising from 0 when $\mu(A) = 1$ (crisp subset) up to 1 when $\mu(A) = 1/2$ (maximum degree of fuzziness).

With respect to considering problem the grade of nonfuzziness of a fuzzy interval NPV can linguistically be interpreted as a risk or uncertainty of obtaining the Net Present Value in interval $[NPV_1, NPV_4]$. Really, the more precise, (more "rectangular") interval obtained, the more a degree of uncertainty and risk. At first glance, this assertion seems to be paradoxical. However, any precise (crisp) interval contains no additional information about relative preference of values placed inside it. Therefore, it contains less useful information than any fuzzy interval being constructed on its basis. In the later case an additional information reducing uncertainty is derived from a membership function of considered fuzzy interval.

2. The second approach is based on the α-cut representation of fuzzy value and the measure of its fuzziness. Let A be fuzzy value and A_r be rectangular fuzzy value defined on the support of A and represented by characteristic function $\eta_A(x) = 1, x \in A; \eta_A(x) = 0, x \notin A$. Obviously, such rectangular value is not a fuzzy value at all, but it is asymptotic limit (object) we obtain when fuzziness of A tends to zero. Hence, it seems quite natural to define a measure of fuzziness of A as its distinction from A_r. To do this we define primarily the measure of non fuzziness as

$$MNF(A) = \int_0^1 f(\alpha)((A_{\alpha 2} - A_{\alpha 1})/(A_{02} - A_{01}))d\alpha,$$

where $f(\alpha)$ is some function of α , e.g, $f(\alpha) = 1$ or $f(\alpha) = \alpha$. Of course, last expression makes sense only for the fuzzy or interval values, i.e., only for non zero width of support $A_{02} - A_{01}$. It is easy to see that if $A \to A_r$ then $MNF(A) \to 1$. Obviously, the measure of fuzziness can be defined as $MF(A) = 1 - MNF(A)$.

We can say that rectangular value A_r defined on the support of A is a more uncertain object than A . Really, only what we know about A_r is that all $x \in A$ belong to A_r with equal degrees, whereas the membership function, $0 \leq \mu(x) \leq 1$, characterizing the fuzzy value A, brings more information to the description and as a consequence, represents a more certain object. Therefore, we can treat the measure of non fuzziness, MNF, as the uncertainty measure. Hence, if some decision is made concerning fuzzy NPV, the uncertainty and, consequently, the risk of such decision can be calculated as $MNF(MPV)$.

3. The authors of [29] proposed approach that can be treated as fuzzy analogue of the sound VAR method [28]. According to this approach the risk associated with fuzzy NPV can presented as

$$Risk = Prob(NPV < G),$$

where G is the fuzzy, interval or real valued effectiveness constrain [29], in other words, G in the low bound on acceptable values of NPV. It is clear the focus of this approach is the method for interval and fuzzy value comparison. In [29], such method based on the geometrical reasoning has been proposed which leads to the resulting formulas nearly the same as earlier were obtained in [31]with a help of probabilistic approach to fuzzy value comparison. In [33] [34], we have presented an overview of existing methods for fuzzy value comparison based on probabilistic approach. It is shown in [33] [34] that analyzed methods have a common drawback- the lack of separate equality relations- leading to the absurd results in the asymptotical cases and some others inconsistencies. The same can be said about of non- probabilistic method proposed in [29]. To solve the problem, in [33] [34] a new method based on the probabilistic approach has been elaborated which generates the complete set of probabilistic interval and fuzzy value relations involving separated equality and inequality relations, comparisons of real numbers with interval or fuzzy values. Let us recall briefly the basics of this approach. There are only two nontrivial situation of intervals setting: the overlapping and inclusion cases (see Fig. 4) are deserved to be considered.

Let $\mathrm{A} = [\mathrm{a}_1, \mathrm{a}_2]$ and $\mathrm{B} = [\mathrm{b}_1, \mathrm{b}_2]$ be independent intervals and $a \in [a_1, a_2]$, $b \in [b_1, b_2]$ be random values distributed on these intervals. As we are dealing with crisp (nonfuzzy) intervals, the natural assumption is that the random values a and b are distributed uniformly. There are some subintervals, which play an important role in our analysis. For example see Fig. 4), falling of random variables $a \in [a_1, a_2]$, $b \in [b_1, b_2]$ in the subintervals $[a_1, b_1]$, $[b_1, a_2]$, $[a_2, b_2]$ may be treated as a set of independent random events. Let us define the events $H_k : a \in A_i, b \in B_j$, for $k = 1$ to n, where A_i and B_j are subintervals formed by the boundaries of compared intervals A and B such that $A = \bigcup_i A_i$, $B = \bigcup_j B_j$. It is easy to see that events H_k form the complete group of events, which describes

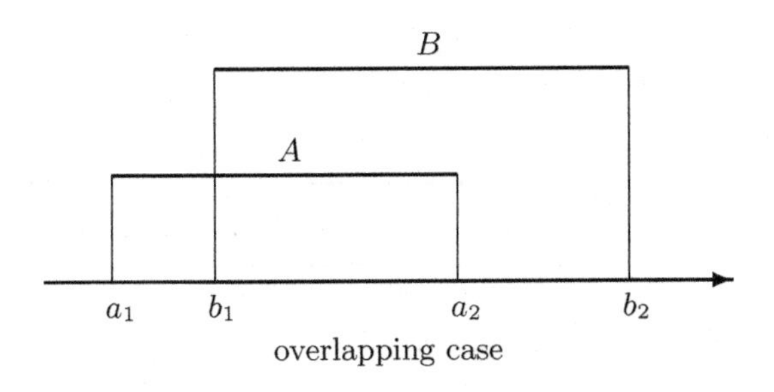

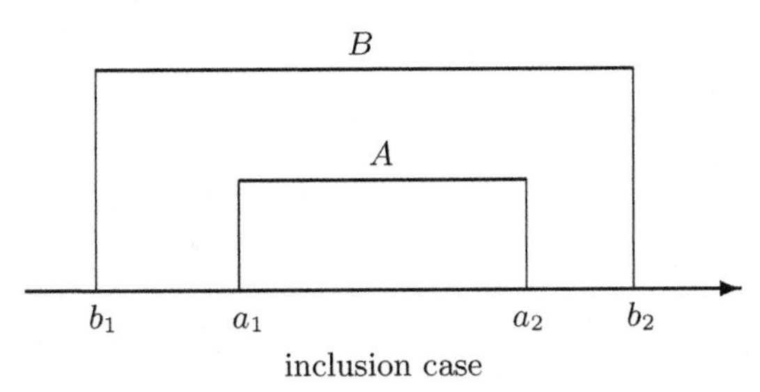

Fig. 4. Examples of interval relations

all the cases of falling random values a and b in the various subintervals A_i and B_j, respectively.

Let $P(H_k)$ be the probability of event H_k, and $P(B > A/H_k)$ be the conditional probability of $B > A$ given H_k. Hence, the composite probability may be expressed as follows:

$$P(B > A) = \sum_{k=1}^{n} P(H_k)P(B > A/H_k)$$

As we are dealing with uniform distributions of the random values a and b in the given subintervals, the probabilities $P(H_k)$ can be easily obtained by simple geometric reasoning. These basic assumptions make it possible to infer the complete set of probabilistic interval relations involving separated equality and inequality relations and comparisons of real numbers and intervals. The complete set of expressions for interval relations is shown in Table 2, obvious cases (without overlapping and inclusion) are omitted. In Table 2, only half of cases that may be realized when considering interval overlapping and including are presented since other three cases, e.q., $b_2 > a_2$ for overlapping and so on, can be easily obtained by changing letter a through b and otherwise in the expressions for the probabilities.

Table 2. The probabilistic interval relations

$P(B > A)$	$P(B < A)$	$P(B = A)$
1. $b_1 > a_1 \wedge b_1 < a_2 \wedge b_1 = b_2$		
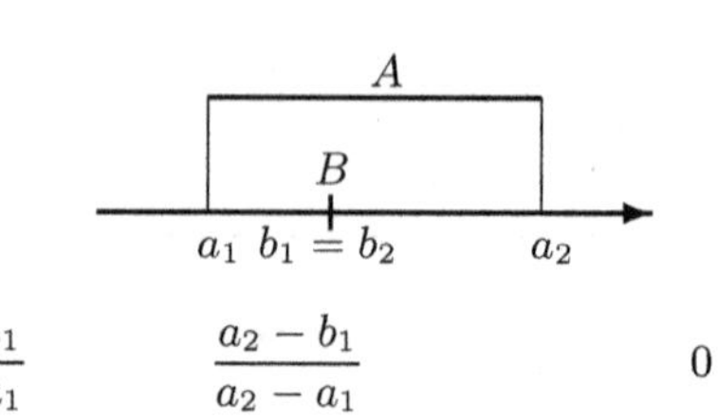		
$\frac{b_1 - a_1}{a_2 - a_1}$	$\frac{a_2 - b_1}{a_2 - a_1}$	0
2. $b_1 \geq a_1 \wedge b_2 \leq a_2$		
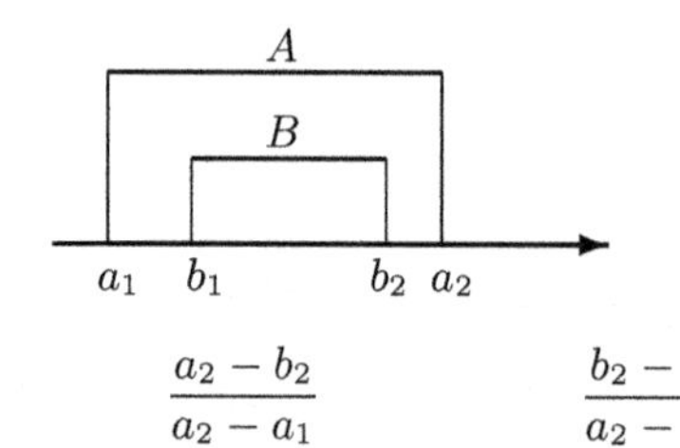		
$\frac{b_1 - a_1}{a_2 - a_1}$	$\frac{a_2 - b_2}{a_2 - a_1}$	$\frac{b_2 - b_1}{a_2 - a_1}$
3. $a_1 \geq b_1 \wedge a_2 \geq b_2 \wedge a_1 \leq b_2$		
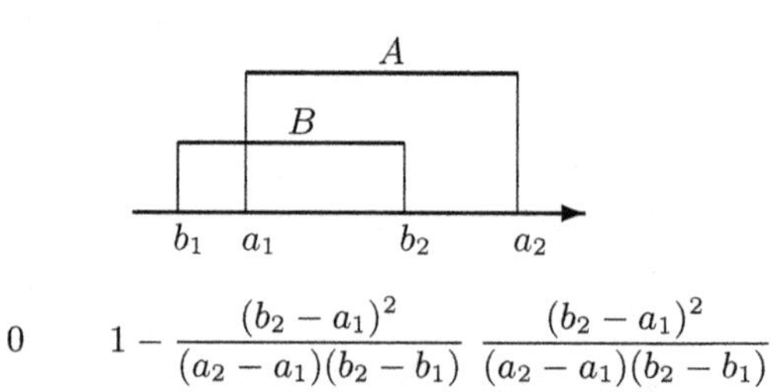		
0	$1 - \frac{(b_2 - a_1)^2}{(a_2 - a_1)(b_2 - b_1)}$	$\frac{(b_2 - a_1)^2}{(a_2 - a_1)(b_2 - b_1)}$

It easy to see that in all cases $P(A < B) + P(A = B) + P(A > B) = 1$. Of course, we can state that $B > A$ if $P(B > A) > max(P(A > B), P(A = B))$, $B = A$ if $P(A = B) > max(P(A > B), P(A < B))$ and $B < A$ if $P(A > B) > max(P(A < B), P(A = B))$. We think that treating an interval equality as an identity ($A = B$ only if $a_1 = b_1, a_2 = b_2$) will not bring good solutions to some practical problems, e.g., when dealing with interval extension of optimization task under equality type restrictions. Obviously, there may be a lot of real-world situations when, from common sense, the intervals $A = [0, 1000.1]$ and $B = [0, 1000.2]$ will be considered as somewhat equal ones. In our approach, equality is not equivalent to identity, since $P(A = B) \leq 1$.

Let $\tilde{A}$ and $\tilde{B}$ be fuzzy numbers on X with corresponding membership functions $\mu_A(x), \mu_B(x) : X \to [0,1]$. We can represent $\tilde{A}$ and $\tilde{B}$ by the sets of α-levels: $\tilde{A} = \bigcup_{\alpha} A_\alpha$, $\tilde{B} = \bigcup_{\alpha} B_\alpha$, where $A_\alpha = \{x \in X : \mu_A(x) \geq \alpha\}$, $B_\alpha = \{x \in X : \mu_B(x) \geq \alpha\}$ are crisp intervals. Then all fuzzy number relations $\tilde{A} rel \tilde{B}$, $rel = \{<,=,>\}$, may be presented by sets of α-cut relations

$$\tilde{A} rel \tilde{B} = \bigcup_{\alpha} A_\alpha \, rel \, B_\alpha .$$

Since A_α and B_α are crisp intervals, the probability $P_\alpha(B_\alpha > A_\alpha)$ for each pair A_α and B_α can be calculated in the way described above. The set of the probabilities $P_\alpha(\alpha \in (0,1])$ may be treated as the support of the fuzzy subset

$$P(\tilde{B} > \tilde{A}) = \{\alpha / P_\alpha(B_\alpha > A_\alpha)\},$$

where the values of α may be considered as grades of membership to fuzzy interval $P(\tilde{B} > \tilde{A})$. In this way, the fuzzy subset $P(\tilde{B} = \tilde{A})$ may also be easily created.

Obtained results are simple enough and reflect in some sense the nature of fuzzy arithmetic. The resulting "fuzzy probabilities" can be used directly. For instance, let $\tilde{A}, \tilde{B}, \tilde{C}$ be fuzzy intervals and $P(\tilde{A} > \tilde{B})$, $P(\tilde{A} > \tilde{C})$ be fuzzy intervals expressing the probabilities $A > \tilde{B}$ and $\tilde{A} > \tilde{C}$, respectively. Hence the probability $P(P(\tilde{A} > \tilde{B}) > P(\tilde{A} > \tilde{C}))$ has a sense of probability's comparison and is expressed in the form of fuzzy interval as well. Such fuzzy calculations may be useful at the intermediate stages of analysis, since they preserve the fuzzy information available. Indeed, it can be shown that in any case $P(\tilde{B} > \tilde{A}) + P(\tilde{B} = \tilde{A}) + P(\tilde{B} < \tilde{A}) =$"near 1", where "near 1" is a symmetrical relative to 1 fuzzy number. It is worth noting here that the main properties of probability are remained in the introduced operations, but in a fuzzy sense. However, a detailed discussion of these questions is out of the scope of this paper.

Nevertheless, in practice, the real-valued number indices are needed for fuzzy interval ordering. For this purpose, some characteristic numbers of fuzzy sets could be used. But it seems more natural to use the defuzzification, which for a discrete set of α-cuts takes the form:

$$\overline{P}(\tilde{B} > \tilde{A}) = \sum_{\alpha} \alpha P_\alpha(B_\alpha > A_\alpha) / \sum_{\alpha} \alpha .$$

Last expression indicates that the contribution of α- level to the overall probability estimation is rising along with the rise in its number. Some typical cases of fuzzy interval comparison are represented in the Fig. 5.

It is easy to see that the resulting quantitative estimations are in a good accordance with our intuition. Obviously, the other approaches to the risk assessment in budgeting can be proposed and can be relevant in the specific

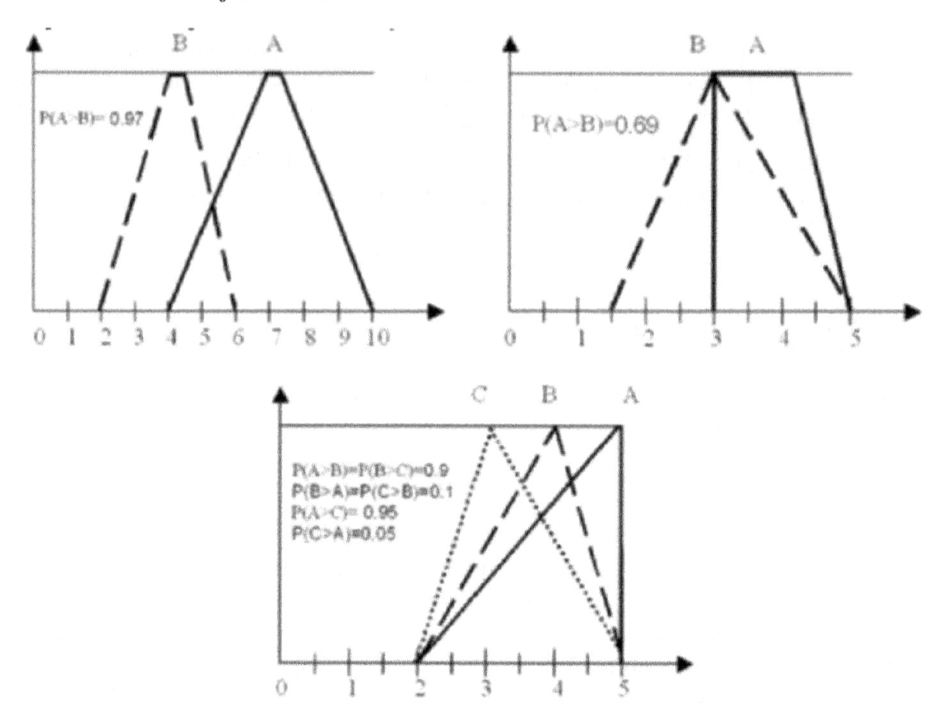

Fig. 5. The typical cases of fuzzy interval ordering.

situations. It is clear they should lead to the different results of investment projects estimation or optimization tasks as they reflect the different decision maker's attitudes to the risk and its importance to the concrete problem considered. Therefore, we think for methodical purposes it is quite enough to consider only one of above approaches. So in further analysis the first model of risk, based on the Exp. (9) will be used. It is important that all considered approaches based on an evaluation of fuzzy NPV inevitably generate two criteria for the estimation of future profits: the fuzzy interval NPV and the degree of its uncertainty (degree of risk).

Therefore, a problem of evaluation of investment efficiency on a base of NPV becomes two-criteria and requires special approach and an appropriate technique. Recently, authors proposed such a technique [32], [14] based on the fuzzy set theory; however, its detailed consideration is out of scope of this paper.

3 The Set of Crisp *IRR* Estimations Based on Fuzzy Cash Flows

In general, the problem of the Internal Rate of Return (IRR) evaluation looks as a fuzzy interval solution of the Eq. (2) with respect to d.

It is proved that a solution of equations with fuzzy parameters (in this case, P_t , KV_t and d) is possible using representation of fuzzy parameters in a form of sets of corresponding α-cuts. For the evaluating IRR, a system of non-linear crisp-interval equations can be obtained:

$$\sum_{t=t_n}^{T} \frac{[P_t]_\alpha}{(1+[d]_\alpha)^t} - \sum_{t=0}^{t_c} \frac{[KV_t]_\alpha}{(1+[d]_\alpha)^t} = [0,0], \tag{10}$$

where $[Pt]_\alpha, [KV_t]_\alpha$ and $[d]_\alpha$ are crisp intervals on corresponding α-cuts.

Of course, it can be claimed that naive assumption, that the degenerated zero interval $[0,0]$ should be placed in the right side of Eq. (10), does not ensure obtaining of adequate outcomes since a non-degenerated interval expression is in the left side of Eq. (10), but this situation needs more thorough consideration.

As the simplest example consider a two-year project when all investments are finished in the first year and all revenues are obtained in the second year. Then each of the equations for α-cuts (10) should be divided on two:

$$\frac{P_{11}}{1+d_2} - KV_{02} = 0, \frac{P_{12}}{1+d_1} - KV_{01} = 0. \tag{11}$$

The formal solution Eq. (11) with respect to d_1 and d_2 is trivial:

$$d_1 = \frac{P_{12}}{KV_{01}} - 1; \quad d_2 = \frac{P_{11}}{KV_{02}} - 1,$$

however it is senseless, as the right boundary of the interval $[d_1, d_2]$ always appears to be less than the left one. This absurd, on a first glance, result is easy to explain from common methodological positions. Really, the rules of the interval mathematics are constructed in such a manner that any arithmetical operation with intervals results in an interval as well. These rules fully coinside with well known common viewpoint stating that any arithmetical operation with uncertainties must increase total uncertainty and the entropy of a system. Therefore, placing the degenerated zero interval in right sides of (10) and (11) is equivalent to the request of reducing uncertainty of the left sides down to zero, which is possible only in case of inverse character of the interval $[d_1, d_2]$, which is in turn can be interpreted as a request to introduce negative entropy into the system.

Thus, the presence of the degenerated zero interval in right sides of interval equations is incorrect. More acceptable approach to solving this problem has been constructed with a help of following reasons. When analysing expressions (11) it is easy to see that for any value d_1 the minimal width of the interval NPV is reached when $d_2 = d_1$. This is in accordance with a common viewpoint: the minimum uncertainty of an outcome (NPV) is reached when uncertainty of all system parameters is minimal. It is clear (see Fig. 6) that the most reasonable decision of "zero" problem is a request for a middle of

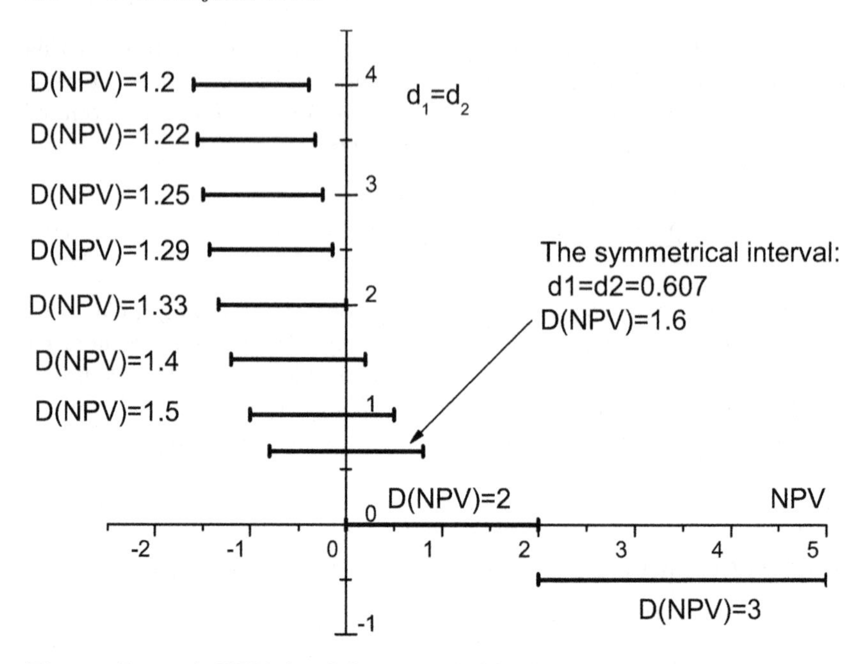

Fig. 6. Interval NPV for different real valued discounts, d, for the case when the investments in the first year is $KV_0 = [1, 2]$, the income in the second year is $P_1 = [2, 3]$, $D(NPV)$ is a width of the interval NPV.

the interval NPV to be placed on a zero point (request of symmetry of the interval against zero). An obvious, on a first glance, intention to minimise the length of interval NPV results in deriving positive or negative intervals of minimum width, but not containing zero point, that does not correspond to a natural definition of zero containing interval. Besides, it can be easily proved that only the request of symmetry of zero containing interval ensures an asymptotically valid outcome when contracting boundaries of all considered intervals to their centres. Thus, the problem is reduced to a search of exact (non-interval) values d that will provide a symmetry of zero resulting intervals NPV on each α-cut in the equations (10), i.e. would guarantee fulfilment of the request $(NPV_1 + NPV_2) = 0$, for each $a = 0, 0.1, 0.2 \ldots, 1$.

Obviously, the problem is solved using numerical methods. To illustrate previous theoretical considerations, compare two investment pro-jects of 4 years duration. Fuzzy cash flows $K_t = P_t - KV_t$ are defined with a help of the four-reference point form described above (see Table 3). It is worth noting that data of the first project are more certain.

The results of calculations for two investment projects with different fuzzy cash flows are also presented in Table 3. It is seen that values of IRR_α obtained for each α-cut can increase or decrease with growth of α. As a result the set of possible crisp values of IRR is obtained for each project. Thus, a problem of

Table 3. The results of IRR_α calculation

Project 1		Project 2	
Year	Cash flow	Year	Cash flow
1	$\{-6.95, -6,95, -7,05, -8.00\}$	1	$\{-6.00, -6.95, -7.50, -8.00\}$
2	$\{4.95, 4.95, 5.05, 6.00\}$	2	$\{4.00, 4.95, 5.50, 6.00\}$
3	$\{3.95, 3.95, 4.05, 5.00\}$	3	$\{3.00, 3.95, 4.50, 5.00\}$
4	$\{1.95, 1.95, 2.05, 3.00\}$	4	$\{1.00, 1.95, 2.50, 3.00\}$

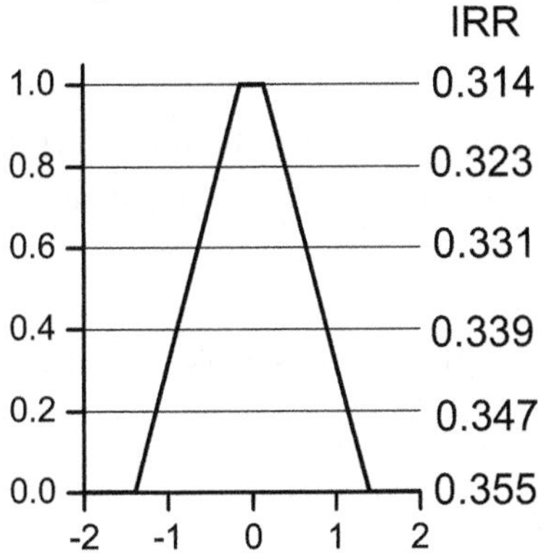

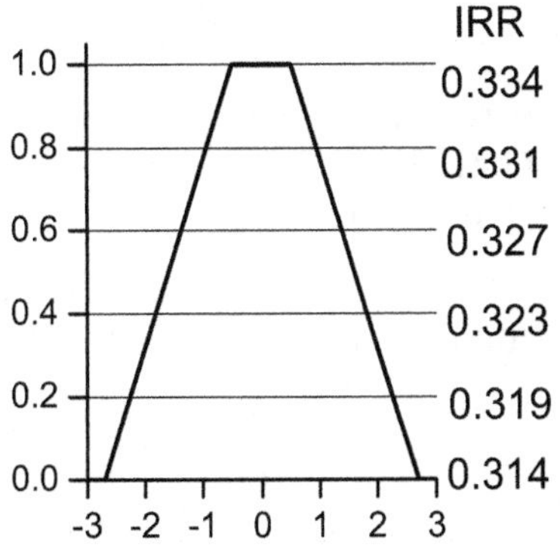

IRR
1.0 0.314
0.8 0.323
0.6 0.331
0.4 0.339
0.2 0.347
0.0 0.355
-2 -1 0 1 2

IRR
1.0 0.334
0.8 0.331
0.6 0.327
0.4 0.323
0.2 0.319
0.0 0.314
-3 -2 -1 0 1 2 3

the results interpretation rises. To solve this problem it is proposed to reduce the sets of IRR_α obtained on each α-cut to a small set of parameters which can be easily interpreted. The first elementary parameter — average value IRR_{m} — is certainly convenient, however it does not take into account that with growth of α the reliability of an outcome increases as well, i.e., IRR_α, obtained on higher α-cuts are more expected than those obtained on lower α-cuts according to the α-cut definition. On the other hand, the width of the crisp interval $[NPV_1, NPV_2]_\alpha$ corresponding to the IRR_α can be considered as, in some sense, a measure of uncertainty for the obtained crisp value IRR_α, since such width quantitatively characterises the difference of the left side of Eq. (10) from the degenerated zero interval $[0, 0]$. This allows to introduce two weighted estimations of IRR on a set IRR_α: least expected (least reliable) $IRR_{\min}$ and most expected (most reliable) $IRR_{\max}$:

$$IRR_{\min} = \frac{\sum\limits_{i=0}^{n-1} IRR_i \left(NPV_{2i} - NPV_{1i}\right)}{\sum\limits_{i=0}^{n-1} \left(NPV_{2i} - NPV_{1i}\right)}, \tag{12}$$

$$IRR_{\max} = \frac{\sum\limits_{i=0}^{n-1} IRR_i \alpha_i}{\sum\limits_{i=0}^{n-1} \alpha_i}, \tag{13}$$

where n is a number of α-cuts.

In a decision making practice it is worth to use all three proposed parameters IRR_{m}, IRR_{min}, IRR_{max} when choosing the best project. An interpretation of length of $[NPV_1, NPV_2]_\alpha$ as an indexes of uncertainty of IRR_α allows to propose a quantitative, expressed in monetary units essessment of financial risk of a project (the degree of uncertainty of the values IRR_{m}, IRR_{min}, IRR_{max} derived from uncertainty of initial data):

$$R_{\mathrm{r}} = \frac{\sum\limits_{i=0}^{n-1} (NPV_{2i} - NPV_{1i})}{n}, \tag{14}$$

Parameter R_{r} can play a key role in project efficiency estimation. THe values of introduced derivative paramters for the considered sample projects are presented in Table 4.

Table 4. The derivative (based on IRR)parameter of sample projects.

Project#	IRR_{min}	IRR_{max}	IRR_{m}	R_{r}
1	0.34	0.327	0.335	1.56
2	0.322	0.329	0.325	3.52

It is seen, the projects have rather the close values of IRR_{m}, IRR_{min}, IRR_{max}. At the same time, the risk R_{r} for the second project is considerably higher than risk of the first one. Hence, the first project is the best one. In addition, some other useful paramters have been proposed: IRR_{mr} — most reliable value of IRR_α — derived from the minimum interval $[NPV_1, NPV_2]_{mr}$ among all $[NPV_1, NPV_2]_\alpha$ and IRR_{lr} — the least reliable value of IRR_α — derived from the maximum interval $[NPV_1, NPV_2]_{\mathrm{lr}}$ among all $[NPV_1, NPV_2]_\alpha$. It is clear, that $[NPV_1, NPV_2]_{\mathrm{mr}}$ and $[NPV_1, NPV_2]_{\mathrm{lr}}$ are the risk estimations for the considering IRR_{mr} and IRR_{lr}. It should be noted (see Table 3) that the difference between values IRR_{mr} for the projects is rather small, but the difference in risk estimations is considerable.

4 A Method for a Numerical Solution of the Project Optimization Problem

Proposed here approach to the optimization problem solving is based on consideration of all initial fuzzy intervals P_t and KV_t as the restrictions on controlled input data, as well as on assumption that d_t is a random parameter describing external, in relation to a considered project, uncertainty. The fact that some preferences for an interval of possible values of d may be expressed by certain membership function $\mu_d(d)$, is also taken into account. Thus, while describing the discount factor one deals with uncertainties of both random and fuzzy nature. The problem is solved in two steps. At first, according to the fuzzy extension principle all parameters P_t , KV_t and d_t in Eq. (1) are substituted for corresponding fuzzy-intervals. As a result the fuzzy-interval NPV is obtained. On the next step, obtained fuzzy-interval NPV is considered as a restriction on a profit when building a local criterion for NPV maximisation. For a mathematical description of local criteria, so-called desirability functions are used. In essence, they can be described as a special interpretation of usual membership functions. Briefly, the desirability function rises from 0 (in area of unacceptable values of its argument) up to 1 (in area of the most preferable values). Thus, a construction of desirability function for NPV is rather obvious: the desirability function $\mu_{NPV}(NPV)$ can be considered only on the interval of possible values restricted by the interval $[NPV_1, NPV_4]$. Hence,the more value of the NPV, the more degree of desirability (see Fig. 7).

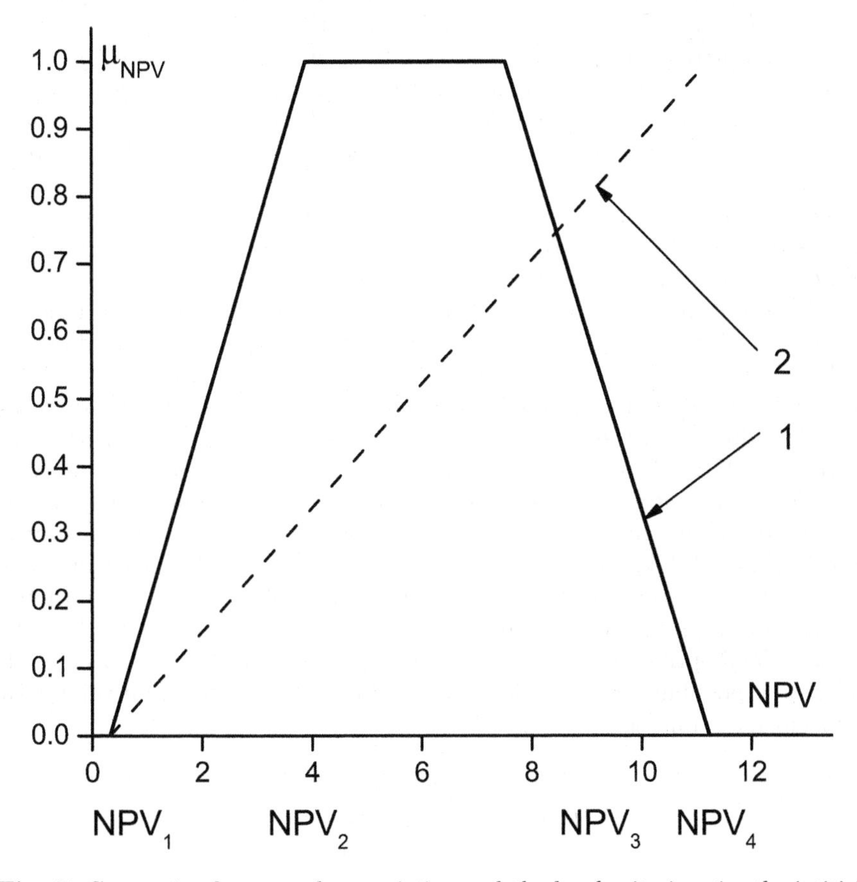

Fig. 7. Connection between the restriction and the local criterion: 1 - the intitial fuzzy interval of NPV (fuzzy restriction); 2 - the desirability function $\mu_{NPV}(NPV)$.

The initial fuzzy intervals P_t and KV_t are also considered as desirability functions $\mu_{P_1}, \mu_{P_2}, \ldots, \mu_{KV_1}, \mu_{KV_2}, \ldots$ describing restrictions on controlled input variables. It is clear that initial intervals were already constructed in such a way that when they are interpreted as desirability functions the more preferable values in intervals of P_t and KV_t appear to be those more realisable (possible). Since these desirability functions are connected with a possibility of realisation of corresponding values of variables P_t and KV_t, they implicitly describe financial risk of the project.

As the result, the general criterion based on the set of all desirability functions has been defined as

$$D(P_t, KV_t, d_t) = \mu_{NPV}^{\alpha_1}(NPV(P_t, KV_t, d_t)) \wedge \\ (\mu_{P_1} \wedge \mu_{P_2} \wedge \ldots \wedge \mu_{KV_1}, \mu_{KV_2} \wedge \ldots)^{\alpha_2}, \tag{15}$$

where α_1 and α_2 are ranks characterising the relative importance of local criteria of profit maximisation and risk minimisation, $\wedge$ is *min* operator, $\mu_{NPV}(NPV(P_t, KV_t, d_t))$ is a desirability function of NPV.

Many different forms of the general criterion are in use. As emphasised in [35],the choice of particular aggregating operator, usually called t-norm, is rather an application dependent problem. However, the choice of *min* operator in Eq. (15) is the most straightforward approach, when a compensation of small values of some criteria by the great values of others is not permited [32], [14]. The problem is reduced to a search for crisp values of $PP_1, PP_2, \ldots, KKV_1, KKV_2, \ldots$ on corresponding fuzzy intervals $P_1, P_2, \ldots, KV_1, KV_2, \ldots$, maximising the general criterion (15).

The problem is complicated by the fact that the discount d is a random parameter, distributed in a specific interval. The solution was carried out as follows.

Firstly, from interval of possible values a fixed value of discount d_i is selected randomly. Further, with a help of the Nollaw-Furst random method an optimum solution is obtained as the best compromise between uncertainty of basic data and intention to maximise profit, i.e., the optimisation problem reduces to maximisation of the general criterion (15). Obtained optimal values PP_t^d and KKV_t^d present the local optimum solution for given discount value. Above procedure is repeated with random discount values until the statistically representative sample of optimum solutions for various d_i is obtained. Final optimum values PP_t^0 , KKV_t^0 are calculated by weighting with degrees of possibility of d_i, which are defined by initial fuzzy interval d with a membership function $\mu_d(d_i)$:

$$PP_t^0 = \frac{\sum_{i=1}^{m} PP_t^d(d_i)\mu_d(d_i)}{\sum_{i=1}^{m} \mu_d(d_i)}, \tag{16}$$

where m is a number of random discount values used for the solution of a problem. Similarly, all KKV_t^0 can be calculated.

It is also possible to take into account the values of the general criterion in optimum points:

$$PP_t^0 = \frac{\sum\limits_{i=1}^{m} \left(PP_t^d(d_i) \left(\mu_d^{\beta_1}(d_i) \wedge D^{\beta_2}(d_i) \right) \right)}{\sum\limits_{i=1}^{m} \left(\mu_d^{\beta_1}(d_i) \wedge D^{\beta_2}(d_i) \right)}, \tag{17}$$

where β_1, β_2 are corresponding weights.

The similar expression can be constructed for KKV_t^0. It is worth noting that last expression gives an ability to take into account, apart from reliability of the values d_i, the degree of compatibility (in other words, the degree of consensus) for each of selected values of discount.

Obtained optimal PP_t^0 and KK_t^0 may be used for a final project's quality estimation. The results of calculation for the first example from the previous Section (Table 3, project 1) are presented in Table 5.

Table 5. The results of optimization

	Expression (16)		Expression (17)	
Years	PP_t^0	KK_t^0	PP_t^0	KK_t^0
0	0.00	2.49	0.00	2.50
1	0.00	0.83	0.00	0.79
2	8.05	0.00	8.04	0.00
3	7.12	0.00	7.09	0.00

Further, with substituting PP_t^0, KK_t^0 and fuzzy interval d in the expression (1) an optimal fuzzy value of NPV was obtained.

For considered example we get the following results:
$NPV_{16} = \{4.057293, 6.110165, 8.073906, 9.454419\}$ using (16)
and
$NPV_{17} = \{4.065489, 6.109793, 8.064094, 9.436519\}$ using (17).

It is clear that there is no great deference between the results obtained using expressions (16) and (17) in this case.

In Fig. 8, the fuzzy NPV_{16} obtained with PP_t^0 and KK_t^0 is compared with the initial one, obtained with the initial fuzzy values P_t and KV_t, without optimisation. It is obvious that in the optimal case the mean value of fuzzy interval NPV is greater.

Using optimal PP_t^0 and KK_t^0 and applying the method described in Section 2, the degree of project risk may be also estimated. This risk can be considered as financial risk of a project as a whole.

For the needs of common accounting practice it is possible to calculate an average weighted value of NPV using following expression:

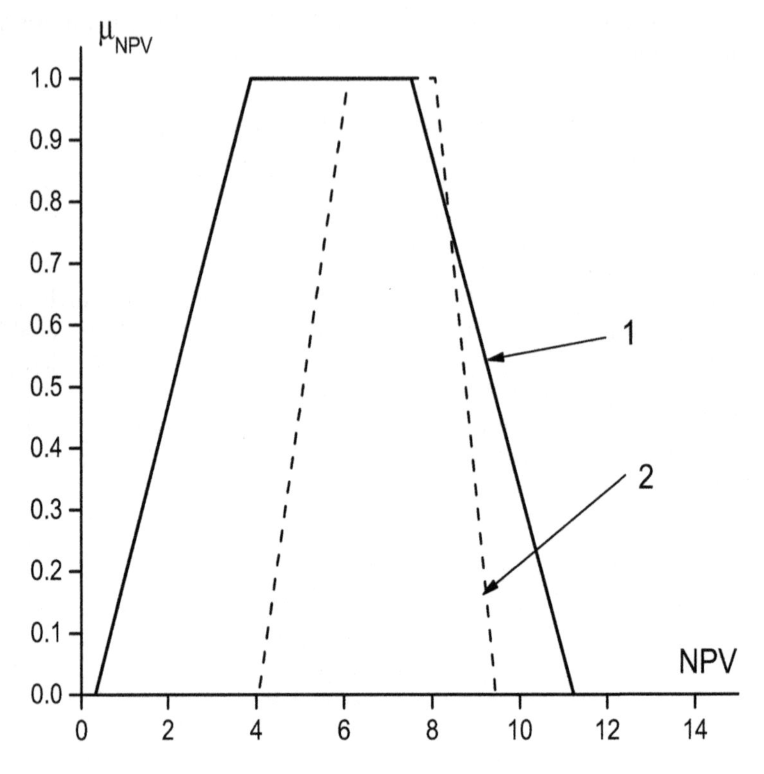

Fig. 8. Comparison of the initial and optimal fuzzy intervals NPV: 1 - initial NPV; 2 - optimal NPV.

$$NPV = \frac{\sum\limits_{i=1}^{m} NPV_i * \mu_{NPV_i}}{\sum\limits_{i=1}^{m} \mu_{NPV_i}}, \tag{18}$$

For the considered example $NPV_{16} = 6.8931$ and $NPV_{17} = 6.8942$ were obtained.

5 Conclusion

The problems of calculation of NPV and IRR and investment project risk assessment i a fuzzy setting are considered. It is shown that the straightforward way of project risk assessment is to consider this risk as a degree of fuzziness of the fuzzy Net Present Value, NPV. Nevertheless, other method for risk astimation on the probability approach to interval and fuzzy value comparison can be relevant in the fuzzy budgeting as well. It is shown that although it is impossible to obtain fuzzy Internal Rate of Return, IRR, the crisp IRR may be obtained as a solution of a fuzzy equation and a set of new useful derivative parameters characterising uncertainty of the problem may be obtained as an additional result.The problem of the multiobjective optimization of a project in a mixed fuzzy and random enviroment is formulated in a form of compromise between local criteria of profit maximisation and risk minimisation. Numerical method for the problem solving is described and tested.

References

1. Belletante B, Arnaud H (1989) Choisir ses investissements. Paris Chotard et Associés Éditeurs
2. Brigham E F (1992) Fundamentals of Financial Management. The Dryden Press, New York
3. Chansa-ngavej Ch, Mount-Campbell CA (1991) Decision criteria in capital budgeting under uncertainties: implications for future research. Int J Prod Economics 23:25–35
4. Liang P, Song F (1994) Computer-aided risk evaluation system for capital investment. Omega 22(4):391–400
5. Bogle HF, Jehenck GK (1985) Investment Analysis: US Oil and Gas Producers Score High in University Survey. In: Proc of Hydrocarbon Economics and Evaluation Symposium. Dallas 234-241
6. Babusiaux D, Pierru A (2001) Capital budgeting, project valuation and financing mix: Methodological proposals. Europian Journal of Operational Research 135:326–337
7. Moore RE (1966) Interval analysis. Englewood Cliffs, Prentice-Hall N J
8. Zadeh LA (1965) Fuzzy sets. Inf.Control 8:338–353
9. Ward TL (1985) Discounted fuzzy cash flow analysis. In: 1985 Fall Industrial Engineering Conference Proceedings 476–481
10. Buckley JJ (1987) The fuzzy mathematics of finance. Fuzzy Sets and Systems 21:257–273
11. Chen S (1995) An empirical examination of capital budgeting techniques: impact of investment types and firm characteristics. Eng Economist 40(2):145–170
12. Chiu CY, Park CS (1994) Fuzzy cash flow analysis using present worth criterion. Eng Economist 39(2):113–138
13. Choobineh F, Behrens A (1992) Use of intervals and possibility distributions in economic analysis. J Oper Res Soc 43(9):907–918
14. Dimova L, Sevastianov P, Sevastianov D (2005) MCDM in a fuzzy setting: investment projects assessment application. International Journal of Production Economics (in press)
15. Li Calzi M (1990) Towards a general setting for the fuzzy mathematics of finance. Fuzzy Sets and Systems 35:265–280
16. Perrone G (1994) Fuzzy multiple criteria decision model for the evaluation of AMS. Comput Integrated Manufacturing Systems 7(4):228–239
17. Chiu CY, Park CS (1994) Fuzzy cash flow analysis using present worth criterion. Eng Economist 39(2):113–138

18. Kahraman C, Tolga E, Ulukan Z (2000) Justification of manufacturing technologies using fuzzy benefit/cost ratio analysis. Int J Product Econom 66(1):45–52
19. Kahraman C, Ulukan Z (1997) Continuous compounding in capital budgeting using fuzzy concept. In: Proc of the 6th IEEE International Conference on Fuzzy Systems 1451–1455
20. Kahraman C, Ulukan Z (1997) Fuzzy cash flows under inflation. In: Proc of the Seventh International Fuzzy Systems Association World Congress (IFSA 97) 4:104–108
21. Sevastianov P, Sevastianov D (1997) Risk and capital budgeting parameters evaluation from the fuzzy sets theory position. Reliable software 1:10–19
22. Dimova L, Sevastianov D, Sevastianov P (2000) Application of fuzzy sets theory, methods for the evaluation of investment efficiency parameters. Fuzzy economic review 5(1):34–48
23. Kuchta D (2000) Fuzzy capital budgeting. Fuzzy Sets and Systems 111:367–385
24. Kahraman C (2001) Fuzzy versus probabilistic benefit/cost ratio analisis for public work projects. Int J Appl Math Comp Sci 11(3):705–718
25. Kahraman C, Ruan D, Tolga E (2002) Capital budgeting techniques using discounted fuzzy versus probabilistic cash flows. Information Sciences 142:57–76
26. Kaufmann A, Gupta M (1985) Introduction to fuzzy-arithmetic theory and applications. Van Nostrand Reinhold, New York
27. Jaulin L, Kieffir M, Didrit O, Walter E (2001) Applied Interval Analysis. Springer-Verlag, London
28. Longerstaey J, Spenser M (1996) RiskMetric-Technical document. RiskMetric Group, J.P. Morgan, New York
29. Nedosekin A, Kokosh A (2004) Investment risk estimation for arbitrary fuzzy factors of investment project. In: Proc. of Int. Conf. on Fuzzy Sets and Soft Computing in Economics and Finance. St. Petersburg 423–437
30. Yager RA (1979) On the measure of fuzziness and negation. Part 1. Membership in the Unit. Interval Int J Gen Syst 5:221–229
31. Yager RR, Detyniecki M, Bouchon-Meunier B (2001) A context-dependent method for ordering fuzzy numbers using probabilities, Information Sciences 138:237–255
32. Sevastianov P, Dimova L, Zhestkova E (2000) Methodology of the multicriteria quality estimation and its software realizing. In: Proc of the Fourth International Conference on New Information Technologies NITe' 1:50–54
33. Sevastianov P, Rog P (2002) A probabilistic approach to fuzzy and interval ordering, Task Quarterly. Special Issue Artificial and Computational Intelligence 7:147–156
34. Sewastianow P, Rog P (2005) Two-objective method for crisp and fuzzy interval comparison in optimization. Computers and Operation Research (in press)
35. Zimmermann HJ, Zysno P (1980) Latest connectives in human decision making. Fuzzy Sets and Systems 4:37–51

Option Pricing Theory in Financial Engineering from the Viewpoint of Fuzzy Logic

Yuji Yoshida

University of Kitakyushu, 4-2-1 Kitagata, Kokuraminami, Kitakyushu 802-8577, Japan

Summary. A mathematical model for European/American options with uncertainty is presented. The uncertainty is represented by both randomness and fuzziness. The randomness and fuzziness are evaluated respectively by probabilistic expectation and fuzzy expectation defined by a possibility measure from the viewpoint of decision-maker's subjective judgment. Prices of European call/put options with uncertainty are presented, and their valuation and properties are discussed under a reasonable assumption. The hedging strategies are also considered for marketability of the European options in portfolio selection. Further, the American options model with uncertainty is discussed by a numerical approach and is compared with the analytical case of the infinite terminal time. The buyer's/seller's rational range of the optimal expected price in each option is presented and the meaning and properties of the optimal expected prices are discussed.

Keywords: American option; European option; Black-Scholes formula; hedging strategy; stopping time; Sugeno integral; Choquet integral; fuzzy goal; fuzzy stochastic process.

1 Introduction and Notations

Option pricing theory in financial market has been developing together with financial engineering based on the famous Black-Scholes model. When we sell or buy stocks in financial market, there sometimes exists a difference between the actual prices and the theoretical value which derived from Black-Scholes method. Actually we cannot utilize timely some of fundamental data regarding the market, and therefore there exists uncertainty which we cannot represent by only probability theory because the concept of probability is constructed on mathematical representation whether something occurs or not in the future. When the market is unstable and changing rapidly, the losses/errors often become bigger between the decision maker's expected price and the actual price. Introducing fuzzy logic to the log-normal stochastic processes designed for the financial market, we present a model with uncertainty of both randomness and

Y. Yoshida: *Option Pricing Theory in Financial Engineering from the Viewpoint of Fuzzy Logic*, StudFuzz **201**, 229–243 (2006)
www.springerlink.com

fuzziness in output, which is a reasonable and natural extension of the original log-normal stochastic processes in Black-Scholes model. To valuate American options, we need to deal with an optimal stopping problem in log-normal stochastic processes (Elliott and Kopp [4], Karatzas and Shreve [5], Ross [10] and so on). We introduce a *fuzzy stochastic process* by fuzzy random variables to define prices in American options, and we evaluate the randomness and fuzziness by probabilistic expectation and fuzzy expectation defined by a possibility measure from the viewpoint of Yoshida [13]. We discuss the following themes on the basis of the results in Yoshida [14, 15, 17].

- American put option in a stochastic and fuzzy environment
  - The case with an expiration date
  - The perpetual option case without expiration dates
- European call/put options in a stochastic and fuzzy environment
  - Option pricing formula
  - Hedging strategies

In the next section, we introduce a fuzzy stochastic process by fuzzy random variables to define prices for American put option with uncertainty. We call the prices *fuzzy prices*. The randomness and fuzziness in the fuzzy stochastic process are evaluated by both probabilistic expectation and fuzzy expectation defined by a possibility measure, taking account of decision-maker's subjective judgment (Yoshida [13]). In Section 2, we deal with two models in American options with uncertainty, the case with an expiration date and the perpetual option case without expiration dates, and it is shown that the optimal fuzzy price is a solution of an optimality equation under a reasonable assumption. In Sections 3 and 4, we consider the optimal expected price in the American put option and we discuss seller's permissible range of expected prices, and we also give an optimal exercise time for the American put option. In Section 5, we give prices in European call/put options with uncertainty and we discuss their valuation and properties under a reasonable assumption. Finally, we give an explicit formula for the fuzzy prices in European options. We consider a rational expected price of the European options and buyer's/seller's permissible range of expected prices. The meaning and properties of rational expected prices are discussed in a numerical example. In the last section, we consider hedging strategies for marketability of the European options.

In the remainder of this section, we describe notations regarding bond price processes and stock price processes. We consider American put option in a finance model where there is no arbitrage opportunities ([4, 5]). Let $(\Omega, \mathcal{M}, P)$ be a probability space, where $\mathcal{M}$ is a σ-field and P is a non-atomic probability measure. $\mathbb{R}$ denotes the set of all real numbers. For a stock, let μ be the appreciation rate and let σ be the volatility ($\mu \in \mathbb{R}$, $\sigma > 0$). Let $\{B_t\}_{t\geq 0}$ be a standard Brownian motion on $(\Omega, \mathcal{M}, P)$. $\{\mathcal{M}_t\}_{t\geq 0}$ denotes a family of nondecreasing right-continuous complete sub-σ-fields of $\mathcal{M}$ such that $\mathcal{M}_t$ is generated by $B_s (0 \leq s \leq t)$. We consider two assets, a bond price and a stock price, where the bond price process $\{R_t\}_{t\geq 0}$ is riskless and the stock price process $\{S_t\}_{t\geq 0}$ is risky. Let r ($r \geq 0$) be a instantaneous interest rate, i.e. a interest factor, on the bond, and the bond price process $\{R_t\}_{t\geq 0}$ is given by $R_t = e^{rt}$ ($t \geq 0$). Let the stock price process $\{S_t\}_{t\geq 0}$ satisfy the following log-normal stochastic differential equation in Black-Scholes model: S_0 is a positive constant, and

$$\mathrm{d}S_t = \mu S_t \,\mathrm{d}t + \sigma S_t \,\mathrm{d}B_t, \tag{1}$$

$t \geq 0$. It is known ([4]) that there exists an equivalent probability measure Q. Under Q, $W_t := B_t - ((r-\mu)/\sigma)t$ is a standard Brownian motion and it holds that $\mathrm{d}S_t = rS_t\mathrm{d}t + \sigma S_t \mathrm{d}W_t$. By Ito's formula, we have $S_t = S_0 \exp((r - \sigma^2/2)t + \sigma W_t)$ ($t \geq 0$). The present stock price is determined by the information regarding the market until the previous time, and the present stock price S_t contains a certain uncertainty since we cannot utilize some of fundamental data actually at the current time t. The uncertainty comes from imprecision of information in the present market and is different from randomness, which is based on whether something occurs or not in the future. In the next section, we introduce fuzzy random variables to represent the uncertainty using fuzzy set theory.

2 Fuzzy Stochastic Processes

Fuzzy random variables, which take values in fuzzy numbers, have been studied by Puri and Ralescu [9] and many authors. It is known that the fuzzy random variable is one of the successful hybrid notions of randomness and fuzziness. First we introduce fuzzy numbers. Let $\mathcal{I}$ be the set of all non-empty bounded closed intervals. A fuzzy number is denoted by its membership function $\tilde{a} : \mathbb{R} \mapsto [0,1]$ which is normal, upper-semicontinuous, fuzzy convex and has a compact support (Zadeh [19], Klir and Yuan [6]). We identify a fuzzy number with its corresponding membership function. $\mathcal{R}$ denotes the set of all fuzzy numbers. The α-cut of a fuzzy number $\tilde{a}(\in \mathcal{R})$ is given by $\tilde{a}_\alpha := \{x \in \mathbb{R} \mid \tilde{a}(x) \geq \alpha\}$ ($\alpha \in (0,1]$) and $\tilde{a}_0 := \mathrm{cl}\{x \in \mathbb{R} \mid \tilde{a}(x) > 0\}$, where cl denotes the closure of an interval. We write the closed intervals as $\tilde{a}_\alpha := [\tilde{a}_\alpha^-, \tilde{a}_\alpha^+]$ for $\alpha \in [0,1]$. We also use a metric δ_∞ on $\mathcal{R}$ defined by $\delta_\infty(\tilde{a}, \tilde{b}) = \sup_{\alpha\in[0,1]} \delta(\tilde{a}_\alpha, \tilde{b}_\alpha)$ for fuzzy numbers $\tilde{a}, \tilde{b} \in \mathcal{R}$, where δ is the Hausdorff metric on $\mathcal{I}$. Hence we introduce a partial order $\succeq$, so called the fuzzy max order, on fuzzy numbers $\mathcal{R}$([6]): Let $\tilde{a}, \tilde{b} \in \mathcal{R}$ be fuzzy numbers. Then $\tilde{a} \succeq \tilde{b}$ means that $\tilde{a}_\alpha^- \geq \tilde{b}_\alpha^-$ and $\tilde{a}_\alpha^+ \geq \tilde{b}_\alpha^+$ for all $\alpha \in [0,1]$. Then $(\mathcal{R}, \succeq)$ becomes a lattice ([13]). For fuzzy numbers $\tilde{a}, \tilde{b} \in \mathcal{R}$, we define the maximum $\tilde{a} \vee \tilde{b}$ with respect to the fuzzy max order $\succeq$ by the fuzzy number whose α-cuts are $(\tilde{a} \vee \tilde{b})_\alpha = [\max\{\tilde{a}_\alpha^-, \tilde{b}_\alpha^-\}, \max\{\tilde{a}_\alpha^+, \tilde{b}_\alpha^+\}]$ ($\alpha \in [0,1]$). An addition, a subtraction and a scalar multiplication for fuzzy numbers are defined as follows: For $\tilde{a}, \tilde{b} \in \mathcal{R}$ and $\lambda \geq 0$, the addition and subtraction $\tilde{a} \pm \tilde{b}$ of $\tilde{a}$ and $\tilde{b}$ and the scalar multiplication $\lambda\tilde{a}$ of λ and $\tilde{a}$ are fuzzy numbers given by $(\tilde{a} + \tilde{b})_\alpha := [\tilde{a}_\alpha^- + \tilde{b}_\alpha^-, \tilde{a}_\alpha^+ + \tilde{b}_\alpha^+]$, $(\tilde{a} - \tilde{b})_\alpha := [\tilde{a}_\alpha^- - \tilde{b}_\alpha^+, \tilde{a}_\alpha^+ - \tilde{b}_\alpha^-]$ and $(\lambda\tilde{a})_\alpha := [\lambda\tilde{a}_\alpha^-, \lambda\tilde{a}_\alpha^+]$ for $\alpha \in [0,1]$.

A fuzzy-number-valued map $\tilde{X} : \Omega \mapsto \mathcal{R}$ is called a fuzzy random variable if $\{(\omega, x) \in \Omega \times \mathbb{R} \mid \tilde{X}(\omega)(x) \geq \alpha\} \in \mathcal{M} \times \mathcal{B}$ for all $\alpha \in [0,1]$, where $\mathcal{B}$ is the Borel σ-field of $\mathbb{R}$. We can find some equivalent conditions in general cases ([9]), however we adopt a simple characterization in the following lemma.

Lemma 1 (Wang and Zhang [12, Theorems 2.1 and 2.2]). *For a map $\tilde{X}$: $\Omega \mapsto \mathcal{R}$, the following (i) and (ii) are equivalent:*

(i) *$\tilde{X}$ is a fuzzy random variable.*
(ii) *The maps $\omega \mapsto \tilde{X}_\alpha^-(\omega)$ and $\omega \mapsto \tilde{X}_\alpha^+(\omega)$ are measurable for all $\alpha \in [0,1]$, where $\tilde{X}_\alpha(\omega) = [\tilde{X}_\alpha^-(\omega), \tilde{X}_\alpha^+(\omega)] := \{x \in \mathbb{R} \mid \tilde{X}(\omega)(x) \geq \alpha\}$.*

From Lemma 1, we obtain the following lemma regarding fuzzy random variables $\tilde{X}$ and their α-cuts $\tilde{X}_\alpha(\omega) = [\tilde{X}_\alpha^-(\omega), \tilde{X}_\alpha^+(\omega)]$.

Lemma 2.

(i) *Let $\tilde{X}$ be a fuzzy random variable. The α-cuts $\tilde{X}_\alpha(\omega) = [\tilde{X}_\alpha^-(\omega), \tilde{X}_\alpha^+(\omega)]$, $\omega \in \Omega$, have the following properties (a) — (c):*
  (a) *$\tilde{X}_\alpha(\omega) \subset \tilde{X}_{\alpha'}(\omega)$ for $\omega \in \Omega$, $0 \leq \alpha' < \alpha \leq 1$.*
  (b) *$\lim_{\alpha' \uparrow \alpha} \tilde{X}_{\alpha'}(\omega) = \tilde{X}_\alpha(\omega)$ for $\omega \in \Omega$, $\alpha > 0$.*
  (c) *The maps $\omega \mapsto \tilde{X}_\alpha^-(\omega)$ and $\omega \mapsto \tilde{X}_\alpha^+(\omega)$ are measurable for $\alpha \in [0,1]$.*
(ii) *Conversely, suppose that a family of interval-valued maps $X_\alpha = [X_\alpha^-, X_\alpha^+]$: $\Omega \mapsto \mathcal{I}$ ($\alpha \in [0,1]$) satisfies the above conditions (a) – (c). Then, a membership function*

$$\tilde{X}(\omega)(x) := \sup_{\alpha \in [0,1]} \min\{\alpha, 1_{X_\alpha(\omega)}(x)\}, \quad \omega \in \Omega,\ x \in \mathbb{R},$$

gives a fuzzy random variable and $\tilde{X}_\alpha(\omega) = X_\alpha(\omega)$ for $\omega \in \Omega$ and $\alpha \in [0,1]$, where $1_{\{\cdot\}}$ denotes the characteristic function of an interval.

Next we need to introduce expectations of fuzzy random variables in order to describe fuzzy-valued European option models in the next section. A fuzzy random variable $\tilde{X}$ is called integrably bounded if both $\omega \mapsto \tilde{X}_\alpha^-(\omega)$ and $\omega \mapsto \tilde{X}_\alpha^+(\omega)$ are integrable for all $\alpha \in [0,1]$. Let $\tilde{X}$ be an integrally bounded fuzzy random variable. The expectation $E(\tilde{X})$ of the fuzzy random variable $\tilde{X}$ is defined by a fuzzy number

$$E(\tilde{X})(x) := \sup_{\alpha \in [0,1]} \min\{\alpha, 1_{E(\tilde{X})_\alpha}(x)\}, \quad x \in \mathbb{R},$$

where $E(\tilde{X})_\alpha := [\int_\Omega \tilde{X}_\alpha^-(\omega)\,\mathrm{d}P(\omega), \int_\Omega \tilde{X}_\alpha^+(\omega)\,\mathrm{d}P(\omega)]$ ($\alpha \in [0,1]$).

Now, we consider a continuous-time fuzzy stochastic process by fuzzy random variables. Let $\{\tilde{X}_t\}_{t\geq 0}$ be a family of integrally bounded fuzzy random variables. We assume that the map $t \mapsto \tilde{X}_t(\omega)(\in \mathcal{R})$ is continuous on $[0,\infty)$ for almost all $\omega \in \Omega$. $\{\mathcal{M}_t\}_{t\geq 0}$ is a family of nondecreasing sub-σ-fields of $\mathcal{M}$

which is right continuous, and fuzzy random variables $\tilde{X}_t$ are $\mathcal{M}_t$-adapted, i.e. random variables $\tilde{X}^-_{r,\alpha}$ and $\tilde{X}^+_{r,\alpha}$ $(0 \le r \le t; \alpha \in [0,1])$ are $\mathcal{M}_t$-measurable. Then $(\tilde{X}_t, \mathcal{M}_t)_{t \ge 0}$ is called a fuzzy stochastic process.

We introduce a valuation method of fuzzy prices, taking into account of decision maker's subjective judgment. Give a fuzzy goal by a fuzzy set $\varphi : [0,\infty) \mapsto [0,1]$ which is a continuous and increasing function with $\varphi(0) = 0$ and $\lim_{x\to\infty} \varphi(x) = 1$. Then we note that the α-cut is $\varphi_\alpha = [\varphi^-_\alpha, \infty)$ for $\alpha \in (0,1)$. For an exercise time T and call/put options with fuzzy values $\tilde{X}_T = \tilde{C}_T$ or $\tilde{X}_T = \tilde{P}_T$, which will be given as fuzzy prices in the next section, we define a fuzzy expectation of the fuzzy numbers $E(\tilde{X}_T)$ by

$$\tilde{E}(E(\tilde{X}_T)) := \fint_{[0,\infty)} E(\tilde{X}_T)(x) \, \mathrm{d}\tilde{m}(x) = \sup_{x \in [0,\infty)} \min\{E(\tilde{X}_T)(x), \varphi(x)\}, \quad (2)$$

where $\tilde{m}$ is the possibility measure generated by the density φ and $\fint \mathrm{d}\tilde{m}$ denotes Sugeno integral ([11]). The fuzzy number $E(\tilde{X}_T)$ means a fuzzy price, and the fuzzy expectation (2) implies the degree of buyer's/seller's satisfaction regarding fuzzy prices $E(\tilde{X}_T)$. Then the fuzzy goal $\varphi(x)$ means a kind of utility function for expected prices x in (2), and it represents a buyer's/seller's subjective judgment from the idea of Bellman and Zadeh [1]. Hence, a real number $x^*(\in [0,\infty))$ is called a rational expected price if it attains the supremum of the fuzzy expectation (2), i.e.

$$\tilde{E}(\tilde{V}) = \sup_{x \in [0,\infty)} \min\{\tilde{V}(x), \varphi(x)\} = \min\{\tilde{V}(x^*), \varphi(x^*)\}, \quad (3)$$

where

$$\tilde{V} := E(\tilde{X}_T)$$

is a fuzzy price of options.

We also consider about an estimation of imprecision regarding fuzzy numbers. One of the methods to evaluate the imprecision regarding a fuzzy number $\tilde{a}$ is given by Choquet integral ([3, 7]):

$$(C) \int \tilde{a}(\cdot) \, \mathrm{d}\tilde{Q}(\cdot) = \int_0^1 \tilde{Q}\{x \in \mathbb{R} | \tilde{a}(x) \ge \alpha\} \, \mathrm{d}\alpha,$$

where $\tilde{Q}$ is a fuzzy measure on $\mathbb{R}$. We take $\tilde{Q}$ Lebesgue measure, then

$$(C) \int \tilde{a}(\cdot) \, \mathrm{d}\tilde{Q}(\cdot) = \int_0^1 (\tilde{a}^+_\alpha - \tilde{a}^-_\alpha) \, \mathrm{d}\alpha.$$

Therefore, the estimation of fuzziness regarding a fuzzy random variable $\tilde{X}_t$ follows

$$(C) \int \tilde{X}_t(\omega)(\cdot) \, \mathrm{d}\tilde{Q}(\cdot) = \int_0^1 (\tilde{X}^+_{t,\alpha}(\omega) - \tilde{X}^-_{t,\alpha}(\omega)) \, \mathrm{d}\alpha \quad (4)$$

for $t \ge 0$, $\omega \in \Omega$.

3 American Put Option in Uncertain Environment

In this section, we introduce American put option with fuzzy prices and we discuss its properties. Let $\{a_t\}_{t\geq 0}$ be an $\mathcal{M}_t$-adapted stochastic process such that the map $t \mapsto a_t(\omega)$ is continuous on $[0,\infty)$ and $0 < a_t(\omega) \leq S_t(\omega)$ for almost all $\omega \in \Omega$. We give a fuzzy stochastic process of the stock prices $\{\tilde{S}_t\}_{t\geq 0}$ by the following fuzzy random variables:

$$\tilde{S}_t(\omega)(x) := L((x - S_t(\omega))/a_t(\omega))$$

for $t \geq 0$, $\omega \in \Omega$ and $x \in \mathbb{R}$, where $L(x) := \max\{1 - |x|, 0\}$ $(x \in \mathbb{R})$ is the triangle-type shape function(Fig. 1) and $\{S_t\}_{t\geq 0}$ is defined by (1). Then, from (4), the fuzziness regarding the fuzzy random variables $\tilde{S}_t$ is estimated by Choquet integral

$$(C)\int \tilde{S}_t(\omega)(\cdot)\,\mathrm{d}\tilde{Q}(\cdot) = \int_0^1 (\tilde{S}^+_{t,\alpha}(\omega) - \tilde{S}^-_{t,\alpha}(\omega))\,\mathrm{d}\alpha = a_t(\omega). \tag{5}$$

Therefore $a_t(\omega)$ means the amount of fuzziness regarding the stock price $\tilde{S}_t(\omega)$ and is the spread of the triangular fuzzy number in Fig. 1.

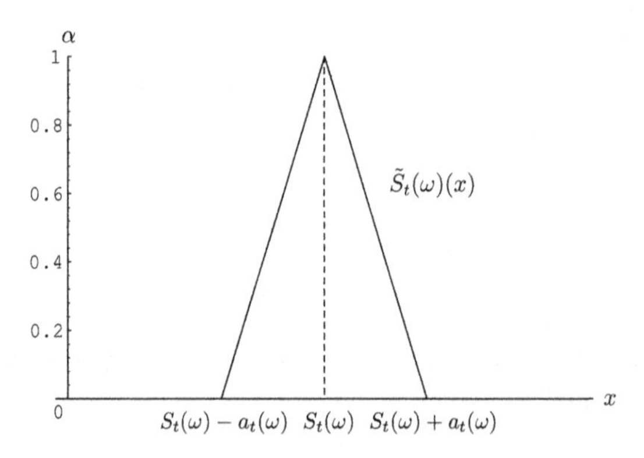

Fig. 1. Fuzzy random variable $\tilde{S}_t(\omega)(x)$.

Let K $(K > 0)$ be a strike price. We define a fuzzy price process by the following fuzzy stochastic process $\{\tilde{P}_t\}_{t\geq 0}$:

$$\tilde{P}_t(\omega) := e^{-rt}(1_{\{K\}} - \tilde{S}_t(\omega)) \vee 1_{\{0\}}$$

for $t \geq 0$, $\omega \in \Omega$, where $\vee$ is the maximum by the fuzzy max order, and $1_{\{K\}}$ and $1_{\{0\}}$ denote the crisp numbers K and zero respectively. By using stopping times τ, we consider a problem to maximize fuzzy price process of American put option. Fix an initial stock price y $(y = S_0 > 0)$. Put the optimal fuzzy price of American put option by

$$\tilde{V} = \bigvee_{\tau\text{: stopping times with values in } \mathbb{T}} E(\tilde{P}_\tau), \tag{6}$$

where $E(\cdot)$ denotes the expectation with respect to the equivalent martingale measure Q, and $\vee$ means the supremum induced from the fuzzy max order.

We consider a valuation method of fuzzy prices, taking into account of decision maker's subjective judgment. From (2), for a stopping time τ, we define a fuzzy expectation of the fuzzy numbers $E(\tilde{P}_\tau)$ by

$$\tilde{E}(E(\tilde{P}_\tau)) = \sup_{x \in [0,\infty)} \min\{E(\tilde{P}_\tau)(x), \varphi(x)\}, \tag{7}$$

where φ is the seller's fuzzy goal. In this section, we discuss the following optimal stopping problem regarding American put option with fuzziness.

Problem P. Find a stopping time τ^* with values in $\mathbb{T}$ such that

$$\tilde{E}(E(\tilde{P}_{\tau^*})) = \tilde{E}(\tilde{V}),$$

where $\tilde{V}$ is given by (6).

Then, τ^* is called an *optimal exercise time* and a real number $x^*(\in [0,\infty))$ is called an *optimal expected price* under the fuzzy expectation generated by possibility measures if it attains the supremum of the fuzzy expectation (7), i.e.

$$\tilde{E}(\tilde{V}) = \min\{\tilde{V}(x^*), \varphi(x^*)\}. \tag{8}$$

The fuzzy random variables $\tilde{Z}_t$ correspond to Snell's envelope in probability theory. Hence, by using dynamic programming approach, we obtain the following optimality characterization for the fuzzy stochastic process regarding the optimal fuzzy price $\tilde{V}$ by fuzzy random variables $\tilde{Z}_t$. Now we introduce a reasonable assumption for computation.

Assumption S. The stochastic process $\{a_t\}_{t \geq 0}$ is represented by

$$a_t(\omega) := cS_t(\omega),$$

$t \geq 0, \omega \in \Omega$, where c is a constant satisfying $0 < c < 1$.

Since (1) can be written as

$$\mathrm{d} \log S_t = \mu \, \mathrm{d}t + \sigma \, \mathrm{d}B_t, \tag{9}$$

$t \geq 0$, one of the most difficulties is estimation of the volatility σ of a stock in actual cases ([10, Sect.7.5.1]). Therefore, Assumption S is reasonable since $a_t(\omega)$ corresponds to the size of fuzziness from (5) and so it is reasonable that $a_t(\omega)$ should depend on the fuzziness of the volatility σ and the stock price $S_t(\omega)$ of the term $\sigma S_t(\omega)$ in (1). In this model, we represent by c the fuzziness

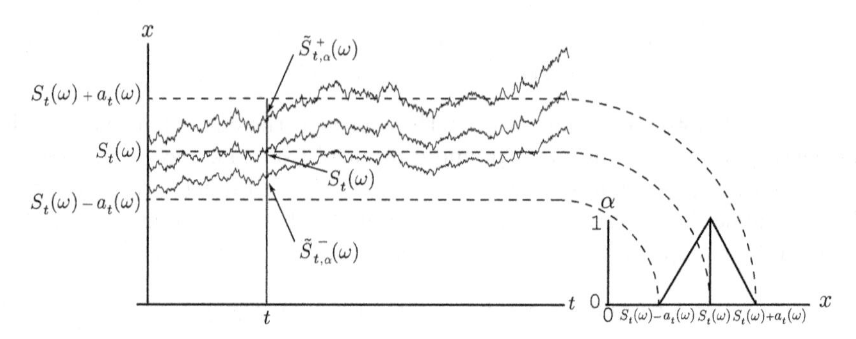

Fig. 2. The stochastic process $\{a_t\}_{t\geq 0}$.

of the volatility σ, and we call c a *fuzzy factor of the stock process.* From now on, we suppose that Assumption S holds. By putting $b^{\pm}(\alpha) := 1 \pm (1-\alpha)c$ ($\alpha \in [0,1]$), from Assumption S we have

$$\tilde{S}^{\pm}_{t,\alpha}(\omega) = b^{\pm}(\alpha) S_t(\omega),$$

$t \geq 0, \omega \in \Omega, \alpha \in [0,1]$, where $\tilde{S}^{\pm}_{t,\alpha}(\omega)$ is the α-cut of $\tilde{S}^{\pm}_t(\omega)$.

Define fuzzy price processes by

$$\tilde{V}^{\pm}_{\alpha}(y,t) := \sup_{\tau \geq t} E(e^{-r(\tau - t)} \max\{K - \tilde{S}^{\mp}_{\tau,\alpha}, 0\} \mid S_t = y)$$

for $y > 0$ and $t \in \mathbb{T}$. Here we consider the following two cases (I) and (II).

Case (I) (American put option with an expiration date T, $\mathbb{T} = [0,T]$). Define an operator

$$\mathcal{L} := \frac{1}{2}\sigma^2 y^2 \frac{\partial^2}{\partial y^2} + ry\frac{\partial}{\partial y} + \frac{\partial}{\partial t}$$

on $[0,\infty) \times [0,T)$. Then we obtain the following optimality conditions by dynamic programming([15]).

Theorem 1 (Free boundary problem). *The fuzzy price* $V(y,t) = \tilde{V}^{\pm}_{\alpha}(y,t)$ *satisfies the following equations:*

$$\begin{aligned}
\mathcal{L}(e^{-rt}V(y,t)) &\leq 0, \\
\mathcal{L}(e^{-rt}V(y,t)) &= 0 \quad \text{on } D, \\
V(y,t) &\geq \max\{K - b^{\mp}(\alpha)y, 0\}, \\
V(y,T) &= \max\{K - b^{\mp}(\alpha)y, 0\},
\end{aligned}$$

where $D := \{(y,t) \in [0,\infty) \times [0,T) \mid \tilde{V}^{\pm}_{\alpha}(y,t) > \max\{K - b^{\mp}(\alpha)y, 0\}\}$. *The corresponding optimal exercise time is*

$$\tau_\alpha(\omega) = \inf\left\{t \in \mathbb{T} \mid \tilde{V}^{\pm}_{\alpha}(S_t(\omega),t) = \max\{K - b^{\mp}(\alpha)S_t(\omega), 0\}\right\}.$$

Case (II) (A perpetual American put option, $\mathbb{T} = [0, \infty)$). The both ends of the α-cuts are

$$\tilde{V}_\alpha^\pm(y,0) := \sup_{\tau \geq 0} E(e^{-r\tau} \max\{K - \tilde{S}_{\tau,\alpha}^\mp, 0\} 1_{\{\tau<\infty\}} \mid S_0 = y)$$

for $y > 0$. Then we obtain the following results ([15]).

Theorem 2. *The fuzzy price $\tilde{V}_\alpha^\pm(y,0)$ is represented by*

$$\tilde{V}_\alpha^\pm(y,0) = \begin{cases} K - b^\mp(\alpha)y & \text{if } y \leq s^\pm(\alpha) \\ (K - b^\mp(\alpha)s^\pm(\alpha))(y/s^\pm(\alpha))^{-\gamma} & \text{if } y > s^\pm(\alpha), \end{cases}$$

where $s^\pm(\alpha) := 2rK/(b^\mp(\alpha)(2r + \sigma^2))$ and $\gamma := 2r/\sigma^2$. The optimal exercise time is

$$\tau_\alpha(\omega) = \inf\left\{t \geq 0 \mid (r - \frac{\sigma^2}{2})t + \sigma W_t(\omega) = \log\left(\frac{s^\pm(\alpha)}{y}\right)\right\}.$$

4 The Optimal Expected Price and the Optimal Exercise Times

Fix an initial stock price $y(> 0)$. In this section, we discuss the optimal expected price of American put option $\tilde{V} := \tilde{V}(y,0)$, which is introduced in the previous section, and we give an optimal exercise time for Problem P. Define a grade α^* by

$$\alpha^* := \sup\{\alpha \in [0,1] | \ \varphi_\alpha^- \leq \tilde{V}_\alpha^+\}, \tag{10}$$

where $\varphi_\alpha = [\varphi_\alpha^-, \infty)$ for $\alpha \in (0,1)$, and the supremum of the empty set is understood to be 0. The following theorem, which is obtained by a modification of the proofs in [13, Theorems 3.1 and 3.2], implies that α^* is the grade of the fuzzy expectation of American put option price $\tilde{V}$ (see (8)).

Theorem 3. *Under the fuzzy expectation generated by possibility measures (8), the following (i) – (iii) hold.*

(i) *The grade of the fuzzy expectation of American put option price $\tilde{V}$ is given by*

$$\alpha^* = \tilde{E}(\tilde{V}) = \sup_{\tau:\ \text{stopping times with values in } \mathbb{T}} \tilde{E}(E(\tilde{P}_\tau)).$$

(ii) *Further, the optimal expected price of American put option is given by*

$$x^* = \varphi_{\alpha^*}^-. \tag{11}$$

(iii)*Define a stopping time*

$$\tau^*(\omega) := \tau_{\alpha^*}(\omega) = \inf\{t \in \mathbb{T} \mid S_t(\omega) \leq s^+(\alpha^*)\}, \tag{12}$$

where the infimum of the empty set is understood to be $\sup \mathbb{T}$. *If* τ^* *is finite, then* τ^* *is an optimal stopping time for Problem P, and it is the optimal exercise time.*

In Theorem 3, we need the assumption *the finiteness of* τ^*, only when $\mathbb{T} = [0, \infty)$. Since the fuzzy expectation (8) is defined by possibility measures, (11) gives an upper bound on optimal expected prices of American put option. Similarly to (10) we can define another grade, which gives a lower bound on optimal expected prices of American put option as follows:

$$x_* = \varphi^-_{\alpha_*}, \tag{13}$$

where α_* is defined by

$$\alpha_* := \sup\{\alpha \in [0,1] | \ \varphi^-_\alpha \leq \tilde{V}^-_\alpha\}.$$

Then, its corresponding stopping time is given by

$$\tau_*(\omega) := \tau_{\alpha_*}(\omega) = \inf\{t \in \mathbb{T} \mid S_t(\omega) \leq s^-(\alpha_*)\}.$$

Hence, from (11) and (13), we can easily check the interval $[x_*, x^*]$ is written as

$$[x_*, x^*] = \{x \in \mathbb{R} \mid \tilde{V}(x) \geq \varphi(x)\},$$

which is the range of prices x such that the reliability degree of the optimal expected price, $\tilde{V}(x)$, is greater than the degree of seller's satisfaction, $\varphi(x)$. Therefore, $[x_*, x^*]$ means *seller's permissible range of expected prices* under his fuzzy goal φ.

Example 1. Consider a perpetual American put option ($\mathbb{T} = [0, \infty)$). Put a fuzzy goal

$$\varphi(x) = \begin{cases} 1 - e^{-0.2x}, & x \geq 0 \\ 0, & x < 0. \end{cases}$$

Then, $\varphi^-_\alpha = -0.2^{-1} \log(1 - \alpha)$ for $\alpha \in (0, 1)$. Put a volatility $\sigma = 0.25$, an interest factor $r = 0.05$, a fuzzy factor $c = 0.1$, an initial stock price $y = 20$ and a strike price $K = 25$. We can easily calculate that the optimal grades are $\alpha_* \approx 0.700468$ and $\alpha^* \approx 0.73281$. From (11) and (13), the *permissible range of expected prices* under seller's fuzzy goal φ is (see Fig. 3) $[x_*, x^*] \approx [6.02767, 6.59897]$. The corresponding optimal exercise times are

$$\tau_*(\omega) = \inf\{t \geq 0 \mid W_t(\omega) - 0.45t = \log(s^-(\alpha_*)/20)\},$$
$$\tau^*(\omega) = \inf\{t \geq 0 \mid W_t(\omega) - 0.45t = \log(s^+(\alpha^*)/20)\}$$

with $s^-(\alpha_*) \approx 14.9372$ and $s^+(\alpha^*) \approx 15.807$.

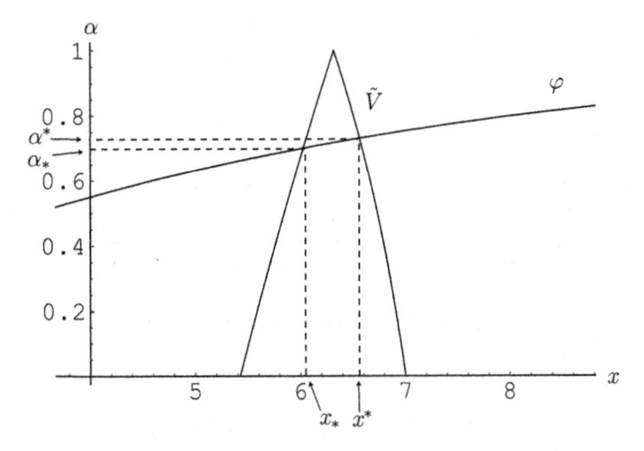

Fig. 3. Optimal fuzzy price $\tilde{V}(x)$ and fuzzy goal $\varphi(x)$.

5 European Options in Uncertain Environment

In this section, we introduce European option with fuzzy prices and we discuss their properties. We define fuzzy stochastic processes of European call/put options by $\{\tilde{C}_t\}_{t\geq 0}$ and $\{\tilde{P}_t\}_{t\geq 0}$:

$$\tilde{C}_t(\omega) := e^{-rt}(\tilde{S}_t(\omega) - 1_{\{K\}}) \vee 1_{\{0\}},$$
$$\tilde{P}_t(\omega) := e^{-rt}(1_{\{K\}} - \tilde{S}_t(\omega)) \vee 1_{\{0\}},$$

$t \geq 0$, $\omega \in \Omega$. We evaluate these fuzzy stochastic processes by the expectations introduced in the pervious sections. Then, the fuzzy price processes of European call/put options are given as follows:

$$\tilde{V}^{\tilde{C}}(y,t) := e^{rt}E(\tilde{C}_T \mid S_t = y)$$
$$\tilde{V}^{\tilde{P}}(y,t) := e^{rt}E(\tilde{P}_T \mid S_t = y)$$

for an initial stock price y $(y > 0)$ and $t \in [0,T]$, where $E(\cdot)$ denotes expectation with respect to the equivalent martingale measure Q. Their α-cuts are

$$\tilde{V}_\alpha^{\tilde{C},\pm}(y,t) = E(e^{-r(T-t)} \max\{\tilde{S}_{T,\alpha}^{\pm} - K, 0\} \mid S_t = y);$$
$$\tilde{V}_\alpha^{\tilde{P},\pm}(y,t) = E(e^{-r(T-t)} \max\{K - \tilde{S}_{T,\alpha}^{\mp}, 0\} \mid S_t = y).$$

Then we obtain the following formulae to calculate fuzzy price in European options ([14]).

Theorem 4 (Black-Scholes formula for fuzzy prices). *Suppose that Assumption S holds. Let $\alpha \in [0,1]$. Let an initial stock price $y(:= S_0 > 0)$.*

(i) *The rational fuzzy price of European call option is given by*

$$\tilde{V}_\alpha^{\tilde{C},\pm}(y,0) = b^{\pm}(\alpha)y\Phi(z_1) - Ke^{-rT}\Phi(z_2), \tag{14}$$

where $\Phi(z) = (1/\sqrt{2\pi})\int_{-\infty}^{z} e^{-w^2/2}dw$ $(z \in \mathbb{R})$ *is the standard normal distribution function, and* z_1 *and* z_2 *are given by*

$$z_1 = \frac{\log b^{\pm}(\alpha) + \log(y/K) + T(r + \sigma^2/2)}{\sigma\sqrt{T}};$$

$$z_2 = \frac{\log b^{\pm}(\alpha) + \log(y/K) + T(r - \sigma^2/2)}{\sigma\sqrt{T}}.$$

(ii) *The rational fuzzy price of European put option is given by the following call-put parity:*

$$\tilde{V}_\alpha^{\tilde{P},\pm}(y,0) = \tilde{V}_\alpha^{\tilde{C},\mp}(y,0) - b^{\mp}(\alpha)y + Ke^{-rT}. \tag{15}$$

Fix an initial stock price $y(>0)$. Next, we discuss the expected price, which is introduced in the previous section, of European call/put options $\tilde{V} = \tilde{V}^{\tilde{C}}(y,0)$ or $\tilde{V} = \tilde{V}^{\tilde{P}}(y,0)$. Define a grade $\alpha^{\tilde{C},+}$ by

$$\alpha^{\tilde{C},+} := \sup\{\alpha \in [0,1] | \ \varphi_\alpha^- \le \tilde{V}_\alpha^{\tilde{C},+}(y,0)\},$$

where $\varphi_\alpha = [\varphi_\alpha^-, \infty)$ for $\alpha \in (0,1)$, and the supremum of the empty set is understood to be 0. From the continuity of φ and $\tilde{V}^{\tilde{C}}$, we can easily check that the grade $\alpha^{\tilde{C},+}$ satisfies

$$\varphi_{\alpha^{\tilde{C},+}}^- = \tilde{V}_{\alpha^{\tilde{C},+}}^{\tilde{C},+}(y,0). \tag{16}$$

The following theorem, which is obtained by a modification of the proofs in [13, Theorems 3.1 and 3.2], implies that $\alpha^{\tilde{C},+}$ is the grade of the fuzzy expectation of European call option $\tilde{V}^{\tilde{C}}(y,0)$.

Theorem 5. *Under the fuzzy expectation generated by possibility measures (2), the following (i) and (ii) hold.*

(i) *The grade of the fuzzy expectation of European call option price* $\tilde{V}^{\tilde{C}}$ *is given by*

$$\alpha^{\tilde{C},+} = \tilde{E}(\tilde{V}^{\tilde{C}}(y,0)) = \tilde{E}(E(\tilde{C}_T)).$$

(ii) *Further, the rational expected price of European call option is given by*

$$x^{\tilde{C},+} = \varphi_{\alpha^{\tilde{C},+}}^-. \tag{17}$$

Since the fuzzy expectation (3) is defined by possibility measures, (17) gives an upper bound on rational expected prices of European call option. Similarly to (16) we can define another grade, which gives a lower bound on rational expected prices of European call option as follows:

$$x^{\tilde{C},-} = \varphi_{\alpha^{\tilde{C},-}}^-, \tag{18}$$

where $\alpha^{\tilde{C},-}$ is defined by

$$\alpha^{\tilde{C},-} := \sup\{\alpha \in [0,1] | \ \varphi_\alpha^- \leq \tilde{V}_\alpha^{\tilde{C},-}(y,0)\}.$$

Hence, from (17) and (18), we can easily check the interval $[x^{\tilde{C},-}, x^{\tilde{C},+}]$ is written as

$$[x^{\tilde{C},-}, x^{\tilde{C},+}] = \{x \in \mathbb{R} \mid \tilde{V}^{\tilde{C}}(y,0)(x) \geq \varphi(x)\},$$

which is the range of prices x such that the reliability degree of the optimal expected price, $\tilde{V}^{\tilde{C}}(y,0)(x)$, is greater than the degree of buyer's satisfaction, $\varphi(x)$. Therefore, $[x^{\tilde{C},-}, x^{\tilde{C},+}]$ means *buyer's permissible range of expected prices* under his fuzzy goal φ. Regarding European put option, similarly we obtain *seller's permissible range of rational expected prices* by $[x^{\tilde{P},-}, x^{\tilde{P},+}]$, where $x^{\tilde{P},-} := \varphi^-_{\alpha^{\tilde{P},-}}$ and $x^{\tilde{P},+} := \varphi^+_{\alpha^{\tilde{P},+}}$, and the grades $\alpha^{\tilde{P},-}$ and $\alpha^{\tilde{P},+}$ are given by $\varphi^-_{\alpha^{\tilde{P},-}} = \tilde{V}^{\tilde{P},-}_{\alpha^{\tilde{P},-}}(y,0)$ and $\varphi^-_{\alpha^{\tilde{P},+}} = \tilde{V}^{\tilde{P},+}_{\alpha^{\tilde{P},+}}(y,0)$.

Example 2. Consider a fuzzy goal

$$\varphi(x) = \begin{cases} 1 - e^{-2x}, & x \geq 0 \\ 0, & x < 0. \end{cases}$$

Then $\varphi_\alpha^- = -2^{-1}\log(1-\alpha)$ for $\alpha \in (0,1)$. Put an exercise time $T = 1$, a volatility $\sigma = 0.25$, an interest factor $r = 0.05$, a fuzzy factor $c = 0.05$, an initial stock price $y = 20$ and a strike price $K = 25$. From Theorem 5, we can easily calculate that the grades of the fuzzy expectation of the fuzzy price are $\alpha^{\tilde{C},-} \approx 0.767815$ and $\alpha^{\tilde{C},+} \approx 0.845076$. These grades means the degree of buyer's satisfaction in pricing. From (17) and (18), the corresponding permissible range of rational expected prices in European call option under his fuzzy goal φ is $[x^{\tilde{C},-}, x^{\tilde{C},+}] \approx [0.730111, 0.84642]$. Consider another fuzzy goal

$$\varphi(x) = \begin{cases} 1 - e^{-0.5x}, & x \geq 0 \\ 0, & x < 0, \end{cases}$$

and take the other parameters in the same as the above. Similarly, in European put option, we can easily calculate the grades, the degree of seller's satisfaction, and the corresponding permissible range of rational expected prices is as follows: $[x^{\tilde{P},-}, x^{\tilde{P},+}] \approx [4.4975, 4.64449]$. Buyer/seller should take into account of *the permissible range of rational expected prices* under their fuzzy goal φ.

6 Hedging Strategies

Finally, we deal with hedging strategies in European call option. Fix any $\alpha \in [0,1]$. A hedging strategy is an $\mathcal{M}_t$-predictable process $\{(\pi_t^{0,\pm}, \pi_t^{1,\pm})\}_{t\geq 0}$

with values in $\mathbb{R} \times \mathbb{R}$, where $\pi_t^{0,\pm}$ means the amount of the bond and $\pi_t^{1,\pm}$ means the amount of the stock at time t, and it satisfies

$$V_{t,\alpha}^{\pm} = \pi_t^{0,\pm} R_t + \pi_t^{1,\pm} \tilde{S}_{t,\alpha}^{\pm}, \quad t \geq 0, \tag{19}$$

where $V_{t,\alpha}^{\pm} := e^{rt} E(\tilde{C}_{T,\alpha}^{\pm} | \mathcal{M}_t)$ is called a wealth process. Then, we obtain the following results ([14]).

Theorem 6. *The minimal hedging strategy* $\{(\pi_t^{0,\pm}, \pi_t^{1,\pm})\}_{t \in [0,T]}$ *for the fuzzy price of European call option is given by*

$$\pi_t^{0,\pm} = \Phi(z_t^{0,\pm}) \quad \text{and} \quad \pi_t^{1,\pm} = -e^{-rT} K \Phi(z_t^{1,\pm})$$

for $t < T$, *where*

$$z_t^{0,\pm} = \log b^{\pm}(\alpha) + \frac{\log(S_t/K) + (T-t)(r + \sigma^2/2)}{\sigma\sqrt{T-t}},$$
$$z_t^{1,\pm} = \log b^{\pm}(\alpha) + \frac{\log(S_t/K) + (T-t)(r - \sigma^2/2)}{\sigma\sqrt{T-t}}.$$

The corresponding wealth process is (19).

7 Concluding Remarks

In this paper, the uncertainty is represented by both randomness and fuzziness. The fuzziness is evaluated by fuzzy expectation defined by a possibility measure from the viewpoint of decision-maker's subjective judgment. We can find other estimation methods instead of fuzzy expectation. For example, mean values by evaluation measures in Yoshida [18] are applicable to this model in the evaluation of fuzzy numbers with the decision maker's subjective judgment.

This paper takes theoretical approach to the option pricing theory under uncertainty. However, in a fuzzy environment, the other approaches from the viewpoint of the option pricing for actual stocks are discussed by Carlsson et al. [2] and Zmeškal [20].

References

1. Bellman RZ, Zadeh LA (1970) Decision-making in a fuzzy environment. Management Sci. Ser. B 17:141-164.
2. Carlsson C, Fullér R, Majlender P (2003) A fuzzy approach to real option valuation. Fuzzy Sets and Systems 139:297-312.
3. Choquet G (1955) Theory of capacities. Ann. Inst. Fourier 5:131-295.
4. Elliott RJ, Kopp PE (1999) Mathematics of financial markets. Springer, New York.

5. Karatzas I, Shreve SE (1998) Methods of mathematical finance. Springer, New York.
6. Klir GJ, Yuan B (1995) Fuzzy sets and fuzzy logic: theory and applications. Prentice-Hall, London.
7. Mesiar R (1995) Choquet-like integrals. J. Math. Anal. Appl. 194:477-488.
8. Pliska SR (1997) Introduction to mathematical finance: discrete-time models. Blackwell Publ., New York.
9. Puri ML, Ralescu DA (1986) Fuzzy random variables. J. Math. Anal. Appl. 114:409-422.
10. Ross SM (1999) An introduction to mathematical finance. Cambridge Univ. Press, Cambridge.
11. Sugeno M (1974) Theory of fuzzy integrals and its applications. Ph.D. thesis, Tokyo Institute of Technology.
12. Wang G, Zhang Y (1992) The theory of fuzzy stochastic processes. Fuzzy Sets and Systems 51:161-178.
13. Yoshida Y (1996) An optimal stopping problem in dynamic fuzzy systems with fuzzy rewards. Computers Math. Appl. 32:17-28.
14. Yoshida Y (2003) The valuation of European options in uncertain environment. Europ. J. Oper. Res. 145:221-229.
15. Yoshida Y (2003) A discrete-time model of American put option in an uncertain environment. Europ. J. Oper. Res. 151:153-166.
16. Yoshida Y (2003) Continuous-time fuzzy decision processes with discounted rewards. Fuzzy Sets and Systems 139:333-348.
17. Yoshida Y (2003) A discrete-time European options model under uncertainty in financial engineering. In: Tanino T, Tanaka T, Inuiguchi M (eds) Multi-objective programming and goal-programming. Springer, Heidelberg, pp.415-420.
18. Yoshida Y (2004) A mean estimation of fuzzy numbers by evaluation measures. In: Negoita MC, Howlett RJ, Jain LC (eds) Knowledge-based intelligent information and engineering systems, part II. Lecture Notes in Artificial Intelligence 3214, Springer, Heidelberg, pp.1222-1229.
19. Zadeh LA (1965) Fuzzy sets. Inform. and Control 8:338-353.
20. Zmeškal Z (2001) Application of the fuzzy-stochastic methodology to appraising the firm value as a European call option. European J. Oper. Res. 135:303-310.

Part III

Fuzzy Group and Multi-criteria Decision-making Techniques

A Multi-granular Linguistic Hierarchical Model to Evaluate the Quality of Web Site Services

F. Herrera[1], E. Herrera-Viedma[1], L. Martínez[2], L.G. Pérez[2], A.G. López-Herrera[1], and S. Alonso[1]

[1] Dept. of Computer Science and Artificial Intelligence. University of Granada, 18071 - Granada, Spain
herrera,viedma@decsai.ugr.es

[2] Dept. of Computer Science. University of Jaén, 23071 - Jaén, Spain
martin,lgonzaga@ujaen.es

Summary. The explosion in the use of Internet has contributed to arise a lot of web sites that offer many kind of services (products, information, etc). At the beginning the quality of these web sites was not too important because the most important fact was that people knew that there was a web site because there was not a big competence. But recently, there are many web sites related to the same topics in Internet and the quality of their services has become a critical factor. Different evaluation approaches for different types of web sites have been developing [2, 22, 35] in which the users provide their opinions in a predefined numerical scale to evaluate their services. Nevertheless, the information provided by users is related to their own perceptions. Usually, human perceptions are subjective and not objective, therefore to assess perceptions with precise information is not very suitable and the results are not accurate. Therefore, in this chapter we propose a linguistic quality evaluation model to evaluate the services offered by the web sites. The use of the fuzzy linguistic approach has provided good results managing human perceptions. Our proposal will consist of a hierarchical model to evaluate the services offered by general purpose web sites, such that, it will choose a few quality dimensions to be evaluated, where each one has different criteria. The users will provide their knowledge about these criteria by means of linguistic assessments. But different users can have different knowledge about the web site's criteria, so the evaluation model should take into account this point. Therefore, our model will be defined in a multi-granular linguistic information context, such that, different users can express their opinions in different linguistic term sets according to their knowledge. In order to develop this evaluation model we shall use different tools and resolution schemes based on decision techniques that are able to deal with multi-granular linguistic information.

Keywords: web quality, evaluation, linguistic variables, aggregation, decision analysis, multi-granular linguistic information, web services.

F. Herrera et al.: *A Multi-granular Linguistic Hierarchical Model to Evaluate the Quality of Web Site Services*, StudFuzz **201**, 247–274 (2006)
www.springerlink.com

1 Introduction

Nowadays, we can assert that the World Wide Web is the largest available repository of data and services with the largest number of visitors. The World Wide Web is a distributed, dynamic, and rapidly growing information resource that has stimulated new and useful developments in areas such as digital libraries, information retrieval, education, commerce, entertainment, government, and health care.

At the beginning, the quality of the web sites did not play a key role for the companies because the most important point was that people knew that the company was in Internet by means of its web site. But recently, due to the fact, there exist a lot of web sites competing in the same area in Internet, the quality of their services has become a critical factor for the competitiveness of the companies. In such a context, Web quality evaluation tools are necessary to filter web resources in order to avoid the bad information and services that users could receive from the web.

When we talk about the quality of a web site services, we want to show how well it meets the consumers necessities and so, it is associated with consumer satisfaction [27]. Companies have noticed that offering quality services has become an essential ingredient for successful competition and they need tools that allow them to evaluate the quality of their services.

One of the first points we must fix is the meaning of quality and satisfaction. Quality can be described as conformance to requirements, while satisfaction has been defined as conformance to expectation. The ideal situation would be that there were no difference between consumer judgement of quality and experienced satisfaction. But, in fact, it is very difficult to meet all the consumers' requirements.

Due to this increasing interest in the evaluation of the services offered by the web sites we can find in the literature different models applied to specific types of web sites [1, 2, 11, 22, 28, 29, 30, 35]. However, the evaluation of the quality offered by a web site is not an easy task because the aspects of the evaluated services are related to the users own perceptions, and usually, most of these perceptions are about subjective aspects. In spite of this fact, most of the evaluation models use precise numerical assessments that it is not very suitable so, sometimes try to hide them labelling the numbers with words or symbols without semantics [22]. Nevertheless, the final results are expressed by means of numbers, so this causes a lack of precision and effectiveness in the evaluation. To overcome these problems, we propose the use of the fuzzy linguistic approach [39] that has provided successful results to manage human perceptions in different topics, such as for example, "information retrieval" [4, 10, 19], "clinical diagnosis" [7], "marketing" [37], "risk in software development" [24], "technology transfer strategy selection" [5], "education" [23], "decision making" [8, 13, 34], etc.

In this chapter, we shall propose a linguistic hierarchical quality evaluation model for general purpose web sites based on decision analysis techniques that could be specialized for specific types of web sites (e-commerce, e-bank, etc, ...). This evaluation model is user centered because it characterizes the quality of the web sites services using judgements provided by different users that surf in those web sites. It takes into account that users are assessing subjective aspects that are related to the quality of the web site services and allow the people to use words (linguistic labels) instead of numbers in their assessments. In addition, we have realized that it should offer to each user the possibility to use a linguistic term set that allow him to express his evaluation values according to their knowledge about the problem or the evaluated aspects (multi-granular linguistic context). Finally, the results generated by our evaluation model will use the same expression domain used by its users and therefore, they will be easier understood by the users and/or companies. To accomplish our aims, the evaluation model will be based on decision analysis techniques and on fuzzy tools that have been used to deal with multi-granular linguistic information [12, 14, 16, 18].

Our proposal for the linguistic multi-granular hierarchical evaluation scheme has the following steps (graphically, Fig. 1):

1. *Evaluation framework:* this model will define an evaluation framework which will be composed by a few number of quality dimensions and their respective criteria that will be evaluated by the users to obtain an evaluation measure of the web site. These dimensions and criteria will be chosen according to their importance and will be chosen taking into account different points of view that we can find in the literature [2, 22, 35]. Once the dimensions and their criteria have been selected our model will gather the information provided by the users offering a multi-granular linguistic context.
2. *Evaluation process:* Once the evaluation framework has been defined, we propose a hierarchical evaluation process based on two steps:
   a) *Quality of each dimension:* the input assessments provided by the users are aggregated to obtain an evaluation assessment for each dimension. The input assessments could be expressed in different linguistic term sets with different granularity or semantics (multi-granular linguistic information). So we shall use fuzzy tools that have been used in the literature to manage and aggregate this type of information [12, 14, 16]
   b) *Global Quality Evaluation:* to compute a global quality measurement of the evaluated web site we shall calculate a satisfaction degree from the quality dimension values obtained in the before phase. This satisfaction degree will be computed by means of a weighting function that will depend on the web site we are evaluating.

This chapter is structured as follows: in the section 2 we revise some linguistic foundations we shall use in our evaluation process. In the section 3, we show in short several evaluation models for different types of web sites, in the section 4 we shall present our proposal for a multi-granular linguistic

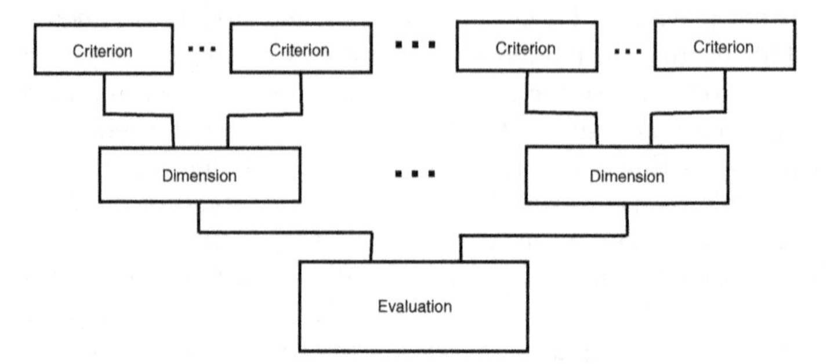

Fig. 1. Hierarchical evaluation model

hierarchical evaluation model for web site services and in the section 5 we shall show an example of this evaluation model. Eventually, some concluding remarks are pointed out.

2 Linguistic Background

In this section, we shall make a brief review of the fuzzy linguistic approach and of the fuzzy linguistic 2-tuple model. Because, we shall use them in the development of our evaluation process to manage human perceptions and multi-granular linguistic information.

2.1 Fuzzy Linguistic Approach

Usually, we work in a quantitative setting, where the information is expressed by means of numerical values. However, many aspects of different activities in the real world cannot be assessed in a quantitative form, but rather in a qualitative one, i.e., with vague or imprecise knowledge. In that case a better approach may be to use linguistic assessments instead of numerical values. The variables which participate in these problems are assessed by means of linguistic terms [39]. This approach is adequate in some situations, for example, when attempting to qualify phenomena related to human perception, we are often led to use words in natural language. This may arise for different reasons. There are some situations where the information may be unquantifiable due to its nature, and thus, it may be stated only in linguistic terms (e.g., when evaluating the "comfort" or "design" of a car, terms like "bad", "poor", "tolerable", "average", "good" can be used [25]. In other cases, precise quantitative information may not be stated because either it is not available or the cost of its computation is too high, then an "approximate value" may be tolerated (e.g., when evaluating the speed of a car, linguistic terms like "fast", "very fast", "slow" are used instead of numerical values). The linguistic approach is less precise than the numerical one, however some advantages may be found using it:

1. The linguistic description is easily understood by human beings even when the concepts are abstract or the context is changing.
2. Furthermore, it diminished the effects of noise since, as it is known the more refined assessment scale is, then more sensitive to noise and consequently the more error facedown it becomes.

In short, the linguistic approach is appropriated for many problems, since it allows a more direct and adequate representation when we are unable to express it with precision. Hence, the burden of qualifying a qualitative concept is eliminated.

The fuzzy linguistic approach represents qualitative aspects as linguistic values by means of linguistic variables:

Definition 1 [39].- *A linguistic variable is characterized by a quintuple (H,T(H),U,G,M) in which H is the name of the variable; T(H) (or simply T) denotes the term set of H, i.e., the set of names of linguistic values of H, with each value being a fuzzy variable denoted generically by X and ranging across a universe of discourse U which is associated with the base variable u; G is a syntactic rule (which usually takes the form of a grammar) for generating the names of values of H; and M is a semantic rule for associating its meaning with each H, M(X), which is a fuzzy subset of U.*

Usually, depending on the problem domain, an appropriate linguistic term set is chosen and used to describe the vague or imprecise knowledge. The number of elements in the term set will determine the granularity of the uncertainty, that is, the level of distinction among different counting of uncertainty. In [3] the use of term sets with an odd cardinal was studied, representing the mid term by an assessment of "approximately 0.5", with the rest of the terms being placed symmetrically around it and the limit of granularity being 11 or no more than 13.

One possibility of generating the linguistic term set consists of directly supplying the term set by considering all terms distributed on scale on which total order is defined [36]. For example, a set of seven terms S, could be given as follows:

$$S = \{s_0 : none, s_1 : verylow, s_2 : low, s_3 : medium, s_4 : high, s_5 : veryhigh, s_6 : perfect\}$$

Usually, in these cases, it is required that in the linguistic term set there exist:

1. A negation operator Neg(s_i) $= s_j$ such that $j = g$-i (g+1 is the cardinality).
2. A max operator: max$(s_i, s_j) = s_i \; if \; s_i \geq s_j$.
3. A min operator: min$(s_i, s_j) = s_i \; if \; s_i \leq s_j$

The semantics of the terms is given by fuzzy numbers. A computationally efficient way to characterize a fuzzy number is to use a representation based on parameters of its membership function [3]. The linguistic assessments given by the users are just approximate ones, some authors consider that linear trapezoidal membership functions are good enough to capture the vagueness of those linguistic assessments. The parametric representation is achieved by the 4-tuple (a, b, d, c), where b and d indicate the interval in which the membership value is 1, with a and c indicating the left and right limits of the definition domain of the trapezoidal membership function [3]. A particular case of this type of representation are the linguistic assessments whose membership functions are triangular, i.e., $b = d$, then we represent this type of membership functions by a 3-tuple (a, b, c). And example may be the following:

$$\begin{array}{ll} P = Perfect = (.83, 1, 1) & VH = Very\_High = (.67, .83, 1) \\ H = High = (.5, .67, .83) & M = Medium = (.33, .5, .67) \\ L = Low = (.17, .33, .5) & VL = Very\_Low = (0, .17, .33) \\ N = None = (0, 0, .17), & \end{array}$$

which is graphically shown in Fig. 2.

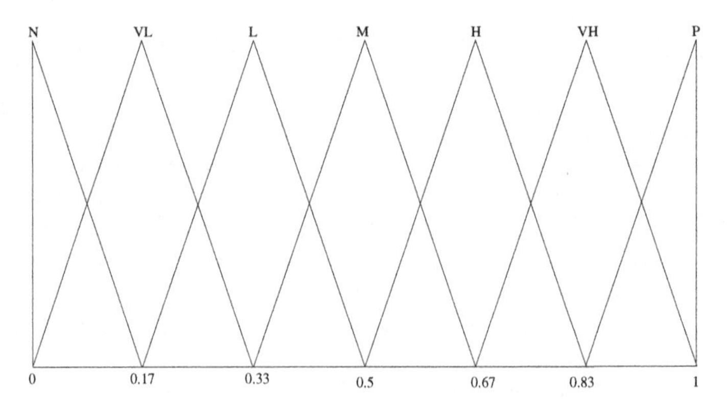

Fig. 2. A Set of Seven Terms with its Semantic

Other authors use a non-parametric representation, e.g., Gaussian functions [4].

The use of linguistic variables implies processes of computing with words such as their fusion, aggregation, comparison, etc. To perform these computations there are different models in the literature:

- *The linguistic computational model based on the Extension Principle*, which allow us to aggregate and compare linguistic terms through computations on the associated membership functions [7].
- *The symbolic method* [9]. This symbolic model makes direct computations on labels, using the ordinal structure of the linguistic term sets.
- *The 2-tuple fuzzy linguistic computational model* [14]. It uses the 2-tuple fuzzy linguistic representation model and its characteristics to make linguistic computations, obtaining as results linguistic 2-tuples. A linguistic

2-tuple is defined by a pair of values, where the first one is a linguistic label and the second one is a real number that represents the value of the symbolic translation.

In the following subsection we shall review the 2-tuple model due to the fact, that it will be the computational model we shall use in our model to deal with multi-granular linguistic information.

2.2 The 2-tuple Fuzzy Linguistic Model

This model has been presented in [14] and has shown itself as useful to deal with heterogeneous information [17, 18], such as the multi-granular linguistic information that we shall use in this paper.

This linguistic model takes as a basis the symbolic aggregation model [9] and in addition defines the concept of Symbolic Translation and uses it to represent the linguistic information by means of a pair of values called linguistic 2-tuple, (s, α), where s is a linguistic term and α is a numeric value representing the symbolic translation.

Definition 2. *Let β be the result of an aggregation of the indexes of a set of labels assessed in a linguistic term set $S = \{s_0, ..., s_g\}$, i.e., the result of a symbolic aggregation operation. $\beta \in [0, g]$, being $g+1$ the cardinality of S. Let $i = round(\beta)$ and $\alpha = \beta - i$ be two values, such that, $i \in [0, g]$ and $\alpha \in [-.5, .5)$ then α is called a Symbolic Translation.*

Graphically, it is represented in Fig. 3.

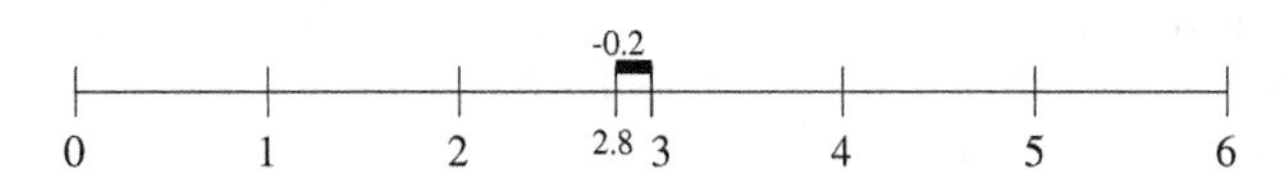

Fig. 3. Example of a Symbolic Translation

From this concept in [14] was developed a linguistic representation model which represents the linguistic information by means of 2-tuples (s_i, α_i), $s_i \in S$ and $\alpha_i \in [-.5, .5)$.

This model defines a set of transformation functions between linguistic terms and 2-tuples, and between numeric values and 2-tuples.

Definition 3.[14] *Let $S = \{s_0, ..., s_g\}$ be a linguistic term set and $\beta \in [0, g]$ a value supporting the result of a symbolic aggregation operation, then the 2-tuple that expresses the equivalent information to β is obtained with the following function:*

$$\Delta : [0, g] \longrightarrow S \times [-0.5, 0.5)$$

$$\Delta(\beta) = \begin{cases} s_i & i = round(\beta) \\ \alpha = \beta - i & \alpha \in [-.5, .5) \end{cases}$$

where round is the usual round *operation, s_i has the closest index label to "β" and "α" is the value of the symbolic translation.*

Proposition 1.[14] *Let $S = \{s_0, ..., s_g\}$ be a linguistic term set and (s_i, α) be a 2-tuple. There is a Δ^{-1} function, such that, from a 2-tuple it returns its equivalent numerical value $\beta \in [0, g] \subset \mathcal{R}$.*

Proof.

It is trivial, we consider the following function:

$$\Delta^{-1} : S \times [-.5, .5) \longrightarrow [0, g]$$
$$\Delta^{-1}(s_i, \alpha) = i + \alpha = \beta$$

Remark 1: From definitions 1 and 2 and from proposition 1, it is obvious that the conversion of a linguistic term into a linguistic 2-tuple consist of adding a value 0 as symbolic translation:

$$s_i \in S \Longrightarrow (s_i, 0)$$

This representation model has associated a computational model that was presented in [14]:

1. **Aggregation of 2-tuples:** The aggregation of linguistic 2-tuples consist of obtaining a value that summarizes a set of values, therefore, the result of the aggregation of a set of 2-tuples must be a linguistic 2-tuple. In [14] we can find several 2-tuple aggregation operators based on classical ones. Here we review the 2-tuple arithmetic mean and the 2-tuple weighted average operators, because we shall use them in our evaluation model:
   Definition 4:*Let $x = \{(r_1, \alpha_1), \cdots, (r_n, \alpha_n)\}$ be a set of 2-tuples, the extended Arithmetic Mean AM^* using the linguistic 2-tuples is computed as,*

$$AM^*\left((r_1, \alpha_1), \ldots, (r_n, \alpha_n)\right) = \Delta\left(\sum_{i=1}^{n} \frac{1}{n} \Delta^{-1}(r_i, \alpha_i)\right) = \Delta\left(\frac{1}{n} \sum_{i=1}^{n} \beta_i\right)$$

   Definition 5:*Let $x = \{(r_1, \alpha_1), \cdots, (r_n, \alpha_n)\}$ be a set of 2-tuples and $W = \{w_1, \cdots, w_n\}$ his associated weights. The 2-tuples weighted mean, W_AM^*, is computed as:*

$$W\_AM^*\left((r_1, \alpha_1), \ldots, (r_n, \alpha_n)\right) = \Delta\left(\frac{\sum_{i=1}^{n} \Delta^{-1}(r_i, \alpha_i) \cdot w_i}{\sum_{i=1}^{n} w_i}\right) =$$
$$= \Delta\left(\frac{\sum_{i=1}^{n} \beta_i \cdot w_i}{\sum_{i=1}^{n} w_i}\right)$$

   More linguistic 2-tuple aggregation operators were defined in [14].
2. **Comparison of 2-tuples:** The comparison of information represented by 2-tuples is carried out according to an ordinary lexico-graphic order.

- if $k < l$ then (s_k, α_1) is smaller than (s_l, α_2)
- if $k = l$ then
  a) if $\alpha_1 = \alpha_2$ then (s_k, α_1), (s_l, α_2) represents the same information
  b) if $\alpha_1 < \alpha_2$ then (s_k, α_1) is smaller than (s_l, α_2)
  c) if $\alpha_1 > \alpha_2$ then (s_k, α_1) is bigger than (s_l, α_2)

3. **Negation Operator of a 2-tuple:** The negation operator over 2-tuples is defined as:
$$Neg\,(s_i, \alpha) = \Delta\left(g - \Delta^{-1}\,(s_i, \alpha)\right)$$
where $g + 1$ is the cardinality of S, $s_i \in S = \{s_0, \ldots, s_g\}$.

3 An Overview of Quality Evaluation Models for Web Services

In this section we shall make a short review about different quality evaluation models applied to evaluate web site services, that we can find in the literature [2, 33, 35]. Our aim is to show their working and their problems to deal with human perceptions, in order to use several concepts of these models and to overcome their limitations in our proposal.

3.1 Evaluation of Web-based Decision Support Systems

Web-based decision support systems are being employed by organizations as decision aids for employees as well as customers. A common usage of web-based DSS (decision support systems) has been to assist customers configure product and service according to their needs. These systems allow each customer to design their own product by choosing from a group of alternatives. Some examples of web-based DSS can be found in www.dell.com, www.ibm.com, www.landsend.com and www.vermontteddybears.com.

In [2] we can see a study about the quality of the services (user satisfaction) of web-based DSS. In this model, the quality of the service is determined by three dimensions where each **dimension** has several *criteria* to evaluate the web-based DSS:

- **System quality**, whose criteria are:
  - *System reliability, Convenient to access, System easy of use, System flexibility.*
- **Information quality**:
  - *Information accuracy, Information completeness, Information relevance, Information content needs, Information timeless.*
- **Information presentation**:
  - *Presentation graphics, Presentation colour, Presentation style, Navigationally efficient.*

The process for reaching a solution in their evaluation problem is as follows:

1. *Data collection:* users provide their opinions about the criteria using numerical scale, in spite of most of them are qualitative aspects.
2. *Evaluation process:* The opinions provided by the users are combined in order to evaluate each dimension and finally a global quality measurement of the web-based DSS is obtained. All the results are expressed by means of numerical values

The results are analyzed using the structural equation model (SEM) approach to study the correlation among the dimensions and the quality, and among the criteria and the quality.

3.2 The Extended Web Evaluation Model (EWAM)

The EWAM (Extended Web Evaluation Model) [33] defines an evaluation grid with a set of criteria for appraising the quality and success of existing e-commerce applications (e.g. assessed some Australian grocery web sites in [22]). The EWAM examines the three classic transaction phases of electronic markets, which include information, agreement, and settlement phases. A fourth element, the community component, is integrated as a link between the actual purchase transaction and the necessary trust relationship in the virtual realm.

In [33] was presented a study based on the EWAM and established the questions about quality, satisfaction or success of a e-commerce application must be allotted to one of the four transaction phases of electronic markets (information, agreement, settlement, and after-sale), to the community component, or to the category "Final Section" which concerns all phases. For instance, in [22] the *criteria* that they used for each **dimension** were:

- **Information phase:** whose criteria are:
  - *Accessibility of the web site, Structure of the contents, Quantity of information, Quality of the content, Passing on price benefits.*
- **Agreement phase:**
  - *Design of the ordering procedure, Models and methods of pricing.*
- **Settlement phase:**
  - *Integration of generic services, Tracking and tracing.*
- **After-sale phase:**
  - *Access to customer support, Performance of customer support.*
- **Community component:**
  - *Sharing opinions.*
- **Final section:**
  - *Availability of the system, The design of the user interface, Increasing productivity by gaining time, The trustworthiness of the web site.*

The dimensions in [33] are formulated in general terms and are valid in every sector but are differentiated by their importance ratings. In order to take into account the differences between the individual sectors, assessor(s) provides weights corresponding to the different sector profiles and their relevance in the sector.

The evaluation of an e-commerce web site with EWAM begins by assigning the concerned web site to a sector. Then, there are two steps involved in the evaluation:

1. *Subjective importance of every dimension:* The assessor(s) declares the subjective importance of a dimension. This importance is recorded on a scale of "unimportant" (-2), "less important" (-1), "important" (+1) and "very important" (+2).
2. *Evaluation of all web site in the concerned sector:* In [22] they used a rating based on a five-point scale: from -2 (very bad) to +2 (very good) to evaluate every question of each criterion. The evaluation process has the following steps:
   a) *Data collection:* EWAM gathers the input information from each user about the satisfaction on each criterion. In EWAM we use a numerical scale although the method offers labels to assess the criteria.
   b) *Evaluation process:* The input information is combined by an aggregation operator and we obtain a global assessment for each dimension. And after according to the importance provided by the assessor(s) to each dimension a global evaluation value is obtained. All the results are expressed by means of numerical values, although seemingly the users provides linguistic information but in fact they are providing numerical values.

3.3 The Servqual Scale Adapted to Electronic Services

In [35] is proposed an "e-satisfaction" model that evaluate the users satisfaction on quality services of information search and purchase web sites based on the Servqual scale [31]. This scale identified six dimensions representing Internet information search satisfaction (reliability, convenience, entertainment, assurance, site design, virtual environment) and three dimensions representing satisfaction with Internet purchase experience (security,product offer,convenience). Each **dimension** is defined by a group of *criteria* that describe the e-satisfaction of the users with that dimension. In [35] are used the following dimensions and criteria:

- For Internet information search satisfaction:
  - **Information reliability,** whose criteria are:
    - *Up-to-date information, Information depth, Search result, Uncluttered web pages, Easy search paths, Easiness in comparing information.*
  - **Convenience:**
    - *Economy of time spent, Effort spent, Easy access, Fast information transmission, Interaction capacity.*

- **Entertainment:**
  - *Interesting places to visit, Pleasant browsing, Entertainment and leisure, Easy browsing, Information diversity.*
- **Assurance:**
  - *Data transmission assurance and Privacy.*
- **Site design:**
  - *Advertising contents and Attractive presentation.*
- **Virtual Environment:**
  - *Capacity of simulating reality and Personal contact absence.*

- For Internet purchase experience satisfaction:
  - **Security:**
    - *Payment security, Trust in supplier, Privacy of purchase, Personal-sales absence, Pleasant way of buying.*
  - **Product Offer:**
    - *Easy to compare products' characteristics, Diversity of products' brands, Product guarantee, Price reduced products, Possibility to return.*
  - **Convenience:**
    - *Fast delivery and Easy way of buying.*

Values on satisfaction are interpreted directly as performance measures (un-weighted) and they evaluate each criterion using a five-point scale (5=High satisfaction,...,1=Low Satisfaction).

The evaluation process is composed by the following phases:

1. *Data collection:* the users provide their opinions about the criteria using numerical values. Each criterion using a five-point scale (5=High satisfaction,...,1=Low Satisfaction).
2. *Evaluation process:* the opinions provide by the users are combined to obtain an evaluation assessment for each dimension. Finally, in this case a global evaluation assessment is obtained using an un-weighted aggregation operator. These results are expressed by means of numerical values.

3.4 Current Web Evaluation Methods: Problems and Working

Reviewing the before web evaluation models we have realized that these models are user centered. Due to the fact that the evaluation of the web site depends on the opinions provided by the users that use the web site.

To evaluate the web site services all the models choose a set of dimensions with several outstanding criteria that have to be assessed by the users according to their perception.

These models present different problems [20], such as, they try to avoid explicit numerical values, because this type of information is not suitable to assess human perceptions. Hence, they try to hide the numbers behind a scale of labels or symbols, but in fact they are using numbers and in addition all the users are forced to use the same scale for all the criteria, despite they can have different knowledge about the criteria or about an specific criterion this causes a lack of expressiveness that means a loss of information and accuracy in the results of the evaluation process. Besides, the final results are expressed by means of numerical values that are far from the user expression domain so sometimes the results are difficult to understand by the users and then a feedback cycle to improve the web site is almost impossible.

We shall propose an linguistic evaluation model to solve the problem of lack of expressiveness using multi-granular linguistic context, in which each user can provide their opinions in a linguistic term set according to his knowledge. In addition, this model will deal with linguistic information and the results will be expressed by means of linguistic values.

This model will have an hierarchical structure to evaluate separately each dimension and from these evaluations to obtain a global evaluation assessment for the web site.

4 A Linguistic Multi-criteria Hierarchical Evaluation Model for Web Sites Services

The aim of this paper is to develop an user centered hierarchical evaluation model for web sites services, in which, the users can express their opinions about the web sites by means of linguistic terms that can be assessed in different linguistic term sets.

In short, we can define mathematically our problem as an evaluation process in which a set of users (experts) $E = \{e_1, \cdots, e_n\}$ will evaluate a web site, W, providing their opinions about a set of quality dimensions, $D = \{d_1, \cdots, d_q\}$, such that each dimension, d_i, has a set of criteria, $C_i = \{c_{1i}, \cdots, c_{ti}\}$, to be evaluated . Therefore, each time a web site is evaluated every expert, e_k, provides his assessments about the different criteria by means of an utility vector:

$$e_k \rightarrow \{u^k_{11}, \ldots, u^k_{t1} \ldots u^k_{1q}, \ldots, u^k_{tq}\},\ u^k_{tq} \in S^k$$

Where u^k_{tq} is the linguistic assessment provided by the expert e_k for the criterion c_{tq} that belongs to the dimension d_q . Due to the fact that our model is defined in a multi-granularity linguistic context, we assume that each user may use a different linguistic term set for each criterion to evaluate the web site services according to their knowledge about the problem. Therefore, each user, e_k, can express his opinions for criterion, c_{tq}, in a linguistic term set $S^k_{tq} = \left\{s_0^{k_{tq}}, \ldots, s_l^{k_{tq}}\right\}$ where $l+1$ is the cardinality of S^k_{tq}.

In the following subsections, we shall present the evaluation framework we shall use to evaluate the web site services, after we shall present our multi-granular linguistic hierarchical evaluation model.

4.1 Evaluation Framework

To develop an evaluation model, first of all we have fix which is the evaluation framework we shall use to evaluate the web sites services.

In the section 3, we have reviewed different evaluation models for different web-based services. All of them chose a few dimensions to evaluate the quality of the web site. Each dimension was evaluated using different criteria, so the dimensions as the criteria depended on the topic related to the evaluated web site. Nevertheless our aim is to define an evaluation model for general purpose web sites, so taking into account the models presented in [1, 2, 6, 21, 22, 26, 33, 35, 38] we have to choose the dimensions and their criteria that our evaluation model will use to define the evaluation framework of our problem.

Our general purpose evaluation model shall use the following **dimensions** and *criteria*:

- **Entertainment**: this criterion is related to amusement and pleasure contents. The criteria we use to assess this dimension are:
  - *Interesting place to visit.*
  - *Pleasant browsing.*
  - *Entertainment and leisure*
  - *Easy browsing.*
  - *Information diversity.*
- **Convenience**: it is related to speed, easy access to information at low cost. The criteria are:
  - *Economy of time spent.*
  - *Effort spent.*
  - *Easy access.*
  - *Fast information transmission.*
  - *Interaction capacity.*
  - *Fast delivery*
  - *Easy way of buying.*
- **Information reliability**: it is about reliability of the information. Information reliability is related with diversity, depth and actuality of information contents. And its criteria are:
  - *Up-to-date information.*
  - *Information depth.*
  - *Search result.*
  - *Uncluttered web pages.*
  - *Easy search paths.*
  - *Easiness in comparing information.*
- **Security and assurance**: security perception and privacy "assurance" are known to have a big impact in user satisfaction of quality web services. Its criteria are:
  - *Payment security.*
  - *Trust in supplier.*

  - *Privacy of purchase.*
  - *Data transmission assurance.*
  - *Privacy.*
- **Site Design**: it depends on functional and attractive elements: ease of browsing, a standard language use, interface design. Its criteria are:
  - *Advertising contents.*
  - *Attractive presentation.*
- **Virtual Environment**: it is used to minimize the absence of human contact and amusement associated to shopping. Its criteria are:
  - *Capacity of simulating reality.*
  - *Personal contact absence.*
  - *Personal-sales absence.*
- **Product Offer**: it is concerning product diversity and available brands. And its criteria are:
  - *Easy to compare products' characteristics*
  - *Diversity of product's brands*
  - *Product guarantee.*
  - *Price reduced products.*
  - *Possibility to return.*

Once the dimensions and criteria to evaluate the web site services have been chosen, the assessments provided by the experts can be expressed in different linguistic term sets S_{tq}^{k} according to their knowledge and the characteristics of the criterion.

This evaluation framework offers several advantages with regards to the evaluation models reviewed in the Section 3:

- The dimensions and criteria used in this framework are general enough to evaluate any kind of web site. Due to the fact we want to develop a general purpose evaluation model for web site services.
- This framework models the information provided by the users by means of linguistic information according to the fuzzy linguistic approach that it is more suitable than numerical values to model human perceptions. In addition, it offers to the users the possibility of using different linguistic term sets to provide their information and in this way avoid lack of expressiveness, lose of information, and probably bad results.

4.2 Evaluation Process: Evaluating the Quality of the Web Services

Once we know the evaluation framework we are going to use in our problem, we are going to present the hierarchical process we propose to evaluate the web site services.

Our proposal to evaluate the web sites services consist of a hierarchical process with the following phases (graphically, Fig. 4) :

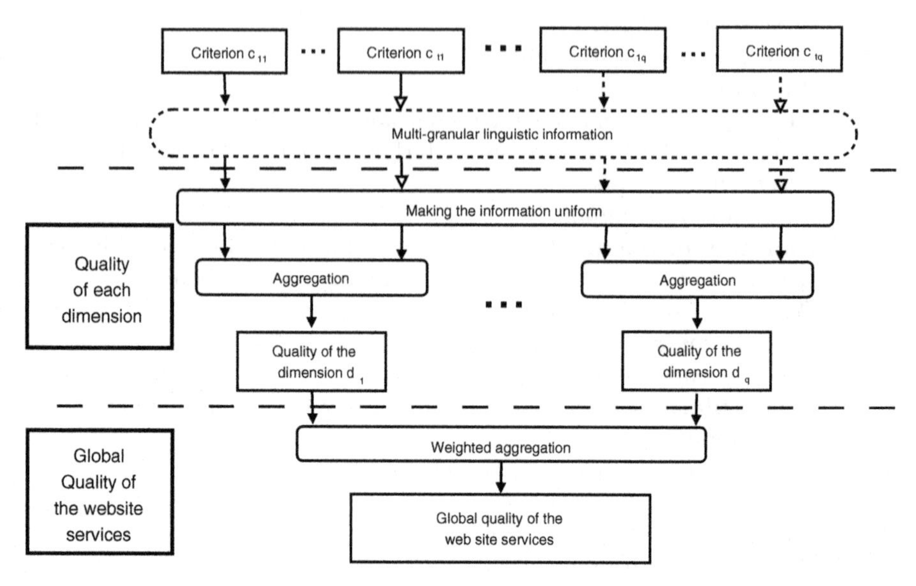

Fig. 4. Evaluation process

1. **Quality of each dimension**: in this phase we want to obtain an evaluation for each dimension of our evaluation framework. To do so, we have to aggregate the criteria belong to the dimension, but the difficulty comes from the multi-granular linguistic context in which it has been defined our framework because there are not standard aggregation operators for this type of information. Therefore this aggregation process will consist of the following steps:
   a) *Making the information uniform:* The input information provided by the users could be expressed in different linguistic term sets with different granularity or semantics (multi-granular linguistic information).Therefore to combine the input assessments, we need to unify the input information, that it is multi-granular, into an unique expression domain. We shall unify this information in a Basic Linguistic Term Set (BLTS) by means of fuzzy sets and afterwards, we shall transform these unified input information into linguistic 2-tuples expressed in the BLTS.
   b) *Aggregation phase:* it combines the unified input assessments provided by the users to obtain a collective value for each dimension.
2. **Global Quality of the web site services**: we want to obtain a global evaluation of the web site services. To do so, we shall aggregate the evaluation assessments obtained for each dimension of quality. In this phase, the aggregation will be carried out by means of a weighting aggregation operator, where the weights assigned to each dimension will depend on the evaluated web site. And with these weights we can annul one or several

dimensions in certain type of web sites. So we can use the same framework for general purpose web sites.

In the next subsections we present each phase of the evaluation model in further detail.

Quality of Each Dimension

We want to obtain a collective assessment on a dimension according to the individual opinions provided by the users regarding the different criteria (assessed in multi-granularity linguistic term sets). We shall aggregate the information according to the following steps:

1. Making the information uniform by means of fuzzy sets.
2. Transforming into 2-tuple.
3. Calculating an evaluation assessment for the dimension.

Now we present the working of each step in detail.

1. Making the Information Uniform

With a view to manage the information we must make it uniform, i.e., the multi-granular linguistic information provided by the users must be transformed into a unified linguistic term set, called BLTS and denoted as S_T.

Before defining a transformation function to unify the multi-granular linguistic information into this BLTS, S_T, we have to decide how to choose S_T. We consider that S_T must be a linguistic term set which allows to express a quality scale easy to understand and maintain the uncertainty degree associated to each expert and the ability of discrimination to express the performance values. So in our case, we propose the following linguistic term set as, BLTS:

$$S_T = \{N, VL, L, M, H, VH, P\},$$

whose semantics has been shown in the Fig. 2.

We shall unify the multi-granular linguistic information by means of fuzzy sets in the BLTS. The process of unifying the information involves the comparison between fuzzy sets. These comparisons are usually carried out by means of a measure of comparison. We focus in measures of comparison which evaluate the resemblance or likeness of two objects (fuzzy sets in our case) [32]. For simplicity, in this paper we shall choose a measure based on a possibility function $S(A, B) = \max_x \min(\mu_a(x), \mu_B(x))$, where μ_A and μ_B are the membership function of the fuzzy sets A and B respectively.

The next step in this process of unifying the information is to define a transformation function that we allow us to express the input information in the BLTS. We shall define a transformation function that will unify the input linguistic multi-granular information by means of fuzzy sets in the BLTS:

Definition 6. *Let* $S = \{l_0, \ldots, l_p\}$ *and* $S_T = \{s_0, \ldots, s_g\}$ *be two linguistic term sets. Then, a linguistic transformation function,* τ_{SS_T}*, is defined as:*

$$\tau_{SS_T} : S \rightarrow F(S_T)$$
$$\tau_{SS_T}(l_i) = \{(s_k, \gamma_k^i) \,/\, k \in \{0, ..., g\}\}, \; \forall l_i \in S$$
$$\gamma_k^i = \max_y \min\{\mu_{l_i}(y), \mu_{s_k}(y)\}$$

where $F(S_T)$ *is the set of fuzzy sets defined in* S_T*, and* $\mu_{l_i}(\cdot)$ *and* $\mu_{s_k}(\cdot)$ *are the membership functions of the fuzzy sets associated with the terms* l_i *and* s_k*, respectively.*

The result of τ_{SS_T} for any linguistic value of S is a fuzzy set defined in the BLTS, S_T. Therefore, after unifying the input information with this transformation function the opinions provided by the experts are expressed by means of fuzzy sets in the BLTS.

Remark 2: In the case that the linguistic term set, S, of the non-homogeneous contexts let be chosen as BLTS, then the fuzzy set that represents a linguistic term will be all **0** except the value correspondent to the ordinal of the linguistic label that will be **1**.

Example. Let $S = \{l_0, l_1, \ldots, l_4\}$ and $S_T = \{s_0, s_1, \ldots, s_6\}$ be two term set, with 5 and 7 labels, respectively, and with the following semantics associated:

$$\begin{array}{ll} l_0 = (0, 0, 0.25) & s_0 = (0, 0, 0.16) \\ l_1 = (0, 0.25, 0.5) & s_1 = (0, 0.16, 0.34) \\ l_2 = (0.25, 0.5, 0.75) & s_2 = (0.16, 0.34, 0.5) \\ l_3 = (0.5, 0.75, 1) & s_3 = (0.34, 0.5, 0.66) \\ l_4 = (0.75, 1, 1) & s_4 = (0.5, 0.66, 0.84) \\ & s_5 = (0.66, 0.84, 1) \\ & s_6 = (0.84, 1, 1) \end{array}$$

The fuzzy set obtained after applying τ_{SS_T} for l_1 is (see Fig. 5):

$$\tau_{SS_T}(l_1) = \{(s_0, 0.39), (s_1, , 0.85), (s_2, 0.85), (s_3, 0.39) \\ (s_4, 0), (s_5, 0), (s_6, 0)\}$$

2. Transforming into 2-tuple

In this phase, we transform the unified information expressed by means of fuzzy sets on the BLTS into linguistic 2-tuples in the BLTS, to facilitate the computation of the satisfaction assessment. In [15] was presented a function χ that transformed a fuzzy set into a numerical value and using Δ we transformed this numerical value in a 2-tuple. In this contribution we have redefined χ in a way that transforms directly a fuzzy set in $F(S_T)$ into a 2-tuple:

$$\chi : F(S_T) \rightarrow S_T x [-0.5, 0.5)$$

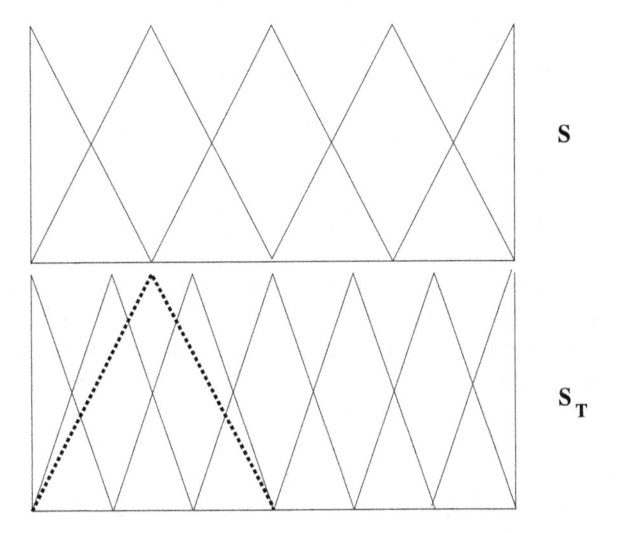

Fig. 5. Transforming $l_1 \in S$ into a Fuzzy Set in S_T

$$\chi\left(F\left(S_T\right)\right)=\chi\left(\left\{\left(s_j,\gamma_j\right),j=0,\ldots,g\right\}\right)=\Delta\left(\frac{\sum_{j=0}^{g} j\gamma_j}{\Sigma_{j=0}^{g}\gamma_j}\right)=$$
$$=\Delta\left(\beta\right)=\left(s,\alpha\right)$$

After applying χ to the fuzzy sets in the BLTS obtained in the before step, we shall obtain the linguistic 2-tuples in the BLTS that express the opinions provided by the users.

Example.

We want to transform the fuzzy set $(0,0,0,.41,1,.19,0)$ in the BLTS to a 2-tuple (Fig. 2):

$$\chi\left((0,0,0,.41,1,.19,0)\right)=\Delta\left(\frac{\sum_{j=0}^{6} j\gamma_j}{\Sigma_{j=0}^{6}\gamma_j}\right)=$$
$$\Delta\left(4.33\right)=\left(H,.33\right)$$

and the result is $(H,.33)$.

3. Calculating an evaluation assessment for each dimension

Our objective is to obtain an evaluation value for each dimension according to the opinions provided by all the users for its criteria. At this moment, these values are expressed by means of linguistic 2-tuples in the BLTS.

Therefore, to reach our objective we follow the next steps:

1. Computing collective values for each criterion: Each dimension, d_i, has a set of criteria, $C_i=\{c_{1i},\cdots,c_{ti}\}$, so first of all we shall compute a collective value for each criterion according to all the users. In this proposal

we shall use a non-weighted aggregation operator as the arithmetic mean for 2-tuples (Definition 4), but weighted operator could considered in the future.

Therefore, the collective value for the criterion (CVC), $c_{ti} \in C_i$, will be computed as:

$$CVC_{ti} = AM^*((u_{ti}^k, \alpha)\,,\; k = 1...n) = (u_{ti}, \alpha)$$

2. Computing an evaluation assessment for each dimension: So far, we have computed a collective value for every criterion that belongs to d_i. Now, we want to obtain an evaluation assessment for each dimension. To do so, we shall aggregate the collective values of its criteria by means of an aggregation operator. As well as before, we shall use a non-weighted aggregation operator as the arithmetic mean for 2-tuples, although could be considered a weighted operator in the future.
   Therefore, to obtain an evaluation assessment for a dimension (ED), d_i, will be computed as:

$$ED_i = AM^*((u_{ji}, \alpha)\,,\; j = 1...t) = (u_i, \alpha)$$

So now, we have an evaluation assessment for each dimension of quality, d_i, of our evaluation model. And we can evaluate separately each dimension to improve just certain drawbacks of our services. In order to improve the global evaluation of our web site services.

Global Quality of the Web Site Services

Our final aim it is to obtain a global evaluation assessment, EAW, for the web sites services we are evaluating. To do so, we shall aggregate the quality assessments obtained for each dimension. In this case we shall use a weighted aggregation operator, because although our model is to evaluate general purpose web site, we think that depends on the specific web site different dimensions could have different importance, even some of them their value can be null in some occasions. So, to obtain the evaluation of the web site services some expert/s provides a weighting vector that indicates the importance of each dimension, $W = \{w_1, ..., w_q\}$. Afterwards we shall apply the 2-tuple linguistic weighting average operator (Definition 5):

$$EAW = W\_AM^*((u_i, \alpha)\,, i = 1, ..., q) = (u, \alpha)$$

We have obtained a global linguistic evaluation for the quality of the web site services that is expressed in the BLTS (linguistic evaluation scale).

In the next section we shall apply this evaluation model to evaluate a lecturer's web site, in order to know the satisfaction of their students regarding the services offered by the web site.

5 Application: Evaluating a Lecturer's Web site

Let us suppose that we want to evaluate Lecturer's web site. We have four students that will provide us their opinions about the web site services. Every student could use a different linguistic term set for each criterion, but, to simplify the problem and show the resolution process easily we shall assume that every student choose a linguistic term set to evaluate all the criteria according to his/her knowledge:

- The first user has chosen the linguistic term set A of 3 labels (Fig. 6).
- The second user has chosen the linguistic term set B of 5 labels (Fig. 7).
- The third user has chosen the linguistic term set C of 7 labels (Fig. 8).
- The fourth user has chosen the linguistic term set D of 9 labels (Fig. 9).

In the table 1 we can see the semantics of the linguistic term sets.

Table 1. Semantic of the linguistic term sets A, B, C and D

Lingustic term set A	Linguistic term set B	Linguistic term set C	Linguistic term set D
$a_0 = (0, 0, 0.5)$	$b_0 = (0, 0, 0.25)$	$c_0 = (0, 0, 0.16)$	$d_0 = (0, 0, 0.12)$
$a_1 = (0, 0.5, 1)$	$b_1 = (0, 0.25, 0.5)$	$c_1 = (0, 0.16, 0.34)$	$d_1 = (0, 0.12, 0.25)$
$a_2 = (0.5, 1, 1)$	$b_2 = (0.25, 0.5, 0.75)$	$c_2 = (0.16, 0.34, 0.5)$	$d_2 = (0.12, 0.25, 0.37)$
	$b_3 = (0.5, 0.75, 1)$	$c_3 = (0.34, 0.5, 0.66)$	$d_3 = (0.25, 0.37, 0.5)$
	$b_4 = (0.75, 1, 1)$	$c_4 = (0.5, 0.66, 0.84)$	$d_4 = (0.37, 0.5, 0.62)$
		$c_5 = (0.66, 0.84, 1)$	$d_5 = (0.5, 0.62, 0.75)$
		$c_6 = (0.84, 1, 1)$	$d_6 = (0.62, 0.75, 0.87)$
			$d_7 = (0.75, 0.87, 1)$
			$d_8 = (0.87, 1, 1)$

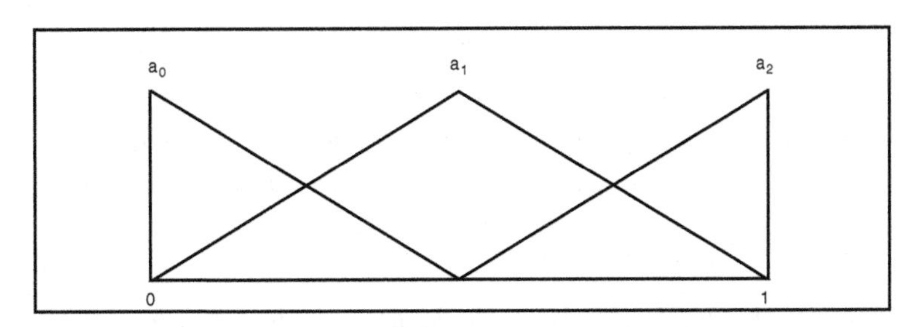

Fig. 6. Semantic of the lingustic term set A

The opinions provided by the students for the dimensions and criteria proposed in the section 4.1 are:

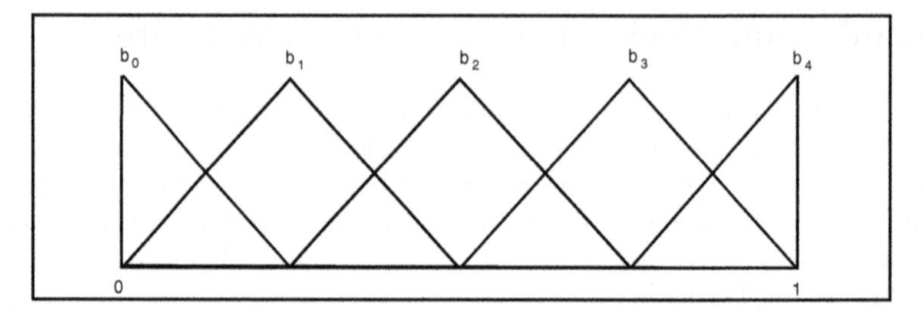

Fig. 7. Semantic of the linguistic term set B

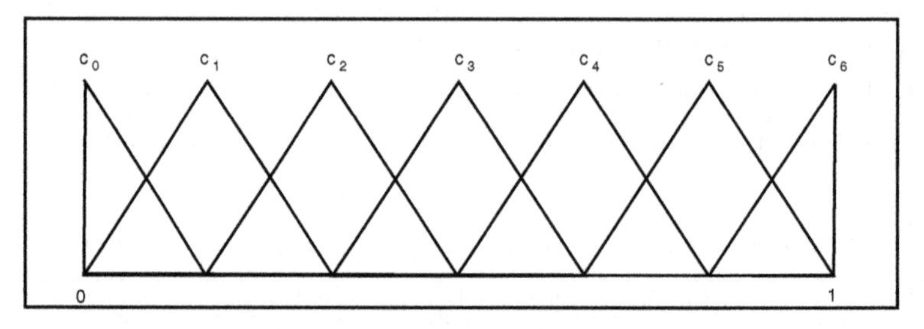

Fig. 8. Semantic of the linguistic term set C

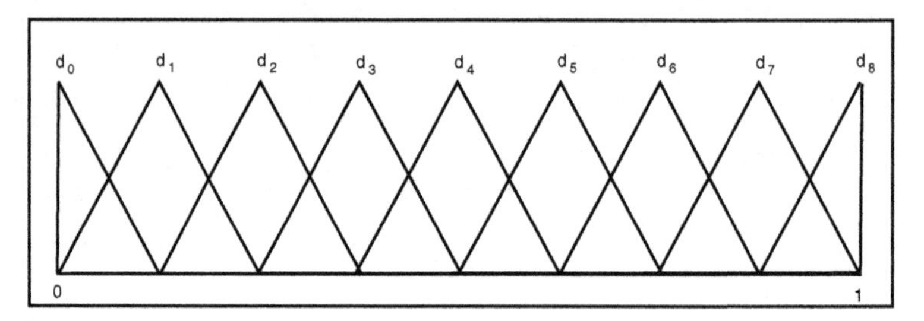

Fig. 9. Semantic of the linguistic term set D

$$e_1 = \{\mathbf{a_2}, \mathbf{a_1}, \mathbf{a_0}, \mathbf{a_2}, \mathbf{a_2}, a_0, a_1, a_1, a_1, a_2, a_0, a_1, a_0, a_0, a_2, a_2, a_2, a_2, a_0, a_0, a_0, a_0, a_0, a_1, a_2, a_2, a_1, a_2, a_0, a_0, a_0, a_0, a_0\}$$

$$e_2 = \{\mathbf{b_1}, \mathbf{b_4}, \mathbf{b_0}, \mathbf{b_0}, \mathbf{b_0}, b_2, b_3, b_4, b_2, b_2, b_3, b_0, b_4, b_0, b_4, b_1, b_4, b_1, b_0, b_0, b_0, b_0, b_0, b_1, b_3, b_3, b_1, b_2, b_0, b_0, b_0, b_0, b_0\}$$

$$e_3 = \{\mathbf{c_4}, \mathbf{c_0}, \mathbf{c_5}, \mathbf{c_1}, \mathbf{c_6}, c_0, c_6, c_2, c_1, c_3, c_2, c_2, c_1, c_1, c_3, c_5, c_1, c_4, c_0, c_0, c_0, c_0, c_0, c_0, c_3, c_3, c_1, c_6, c_0, c_0, c_0, c_0, c_0\}$$

$$e_4 = \{\mathbf{d_7}, \mathbf{d_7}, \mathbf{d_1}, \mathbf{d_2}, \mathbf{d_8}, d_4, d_3, d_7, d_2, d_2, d_0, d_3, d_7, d_4, d_6, d_5, d_0, d_1, d_0, d_0, d_0, d_0, d_0, d_4, d_2, d_3, d_0, d_3, d_0, d_0, d_0, d_0, d_0\}$$

We shall resolve this evaluation problem but for a better comprehension of the evaluation process we just show the main operations over the dimension Entertainment (the first five assessments of every student that are in bold).

Once we have obtained the opinions from the students we have to apply the evaluation model to obtain: firstly, the quality of each dimension and afterwards we shall compute the global quality of the web site.

Quality of each dimension

We want to remark that in this case the BLTS we have chosen for our model is the linguistic tern set C of the multi-granular linguistic context used in this example.

To obtain the quality value for each dimension we shall apply the process presented in the section 4.2:

1. Make the information uniform: We unify the input information by means of fuzzy sets in the BLTS and afterwards, we shall transform these unified information into linguistic 2-tuples expressed in the BLTS.
   a) Transforming into the BLTS: We use the functions τ_{AC}, τ_{BC}, τ_{CC} and τ_{DC} (definition 6). The results obtained for the criteria belonging to the dimension entertainment are:

$$e_1 = \{(\mathbf{0,0,0,0.25,0.5,0.5,0}), (\mathbf{0.25,0.5,0.5,0,0.5,0.5,0.25}), (\mathbf{0,0.5,0.5,0.25,0,0,0}), (\mathbf{0,0,0,0.25,0.5,0.5,0}), (\mathbf{0,0,0,0.25,0.5,0.5,0}), \ldots\}$$

$$e_2 = \{(\mathbf{0.4,0.8,0.8,0.4,0,0,0}), (\mathbf{0,0,0,0,0.2,0.6,0}), (\mathbf{0,0.6,0.2,0,0,0,0}), (\mathbf{0,0.6,0.2,0,0,0,0}), (\mathbf{0,0.6,0.2,0,0,0,0}), \ldots\}$$

$$e_3 = \{(\mathbf{0,0,0,0,1,0,0}), (\mathbf{1,0,0,0,0,0,0}), (\mathbf{0,0,0,0,0,1,0}), (\mathbf{0,1,0,0,0,0,0}), (\mathbf{0,0,0,0,0,0,1}), \ldots\}$$

$$e_4 = \{(\mathbf{0,0,0,0,0.28,0.85,0.57}), (\mathbf{0,0,0,0,0.28,0.85,0.57}), (\mathbf{0.57,0.85,0.28,0,0,0,0}), (\mathbf{0.14,0.71,0.71,0.14,0,0,0}), (\mathbf{0,0,0,0,0,0.42,0}), \ldots\}$$

   Where, for example, the corresponding fuzzy set for c_{11} of the e_1 is obtaining as:

$$\tau_{AC}(a_2) = (0, 0, 0, 0.25, 0.5, 0.5, 0)$$

   b) Transforming into 2-tuples. The fuzzy sets are transformed into linguistic 2-tuples by means of the function χ. The results for the dimension Entertainment are:

$$e_1 = \{(\mathbf{c_4}, \mathbf{0.2}), (\mathbf{c_3}, \mathbf{0}), (\mathbf{c_2}, \mathbf{-0.2}), (\mathbf{c_4}, \mathbf{0.2}), (\mathbf{c_4}, \mathbf{0.2}), \cdots\}$$

$$e_2 = \{(\mathbf{c_2}, \mathbf{-0.5}), (\mathbf{c_5}, \mathbf{-0.25}), (\mathbf{c_1}, \mathbf{0.25}), (\mathbf{c_1}, \mathbf{0.25}), (\mathbf{c_1}, \mathbf{0.25}), \cdots\}$$

$$e_3 = \{(\mathbf{c_4}, \mathbf{0}), (\mathbf{c_0}, \mathbf{0}), (\mathbf{c_5}, \mathbf{0}), (\mathbf{c_1}, \mathbf{0}), (\mathbf{c_6}, \mathbf{0}), \cdots\}$$

$$e_4 = \{(\mathbf{c_5}, \mathbf{0.16}), (\mathbf{c_5}, \mathbf{0.16}), (\mathbf{c_1}, \mathbf{-0.16}), (\mathbf{c_2}, \mathbf{-0.5}), (\mathbf{c_5}, \mathbf{0}), \cdots\}$$

For example, the transformation into a 2-tuple of c_{11}^1 is computed as:

$$\chi((0, 0, 0, 0.25, 0.5, 0.5, 0)) = \Delta(4.02) = (c_4, 0.2)$$

2. Aggregation phase: we shall combine the unified input assessments.
   a) Computing collective values for each criterion. We shall compute a collective value for each criterion according to all the users. We shall use the arithmetic mean for 2-tuples(Definition 4):

$$CVC_{1i} \Rightarrow \{(\mathbf{c_4}, \mathbf{-0.28}), (\mathbf{c_3}, \mathbf{0.22}), (\mathbf{c_2}, \mathbf{0.22}), (\mathbf{c_2}, \mathbf{-0.01}), (\mathbf{c_4}, \mathbf{0.11}), \cdots\}$$

   Where, for example the first assessment is computed as:

$$CVC_{11} = AM^*((c_4, 0.2), (c_2, -0.5), (c_4, 0), (c_5, 0.16)) = (c_4, -0.28)$$

   b) Computing an evaluation assessment for each dimension. Now, we want to obtain an evaluation assessment for each dimension. We shall aggregate the collective values of its criteria by means of the arithmetic mean for 2-tuple (Definition 4):

$$ED_i \Rightarrow \{(\mathbf{c_3}, \mathbf{0.05}), (c_3, -0.27), (c_3, 0), (c_1, 0.01), (c_3, -0.41), (c_3, -0.02), (c_1, 0.01)\}$$

   where, the first assessment is obtained according to this expression:

$$ED_1 = AM^*((c_4, -0.28), (c_3, 0.22), (c_2, 0.22), (c_2, -0.01), (c_4, 0.11)) = (c_3, 0.05)$$

Global Quality of the web site services.

Depending on the kind of web site services we are assessing, the importance of every dimension is different. In this case we have established that Entertainment, Convenience and Information reliability are more importance. The weighted vector we shall use is $= W\{0.2, 0.2, 0.2, 0, 0.2, 0.2, 0\}$ because, we are evaluating an educational web site, so the dimensions *security and assurance* and *product offer* are not crucial at all, hence the users don't provide proper assessments about them. We shall apply the 2-tuple linguistic weighting average operator (Definition 5) to compute the EAW and finally we shall obtain:

$$EAW = W\_AM^{*}\left(\left(c_3, 0.05\right), \left(c_3, -0.27\right), \left(c_1, 0.01\right), \left(c_3, 0.12\right), \left(c_3, -0.41\right), \left(c_3, -0.02\right), \left(c_1, 0.01\right)\right)$$

$$= \left(c_3, -0.06\right)$$

where $(c_3, -0.06)$ is the global linguistic evaluation for the quality of the lecturer's web site.

6 Concluding Remarks

The evaluation of web site services have become a critical factor for users and companies in order to improve their commercial exchanges. So recently, different evaluation methods for this topic have arisen. The evaluation of these services is user centered because it depends on the opinions provided by the users according to their perceptions.

Current evaluation methods use numerical information to model users opinions. This modelling is not suitable for human perceptions. Therefore, we have proposed a hierarchical evaluation model that models the users opinions by means of linguistic information and these opinions could be assessed in different linguistic term sets to offer a greater flexibility to the users that take part in the evaluation process.

To manage the multi-granular linguistic information of the evaluation framework we have used fuzzy tools and the linguistic 2-tuples representation model.

Acknowledgments

This work has been partially supported by the Research Projects TIC 2003-03348 and TIC 2002-11942-E

References

1. M. Banâtre, V. Issarny, F. Leleu, and B. Charpiot. Providing quality of service over the web: A newspaper-based approach. *Computer Networks and ISDN Systems*, (29):1457–1465, 1997.
2. P. Bharati and A. Chaudhury. An empirical investigation of decision-making satisfaction in web-based decision support systems. *Decision Support Systems*, (37):187–197, 2004.
3. P.P. Bonissone and K.S. Decker. *Selecting Uncertainty Calculi and Granularity: An Experiment in Trading-Off Precision and Complexity.* In L.H. Kanal and J.F. Lemmer, Editors., Uncertainty in Artificial Intelligence. North-Holland, 1986.

4. G. Bordoga and G. Pasi. A fuzzy linguistic approach generalizing boolean information retrieval: A model and its evaluation. *Journal of the American Society for Information Science*, (44):70–82, 1993.
5. P. Chang and Y. Chen. A fuzzy multicriteria decision making method for technology tranfer strategy selection in biotechnology. *Fuzzy Sets and Systems*, 63:131–139, 1994.
6. L.-D. Chen and J. Tan. Technology adaption in e-commerce: Key determinants of virtual stores acceptance. *European Management Journal*, 22(1):74–86, 2004.
7. R. Degani and G. Bortolan. The problem of linguistic approximation in clinical decision making. *International Journal of Approximate Reasoning*, 2:143–162, 1988.
8. M. Delgado, J.L. Verdegay, and M.A. Vila. Linguistic decision making models. *International Journal of Intelligent Systems*, 7:479–492, 1993.
9. M. Delgado, J.L. Verdegay, and M.A Vila. On aggregation operations of linguistic labels. *International Journal of Intelligent Systems*, 8:351–370, 1993.
10. E. Herrera-Viedma, O. Cordon, M. Luque, A.G. Lopez, and A.N. Muñoz. A model of fuzzy linguistic irs based on multi-granular linguistic information. *International Journal of Approximate Reasoning*, 34(3):221–239, 2003.
11. M. Gertz, M.T. Ozsu, G. Saake, and K.U. Sattler. Report on the dagstuhl seminar "data quality on the web". *Sigmod Record*, 33(1):127–132, 2004.
12. F. Herrera, E. Herrera-Viedma, and L. Martinez. A fusion approach for managing multigranularity linguistic term sets in decision making. *Fuzzy Sets and Systems*, (114):43–58, 2000.
13. F. Herrera, E. Herrera-Viedma, and J.L. Verdegay. A sequential selection process in group decision making with linguistic assessment. *Information Sciences*, 85:223–239, 1995.
14. F. Herrera and L. Martínez. A 2-tuple fuzzy linguistic representation model for computing with words. *IEEE Transactions on Fuzzy Systems*, 8(6):746–752, 2000.
15. F. Herrera and L. Martínez. An approach for combining linguistic and numerical information based on 2-tuple fuzzy representation model in decision-making. *International Journal of Uncertainty, Fuzziness and Knowledge-Based Systems*, 8(5):539–562, 2000.
16. F. Herrera and L. Martínez. The 2-tuple linguistic computational model. Advantages of its linguistic description, accuracy and consistency. *International Journal of Uncertainty, Fuzziness and Knowledge-Based Systems*, 9(Suppl.):33–49, 2001.
17. F. Herrera and L. Martínez. A model based on linguistic 2-tuples for dealing with multigranularity hierarchical linguistic contexts in multiexpert decision-making. *IEEE Transactions on Systems, Man and Cybernetics. Part B: Cybernetics*, 31(2):227–234, 2001.
18. F. Herrera, L. Martínez, and P.J. Sánchez. Managing non-homogeneus information in group decision making. *European Journal of Operational Research*, (To appear), 2004.
19. E. Herrera-Viedma. Modeling the retrieval process of an information retrieval system using an ordinal fuzzy linguistic approach. *Journal of the American Society for Information Science and Technology (JASIST)*, 52(6):460–475, 2001.
20. E. Herrera-Viedma and G. Pasi. Fuzzy approaches to access information on the web: Recent developments and research trends. In *Proceedings Interna-*

tional Conference on Fuzzy Logic and Technology (EUSFLAT03), pages 25–31, Zittau(Germany), 2003.

21. J. Van Iwaarden, T. Van der Wiele, L. Ball, and R. Millen. Perceptions about the quality of web sites: A survey amongst students at northeastern university and erausmus university. *Information and Management*, (41):947–959, 2004.
22. S. Kurnia and P. Schubert. An assessment of australian web sites in the grocery sector. pages 229–236. IADIS International Conference e-Commerce 2004.
23. C.K. Law. Using fuzzy numbers in educational grading system. *Fuzzy Sets and Systems*, (83):311–323, 1996.
24. H.M. Lee. Group decison making using fuzzy sets theory for evaluating the rate of aggregative risk in software development. *Fuzzy Sets and Systems*, 80:261–271, 1996.
25. E. Levrat, A. Voisin, S. Bombardier, and J. Bremont. Subjective evaluation of car seat comfort with fuzzy techniques. *International Journal of Intelligent Systems*, (12):891–913, 1997.
26. C. Liu and K.P. Arnett. Exploring the factors associated with web site success in the context of electronic commerce. *Information and Management*, (38):23–33, 2000.
27. P.J.A. Nagel and W.W. Cilliers. Customer satisfaction: A comprehensive approach. *International Journal of Physical Distribution and Logistic Management*, 20(6):2–46, 1990.
28. F. Naumann. Quality-driven query answering for integrated information systems. *LNCS*, 2261, 2002.
29. S. Negash, T. Ryan, and M. Igbaria. Quality and effectiveness in web-based customer support systems. *Information and Management*, (40):757–768, 2003.
30. L. Olsina and G. Rossi. Measuring web application quality with WebQEM. *IEEE Mul-timedia*, October-December:20–29, 2002.
31. A. Parasuraman, L.L. Berry, and V.A. Zeithaml. Servqual:a multiple-item scale for measuring customer perceptions of service quality. *In Journal of Retailing*, 1988.
32. M. Rifqi, V. Berger, and B. Bouchon-Meunier. Discrimination power of measures of comparison. *Fuzzy Sets and Systems*, (110):189–196, 2000.
33. P. Schubert. Extended web assessment method (EWAM) - evaluation of electronic commerce applications from the customer's viewpoint. *International Journal of Electronic Commerce*, 7(2):51–81, 2003.
34. M. Tong and P.P. Bonissone. A linguistic approach to decision making with fuzzy sets. *IEEE Transactions on Systems, Man and Cybernetics*, 10:716–723, 1980.
35. A. I. Torres and F. Vitorino. An e-satisfaction model - application to internet information search and purchase. pages 313–319. AIDIS International Conference e-Commerce, 2004.
36. R.R. Yager. An approach to ordinal decision making. *International Journal of Approximate Reasoning*, (12):237–261, 1995.
37. R.R. Yager, L.S. Goldstein, and E. Mendels. Fuzmar: an approach to aggregating market research data based on fuzzy reasoning. *Fuzzy Sets and Systems*, (68):1–11, 1994.

38. Z. Yang, S. Cai, Z. Zhou, and N. Zhou. Development and validation of an instrument to measure user perceived service quality of information presenting web portals. *Information and management*, (Article in press), 2004.
39. L.A. Zadeh. The concept of a linguistic variable and its applications to approximate reasoning. *Information Sciences, Part I, II, III*, 8,8,9:199–249,301–357,43–80, 1975.

Optimization and Decision Making under Interval and Fuzzy Uncertainty: Towards New Mathematical Foundations

Hung T. Nguyen[1] and Vladik Kreinovich[2]

[1] New Mexico State University
hunguyen@nmsu.edu
[2] University of Texas at El Paso
vladik@cs.utep.edu

Summary. In many industrial engineering problems, we must select a design, select parameters of a process, or, in general, make a decision. Informally, this decision must be *optimal*, the best for the users. In traditional operations research, we assume that we know the *objective function* $f(x)$ whose values describe the consequence of a decision x for the user. Optimization of well-defined functions is what started calculus in the first place: once we know the objective function $f(x)$, we can use differentiation to find its maximum, e.g., as the point x at which the derivative of f with respect to x is equal to 0.

In real life, we often do not know the exact consequence $f(x)$ of each possible decision x, we only know this consequence with uncertainty. The simplest case is when we have *tolerance-type (interval) uncertainty*, i.e., when all we know is that the deviation between the the actual (unknown) value $f(x)$ and the approximate (known) value $f_0(x)$ cannot exceed the (known) bound $\Delta(x)$. In precise terms, this means that $f(x)$ belongs to the interval $[f_0(x) - \Delta(x), f_0(x) + \Delta(x)]$. In other situations, in addition to the interval that is guaranteed to contain $f(x)$, experts can also provide us with narrower intervals that contain $f(x)$ with certain degree of confidence α; such a nested family of intervals is equivalent to a more traditional definition of fuzzy set. To solve the corresponding optimization problems, in this paper, we extend differentiation formalisms to the cases of interval and fuzzy uncertainty.

Key words: optimization, interval uncertainty, fuzzy uncertainty, extending differentiation

1 Introduction

Optimization and decision making are important. In many industrial engineering problems, we must select a design, select parameters of a process, or, in general, make a decision. Informally, this decision must be *optimal*, the

H.T. Nguyen and V. Kreinovich: *Optimization and Decision Making under Interval and Fuzzy Uncertainty: Towards New Mathematical Foundations*, StudFuzz **201**, 275–290 (2006)
www.springerlink.com

best for the users. In traditional operations research, we assume that we know the *objective function* $f(x)$ whose values describe what is best for the users. For example, for a chemical plant, this function $f(x)$ may represent the profit resulting from using parameters x.

In such situations, the problem is to find a design (parameters, decision) x that optimizes (e.g., maximizes) the given function $f(x)$ on the given range X.

Deterministic case: traditional approaches. Optimization of well-defined functions is what started calculus in the first place: once we know the objective function $f(x)$, we can use differentiation to find its maximum, e.g., as the point x at which the derivative of f with respect to x is equal to 0.

Sometimes, this equation $f'(x) = 0$ can be solved directly; sometimes, it is difficult to solve it directly, so we use gradient-based (i.e., derivatives-based) techniques to either solve this equation or to optimize the original objective function $f(x)$.

Case of probabilistic uncertainty. Often, we need to make decisions under uncertainty. In this case, we cannot predict the exact outcome $f(x)$ of a decision x; this outcome depends on the unknown factors. If our description of possible factors is reasonably complete, then, for each value v of these unknown factors and for each decision x, we can predict the outcome $f(x, v)$ of the decision x under the situation v. In the traditional approach to decision making, we assume that we can estimate the probability $p(v)$ of each situation v. In this case, it is reasonable to select a decision x for which the *expected utility* $f(x) \stackrel{\text{def}}{=} \sum\limits_{v} p(v) \cdot f(x, v)$ is the largest possible.

In many decision-making problems, instead of *finitely many* situations v, we have a *continuum* of possible situations – e.g., we may have one or several continuous variables like the percentage of sulfur in the oil, the oil price, the outdoors temperature, etc., that describe possible situations. In such problems, instead of finitely many probabilities $p(v)$, we have a probability distribution with a probability density $\rho(v)$, and instead of a sum, we represent the expected value as an integral $f(x) = \int_v \rho(v) \cdot f(x, v)\, dv$.

Real-life situations: beyond probabilistic uncertainty. In real life, we often do not know the probabilities of different possible situations – or at least we only have partial knowledge about these probabilities.

Interval uncertainty. The simplest case is when we have *tolerance-type (interval) uncertainty*, i.e., when all we know is that the deviation of actual value x of the parameter (e.g., thickness of a beam) from the nominal value x_0 cannot exceed the given tolerance Δ. In precise terms, this means that x belongs to the interval $[x_0 - \Delta, x_0 + \Delta]$, an we have no information about the probabilities within this interval.

Fuzzy uncertainty. In other situations, in addition to the interval that is guaranteed to contain x, experts can also provide us with narrower intervals that contain x with certain degree of confidence α. Such a nested family of intervals is also called a *fuzzy set*, because it turns out to be equivalent to a more traditional definition of fuzzy set [3, 12, 21, 24, 26] (if a traditional fuzzy set is given, then different intervals from the nested family can be viewed as α-cuts corresponding to different levels of uncertainty α).

Dealing with interval and fuzzy uncertainty is computationally difficult. The resulting interval and fuzzy computations techniques have been well developed, and they are still actively used in many application areas.

For foundations and applications of fuzzy techniques, see, e.g., [3, 4, 12, 26]. For applications of interval computations techniques, see, e.g., [9, 10, 11, 20].

Solving the corresponding data processing and optimization problems is often computationally difficult (NP-hard), even in the simplest case of interval uncertainty; see, e.g., [14].

At present, mostly heuristic methods are used. As a result of the above-mentioned computational difficulty, to handle optimization under interval and fuzzy uncertainty, researchers mainly use heuristic techniques, techniques that often modify the techniques used in traditional optimization.

What is the problem with using heuristic techniques. Often, as a result, we get very heuristic techniques, with no clear understanding of what they are optimizing and why this particular objective function is selected to represent the original uncertain situation.

Towards new mathematical foundations. It is therefore desirable, instead of modifying *techniques*, to first modify the basic *foundations*, the main mathematical methods behind these techniques – and hopefully, justified methods will come out not just heuristic ones. In this paper, we overview the preliminary results of this research, and describe some useful mathematical techniques in detail.

Specifically:

- to optimize imprecisely defined objective functions, we must extend differential formalism to interval-valued and fuzzy-valued functions; our preliminary results are described in [15];
- to find out the utility function, we must extract a function from the data; here, we usually have linear, quadratic function, Taylor series etc. - a natural part of calculus; in uncertain case, it is often more beneficial to extract rules than expressions; see, e.g., [28];
- we must extend the expected utility theory to decision making under partially known preferences; our preliminary results are given in [7, 33];
- finally, we must extend foundations of probability to the case when we only have partial information about probabilities; this topic is covered in many monographs and research papers starting with [36]; our research emphasizes computational-related aspects of these problems; see, e.g., [13, 16, 17].

What we are planning to do. In this paper, we concentrate on extending differentiation formalisms to the interval-valued and fuzzy-valued cases.

As we have mentioned, the results of this paper were previously presented at a conference [15].

2 Extending Differentiation to Interval- and Fuzzy-Valued Functions: A Problem

In many real-life problems, we want to know the values of the derivatives. In many areas of science and engineering, we are interested in slopes. For example, a 1-D landscape is described as a dependence of the altitude y on the coordinate x; different 1-D landscape features are defined by different values of the slope dy/dx of this dependence: low values of this slope correspond to a plain, high values to steep mountains, and medium values to a hilly terrain.

In industrial engineering, we often want to make sure that a certain parameter of a plant stays within the given range. As we monitor the value of this parameter, we would like not only to check that this value is within the range, but we also would like to look at the trend (slope) of this dependence, to be able to predict what the value of this parameter will be in the future and thus, to take preventive measures if necessary.

Interval uncertainty. In the ideal situation, when we know the exact values of $y(x)$ for every x, we can simply differentiate the corresponding dependence. In practice, however, the information on y comes from measurements, and measurements are never exact. E.g., in the landscape example, we measure the altitudes $y_1, \ldots, y_n$ at different points $x_1 < \ldots < x_n$. Since the measurements are not exact, the measured values $\widetilde{y}_i$ are, in general, slightly different from the the (unknown) actual altitudes y_i.

For measuring instruments, we usually have an upper bound Δ_i on the measurement error $\Delta y_i \stackrel{\text{def}}{=} \widetilde{y}_i - y_i$. This upper bound is usually provided by the manufacturer of this instrument: $|\Delta y_i| \le \Delta_i$. Thus, after the measurement, the only information that we have about the actual (unknown) value y_i is that this value belongs to the interval $\mathbf{y}_i = [\underline{y}_i, \overline{y}_i]$, where $\underline{y}_i \stackrel{\text{def}}{=} \widetilde{y}_i - \Delta_i$ and $\overline{y}_i \stackrel{\text{def}}{=} \widetilde{y}_i + \Delta_i$ (for a more detailed description of interval uncertainty, see, e.g., [9, 10, 11, 20]).

Thus, the only information that we have about the actual dependence $y = f(x)$ of y on x is that the (unknown) function $f(x)$ belongs to the class

$$F \stackrel{\text{def}}{=} \{f(x) \mid f(x_i) \in \mathbf{y}_i \text{ for all } i = 1, \ldots, n\}. \quad (1)$$

We also know that the (unknown) function $f(x)$ is smooth (differentiable) – because otherwise, the notion of a slope does not make sense.

In many practical applications, the derivative has a physical meaning, and this meaning implies that it is itself a continuous (or even differentiable) function. For example, when we monitor the locations y_i of a particle at different moments of time x_i, then the derivative dy/dx is a velocity; when we monitor the values y_i of the velocity, then the derivative dy/dx is the acceleration, etc. Thus, we can assume that the function f is continuously differentiable.

How can we determine the slopes under such interval uncertainty?

Toward a formal definition. Let us assume that we look for areas where the slope takes a given value s. In a simplified example, we monitor the location y_i of a car on a highway at different moments of time, and we want to find out where the car was driving at the maximal allowed speed s (or, alternatively, where it was driving at an excessive speed s).

Since we only know the values of the unknown function $f(x)$ at finitely many points $x_1 < \ldots < x_n$, it is always possible that the derivative of the (unknown) function $f(x)$ attains the desired value s at some point between x_i and x_{i+1}. For example, if we are checking for the areas where the car was overspeeding, it is always possible that the car was going very fast when no one was looking (i.e., in between x_i and x_{i+1}), for a short period of time, just for fun, so that the overall traveled distance was not affected.

In other words, for every interval $[a, b]$ $(a < b)$, it is always possible to have a function f within the class F (defined by the formula (1)) for which $f'(x) = s$ for some $s \in [a, b]$.

What we are really interested in is not whether it is *possible* that somewhere, the slope is equal to s (it is always possible), but whether the data *imply* that somewhere, the slope was indeed equal to x. This "implies" means that whatever function $f \in F$ we take, there always is a point $x \in [a, b]$ for which $f'(x) = s$ (this point may be different for different functions $f \in F$).

In other words, we say that the slope is guaranteed to attain a given value s somewhere on a given interval $[a, b]$ if for every function $f \in F$, the range $f'([a, b])$ of its derivative $f'(x)$ contains the value s. In mathematical terms, this means that the value s belongs to the *intersection* of the ranges $f'([a, b])$ corresponding to all $f \in F$.

This intersection thus describes the "range of the derivative" of the interval function F on the given interval $[a, b]$. In other words, we arrive at the following definitions.

From interval to fuzzy uncertainty. As we have mentioned, interval uncertainty is just the simplest possible case of non-probabilistic uncertainty. In many real life situations, instead of an interval, we have a fuzzy set – i.e., nested family of intervals corresponding to different levels of certainty α.

In this case, instead of an "interval function", i.e., a finite sequence of pairs $\langle x_i, \mathbf{y}_i \rangle$ $(i = 1, 2, \ldots, n)$, where x_i is a real number and $\mathbf{y}_i$ is an interval, we have a "fuzzy function", i.e., a finite sequence of pairs $\langle x_i, Y_i \rangle$ $(i = 1, 2, \ldots, n)$, where Y_i is a fuzzy number.

Once we define the derivative of an interval-valued function, we can naturally extend this definition to derivatives of fuzzy-valued function: Namely, for every degree of certainty α, we consider an interval function formed by the α-cuts of Y_i. We can then compute the "derivative" of this interval function. This derivative, as we will see, is, by itself, also an interval.

So, for each level α, we have a derivative interval corresponding to this level α. The nested family of these derivative intervals forms a fuzzy number – which can be thus viewed as a "derivative" of the original fuzzy function.

In view of this comment, once we know how to define and compute derivatives of interval-valued functions, we can naturally extend this definition to fuzzy-valued functions as well. Because of this fact, in this text, we will concentrate on definitions and algorithms corresponding to interval-valued functions.

3 Precise Formulation of the Problem

Definition 1. *By an* interval function F, *we mean a finite sequence of pairs* $\langle x_i, \mathbf{y}_i \rangle$ $(i = 1, 2, \ldots, n)$, *where for each* i, x_i *is a real number,* $\mathbf{y}_i$ *is a non-degenerate interval, and* $x_1 < x_2 < \ldots < x_n$.

Definition 2. *We say that a function* $f : R \to R$ *from reals to reals* belongs *to an interval function* $F = \{\langle x_1, \mathbf{y}_1 \rangle, \ldots, \langle x_n, \mathbf{y}_n \rangle\}$ *if* $f(x)$ *is continuously differentiable and for every* i *from 1 to* n, *we have* $f(x_i) \in \mathbf{y}_i$.

Definition 3. *Let* F *be an interval function, and let* $[a, b]$ *be an interval. By a* derivative $F'([a, b])$, *we mean the intersection*

$$F'([a,b]) \stackrel{\text{def}}{=} \bigcap_{f \in F} f'([a,b]),$$

where $f'(x)$ *denotes the derivative of a differentiable function* $f(x)$, *and* $f'([a, b]) \stackrel{\text{def}}{=} \{f'(x) \mid x \in [a, b]\}$ *is the range of the derivative* $f'(x)$ *over the interval* $[a, b]$.

Comment. The notation $F'([a, b])$ looks like the notation of a range for a real-valued function, but it is not a range: in contrast to range, if an interval is narrow enough, we can have $F'([a, b]) = \emptyset$ (see examples below).

This newly defined derivative does share some properties of the range. For example, it is well known that the range is *inclusion-monotonic* – in the sense that $[a, b] \subseteq [c, d]$ implies $f'([a, b]) \subseteq f'([c, d])$. From this property of the range, we can conclude that $[a, b] \subseteq [c, d]$ implies $F'([a, b]) \subseteq F'([c, d])$ – i.e., that the newly defined derivative is also inclusion-monotonic. Thus, if the union $A \cup B$ of two intervals is also an interval, we have $F'(A \cup B) \supseteq F'(A) \cup F'(B)$.

Formulation of the problem. How can we compute the derivative of an interval function? The above definition, if taken literally, requires that we consider all (infinitely many) functions $f \in F$ – which is computationally excessive. Thus, we must find an efficient algorithm for computing this derivative. This is what we will do in this paper.

We will try our best to make sure that these algorithms are not simply tricks, that the ideas behind these algorithms are clear and understandable.

Therefore, instead of simply presenting the final algorithm, we will, instead, present our reasoning in a series of auxiliary results that eventually leads to the asymptotically optimal algorithms for computing the desired derivative $F'([a,b])$.

Previous work. In our research, we were guided by results from two related research directions:

First, we were guided by different definitions of differentiation of an interval function that have been proposed by interval computations community [2, 10, 19, 22, 23, 27, 29, 30, 31, 32]. The main difference from our problem is that most of these papers assume that we have intervals $\mathbf{y}$ for all x, while we consider a more realistic situation when the interval bounds on $f(x)$ are only known for finitely many values $x_1, \ldots, x_n$.

Second, we were guided by a paper [34] in which an algorithm was developed to check for local maxima and minima of an interval function f. This result has been applied to detecting geological areas [1, 5, 6] and to financial analysis [8]. This result can be viewed as detecting the areas where the derivative is equal to 0 – and, in this sense, as a particular case of our current problem.

4 First Auxiliary Result: Checking Monotonicity

Definition 4. *We say that a function $f(x)$ is* strongly increasing *if $f'(x) > 0$ for all x.*

Comment. Every strongly increasing function is strictly increasing, but the inverse is not necessarily true: the function $f(x) = x^3$ is strictly increasing but not strongly increasing.

Proposition 1. *For every interval function F, the existence of a strongly increasing function $f \in F$ with $f'(x) > 0$ is equivalent to*

$$\underline{y}_i < \overline{y}_j \text{ for all } i < j. \tag{2}$$

Proof. If $f \in F$ and $f(x)$ is strongly increasing, then it is also strictly increasing hence for every $i < j$, the inequality $x_i < x_j$ implies that $f(x_i) < f(x_j)$. Since $f \in F$, we have $f(x_i) \in \mathbf{y}_i = [\underline{y}_i, \overline{y}_i]$ and $f(x_j) \in \mathbf{y}_i = [\underline{y}_j, \overline{y}_j]$. Thus, from $\underline{y}_i \le f(x_i) < f(x_j) \le \overline{y}_j$, we conclude that $\underline{y}_i < \overline{y}_j$, which is exactly the inequality (2).

Vice versa, let us assume that the inequalities (2) are satisfied, and let us design the corresponding strictly increasing function $f \in F$. We will first build a piece-wise linear strictly increasing function $f_0(x)$ for which $f_0(x_i) \in \mathbf{y}_i$, and then we will show how to modify $f_0(x)$ into a continuously differentiable strongly increasing function $f \in F$.

According to the inequalities (2), all the differences $\overline{y}_j - \underline{y}_i$ $(i < j)$ are positive. Since all intervals are non-degenerate, the differences $\overline{y}_i - \underline{y}_i$ are also positive. Let us denote the smallest of these positive numbers by Δ. For every i, let us denote

$$y_i \stackrel{\text{def}}{=} \max(\underline{y}_1, \ldots, \underline{y}_i) + \frac{i}{2n} \cdot \Delta. \quad (3)$$

We will then design $f_0(x)$ as a piece-wise linear function for which $f_0(x_i) = y_i$. To show that $f_0(x)$ is the desired piece-wise linear function, we must show that for every i, $y_i \in \mathbf{y}_i$, and that this function is strictly increasing, i.e., that $i < j$ implies $y_i < y_j$.

That $i < j$ implies $y_i < y_j$ is clear: the first (maximum) term in the formula (3) can only increase (or stay the same) when we replace i by j, and the second term increases. Thus, it is sufficient to prove that $y_i \in \mathbf{y}_i = [\underline{y}_i, \overline{y}_i]$, i.e., that $\underline{y}_i \le y_i$ and $y_i \le \overline{y}_i$. We will actually prove a stronger statement: that $\underline{y}_i < y_i$ and $y_i < \overline{y}_i$.

The first inequality $\underline{y}_i < y_i$ follows directly from the formula (3): by definition of a maximum, $\max(\underline{y}_1, \ldots, \underline{y}_i) \ge \underline{y}_i$, and when we add a positive number to this maximum, the result only increases. So, y_i is actually larger than $\underline{y}_i$.

Let us now prove that $y_i < \overline{y}_i$. Indeed, by definition of Δ, for all $k \le i$, we have $\underline{y}_k + \Delta \le \overline{y}_i$, hence (since $(i/2n) \cdot \Delta < \Delta$) $\underline{y}_k + (i/2n)\Delta < \overline{y}_i$. Thus, y_i – which is the largest of the values $\underline{y}_k + (i/2n)\Delta$ – is also smaller than $\overline{y}_i$. So, the desired $f_0(x)$ is designed.

Let us now show how to build the corresponding continuously differentiable function $f(x)$. For the piece-wise linear function $f_0(x)$, the first derivative $f_0'(x)$ is piece-wise constant; since the function $f_0(x)$ is strictly increasing, the values $f_0'(x)$ are all positive. Around each discontinuity point x_i, replace the abrupt transition with a linear one; as we integrate the resulting function, we get a new function $f(x)$ that is continuously differentiable and – since the new values of the derivative are still everywhere positive – strongly increasing. When the replacement is fast enough, the change in the value $f(x_i)$ is so small that $f(x_i)$ is still inside the desired interval $\mathbf{y}_i$. The proposition is proven.

Similarly, we can prove the following results:

Definition 5. *We say that a function $f(x)$ is* strongly decreasing *if $f'(x) < 0$ for all x.*

Proposition 2. *For every interval function F, the existence of a strongly decreasing function $f \in F$ is equivalent to*

$$\overline{y}_i > \underline{y}_j \text{ for all } i < j. \quad (4)$$

Proposition 3. *For every interval function F and for every interval $[a, b]$, the existence of a function $f \in F$ that is strongly increasing on the interval $[a, b]$ is equivalent to*

$$\underline{y}_i < \overline{y}_j \text{ for all } i < j \text{ for which } x_i, x_j \in [a, b]. \quad (5)$$

Proposition 4. *For every interval function F and for every interval $[a,b]$, the existence of a function $f \in F$ that is strongly decreasing on the interval $[a,b]$ is equivalent to*

$$\overline{y}_i > \underline{y}_j \text{ for all } i < j \text{ for which } x_i, x_j \in [a,b]. \tag{6}$$

5 Second Auxiliary Result: Checking Whether $0 \in F'([a,b])$

Proposition 5. *For every interval function F and for every interval $[a,b]$, $0 \in F'([a,b])$ if and only if neither conditions (5) not conditions (6) are satisfied.*

Proof. Let us first show that if either the conditions (5) or the conditions (6) are satisfied, then $0 \notin F'([a,b])$.

Indeed, according to Proposition 3, if the conditions (5) are satisfied, then there exists a function $f \in F$ that is strongly increasing on $[a,b]$. For this function, $f'(x) > 0$ for all $x \in [a,b]$; therefore, $f'([a,b]) \subseteq (0,\infty)$. Since $F'([a,b])$ is defined as the intersection of such range sets, we have $F'([a,b]) \subseteq f'([a,b]) \subseteq (0,\infty)$ hence $0 \notin F'([a,b])$.

Similarly, if the conditions (6) are not satisfied, then $0 \notin F'([a,b])$.

Vice versa, let us assume that neither the conditions (5) nor the conditions (6) are satisfied, and let us show that then $0 \in F'([a,b])$. Indeed, let $f \in F$ be an arbitrary function from the class F. Since the conditions (5) are not satisfied, the function $f(x)$ cannot be strongly increasing; therefore, there must be a point $x_1 \in [a,b]$ for which $f'(x_1) \le 0$. Similarly, since the conditions (6) are not satisfied, the function $f(x)$ cannot be strongly decreasing; therefore, there must be a point $x_2 \in [a,b]$ for which $f'(x_2) \ge 0$.

Since the function $f(x)$ is continuously differentiable, the continuous derivative $f'(x)$ must attain the 0 value somewhere on the interval $[x_1,x_2] \subseteq [a,b]$. In other words, $0 \in f'([a,b])$ for all $f \in F$. Thus, 0 belongs to intersection $F'([a,b])$ of all possible ranges $f'([a,b])$. The proposition is proven.

6 Third Auxiliary Result and Final Description of $F'([a,b])$

Definition 6. *Let $F = \{\langle x_1, \mathbf{y}_1\rangle, \ldots, \langle x_n, \mathbf{y}_n\rangle\}$ be an interval function, and let v be a real number. Then, we define a new interval function $F - v \cdot x$ as follows:*

$$F - v \cdot x = \{\langle x_1, \mathbf{y}_1 - v \cdot x_1\rangle, \ldots, \langle x_n, \mathbf{y}_n - v \cdot x_n\rangle\},$$

where, for an interval $\mathbf{y} = [\underline{y}, \overline{y}]$ and for a real number c, the difference $\mathbf{y} - c$ is defined as $[\underline{y} - c, \overline{y} - c]$.

It is easy to prove the following auxiliary result:

Proposition 6. *For every interval function F and for every interval $[a,b]$, $v \in F'([a,b])$ if and only if $0 \in (F - v \cdot x)'([a,b])$.*

This results leads to the following description of the derivative $F'([a,b])$:

Proposition 7. *For every interval function F and for every interval $[a,b]$, let i_0 and j_0 be the first and the last index of the values x_i inside $[a,b]$. Then $F'([a,b]) = [\underline{F}_{i_0,j_0}, \overline{F}_{i_0 j_0}]$, where*

$$\underline{F}_{i_0 j_0} \stackrel{\text{def}}{=} \min_{i_0 \le i < j \le j_0} \overline{\Delta}_{ij}, \quad \overline{F}_{i_0 j_0} \stackrel{\text{def}}{=} \max_{i_0 \le i < j \le j_0} \underline{\Delta}_{ij}, \tag{7}$$

$$\underline{\Delta}_{ij} \stackrel{\text{def}}{=} \frac{\underline{y}_i - \overline{y}_j}{x_j - x_i}, \quad \overline{\Delta}_{ij} \stackrel{\text{def}}{=} \frac{\overline{y}_i - \underline{y}_j}{x_j - x_i}, \tag{8}$$

and $[p,q] \stackrel{\text{def}}{=} \{x \,|\, p \le x \,\&\, x \le q\}$ – so when when $p > q$, the interval $[p,q]$ is the empty set.

Comment. The above expression is rather intuitively reasonable because the ratios $\underline{\Delta}_{ij}$ and $\overline{\Delta}_{ij}$ are finite differences – natural estimates for the derivatives.

Comment. As a corollary of this general result, we can conclude that if the interval $[a,b]$ contains a single point x_i (or no points at all), then

$$F'([a,b]) = \emptyset.$$

Mathematically, this conclusion follows from our general result because in this case, there is no pair $i < j$, so the minimum and the maximum are taken over an empty set. By definition, the minimum of an empty set is infinite, so $\underline{F}_{i_0 j_0} = +\infty$; similarly, $\overline{F}_{i_0 j_0} = -\infty$. Here, $\underline{F}_{i_0 j_0} > \underline{F}_{i_0 j_0}$, so the interval is empty. Intuitively, however, this conclusion can be understood without invoking minima and maxima over an empty set.

Indeed, let us assume that the given interval $[a,b]$ contains only one point x_i from the original list $x_1, \ldots, x_n$. Then, for any real number s, we can take, as $f \in F$, a function that takes an arbitrary value $y_i \in \mathbf{y}_i$ for $x = x_i$ and that is linear with a slope s on $[a,b]$ – i.e., the function

$$f(x) = y_i + s \cdot (x - x_i).$$

For this function $f(x)$, the range $f'([a,b])$ of the derivative $f'(x)$ on the interval $[a,b]$ consists of a single point s. Thus, if we take two such functions corresponding to two different values of s, then the intersection of their ranges is empty. Therefore, the range $F'([a,b])$ – which is defined (in Definition 3) as the intersection of all such ranges $f'([a,b])$ – is also empty.

Proof. The fact that conditions (5) are not satisfied means that there exist value $i_0 \leq i < j \leq j_0$ for which $\underline{y}_i \geq \overline{y}_j$. The fact that the conditions (6) are not satisfied means that there exist values $i_0 \leq i' < j' \leq j_0$ for which $\overline{y}_{i'} \leq \underline{y}_{j'}$.

Similarly, the fact that the conditions (5) and (6) are not satisfied for the interval function $F - v \cdot x$ mean that

$$\exists i, j\, (i_0 \leq i < j \leq j_0 \,\&\, \underline{y}_i - v \cdot x_i \geq \overline{y}_j - v \cdot x_j) \tag{9}$$

and

$$\exists i', j'\, (i_0 \leq i' < j' \leq j_0 \,\&\, \overline{y}_{i'} - v \cdot x_{i'} \leq \underline{y}_{j'} - v \cdot x_{j'}). \tag{10}$$

The inequality $\underline{y}_i - v \cdot x_i \geq \overline{y}_j - v \cdot x_j$ can be described in the equivalent form $v \cdot (x_j - x_i) \geq \overline{y}_j - \underline{y}_i$, i.e., since $x_i < x_j$, in the form $v \geq \overline{\Delta}_{ij}$. Thus, the existence of i and j as expressed by the formula (9) can be described as the existence of i and j for which v is larger than the corresponding value $\overline{\Delta}_{ij}$, i.e., as

$$v \geq \min_{i_0 \leq i < j \leq j_0} \overline{\Delta}_{ij}.$$

Similarly, the condition (10) is equivalent to

$$v \leq \max_{i_0 \leq i < j \leq j_0} \underline{\Delta}_{ij}.$$

The proposition is proven.

7 Towards a Faster Algorithm

Proposition 7 provides an explicit formula for computing $F'([a, b])$ for each interval $[a, b]$. For each $[a, b]$, we need to compute $O(n^2)$ values of $\underline{\Delta}_{ij}$ and $\overline{\Delta}_{ij}$.

In problem like locating landscape features, we are not so much interested in knowing whether a given type of landscape exists in a given zone, but rather in locating all types of landscape. In other words, we would like to be able to find the values $F'([a, b])$ for all possible intervals $[a, b]$. According to Proposition 7, it is sufficient to find all the values $F'([x_{i_0}, x_{j_0}])$ for all $i_0, j_0 = 1, \ldots, n$ for which $i_0 < j_0$. There are $n \cdot (n+1)/2 = O(n^2)$ such values. If we use the formula from Proposition 7 – that takes $O(n^2)$ computational steps – to compute each of these $O(n^2)$ values, we will need an overall of $O(n^2) \cdot O(n^2) = O(n^4)$ steps.

For large n – e.g., for $n \approx 10^6$ – we need $n^4 \approx 10^{24}$ computational steps; this is too long for even the fastest computers. Let us show that we can compute the interval derivative faster, actually in $O(n^2)$ time. Since we must return $O(n^2 0$ results, we cannot do it in less than $O(n^2)$ computational steps – so this algorithm is (asymptotically) optimal.

Proposition 8. *There exists an algorithm that, given an interval function* $F = \{\langle x_1, \mathbf{y}_1\rangle, \ldots, \langle x_n, \mathbf{y}_n\rangle\}$, *computes all possible values of the derivative* $F'([a,b])$ *in* $O(n^2)$ *computational steps.*

Proof. At first, we compute $O(n^2)$ values $\underline{\Delta}_{ij}$ and $\overline{\Delta}_{ij}$ by using the formulas (8); this requires $O(n^2)$ steps.

Let us now show how to compute all n^2 values $\overline{F}_{i_0 j_0}$ in $O(n^2)$ steps.

First, for each i, we sequentially compute the "vertical" maxima $\overline{v}_{ij} \stackrel{\text{def}}{=} \max(\underline{\Delta}_{i,i+1}, \ldots, \underline{\Delta}_{ij})$ corresponding to $j = i+1, i+2, \ldots, n$ as follows: $\overline{v}_{i,i+1} = \underline{\Delta}_{i,i+1}$ and $\overline{v}_{ij} = \max(\overline{v}_{i,j-1}, \underline{\Delta}_{ij})$ for $j > i+1$. For each $i = 1, \ldots, n$, to compute all these values, we need $\leq n$ computational steps. Thus, to compute all such values $\overline{v}_{ij}$ for all i and j, we need $\leq n \cdot n = O(n^2)$ computational steps.

Then, for every j_0, we sequentially compute the values $\overline{F}_{i_0 j_0}$ for $i_0 = j_0 - 1, j_0 - 2, \ldots, 1$ as follows: $\overline{F}_{j_0-1,j_0} = \underline{v}_{j_0-1,j_0}$ and $\overline{F}_{i_0,j_0} = \max(\overline{F}_{i_0+1,j_0}, \overline{v}_{i_0,j_0})$ (it is easy to see that this formula is indeed correct). For each $j_0 = 1, \ldots, n$, to compute all these values, we need $\leq n$ computational steps. Thus, to compute all such values $\overline{F}_{ij}$ for all i_0 and j_0, we need $\leq n \cdot n = O(n^2)$ computational steps.

Similarly, by using $\overline{\Delta}_{ij}$ instead of $\underline{\Delta}_{ij}$ and min instead of max, we can compute all n^2 values $\underline{F}_{i_0 j_0}$ in $O(n^2)$ steps. The proposition is proven.

8 This Same Differential Formalism Also Serves an Alternative Definition of Zones

In some practical problems, a zone is defined not by an exact value of the derivative v, but an interval $\mathbf{v} = [\underline{v}, \overline{v}]$ of possible values. In this case, it makes sense to say that an interval $[a,b]$ contains a zone if for every function $f \in F$, there is at least one point $x \in [a,b]$ for which $f'(x) \in \mathbf{v}$. In other words, we say that the interval $[a,b]$ contains a zone of a given type if $f'([a,b]) \cap \mathbf{v} \neq \emptyset$ for all functions $f \in F$.

It turns out that the above notion of a derivative can help us detect such zones as well. Namely, the following statement is true:

Proposition 9. *For every interval function* F *and for every two intervals* $[a,b]$ *and* $\mathbf{v}$, *the following properties are equivalent to each other:*

- *for every function* $f \in F$, *we have* $f'([a,b]) \cap \mathbf{v} \neq \emptyset$;
- $\underline{F}_{i_0 j_0} \leq \overline{v}$ *and* $\overline{F}_{i_0 j_0} \geq \underline{v}$.

Proof. We will prove the equivalence of the two opposite statements:

- there exists a function $f \in F$ for which $f'([a,b]) \cap \mathbf{v} = \emptyset$;
- $\underline{F}_{i_0 j_0} > \overline{v}$ or $\overline{F}_{i_0 j_0} < \underline{v}$.

Indeed, let us assume that there exists a function $f \in F$ for which $f'([a,b]) \cap \mathbf{v} = \emptyset$. Since every function $f \in F$ is continuously differentiable, its derivative $f'(x)$ is a continuous function, hence the range $f'([a,b])$ is an interval. There are two possible situations when this interval range does not intersect with $\mathbf{v}$:

- either all the values from this range are $> \overline{v}$,
- or all the values from this range are $< \underline{v}$.

In the first case, we have $f'(x) > \overline{v}$ for all $x \in [a,b]$. Therefore, for the function $g(x) \stackrel{\text{def}}{=} f(x) - \overline{v} \cdot x$, we get $g'(x) > 0$ for all x, i.e., the function $g(x)$ is strongly increasing. Since $f \in F$, we have $g \in G \stackrel{\text{def}}{=} F - \overline{v} \cdot x$. Due to Proposition 1, the existence of a strongly increasing function $g \in G$ means that $\underline{y}_i - \overline{v} \cdot x_i < \overline{y}_j - \overline{v} \cdot x_j$ for all $i < j$. This inequality, in its turn, means that $\overline{\Delta}_{ij} > \overline{v}$ for all $i < j$. Thus, $\overline{v}$ is smaller than the smallest of the values $\overline{\Delta}_{ij}$, i.e., smaller than $\underline{F}_{i_0 j_0}$.

Similarly, in the second case, we have $f'(x) < \underline{v}$ for all $x \in [a,b]$, hence $\overline{F}_{i_0 j_0} < \underline{v}$.

Vice versa, let $\underline{F}_{i_0 j_0} > \overline{v}$. By definition of $\underline{F}_{i_0 j_0}$ as the minimum, this means that $\overline{\Delta}_{ij} > \overline{v}$ for all i, j for which $i_0 \le i < j \le j_0$. Substituting the definition of $\overline{\Delta}_{ij}$, multiplying both sides of the inequality by a positive term $x_j - x_i$ and moving terms to another side, we conclude that $\underline{y}_i - \overline{v} \cdot x_i < \overline{y}_j - \overline{v} \cdot x_j$ for all $i < j$. This inequality, in its turn, means that for the interval function $G \stackrel{\text{def}}{=} F - \overline{v} \cdot x$, formula (2) holds and thus, due to Proposition 1, there exist a strongly monotonic function $g \in G$ for which $g'(x) > 0$ for all x. Then, for the function $f(x) \stackrel{\text{def}}{=} g(x) + \overline{v} \cdot x$, we have $f \in F$ and $f'(x) = g'(x) + \overline{v} > \overline{v}$ for all x – hence, $f'([a,b]) \cap \mathbf{v} = \emptyset$.

Similarly, if $\overline{F}_{i_0 j_0} < \underline{v}$, there exists a function $f \in F$ for which $f'([a,b]) \cap \mathbf{v} = \emptyset$. The proposition is proven.

9 Open Problems

What if we take into consideration uncertainty in measuring x? In the above text, we took into consideration the uncertainty of measuring y, but assumed that we know x exactly. In real life, there is also some uncertainty in measuring x as well. How can we take this uncertainty into consideration?

For the problem of finding local minima and maxima, this uncertainty was taken into consideration in [18]. It is desirable to extend this approach to finding the range of the derivatives.

Parallelization. In the above text, we described how to compute the derivative of an interval function in time $O(n^2)$, where n is the number of observations, and showed that this algorithm is (asymptotically) optimal in the sense that no algorithm can compute this derivative faster.

For reasonable n, e.g., for $n \approx 10^3$, n^2 computational steps means a million steps; it is quite doable on modern computers. However, for large n, e.g., for

$n \approx 10^6$, n^2 computational steps is 10^{12} steps, so on a modern Gigaherz machine, the corresponding computations will take 10^3 sec – almost an hour.

How can we further speed up the corresponding computations? Our optimality result shows that we cannot achieve a drastic speed-up if we use sequential computers. Thus, the only way to speed up the corresponding computations is to use *parallel* computers.

For the problem of finding local minima and local maxima, parallel computers can indeed speed up the corresponding computations; see, e.g., [35]. An important question is therefore: How can speed up the computation of the corresponding derivative by using parallel computers?

Acknowledgments

This work was supported in part by NASA under cooperative agreement NCC5-209, by NSF grants EAR-0112968, EAR-0225670, and EIA-0321328, and by NIH grant 3T34GM008048-20S1. The authors are also thankful to participants of the interval computations mailing list, especially to R. Baker Kearfott, Svetoslav Markov, Arnold Neumaier, and Andrei Sobolevskii, for valuable comments.

References

1. Aguiar MS, Costa ACR, Dimuro GP (2004) ICTM: an interval tessellation-based model for reliable topographic segmentation. Numerical Algorithms 37:3–11.
2. Angelov R, Markov S (1981) Extended Segment Analysis. Freiburger Intervall-Berichte 81(10):1–63.
3. Bojadziev G, Bojadziev M (1995). Fuzzy sets, fuzzy logic, applications, World Scientific, Singapore.
4. Booker J, Parkinson J, Ross TJ, eds. (2002) Combined fuzzy logic and probability applications. SIAM, Philadelphia.
5. Coblentz DD, Kreinovich V, Penn BS, Starks SA (2000) Towards reliable subdivision of geological areas: interval approach. In: Proc. NAFIPS'2000, Atlanta, Georgia, July 13–15, 2000, 368–372.
6. Coblentz DD, Kreinovich V, Penn BS, Starks SA (2003) Towards reliable subdivision of geological areas: interval approach, In: Reznik L, Kreinovich V (eds.), Soft Computing in Measurements and Information Acquisition, Springer-Verlag, Berlin-Heidelberg, 223–233.
7. de la Mora C, Wojciechowski P, Kreinovich V, Starks SA, Tanenbaum P, Kuzminykh A (2003) Robust methodology for characterizing system response to damage: a subjective (fuzzy) partial ordered modification of the traditional utility-probability scheme. In: Proceedings of the 22nd International Conference of the North American Fuzzy Information Processing Society NAFIPS'2003, Chicago, Illinois, July 24–26, 2003, 413–419.

8. Deboeck GJ, Villaverde K, Kreinovich K (1995) Interval methods for presenting performance of financial trading systems. Reliable Computing, Supplement (Extended Abstracts of APIC'95: International Workshop on Applications of Interval Computations, El Paso, TX, Febr. 23–25, 1995), 67–70.
9. Jaulin L, Kieffer M, Didrit O, Walter E (2001) Applied interval analysis: with examples in parameter and state estimation, robust control and robotics. Springer, London.
10. Kearfott RB (1996) Rigorous global search: continuous problems. Kluwer, Dordrecht.
11. Kearfott Rb, Kreinovich V., eds (1996) Applications of interval computations. Kluwer, Dordrecht.
12. Klir G, Yuan, B (1995) Fuzzy sets and fuzzy logic: theory and applications. Prentice Hall, Upper Saddle River, New Jersey.
13. Kreinovich V (2004) Probabilities, intervals, what next? optimization problems related to extension of interval computations to situations with partial information about probabilities. Journal of Global Optimization 29(3):265–280.
14. Kreinovich V, Lakeyev A, Rohn J, Kahl P (1997) Computational complexity and feasibility of data processing and interval computations. Kluwer, Dordrecht.
15. Kreinovich V, Nguyen HT, Dimuro GP, da Rocha Costa AC, Bedregal BRC (2003) A new differential formalism for interval-valued functions and its potential use in detecting 1-D landscape features. In: Proceedings of the International Conference on Information Technology InTech'03, Chiang Mai, Thailand, December 17–19, 2003, 491–498.
16. Kreinovich V, Nguyen HT, Ferson S, Ginzburg L (2003) From computation with guaranteed intervals to computation with confidence intervals: a new application of fuzzy techniques. In: Proceedings of the 21st International Conference of the North American Fuzzy Information Processing Society NAFIPS'2002, New Orleans, Louisiana, June 27–29, 2002, 418–422.
17. Kreinovich V, Nguyen HT, Wu B (to appear) On-line algorithms for computing mean and variance of interval data, and their use in intelligent systems, Information Sciences.
18. Lorkowski J, Kreinovich V (1996) If we measure a number, we get an interval. What if we measure a function or an operator? Reliable Computing 2(3):287–298.
19. Markov S (1979) Calculus for interval functions of a real variable. Computing 22:325–337.
20. Moore RE (1979) Methods and applications of interval analysis. SIAM, Philadelphia.
21. Moore RE, Lodwick WA (2003) Interval analysis and fuzzy set theory. Fuzzy Sets and Systems 135(1):5–9.
22. Muñoz H, Kearfott RB (2004) Slope intervals, generalized gradients, semigradients, and csets. Reliable Computing 10(3):163–193.
23. Neumaier A (2001) Introduction to numerical analysis. Cambridge Univ. Press, Cambridge.
24. Nguyen HT, Kreinovich V (1996) Nested intervals and sets: concepts, relations to fuzzy sets, and applications, In: Kearfott, RB, Kreinovich, V (eds.), Applications of interval computations. Kluwer, Dordrecht, 245–290.
25. Nguyen HT, Kreinovich V (1999) How to divide a territory? A new simple differential formalism for optimization of set functions. International Journal of Intelligent Systems 14(3):223–251.

26. Nguyen HT, Walker, EA (1999) First course in fuzzy logic, CRC Press, Boca Raton, Florida.
27. Ratz D (1998) Automatic slope computation and its application in nonsmooth global optimization, Shaker-Verlag, Aachen.
28. Salvatore AP, Biswas A, Kreinovich V, Manriquez B, Cannito MP, Sinard RJ (2004) Expert system-type approach to voice disorders: scheduling botulinum toxin treatment for adductor spasmodic dysphonia. In: Proceedings of the Fifth International Conference on Intelligent Technologies InTech'04, Houston, Texas, December 2–4, 2004.
29. Sendov B (1977) Segment arithmetic and segment limit. C. R. Acad. Bulg. Sci. 30:955–968.
30. Sendov B (1977) Segment derivatives and Taylor's formula. C. R. Acad. Bulg. Sci. 30:1093–1096.
31. Sendov B (1980) Some topics of segment analysis. In: Nickel K (ed.), Interval mathematics'80, Academic Press, N.Y., 236–245.
32. Sendov B (1990) Hausdorff approximations. Kluwer, Dordrecht.
33. Tanenbaum PJ, de la Mora C, Wojciechowski P, Kosheleva O, Kreinovich V, Starks SA, Kuzminykh A (2004) Robust methodology for characterizing system response to damage: approach based on partial order. In: Lirkov I, Margenov S, Wasniewski J, Yalamov P (eds.), Large-Scale Scientific Computing, Proceedings of the 4-th International Conference LSSC'2003, Sozopol, Bulgaria, June 4–8, 2003, Springer Lecture Notes in Computer Science 2907:276–283.
34. Villaverde K, Kreinovich V (1993) A linear-time algorithm that locates local extrema of a function of one variable from interval measurement results. Interval Computations, No. 4:176–194.
35. Villaverde K, Kreinovich V (1995) Parallel algorithm that locates local extrema of a function of one variable from interval measurement results. Reliable Computing, Supplement (Extended Abstracts of APIC'95: International Workshop on Applications of Interval Computations, El Paso, TX, Febr. 23–25, 1995), 212–219.
36. Walley P (1991) Statistical reasoning with imprecise probabilities, Chapman & Hall, New York.

Interval Evaluations in DEA and AHP

Tomoe Entani[1], Kazutomi Sugihara[2], and Hideo Tanaka[3]

[1] Department of Economics and Social Sciences, Kochi University, 2-5-1 Akebono, Kochi, 780-8520 Japan, entani@cc.kochi-u.ac.jp
[2] Department of Management Science, Fukui University of Technology 3-6-1 Gakuen Fukui 910-8505 Japan, sugihara@ccmails.fukui-ut.ac.jp
[3] Department of Kansei Information, Hiroshima International University, Gakuendai 555-36, Kurose, Hiroshima, 724-0695 Japan, h-tanaka@he.hirokoku-u.ac.jp

Summary. Even if the given data are crisp, there exists uncertainty in decision making process and inconsistency based on human judgements. The purpose of this paper is to obtain the evaluations which reflect such an uncertainty and inconsistency of the given information. Based on the idea that intervals are more suitable than crisp values to represent evaluations in uncertain situations, we introduce this interval analysis concept into two well-known decision making models, DEA and AHP. In the conventional DEA, the relative efficiency values are measured and in the proposed interval DEA, the efficiency values are defined as intervals considering various viewpoints of evaluations. In the conventional AHP, the priority weights of alternatives are obtained and in the proposed interval AHP, the priority weights are also defined as intervals reflecting the inconsistency among the given judgements.

Key words: Decision making, Uncertain information, Efficiency interval, Interval priority weight

1 Introduction

In the decision making problem involving human judgements, usually the information is uncertain, even if the data are given as crisp values. Through most of the conventional decision making models, the results such as evaluations from the given data are obtained as crisp values. However, there exists uncertainty in the decision making process involved in different viewpoints, human intuitive judgements and fuzzy environments. It seems to be suitable to obtain the evaluations as intervals in order to reflect various uncertainty in the given data and evaluating process. In this viewpoint, the concept of interval analysis is introduced into DEA(Data Envelopment Analysis) and AHP(Analytic Hierarchy Process), which are well-known evaluation models. DEA is relative evaluation model to measure the efficiency of DMUs (Decision Making Units) with common input and output terms. In Section 2, we propose

T. Entani et al.: *Interval Evaluations in DEA and AHP*, StudFuzz **201**, 291–304 (2006)
www.springerlink.com

Interval DEA, where the efficiency value calculated from various viewpoints for each DMU are considered and efficiency intervals are obtained. AHP is the useful method to obtain the priority weight of each item in multiple criteria decision making problems. In Section 3, we propose Interval AHP, where inconsistency based on human intuition in the given pairwise comparisons are considered and interval weights are obtained. In this paper, evaluations as results with crisp data through models are obtained as intervals reflecting the uncertainty of the given data and the decision making process. Interval evaluations are more useful information for decision making and helpful for decision makers than crisp evaluations, since the former can consider various viewpoints and inconsistency in the given data.

2 Interval DEA

2.1 Efficiency Value by Conventional DEA

DEA (Data Envelopment Analysis) is a non-parametric technique for measuring the efficiency of DMUs (Decision Making Units) with common input and output terms [1, 2]. In DEA, the efficiency for DMU_o which is the analyzed object is evaluated by the following basic fractional model.

$$\begin{aligned} \theta_o^{E^*} &= \max_{\boldsymbol{u},\boldsymbol{v}} \frac{\boldsymbol{u}^t \boldsymbol{y}_o}{\boldsymbol{v}^t \boldsymbol{x}_o} \\ \text{s.t. } & \frac{\boldsymbol{u}^t \boldsymbol{y}_j}{\boldsymbol{v}^t \boldsymbol{x}_j} \leq 1 \quad \forall j \\ & \boldsymbol{u} \geq \boldsymbol{0} \\ & \boldsymbol{v} \geq \boldsymbol{0} \end{aligned} \tag{1}$$

where the decision variables are the weight vectors $\boldsymbol{u}$ and $\boldsymbol{v}$, $\boldsymbol{x}_j \geq \boldsymbol{0}$ and $\boldsymbol{y}_j \geq \boldsymbol{0}$ are the given input and output vectors for DMU_j and the numbers of inputs, outputs and DMUs are m, k and n, respectively.

The efficiency is obtained by maximizing the ratio of weighted sum of outputs to that of inputs for DMU_o under the condition that the ratios for all DMUs are less than or equal to one. To deal with many inputs and outputs, the weighted sum of inputs and that of outputs are considered as a hypothetical input and a hypothetical output, respectively. The maximum ratio of this output to this input is assumed as the efficiency which is calculated from the optimistic viewpoint for each DMU. The efficiency for DMU_o is evaluated relatively by the other DMUs.

This fractional programming problem is replaced with the following linear programming (LP) problem, which is the basic DEA model called CCR (Charnes Cooper Rhodes) model, by fixing the denominator of the objective function to one.

$$\begin{aligned} \theta_o^{E*} &= \max_{\boldsymbol{u},\boldsymbol{v}} \boldsymbol{u}^t \boldsymbol{y}_o \\ \text{s.t. } & \boldsymbol{v}^t \boldsymbol{x}_o = 1 \\ & \boldsymbol{u}^t \boldsymbol{y}_j - \boldsymbol{v}^t \boldsymbol{x}_j \leq 0 \quad \forall j \\ & \boldsymbol{u} \geq \boldsymbol{0} \\ & \boldsymbol{v} \geq \boldsymbol{0} \end{aligned} \tag{2}$$

θ_o^{E*} is obtained with the superior inputs and outputs of DMU_o by maximizing the objective function in (2) with respect to the weight variables. Therefore, it can be said that it is the evaluation from the optimistic viewpoint for DMU_o.

When the optimal value of objective function is equal to one, DMU_o is rated as efficient and otherwise it is not rated as efficient. Precisely speaking, the word "efficient" which we use in this paper is called "weak efficient". In this model the production possibility set is assumed as follows.

$$P = \{(\boldsymbol{x}, \boldsymbol{y}) | \boldsymbol{x} \geq X\boldsymbol{\lambda}, \boldsymbol{y} \leq Y\boldsymbol{\lambda}, \boldsymbol{\lambda} \geq \boldsymbol{0}\} \tag{3}$$

where $X \in \Re^{m \times n}$ is an input matrix consisting of all input vectors, $Y \in \Re^{k \times n}$ is an output matrix consisting of all output vectors. (3) means that the more inputs, smaller outputs or both than those of given data can be productive.

On the other hand, the inefficiency measure has defined by using inverse relation to the ratio defined in DEA in [3], that is the ratio of weighted sum of inputs to that of outputs. Thus, the inefficiency model is called "Inverted DEA". However the ratios considered in DEA and Inverted DEA are different each other so that there is no mathematical relation between the efficiency by DEA and the inefficiency by Inverted DEA. In the literature [4], the maximum and minimum efficiency values for a new DMU have been proposed using the benchmarks obtained by DEA. This approach is called DEA-based benchmarking model and it is an effective measure as an interval for a new DMU, considering a set of the benchmark frontier by DEA. In this paper, the proposed minimum efficiency for each DMU has been defined by using all the given DMUs. We propose Interval DEA [5], where the efficiency intervals are obtained so as to reflect uncertainty in evaluating viewpoints. The following points should be noted: 1) the proposed approach has the same mathematical structures for the maximum and minimum efficiency values, and 2) efficiency interval is obtained by all the given DMUs. These are different from Inverted DEA [3] and DEA-based benchmarking model [4].

2.2 Efficiency Interval

The relative efficiency can be obtained from various viewpoints. In this section, we propose Interval DEA model to obtain the efficiency interval [5]. The efficiency interval is denoted as its upper and lower bounds. Then, they are obtained by solving two optimization problems such that the relative ratio of the analyzed DMU to the others is maximized and minimized with respect to input and output weights, respectively. In both models the same ratios are considered to be maximized and minimized respectively. The upper and lower bounds of efficiency interval denote the evaluations from the optimistic and pessimistic viewpoints, respectively.

Since the conventional DEA can be regarded as the evaluation from the optimistic viewpoint, the upper bound of efficiency interval for DMU_o can be obtained by the conventional CCR model in [1, 8]. Considering the original CCR model formulated as a fractional programming problem (1), the problem to obtain the upper bound of efficiency interval is formulated as follows.

$$\begin{aligned} \theta_o^{E*} &= \max_{\boldsymbol{u},\boldsymbol{v}} \frac{\dfrac{\boldsymbol{u}^t\boldsymbol{y}_o}{\boldsymbol{v}^t\boldsymbol{x}_o}}{\max\limits_j \dfrac{\boldsymbol{u}^t\boldsymbol{y}_j}{\boldsymbol{v}^t\boldsymbol{x}_j}} \\ \text{s.t. } \boldsymbol{u} &\geq \boldsymbol{0} \\ \boldsymbol{v} &\geq \boldsymbol{0} \end{aligned} \tag{4}$$

It should be noted that the denominator in (4) plays an important role of normalizing efficiency value. The ratio of the weighted sum of outputs to that of inputs for DMU_o is compared to the maximum ratio of all DMUs. In (1), the ratios of the weighted sum of outputs to that of inputs for all DMUs are constrained to be less than one for normalization. Furthermore, formulating the upper bound of efficiency interval as (4) is very useful for defining the lower bound of efficiency interval.

When the denominator of the objective function is fixed to one, (4) can be reduced to the following problem.

$$\begin{aligned} \theta_o^{E*} &= \max_{\boldsymbol{u},\boldsymbol{v}} \frac{\boldsymbol{u}^t\boldsymbol{y}_o}{\boldsymbol{v}^t\boldsymbol{x}_o} \\ \text{s.t. } \max_j \frac{\boldsymbol{u}^t\boldsymbol{y}_j}{\boldsymbol{v}^t\boldsymbol{x}_j} &= 1 \\ \boldsymbol{u} &\geq \boldsymbol{0} \\ \boldsymbol{v} &\geq \boldsymbol{0} \end{aligned} \tag{5}$$

Comparing with (5) and (1), the conditions of (5) is stricter than that of (1). However, the optimization problem (5) is equal to (2), which is the original CCR model described as LP problem (see [5]).

On the other hand, by minimizing the objective function in (4) with respect to the weight variables, the lower bound of efficiency interval is obtained by the following problem.

$$\begin{aligned} \theta_{o*}^{E} &= \min_{\boldsymbol{u},\boldsymbol{v}} \frac{\dfrac{\boldsymbol{u}^t\boldsymbol{y}_o}{\boldsymbol{v}^t\boldsymbol{x}_o}}{\max_j \dfrac{\boldsymbol{u}^t\boldsymbol{y}_j}{\boldsymbol{v}^t\boldsymbol{x}_j}} \\ \text{s.t. } \boldsymbol{u} &\geq \boldsymbol{0} \\ \boldsymbol{v} &\geq \boldsymbol{0} \end{aligned} \tag{6}$$

θ^E_{o*} is obtained with inferior inputs and outputs of DMU_o. Therefore, it can be said that it is the evaluation from the pessimistic viewpoint considering all DMUs. The optimization problem (6) can be reduced to the following problem (see [5]).

$$\theta^E_{o*} = \min_{p,r} \frac{\dfrac{y_{op}}{x_{or}}}{\max\limits_j \dfrac{y_{jp}}{x_{jr}}} \tag{7}$$

where the rth element of input weight vector $\boldsymbol{v}$ and the pth element of output vector $\boldsymbol{u}$ are one and the other elements are all zero in (6). Only the rth input and pth output are used to determine the lower bound of efficiency interval. These are inferior input and output of DMU_o relatively to the others.

The efficiency interval denoted as $[\theta^E_{o*}, \theta^{E*}_o]$ illustrates all the possible evaluations for DMU_o from various viewpoints. Thus, Interval DEA gives a decision maker all the possible efficiency values that reflect different perspectives. The efficiency intervals are important and useful information to a decision maker in a sense of perspectives.

In order to improve the efficiency, in the conventional DEA the efficiency value is obtained as a real value and the inputs and outputs are adjusted to make the efficiency value be one. Several approaches to improvement for the conventional efficiency value by adjusting inputs and outputs have been proposed in [4, 6, 7, 8, 9] On the contrary, in Interval DEA the efficiency is obtained as an interval. The given inputs and outputs are adjusted so that the efficiency interval with the adjusted ones become larger than one before the improvement. The approach to improve the efficiency interval can be described in [10]. It is done with the following way: the upper bound of efficiency interval becomes one and the lower one becomes as large as possible. It can be said that the superior inputs and outputs are shown as a target for each improved DMU.

2.3 Numerical Example

We calculate the efficiency interval by using one-input and two-output data in Table 1. Efficiency intervals determined by (2) and (7) are shown in Table 1 and Fig. 1. The conventional efficiency value and the upper bound of efficiency intervals are the same, since both of them are obtained from the optimistic viewpoint by (2).

Although the upper bounds of efficiency intervals for A and J are equal to 1, their lower bounds are small. Their ranges of efficiency intervals are large, therefore, they are called as peculiar. Peculiar DMUs have some inferior data so that the interval ranges are large, while the upper bounds are one.

The interval order relation is defined as follows in [11].

Definition 1. *Interval order relation*
$A = [\underline{a}, \overline{a}] \succ B = [\underline{b}, \overline{b}]$ *holds if and only if* $\underline{b} \leq \underline{a}$ *and* $\overline{b} \leq \overline{a}$.

Using Definition 1, the relations between DMUs by the efficiency intervals are illustrated in Fig. 2. By the obtained efficiency intervals, E and G do not have any DMUs whose efficiency intervals are greater than those of them. Then they are picked out as non-dominated DMUs and rated as efficient in Interval DEA. Peculiar DMUs such as A and J are not rated as efficient. Considering all the possible viewpoints of evaluations by Interval DEA, the partial order relation of DMUs is obtained. Efficiency intervals reflect uncertainty on perspectives of evaluations so that they are similar to our natural evaluation and give more useful information than crisp efficiency values do.

Table 1. Given crisp data and efficiency intervals

DMU	input	output1	output2	efficiency interval
A	1	1	8	[0.143,1.000]
B	1	2	3	[0.286,0.522]
C	1	2	6	[0.286,0.824]
D	1	3	3	[0.375,0.652]
E	1	3	7	[0.428,1.000]
F	1	4	2	[0.250,0.696]
G	1	4	5	[0.571,0.957]
H	1	5	2	[0.250,0.826]
I	1	6	2	[0.250,0.957]
J	1	7	1	[0.125,1.000]

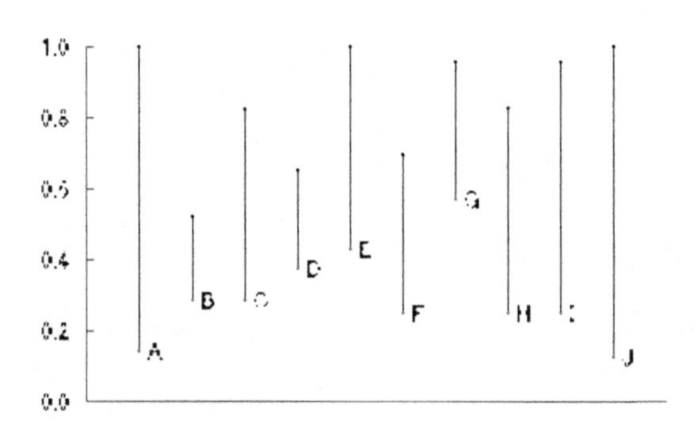

Fig. 1. Efficiency intervals

3 Interval AHP

3.1 Crisp Weights by Conventional AHP

AHP (Analytic Hierarchy Process) is useful in multi-criteria decision making problems. AHP is a method to deal with the priority weights with respect

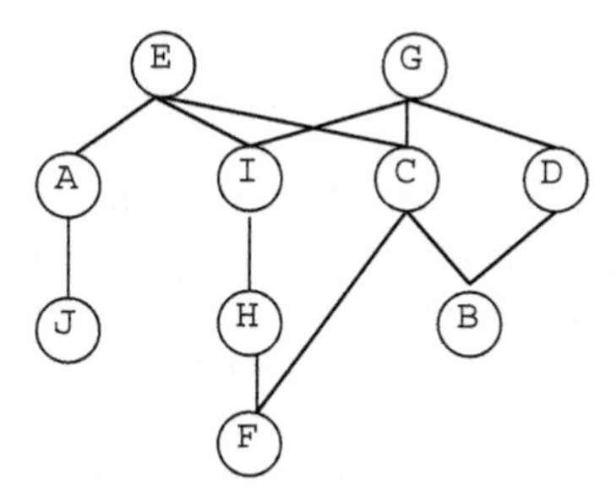

Fig. 2. Partial order of DMUs by efficiency intervals

to many items and proposed to determine the priority weight of each item [12]. When there are n items, a decision maker compares a pair of items for all possible pairs then we can obtain a comparison matrix A as follows. The elements of the matrix called pairwise comparisons are relative measurements and given by a decision maker.

$$A = [a_{ij}] = \begin{pmatrix} 1 & \cdots & a_{1n} \\ \vdots & a_{ij} & \vdots \\ a_{n1} & \cdots & 1 \end{pmatrix}$$

where a_{ij} shows the priority ratio of item i comparing to item j.

The elements of pairwise comparison matrix satisfy the following relations. The decision maker gives $n(n-1)/2$ pairwise comparisons in case of n items.

$$\begin{array}{ll} \text{Diagonal elements} & a_{ii} = 1 \\ \text{Reciprocal elements} & a_{ij} = 1/a_{ji} \end{array} \tag{8}$$

From the given comparison matrix, the priority weights w_i^* are obtained by the well-known eigenvector method. The eigenvector problem is as follows.

$$A\boldsymbol{w} = \lambda \boldsymbol{w} \tag{9}$$

where λ is an eigenvalue and $\boldsymbol{w}$ is a corresponding eigenvector. By (9), the eigenvector $\boldsymbol{w}^* = (w_1^*, \ldots, w_n^*)^t$ corresponding to the principal eigenvalue λ_{max} is obtained as the weight vector. It is noted that the sum of the obtained weights w_i^* is normalized to be one; $\sum_i w_i^* = 1$. The obtained weights from the given comparison matrix can reflect his/her attitude in the actual decision problem.

The weights obtained by the conventional AHP lead to a linear order of items. Uncertainty of an order of items in AHP is discussed in [13]. However, there exists a problem that pairwise comparisons might be inconsistent with each other because they are based on human intuition. The approach for dealing with interval comparisons has been proposed in [14]. It is easier for a decision maker to give interval comparisons than crisp ones. This approach is rather complex comparing to our approach shown in this paper in view of solving problems on all vertices for obtaining interval weights. In the similar setting to [14], the approaches for dealing with decision maker's preference statements instead of pairwise comparisons have been described in [15]. This seems to be very practical, but obtaining the upper and lower bounds of interval weights has been proposed without defining the interval weights. We propose Interval AHP where interval weights are obtained so as to reflect inconsistency among the given crisp comparisons.

3.2 Interval Priority Weight

It is assumed that the estimated weights are intervals to reflect inconsistency of pairwise comparisons. Since the decision maker's judgements are usually inconsistent [16, 17]. We obtain the interval weights so as to include all the given pairwise comparisons and minimize the widths. We formulate the approach for obtaining interval weights as a LP problem, instead of the eigenvector problem in the conventional AHP. This concept is similar to interval regression analysis [18]. The width of the obtained interval weight represents inconsistency of the pairwise comparisons. A decision maker always gives inconsistent information since his/her judgements on each item's weight are uncertain. Then, such inconsistency in the item's weight can be denoted as the widths of interval weights. The given pairwise comparison a_{ij} is approximated by the ratio of priority weights, w_i and w_j, symbolically written as follows.

$$a_{ij} \approx w_i / w_j$$

It is noted that the consistent comparison matrix satisfy the following relations.

$$a_{ij} = a_{ik} a_{kj} \quad \forall (i, j, k) \tag{10}$$

In usual cases such that comparisons are based on the decision maker's intuitive judgements, the relation (10) is not satisfied. Therefore, there is some inconsistency in the given matrix.

Assuming the priority weight as an interval W_i, the interval priority weights are denoted as $W_i = [\underline{w}_i, \overline{w}_i]$. Then, the approximated pairwise comparison with the interval weights is defined as the following interval.

$$\frac{W_i}{W_j} = \left[\frac{\underline{w}_i}{\overline{w}_j}, \frac{\overline{w}_i}{\underline{w}_j} \right]$$

where the upper and lower bounds of the approximated comparison are defined as the maximum range considering all the possible values.

Interval Weights Normalization

While the sum of weights obtained by AHP is normalized to be one, interval probability proposed in [19] can be regarded as a normalization of interval weights. The normalization for interval weights is defined as follows.

Definition 2. *Interval normalization*
Interval weights $(W_1, ..., W_n)$ *are called interval probability if and only if*

$$\begin{array}{l} \sum_{i \neq j} \overline{w}_i + \underline{w}_j \geq 1 \quad \forall j \\ \sum_{i \neq j} \underline{w}_i + \overline{w}_j \leq 1 \quad \forall j \end{array} \tag{11}$$

where $W_i = [\underline{w}_i, \overline{w}_i]$.

It can be said that the conventional normalization is extended to the interval normalization by using the above conditions. In order to explain interval normalization, we use the following example intervals which do not satisfy the conditions (11),

$$W_1 = [0.3, 0.6], \quad W_2 = [0.2, 0.4], \quad W_3 = [0.1, 0.2].$$

Assuming the value $w_1^* = 0.3$ in W_1, there do not exist the values, w_2^* and w_3^*, in W_2 and W_3 whose sum is one, $w_1^* + w_2^* + w_3^* = 1$. Transforming W_1 into $W_1' = [0.5, 0.6]$, these intervals satisfy the conditions for interval normalization (11) and the sum of values in the intervals can be one. Definition 2 is effective to reduce redundancy under the condition that the sum of crisp weights in the interval weights is equal to one.

Approximation of Crisp Pairwise Comparison Matrix

The model to obtain the interval weights is determined so as to include the given interval comparisons [16]. The obtained interval weights satisfy the following inclusion relations.

$$a_{ij} \in \frac{W_i}{W_j} = \left[\frac{\underline{w}_i}{\overline{w}_j}, \frac{\overline{w}_i}{\underline{w}_j}\right] \quad \forall (i, j)$$

It is denoted as the following two inequalities.

$$\frac{\underline{w}_i}{\overline{w}_j} \leq a_{ij} \leq \frac{\overline{w}_i}{\underline{w}_j} \Leftrightarrow \begin{cases} \underline{w}_i \leq a_{ij} \overline{w}_j & \forall (i, j) \\ \overline{w}_i \geq a_{ij} \underline{w}_j & \forall (i, j) \end{cases} \tag{12}$$

The interval weights include the given inconsistent comparisons. In order to obtain the least interval weights, the width of each weight must be minimized. The problem for obtaining interval weights is formulated as the following LP problem.

$$\begin{aligned}
\min \quad & \sum_i (\overline{w}_i - \underline{w}_i) \\
\text{s.t.} \quad & \underline{w}_i \leq a_{ij}\overline{w}_j \quad \forall(i,j) \\
& \overline{w}_i \geq a_{ij}\underline{w}_j \quad \forall(i,j) \\
& \textstyle\sum_{i \neq j} \overline{w}_i + \underline{w}_j \geq 1 \quad \forall j \\
& \textstyle\sum_{i \neq j} \underline{w}_i + \overline{w}_j \leq 1 \quad \forall j \\
& \overline{w}_i \geq \underline{w}_i \geq \varepsilon \quad \forall i
\end{aligned} \tag{13}$$

where ε is a small positive value and the first two and the next two conditions show the inclusion relations (12) and interval normalization (11), respectively.

The width of the interval weight represents uncertainty of each weight and the least uncertain weights are obtained by this model (13). When a decision maker has some information over uncertainties of the items' priority weights, he/she gives them as the uncertainty weights $p_i \forall i$. Then, the weighted sum of widths $\sum_i p_i(\overline{w}_i - \underline{w}_i)$ can be minimized. However, it is not easy for a decision maker to give the weight of each width. Simply the sum of widths of all weights is minimized as in (13) without information over uncertainties of items.

Since the proposed Interval AHP is the ratio model, its concept is similar to interval regression analysis in view of the least approximation. The proposed model (13) is formulated as LP problem, the following inequality should be satisfied.

$$\frac{n(n-1)}{2} \geq 2n \tag{14}$$

where n is the number of items. (14) requires that the number of given comparison data should be larger than that of decision variables.

If (14) is satisfied, whatever the given comparison matrix is, there exist an optimal solution that minimizes the objective function in (13). If the optimal value of the objective function is equal to zero; $\sum_i(\overline{w}_i^* - \underline{w}_i^*) = 0$, it can be said that the given comparison matrix is perfectly consistent. The weights are obtained as crisp values and they are the same as those by the conventional eigenvector method (9). In the conventional AHP, the consistency index is defined considering the eigenvector corresponding to the principal eigenvalue of the given matrix. If it is equal to 0, the elements of the matrix satisfy the relations (10), that is, the matrix is perfectly consistent. Experimentally it can seem to be consistent in case where the index is less than 0.1. In the proposed LP method, the consistency of the matrix is represented as the optimal value of the objective function that is the sum of widths of the obtained interval weights. The optimal value becomes small for consistent matrix.

The decision problem in AHP is structured hierarchically as criteria $(C_1, ..., C_k)$ and alternatives $(A_1, ..., A_n)$ as in Fig.3. In Fig. 3 the criteria are at one layer, however, it is possible to construct several layers of criteria.

The criterion weights and alternative scores with respect to the criteria are obtained from the corresponding pairwise comparison matrices. Concerning all criteria, the overall priority of each alternative is obtained as the sum of

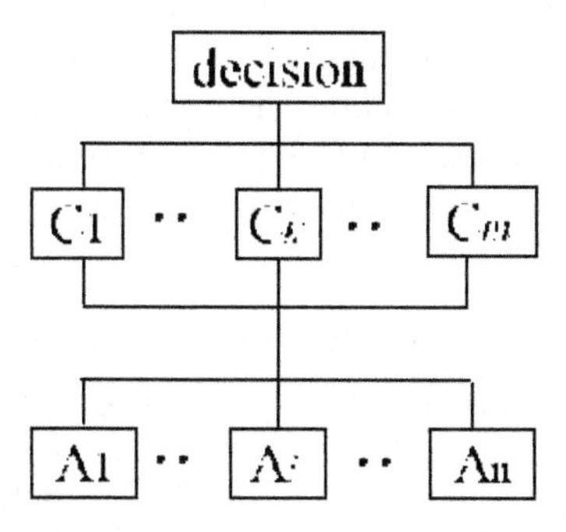

Fig. 3. Structure of decision problem in AHP

the products of criterion weights and corresponding alternative scores. By the proposed model, the criterion weights and alternative scores are obtained as intervals. Then, by interval arithmetic the overall priority is also obtained as an interval.

3.3 Numerical Example

The following pairwise comparison matrix with five items is given by a decision maker. The decision maker gives 10 comparisons marked * and the other elements are filled by (8).

$$A = [a_{ij}] = \begin{pmatrix} 1 & 2^* & 3^* & 5^* & 7^* \\ 1/2 & 1 & 2^* & 2^* & 4^* \\ 1/3 & 1/2 & 1 & 1^* & 1^* \\ 1/5 & 1/2 & 1 & 1 & 1^* \\ 1/7 & 1/4 & 1 & 1 & 1 \end{pmatrix}$$

The crisp weights obtained by conventional eigenvector method (9) are shown in the right column of Table 2. The linear order relation of items is $1 > 2 > 3 > 4 > 5$. The elements of this comparison matrix satisfy $a_{ik} \geq a_{jk}$ for all k and such a comparison matrix is in row dominance relation. In case of row dominance relation, it is assured that the obtained weights by eigenvector method satisfy the order relation $w_i \geq w_j$. It can be seen from the results of this example.

The interval weights obtained by the proposed model (13) is shown in Table 2 and Fig. 4. With the obtained interval weights, by Definition 1, the order relation of items is $1 \succ 2 \succ 3 \succ 4 \succ 5$. Although the linear order relation is obtained in this example, it is noted that the partial order relation is often obtained because of interval weights. The obtained weights of items 1 and 2 are crisp values and items 3, 4 and 5 are intervals. From the given comparison matrix, it is estimated that item 1 is prior to item 2 and both of them are apparently prior to items 3, 4 and 5. However, the relations among items 3, 4 and 5 are not easily estimated, since the comparisons over them are contradicted each other. The obtained interval weights, W_3, W_4 and W_5, by the proposed Interval AHP reflect inconsistency among the given crisp comparisons. Since the decision maker gives comparisons of all pairs of items intuitively, it is natural to consider that the obtained weights are intervals reflecting the uncertainty.

Table 2. Interval weights with crisp comparison matrix

item	interval weights (13)	width	eigenvector (9)
1	0.453	0.000	0.464
2	0.226	0.000	0.241
3	[0.104, 0.151]	0.047	0.112
4	[0.091, 0.113]	0.023	0.100
5	[0.057, 0.104]	0.047	0.083

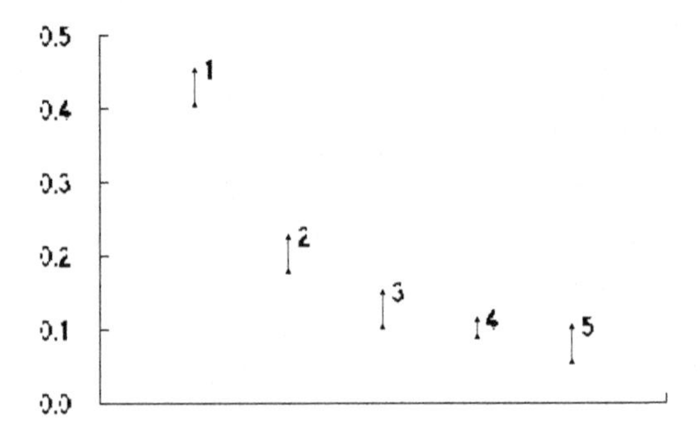

Fig. 4. Interval weights

4 Conclusion

The decision problems usually include uncertainty since humans are involved in the process. The given information and evaluations based on human intuitive judgements might be from various viewpoints and inconsistent. In order to deal with the uncertainty in the given information, interval evaluations have been introduced. Reflecting the uncertainty of the given data and evaluating process, the results obtained by the proposed interval evaluation models are intervals even in case of crisp given data. Interval evaluations are more suitable for our natural judgements than crisp evaluations.

By the proposed Interval DEA, the efficiency interval is obtained so as to include all the possible efficiency values from various viewpoints. By the proposed Interval AHP, the interval weights are obtained so as to include inconsistency among the given comparisons. The proposed interval evaluation models deal with the uncertainty by human judgements. In view of considering all the possibility of data and perspectives of evaluations, the proposed interval models are a kind of possibility analysis [18]. They can give useful and helpful information for decision makers, since various viewpoints and inconsistency in the given data are considered.

References

1. Charnes A, Cooper W.W, Rhodes E (1978) Measuring the efficiency of decision making units. European Journal of Operational Research 2:429-444
2. Charnes A, Cooper W.W, Rhodes E (1981) Evaluating program and managerial efficiency: An application of data envelopment analysis to program follow through. Management Science 27(6):668-734
3. Yamada Y, Matui T, Sugiyama M (1994) New analysis of efficiency based on DEA. Journal of the Operations Research Society of Japan 37(2):158-167 (in Japanese)
4. Zhu J (2002) Quantitative Models for Performance Evaluation and Benchmarking : Data Envelopment Anaylsis with Spreadsheets. Kluwer Academic Publishers, Boston
5. Entani T, Maeda Y, Tanaka H (2002) Dual models of Interval DEA and its extension to interval data. European Journal of Operational Research 136(1):32-45
6. Golany B, Tamir E (1995) Evaluating efficiency - effectiveness - equality tradeoffs: A data envelopment analysis approach. Management Science 41(7):1172-1184
7. Golany B, Phillips F.Y, Rousseau J.J (1993) Models for improved effectiveness based on DEA efficiency results. IIE Transactions 25: 2-9
8. Ramanathan R (1996) An Introduction to Data Envelopment Analysis. Sage Publications
9. Takeda E (2000) An extened DEA model: Appending an additional input to make all DMUs at least weakly efficient. European Journal of Operational Research 125(1):25-33.
10. Entani T, Tanaka H (2005) Improvement of Efficiency Intervals Based on DEA by Adjusting Inputs and Outputs. European Journal of Operational Research (in press)
11. Dubious D, Prade H (1980) Systems of Linear Fuzzy Constraints. Fuzzy Sets and Systems 3:37-48
12. Saaty T.L (1980) The Analytic Hierarchy Process. McGraw-Hill
13. Saaty T.L, Vergas L.G (1987) Uncertainty and Rank Order in the Analytic Hiearchy Process. European Journal of Operations Research 32:107-117
14. Arbel A (1989) Approximation Articulation of Preference and Priority Deviation. European Journal of Operations Research 43:317-326
15. Solo A.A, Hamalainen R.P (1995) Prefrence Performming through Approximate Ratio Comparisons. European Journal of Operations Research 82:458-475
16. Sugihara K, Tanaka H (2001) Evaluation in the Analytic Hierarchy Process by Possibility Analysis. International Journal of Computational Intelligence 17(3):567-579

17. Sugihara K, Ishii H, Tanaka H (2004) Interval priorities in AHP by interval regression anaalysis. European Journal of Operations Research 158:745-754
18. Tanaka H, Guo P(1999) Possibilistic Data Analysis for Operation Research. Physica-Verlag, A Springer Verlag Company
19. Tanaka H, Sugihara K, Maeda Y (2004) Non-additive Measures by Interval Probability Functions. Information Sciences 164:209-227

A General Form of Fuzzy Group Decision Making Choice Functions under Fuzzy Preference Relations and Fuzzy Majority

Janusz Kacprzyk and Sławomir Zadrożny

Systems Research Institute, Polish Academy of Sciences
Ul. Newelska 6, 01-447 Warsaw, Poland
{kacprzyk,zadrozny}@ibspan.waw.pl

Summary. A general form of a collective choice rule in group decision making under fuzzy preferences and a fuzzy majority is proposed. It encompasses some well-known choice rules. Our point of departure is the fuzzy majority based linguistic aggregation rule (solution concept) proposed by Kacprzyk [11–13]. This rule is viewed here from a more general perspective, and the fuzzy majority – meant as a fuzzy linguistic quantifier – is dealt with by using Yager's [42] OWA operators. The particular collective choice rules derived via the general scheme proposed are shown to be applicable in the case of nonfuzzy preferences too. Moreover, a relation to Zadeh's concept of a protoform is mentioned in this context.

Key words: Group decision-making; preference relations; fuzzy majority; OWA operators; linguistic quantifiers; collective choice rules

1 Introduction

We consider group decision making stated as follows: we have a set of M options, $S = \{s_1, ..., s_M\}$, and a set of N individuals, $X = \{x_1, ..., x_N\}$. Each individual x_k from X provides his or her preferences over S assumed to be given by *individual fuzzy preference relations* (see, e.g., Kacprzyk and Fedrizzi [14–16], Fodor and Roubens [7], Ovchinnikov [36], etc.).

A *fuzzy preference relation*, R_k, is given by its membership function $\mu_{R_k}: S \times S \rightarrow [0,1]$ such that $\mu_{R_k}(\cdot,\cdot) \in [0,1]$ denotes the strength of preference. If card S is small enough, R_k may be represented by a matrix $[r_{ij}^k]$, $r_{ij}^k = \mu_{R_k}(s_i, s_j)$; $i, j = 1,...,M$; $k = 1,..., N$.

While deriving solution concepts, two lines of reasoning may be followed here (cf. Kacprzyk [11–13]):

- a direct approach: $\{R_1, ..., R_N\} \rightarrow$ solution

J. Kacprzyk and S. Zadrozny: *A General Form of Fuzzy Group Decision Making Choice Functions under Fuzzy Preference Relations and Fuzzy Majority*, StudFuzz **201**, 305–319 (2006)
www.springerlink.com

- an indirect approach: $\{R_1, ..., R_N\} \rightarrow R$ (a social fuzzy preference relation)$\rightarrow$ solution.

A solution is here not clearly understood. For instance, Kacprzyk [11-13] introduced the *core* for the direct approach and the *consensus winner* for the indirect approach, using a fuzzy majority represented by a linguistic quantifier. For more on solution concepts, see, e.g., Nurmi [28,29].

Notice that the direct and indirect approach are equivalent to the determination of a (collective) choice function (cf. Aizerman and Aleskerov [1]) acting from S to a family of all nonfuzzy or fuzzy subsets of S.

We present a general scheme of collective choice rule that covers a number of well-known rules. We start from Kacprzyk's rule [11–13], using the OWA operators instead of the originally employed linguistic quantifiers. This is clearly for convenience in the future discussion. Then, we present the idea of Zadeh's protoform [50] and show that this collective choice rule can be viewed as a protoform for a wide variety of both fuzzy and non-fuzzy choice rules known from the literature.

2 Fuzzy Preference Relations

Let $S = \{s_1,...,s_M\}$ be a finite set of options and $X = \{x_1,...,x_N\}$ the set of individuals. A fuzzy preference relation R is a fuzzy subset of $S \times S$ characterized by the membership function:

$$\mu_R(s_i,s_j) = \begin{cases} 1 & \text{definite preference} \\ c \in (0.5,1) & \text{preference to some extent} \\ 0.5 & \text{indifference} \\ d \in (0,0.5) & \text{preference to some extent} \\ 0 & \text{definite preference} \end{cases}$$

represented as a matrix $[r_{ij}^k] = [\mu_{R_k}(s_i,s_j)], \forall i,j,k$.

The degree of preference, $\mu_R(s_i,s_j)$, is here interpreted in a continuous manner, i.e. when the value of $\mu_R(s_i,s_j)$ function changes from the one slightly below 0.5 to the one slightly above 0.5, there is no abrupt change of its meaning - both values more or less correspond to the indifference. In the other words, the particular values of a membership function $\mu_R(s_i,s_j)$ express some uncertainty as to the actual preferences, highest in the case of 0.5 and lowest in the case of 1.0 and 0.0. The particular values of a membership function $\mu_R(s_i,s_j)$ may be interpreted in a different way. For example, Nurmi [24] assumes, that $\mu_R(s_i,s_j) > 0.5$ means a definite preference of s_i over s_j and the particular values from the (0.5,1] interval express the intensity of this preference. In what follows, we refer to this interpretation as Nurmi's interpretation.

Usually, the fuzzy preference relation is assumed to meet certain conditions, most often those of reciprocity and transitivity. For our main result, none of these properties are directly relevant. Nevertheless, for some parts of our presentation it is suitable to assume the reciprocity property, i.e., we will assume the following condition holds

$$\mu_R(s_i,s_j)+\mu_R(s_j,s_i)=1, \forall i \neq j$$

Such relations are known as fuzzy tournaments; cf. Nurmi and Kacprzyk [34]. Assuming a reasonable (small) cardinality of S, it is convenient to represent a preference relation R_k of the individual k as a matrix (table):

$$[r_{ij}^k]=[\mu_{R_k}(s_i,s_j)], \forall i,j,k$$

3 Fuzzy Majority, Linguistic Quantifiers, and the OWA Operators

Fuzzy majority constitutes a natural generalization of the concept of majority in the case of a fuzzy setting within which a group decision making problem is considered. A fuzzy majority was introduced into group decision making under fuzziness by Kacprzyk [11–13], and then considerable extended in the works of Fedrizzi, Herrera, Herrera-Viedma, Kacprzyk, Nurmi, Verdegay, Zadrozny etc. (see, e.g., a review by Kacprzyk and Nurmi [19], and papers cited in the bibliography).

Basically, Kacprzyk's [11–13] idea was to equate a fuzzy majority with a fuzzy linguistic quantifiers which often appear in a natural language discourse. Linguistic (fuzzy) quantifiers exemplified by expressions like "most", "almost all", etc. allow for a more flexible quantification than the classic general and existential quantifiers.

There exist a few approaches to the linguistic quantifiers modeling. Basically, we are looking for the truth of a proposition:

"Most objects posses a certain property"

that may be formally expressed as follows:

$$\underset{x \in X}{\mathrm{Q}} \; P(x) \tag{1}$$

where Q denotes a fuzzy linguistic quantifier (in this case "most"), $X = \{x_1, \ldots x_N\}$ is a set of objects, $P(\cdot)$ corresponds to the property. It is assumed that the property P is fuzzy and its interpretation may be informally equated with a fuzzy set and its membership function, i.e.:

$$\mathrm{truth}(P(x_i)) = \mu_P(x_i)$$

The first approach, proposed by Zadeh [45, 46], is called a calculus of linguistically quantified propositions. Here, a linguistic quantifier is represented as a fuzzy set $Q \in \mathrm{F}([0,1])$, where F(A) denotes the family of all fuzzy sets defined on A. For our purposes and for practical reasons, its membership function is usually assumed piece-wise linear. Thus the fuzzy set corresponding to the fuzzy quantifier Q ("most") may be defined by, e.g., the following membership function:

$$\mu_Q(y) = \begin{cases} 1 & \text{for } \mathrm{y} \geq 0.8 \\ 2y - 0.6 & \text{for } 0.3 < \mathrm{y} < 0.8 \\ 0 & \text{for } \mathrm{y} \leq 0.3 \end{cases} \tag{3}$$

The truth of the proposition (1) is determined as:

$$\mathrm{truth}(\mathrm{Q}P(X)) = \mu_Q(\sum_{i=1}^{N} \mu_P(x_i) / N) \tag{3}$$

where N=card(X).

We can also add importance to a linguistically quantified proposition, for instance

"Most important objects posses a certain property"

that may be formally expressed as follows:

$$\underset{x \in X}{QR} \; P(x) \tag{4}$$

with R representing importance and being a fuzzy set in X = $\{x\}$, $R \in \mathrm{F}(X)$, which yields:

$$\text{truth}(QR\ P(X)) = \mu_Q[\sum_{i=1}^{N}(\mu_P(x_i) \wedge (\mu_R(x_i)) / \sum_{i=1}^{N}\mu_R(x_i)] \tag{5}$$

where "∧" denotes the minimum.

Zadeh's calculus works well but for linguistic quantifiers that are "fuzzy enough". Since we wish to encompass in our approach both fuzzy and non-fuzzy quantifiers (corresponding to fuzzy and non-fuzzy majorities and other choice rationality conditions), and also to have a higher notational uniformity, we will use here another approach to the modeling of fuzzy linguistic quantifiers by using Yager's OWA (ordered weighted averaging) operators [42, 43] (see also Yager and Kacprzyk's [44] volume).

An OWA operator O of dimension n is defined as:

$$O : \Re^n \to \Re$$

$$W = [w_1, \ldots, w_n], \quad w_i \in [0,1], \quad \sum_{i=1}^{n} w_i = 1$$

$$O(a_1, \ldots a_n) = \sum_{j=1}^{n} w_j b_j, \quad b_j \text{ is } j\text{-th largest of the } a_i\text{'s}$$

Thus, an OWA operator is fully defined by its vector of weights W. The correspondence between an OWA operator (its vector of weights) and a fuzzy linguistic quantifier in Zadeh's sense is often given by the well-known formula:

$$w_i = \mu_Q(i/n) - \mu_Q((i-1)/n) \tag{6}$$

Basically, using this formula we may define an OWA operator that behaves (in the sense of its aggregating bahavior) similarly to a Zadeh's linguistic quantifier given by the membership function $\mu_Q(.)$.

The OWA operators provides us with a convenient, compact and simple, representation of classical quantifiers, i.e. the general and existential, respectively:

$$\forall \to W = [0, \ldots, 0, 1] \qquad O_\forall \tag{7}$$

$$\exists \to W = [1, 0, \ldots, 0] \qquad O_\exists \tag{8}$$

For our purposes, related to group decision making, the following vectors of weights define some other OWA operators that correspond to:

- ***the classic crisp majority*** (at least a half) O_{maj}

$$W = [0,\ldots,0,1,0,\ldots,0] \quad w_{\lfloor n/2 \rfloor+1} \text{ or } w_{\lceil n+1 \rceil/2} = 1 \tag{9}$$

for n odd and even, respectively

- ***the average*** O_{avg}

$$W = [1/n,\ldots,1/n] \tag{10}$$

- ***most*** (fuzzy majority) O_{most}

and the weight vector may be, e.g., calculated by using (6).

It may be noticed that the concept of a protoform in the sense of Zadeh [47] is highly relevant in this context. First of all, a protoform is defined as an abstract prototype, or summary. So, in our context, of propositions (statements) about objects (with importance or not) that possess some properties, say:

$$\text{"}X\text{'s are } P\text{"} \tag{11}$$

$$\text{"}RX\text{'s are } P\text{"} \tag{12}$$

by addition a "natural" linguistic quantifier "most" we can obtain:

$$\text{"Most } X\text{'s are } P\text{"} \tag{13}$$

$$\text{"Most } RX\text{'s are } P\text{"} \tag{14}$$

Evidently, as protoforms may form a hierarchy, we can define higher level (more abstract) protoforms, for instance replacing *most* by a general linguistic quantifier Q, we obtain, respectively:

$$\text{"}QX\text{'s are } P\text{"} \tag{15}$$

$$\text{"}QRX\text{'s are } P\text{"} \tag{16}$$

We will show that when we start with a general group decision choice rule, we can view it as a protoform which can represent a wide variety of both non-fuzzy and fuzzy group choice rules.

4 Collective Choice Rules under Fuzzy Majority

A collective choice rule, as meant here, describes how to determine a set of preferred options starting from the set of individual preference relations. Thus, it may be informally represented as follows:

$$\{R_1,\ldots,R_N\} \to 2^S$$

Notice that this expression reflects the direct approach to the determination of a solution. In fact, for our discussion later on its is not important if we assume the collective choice function to be derived directly as above, i.e. via $\{R_1,\ldots,R_N\} \to 2^S$ or indirectly, i.e. via the indirect approach $\{R_1,\ldots,R_N\} \to R \to 2^S$. It is only important that the individual preferences should somehow be aggregated so as to produce a set of options satisfying preferences of all involved parties according to some rationality principles. Here, we do not care if there are some intermediate steps in the process of choice. For example, the rule may first require creation of a group (collective) preference relation and only then – using this relation – select a set of options (via the indirect approach). Moreover, some interesting and popular rules are meant just for producing group preference relations leaving the choice of a "best" options as irrelevant or obvious (e.g., social welfare functions – cf. Sen [35]). In cases where the group preference relations are required to be linear orderings we will assume that the option(s) that is (are) first in that ordering are selected.

One of the most popular rules of aggregation is the simple majority rule (known also as the Condorcet rule) – cf. Nurmi [30]. Basically, it is assumed to work for linear orderings and produce group linear ordering (what is not always possible, in general). Thus, this rule may be described by the following formulas:

$$R(s_i,s_j) \Leftrightarrow \text{Card}\{k : R_k(s_i,s_j)\} \ge \text{Card}\{k : R_k(s_j,s_i)\} \tag{17}$$

$$S_0 = \{s_i \in S : \underset{i \neq j}{\forall} R(s_i,s_j)\} \tag{18}$$

where Card$\{A\}$ denotes cardinality of the set A, R_i, R are individual and group preference relations, respectively and S_0 is the set of collectively preferred options.

As a counterpart for this rule in the fuzzy case Nurmi [28] proposed the following rule:

$$R(s_i, s_j) \Leftrightarrow \text{Card}\,\{k : R_k(s_i, s_j) > \alpha\} \geq \textit{threshold} \tag{19}$$

$$S_0 = \{s_i \in S : \neg \exists_j R(s_j, s_i)\} \tag{20}$$

Therefore, Nurmi [28] restated (17) adapting it to the case of a fuzzy relations R_i and R and employing a more flexible concept of majority defined by a threshold. Notice that in (20) still the strict quantifying is used (referring here to the concept of the non-domination).

Kacprzyk [11–13] interpreted the rule (19) – (20) employing the concept of a fuzzy majority equated with a linguistic quantifier. He introduced the concept of a Q-core that may be informally stated in a slightly modified version, as the $Q1/Q2$-core (cf. Zadro ny [48]) as:

$CC_{Q1,Q2}$: *Set of options, which are*
"better" than most (Q1) of the rest of options from the set S for most (Q2) of individuals

$CC_{Q1,Q2} \in F(S)$

$$\mu_{CC_{Q1,Q2}}(s_i) \rightarrow \underset{s_j}{Q1} \ \underset{x_k \in X}{Q2} \ R_k(s_i, s_j) \tag{21}$$

Then, using Zadeh's fuzzy linguistic quantifiers, we obtain:

$$h_i^j = \frac{1}{N}\sum_{k=1}^{N} r_{ij}^k \quad h_i = \frac{1}{M-1}\sum_{\substack{j=1 \\ j \neq i}}^{M} \mu_{Q2}(h_i^j) \tag{22}$$

$$\mu_{CC_{Q1,Q2}}(s_i) = \mu_{Q1}(h_i)$$

where h_i^j denotes the degree to which, option s_i is better than option s_j in the opinion of all individuals,; h_i denotes the degree to which option s_i is better than all other options, in the opinion of most ($Q2$) individuals,; $\mu_{Q1}(h_i)$ denotes the degree (to be determined) to which option s_i is better than most ($Q1$) other options, in the opinion of most ($Q2$) individuals.

Formula (21) serves as a prototype for our generic collective choice rule proposed in the next section.

5 Classification of Collective Choice Rules

It turns out that the $Q1/Q2$-core rule given by (21) and (22) may be viewed as a generic form for many well-known aggregation rules that employ, more or less explicitly, only classic (non-fuzzy) quantifiers. Thus, in order to cover them by our generic rule given by (21) and (22), we would rather use OWA operators instead of linguistic quantifiers in Zadeh's sense.

Thus, using the notation from the previous section, we first transform (21) into:

$$\underset{s_j}{Q1}\ \underset{x_k \in X}{Q2}\ R_k(s_i, s_j) \rightarrow O^j_{most}\, O^k_{most}\, R_k(s_i, s_j)$$

In our next discussion j and k will be indexing the set of options and individuals, respectively. Thus, O^j_{most} (O^k_{most}) denotes an OWA operator aggregating some values for all options (individuals) and governed by the weight vector indicated by the lower index, i.e. corresponding to the linguistic quantifier *most* with weights determined by (6).

Now, the generic collective choice rule (CCR) proposed in his paper may be expressed as follows:

$$\mu_{CCR}(s_i) = O_1\, O_2\, R_k(s_p, s_q)$$

This form has a number of "degrees of freedom". Namely, specific collective choice rules may be recovered by specifying:

1. what are the upper indexes of the OWA operators, i.e., if we first aggregate over the individuals and then over the options or in the opposite way,
2. what are weight vectors of both the OWA operators,
3. whether the pair of option indexes (p, q) corresponds to (i, j) or to (j, i)

Therefore, we can basically distinguish four types of collective choice rules:

I. $\mu_{CCR}(s_i) = O^k_1\, O^j_2\, R_k(s_i, s_j)$

II. $\mu_{CCR}(s_i) = O^j_1\, O^k_2\, R_k(s_i, s_j)$

III. $\mu_{CCR}(s_i) = O^k_1\, O^j_2\, R_k(s_j, s_i)$

IV. $\mu_{CCR}(s_i) = O^j_1\, O^k_2\, R_k(s_j, s_i)$

In order to identify the classical rules covered by this generic scheme, which are meant to provide a non-fuzzy set of options as a solution, we have to propose a way to determine such a non-fuzzy set of preferred options having a fuzzy set represented by the membership function μ_{CCR}.

This may be done in the following way:

- for type I and II rules choose s_i such that

$$\mu_{CCR}(s_i) = \max_j \mu_{CCR}(s_j)$$

- for type III and IV rules choose s_i such that

$$\mu_{CCR}(s_i) = \min_j \mu_{CCR}(s_j)$$

Now we can mention some well-known rules covered by our generic form of a collective choice rule. In the sequel we will use some specific OWA operators as defined at the end Section 3. Most of these rules assume the individual preferences in the form of linear orderings and we will comment upon some of them in these terms.

First, let us list some of the rules which may be classified as type I as well as type II rules:

1. $O_{\forall}O_{\forall}$ - a "consensus solution"
2. $O_{avg}O_{avg}$ - Borda's rule

Here we have a full, direct correspondence in case of non-fuzzy preference relations. The first rule does not require any comment (maybe except for that it works also for fuzzy preferences). In its simplest version the Borda rule scores each option by the number of options that are dominated by (non-dominating) this option in the orderings of all individuals. Our rule mimics this behavior. Counting of the dominated/non-dominating options is secured by using a simple averaging version of OWA operator given by (10), which additionally normalizes the number of options, but that does not change the order of the obtained membership degrees of the particular options to S_0. It is worth noticing again that the rule expressed in this form is applicable to any fuzzy and crisp preference relation, thus providing a counterpart of the Borda rule for preference representations that are not orderings.

The following rule may be classified as type III or IV:

3. $O_{\exists}O_{\exists}$ - the minimax degree set (Nurmi) [28]

Now, some type I rules are:

4. $O^k_{avg} O^j_{\forall}$ - the plurality voting

In the crisp case, the classical general quantifier over options picks up for each individual an option that is the first one in its ordering. Then, the options are assigned membership degrees to S_0 proportional to the number of first places in the orderings of individuals, due to the averaging OWA operator over the individuals. Thus, an option(s) with the highest membership degrees is (are) identical with those chosen due to the original plurality voting procedure.

5. $O^k_{maj} O^j_{\forall}$ - the qualified plurality voting

In this variant of plurality voting, an option to be chosen have to be first in the orderings of most individuals. These is secured in our rule by the use of the classic crisp majority OWA operator (9) instead of simple averaging.

6. $O^k_{avg} O^j_{maj}$ - the approval voting-like

provided that: O^j_{maj} models the individuals' behavior, i.e., the selection of options they approve; O^j_{most} leads to a cumulative variant

7. $O^k_{\forall} O^j_{maj}$ - the "consensus+approval voting"

Some examples of type II rules are:

8. $O^j_{\forall} O^k_{maj}$ - the simple majority (Condorcet's rule)

9. $O^j_{\forall} O^k_{\exists}$ - the Pareto rule

10. $O^j_{avg} O^k_{maj}$ - Copeland's rule

An example of a type III rule is:

11. $O^k_{most} O^j_{avg}$ - Kacprzyk's Q-minimax set [10–12]

And finally, some type IV rules are:

12. $o_{\exists}^{j} o_{avg}^{k}$ - the minimax set (cf. Nurmi [24])

13. $o_{\forall}^{j} o_{maj}^{k}$ - the Condorcet looser

14. $o_{\exists}^{j} o_{\forall}^{k}$ - the Pareto inferior options

Thus, the generic scheme proposed in this paper covers some classical rules, particularly well-known in the context of voting (cf. Nurmi [28]). Some of those rules are not, strictly speaking, collective choice rules. For example, rules 13 and 14 produce sets of options that may be viewed as being collectively rejected rather than collectively selected.

6 Concluding Remarks

We proposed a generic form of a collective choice rule. Using the OWA operators that can easily represent both fuzzy and nonfuzzy majorities, the generic collective choice rule can represent both choice rules derived in the context of group decision making under fuzziness (cf. Kacprzyk and Nurmi [19] for a comprehensive review) and also many classic collective choice rules (cf. Nurmi [28]).

We showed that the above rules can be in a sense "derived" from Kacprzyk's *Q*1/*Q*2-core [11-12] (cf. Zadrożny [48] for its redefinition that is more suitable for the present analysis) so that it can be viewed as a protoform for the other rules defined.

References

1. Aizerman M., Aleskerov F. (1995) Theory of Choice. North-Holland, Amsterdam.
2. Barrett C.R, Pattanaik PK, Salles M. (1990) On choosing rationally when preferences are fuzzy. Fuzzy Sets and Systems 34: 197-212.
3. Billot A. (1991) Aggregation of preferences: the fuzzy case, Theory and Decision 30: 51-93.
4. Fedrizzi M., Kacprzyk J., Nurmi H. (1993) Consensus degrees under fuzzy majorities and fuzzy preferences using OWA (ordered weighted average) operators. Control and Cybernetics 22: 71-80.
5. Fedrizzi M., Kacprzyk J. Nurmi H. (1996) How different are social choice functions: a rough sets approach, Quality and Quantity 30: 87-99.

6. Fedrizzi M., Kacprzyk J., Zadrożny S. (1988) An interactive multi-user decision support system for consensus reaching processes using fuzzy logic with linguistic quantifiers, Decision Support Systems 4: 313-327.
7. Fodor J.C., Roubens M. (1994) Fuzzy Preference Modelling and Multicriteria Decision Support. Kluwer, Dordrecht.
8. Herrera F., Herrera-Viedma E. (1997) Aggregation operators for linguistic weighted information, IEEE Transactions on Systems, Man and Cybernetics SMC-27: 646-656.
9. Herrera F., Herrera-Viedma E. (2000) Choice functions and mechanisms for linguistic preference relations, European Journal of Operational Research 120: 144-161.
10. Herrera F., Herrera-Viedma E., Verdegay J.L. (1996) Direct approach processes in group decision making using linguistic OWA operators, Fuzzy Sets and Systems 79: 175-190.
11. Kacprzyk J. (1985) Group decision making with a fuzzy majority via linguistic quantifiers. Part I: A consensory - like pooling. Cybernetics and Systems: an Int. Journal 16: 119-129.
12. Kacprzyk J. (1985) Group decision making with a fuzzy majority via linguistic quantifiers. Part II: A competitive - like pooling. Cybernetics and Systems: an Int. Journal 16: 131-144.
13. Kacprzyk J. (1986) Group decision making with a fuzzy majority, Fuzzy Sets and Systems 18: 105-118.
14. Kacprzyk J., Fedrizzi M. (1986) 'Soft' consensus measures for monitoring real consensus reaching processes under fuzzy preferences, Control and Cybernetics 15: 309-323.
15. Kacprzyk J., Fedrizzi M. (1988) A `soft' measure of consensus in the setting of partial (fuzzy) preferences, European Journal of Operational Research 34: 316-325.
16. Kacprzyk J., Fedrizzi M. (1989) A 'human-consistent' degree of consensus based on fuzzy logic with linguistic quantifiers, Mathematical Social Sciences 18: 275-290.
17. Kacprzyk J., Fedrizzi M. (eds.) (1990) Multiperson Decision Making Problems Using Fuzzy Sets and Possibility Theory, Kluwer, Dordrecht/Boston/London.
18. Kacprzyk J., Fedrizzi M., Nurmi H. (1992) Group decision making and consensus under fuzzy preferences and fuzzy majority, Fuzzy Sets and Systems 49: 21-31.
19. Kacprzyk J., Nurmi H. (1988) Group decision making under fuzziness, in R. Słowiński (ed.): Fuzzy Sets in Decision Analysis, Operations Research and Statistics, Kluwer, Boston, pp. 103-136.
20. Kacprzyk J., Nurmi H., Fedrizzi M. (eds.) (1996) Consensus under Fuzziness, Kluwer, Boston.
21. Kacprzyk J., Nurmi H., Fedrizzi M. (1999) Group Decision Making and a Measure of Consensus under Fuzzy Preferences and a Fuzzy Linguistic Majority, in: LA. Zadeh and J. Kacprzyk (eds.): Computing with Words in

Information/Intelligent Systems. Part 2. Foundations. Physica-Verlag, Heidelberg and New York, pp. 233-243.
22. Kacprzyk J., Roubens M. (eds.) (1988) Non-conventional Preference Relations in Decision Making, Springer - Verlag, Berlin and New York.
23. Kacprzyk J., Zadrożny S. (2000) Collective choice rules under linguistic preferences: an example of the computing with words/perceptions paradigm, Proceedings of 9th IEEE International Conference on Fuzzy Systems (FUZZ-IEEE'2000), San Antonio, USA, pp. 786-791.
24. Kacprzyk J., Zadrożny S. (2001) Computing with words in decision making through individual and collective linguistic choice rules, International Journal of Uncertainty, Fuzziness and Knowledge-Based Systems 9: 89-102.
25. Kacprzyk J., Zadrożny S. (2002) Collective choice rules in group decision making under fuzzy preferences and fuzzy majority: a unified OWA operator based approach, Control and Cybernetics 31(4): 937-948.
26. Kacprzyk J., Zadrożny S., Fedrizzi M. (1997) An interactive GDSS for consensus reaching using fuzzy logic with linguistic quantifiers, in D. Dubois, H. Prade and R.R. Yager (eds.): Fuzzy Information Engineering - A Guided Tour of Applications, Wiley, New York, pp. 567-574.
27. Kitainik L. (1993). Fuzzy Decision Procedures with Binary Relations: Towards a Unified Theory. Kluwer Academic Publishers, Boston/Dordrecht/London.
28. Nurmi H. (1981) Approaches to collective decision making with fuzzy preference relations. Fuzzy Sets and Systems 6: 249-259.
29. Nurmi H. (1983) Voting procedures: a summary analysis. British Journal of Political Science 13: 181-208.
30. Nurmi H. (1987) Comparing Voting Systems. Reidel, Dordrecht.
31. Nurmi H. (1988) Assumptions on individual preferences in the theory of voting procedures. In J. Kacprzyk and M. Roubens (eds.): Non-Conventional Preference Relations in Decision Making, Springer-Verlag, Heidelberg, pp. 142-155.
32. Nurmi H., Fedrizzi M., Kacprzyk J. (1990) Vague notions in the theory of voting. In J. Kacprzyk and M. Fedrizzi (eds.): Multiperson Decision Making Models Using Fuzzy Sets and Possibility Theory. Kluwer, Dordrecht, pp. 43-52.
33. Nurmi H., Kacprzyk J. (2000) Social choice under fuzziness: a perspective, in J. Fodor, B. De Baets and P. Perny (eds.): Preferences and Decisions under Incomplete Knowledge. Physica-Verlag (Springer-Verlag), Heidelberg and New York, pp. 107-130.
34. Nurmi H., Kacprzyk J. (1991) On fuzzy tournaments and their solution concepts in group decision making. European Journal of Operational Research 51: 223-232.
35. Nurmi H., Kacprzyk J., Fedrizzi M. (1996) Probabilistic, fuzzy and rough concepts in social choice, European Journal of Operational Research 95: 264-277.

36. Ovchinnikov S. (1990) Modelling valued preference relations, In J. Kacprzyk and M. Fedrizzi (eds.): Multiperson Decision Making Models Using Fuzzy Sets and Possibility Theory, Kluwer, Dordrecht, pp. 64-70.
37. Roubens M. (1989) Some properties of choice functions based on valued binary relations, European Journal of Operational Research 40: 309-321.
38. Schwartz T. (1986) The Logic of Collective Choice. Columbia University Press, New York.
39. Sen A. K. (1970) Collective Choice and Social Welfare. Oliver and Boyd, Edinburgh.
40. Świtalski Z. (1988) Choice functions associated with fuzzy preference relations, in J. Kacprzyk, M. Roubens (eds.): Non - conventional Preference Relations in Decision Making, Springer-Verlag, Berlin, pp. 106-118.
41. Van de Walle B., De Baets B., Kerre E. E. (1998). A plea for the use of Łukasiewicz triplets in fuzzy preference structures. Part 1: General argumentation. Fuzzy Sets and Systems 97: 349-359.
42. Yager R. R. (1988) On ordered weighted averaging aggregation operators in multi-criteria decision making. IEEE Transactions on Systems, Man and Cybernetics SMC-18: 183-190.
43. Yager R. R. (1994) Interpreting linguistically quantified propositions. International Journal of Intelligent Systems 9: 541-569.
44. Yager R. R., Kacprzyk, J. (eds.) (1997) The Ordered Weighted Averaging Operators: Theory and Applications, Kluwer, Boston.
45. Zadeh L A. (1983) A computational approach to fuzzy quantifiers in natural languages. Comp. and Maths. with Appls. 9: 149-184.
46. Zadeh L. A. (1987) A computational theory of dispositions. International Journal of Intelligent Systems 2: 39-64.
47. Zadeh L. A. (2002) A prototype-centered approach to adding deduction capabilities to search engines – the concept of a protoform. BISC Seminar, 2002, University of California, Berkeley.
48. Zadrożny S (1996) An approach to the consensus reaching support in fuzzy environment, in J. Kacprzyk, H. Nurmi and M. Fedrizzi (eds.): Consensus under Fuzziness. Kluwer, Boston, pp. 83–109.
49. Zadrożny S., Kacprzyk J. (1999) Collective choice rules: a classification using the OWA operators. Proceedings of EUSFLAT-ESTYLF Joint Conference, Palma de Mallorca, Spain, pp. 21-24.
50. Zadrożny S., Kacprzyk J. (2000) An approach to individual and collective choice under linguistic preferences, Proceedings of 8th International Conference on Information Processing and Management of Uncertainty in Knowledge-based Systems IPMU 2000, Madrid, pp. 462-469.

Part IV

Fuzzy Techniques in Human Factors Engineering and Ergonomics

The Application of Neural Fuzzy Approaches to Modeling of Musculoskeletal Responses in Manual Lifting Tasks *

Yanfeng Hou[1], Jacek M. Zurada[2], Waldemar Karwowski[3], and William S. Marras[4]

[1] Department of Electrical and Computer Engineering, University of Louisville y0hou002@louisville.edu
[2] Department of Electrical and Computer Engineering, University of Louisville jacek.zurada@louisville.edu
[3] Center for Industrial Ergonomics, Department of Industrial Engineering, University of Louisville karwowski@louisville.edu
[4] Biodynamics Laboratory, Institute for Ergonomics, The Ohio State University marras.1@osu.edu

Summary. Electromyographic signals and spinal forces in trunk muscles during lifting motion can be used to describe the dynamics of muscular activities. Yet spinal forces can not be measured directly, and EMG signals are often difficult to measure in industry due to environmental hostilities. EMG waves, however, can be treated and analyzed as responses of a system that takes kinematics measurements and other auxiliary factors as inputs. By establishing the kinematics-EMG-force relationship using neural and fuzzy approaches, we propose models for EMG and spinal force estimation. Key variables affecting EMG and forces in lifting tasks are identified using fuzzy average with fuzzy cluster distribution method. An EMG signal estimation model with a novel structure of feedforward neural network is then built. And the spinal forces are estimated by a recurrent fuzzy neural network model. The proposed neural and fuzzy approaches can prune the input variables and estimate EMG and forces effectively.

Key words: EMG, Spinal forces, Neural networks, Fuzzy logic

1 Introduction

Manual materials handling tasks performed in industry have been related to the onset of low back disorders [1]. Since electromyography (EMG) response is a direct reflection of muscular activity [2], it is important to study the EMG signals generated during lifting motion of the human body. EMG signals provide useful information about the levels of physical exertion. Studying the forces applied to the spine is also

* This study was conducted under a research grant on the "Development of a Neuro-Fuzzy System to Predict Spinal Loading as a Function of Multiple Dimensions of Risk", sponsored by the National Institute for Occupational Safety and Health (DHHS).

Y. Hou et al.: *The Application of Neural Fuzzy Approaches to Modeling of Musculoskeletal Responses in Manual Lifting Tasks*, StudFuzz **201**, 323–337 (2006)
www.springerlink.com

fundamental to the understanding of low back injury [3]. The forces on the spine during manual lifting are very useful to estimate whether a given lifting task would be safe. The clear understanding of the EMG and spinal forces in manual lifting plays an important role for guiding the reduction of musculoskeletal loading in heavy work situations [4].

However, the spinal forces can not be measured directly. Thus the low back biomechanical models are often used to estimate the loads on the lumbar spine using variety of human and environment related variables. Most biomechanical models rely on the EMG data because the internal behavior of muscles is usually described with EMG signals. However, the measuring of EMG is often difficult to perform in industrial environments. Since EMG signals are related to the kinematic characteristics, evaluating EMG from kinematics measurements and other auxiliary factors becomes a better choice [5].

In view of the above, the kinematics-EMG-force relationship could be found in the load evaluation system. The spinal forces are connected with kinematic variables through EMG signals. So models can be developed to express the relationships and estimate EMG and forces on lumbar spine [5, 6, 7]. These models do not need the measuring of EMG signals and the use of biomechanics model.

The information obtained for the evaluation of body stresses in manual lifting activities is normally uncertain, imprecise, and noisy. The muscle activities are influenced by multiple factors, without much knowledge of their underlying dynamics. Since the exact relationships between the multiple variables are not clear in many situations, neural networks and fuzzy logic are appropriate methods in this situation.

The neural and fuzzy approaches have been successfully applied to many complex and uncertain systems. They have played an important role in solving many engineering problems. Neural networks can model the nonlinear relationship between the input and output by extracting information from examples, while fuzzy systems provide an approximate human reasoning capability in a fuzzy inference system. Fuzzy systems are good at dealing with imprecise information and it is clear how they determine their output. However, it is usually hard to determine the membership functions in the fuzzy inference system. This problem can be solved by combining neural network with the fuzzy logic. The neural network can make up membership functions and extract fuzzy rules from numeric data. This hybrid method combines the advantages of the neural network and fuzzy logic approaches.

Different neural and fuzzy models can be developed to estimate the EMG signals and spinal forces due to manual lifting tasks. The input-output relationship of the EMG and force prediction systems, however, is not well understood. It is important to find out which variables have significant influence on the forces during the lifting motion, so that they can be selected as input variables. The kinematic variables such as velocities, accelerations, and angles affect the spinal forces. Furthermore, the differences between subjects also affect the EMG responses. Different people produce different patterns of EMG even though the kinematics data may be similar in doing the same task. Subject variables include body weight, height, arm length, etc. From a lot of input candidates, if we can remove those have little or no influence on the output and put emphasis on the important variables, a more parsimonious and more effective model could be built. So it is important to identify those important input variables before building the models for EMG and force evaluation.

The modelling of the kinematics-EMG-force dynamics can be divided into three parts. In the first part the key input variables of the models are identified using a fuzzy method called Fuzzy Average with Fuzzy Cluster Distribution (FAFCD). In the second part a neural network model is built to translate kinematics data into EMG signals under different task conditions. In the third part a hybrid neuro-fuzzy model is developed to predict the forces on the lumbar spine.

2 Identification of Input Variables using FAFCD

Since we do not know how significantly each input variable affects the output of the EMG and forces, all the associated kinematic variables and subject variables are recorded. The twelve kinematic variables are dynamic variables which change their values during the motion. While the fifteen subject variables are static variables which are the anthropometric characteristics of the subjects and they are the same during a motion for a particular subject. If we take all the kinematic variables and subject variables as input of the model, the dimension of the input space will be very high. It is important to identify the influence of the variables and select only the key variables as inputs of the model.

2.1 The Fuzzy Curve Method and Its Improvements

In [8] and [9], Lin et al. proposed their "fuzzy curves" method. There are m sampling data points obtained for a nonlinear system with one output variable and n associated input variables. For each input variable x_i, the m data points are plotted in the $x_i - y$ space. A fuzzy rule is defined according to each sampling data point (x_i^j, y^j) $(i = 1, 2, ..., n, j = 1, 2, ..., m)$ in the following form:

$$R_i^j : \text{IF } x_i \text{ is } \mu_{ij}(x_i) \text{ THEN } y \text{ is } y^j;$$

where $\mu_{ij}(x_i)$ is a Gaussian membership function of x_i^j. From m data points, m fuzzy rules can be obtained. The fuzzy membership functions for input variable x_i are Gaussian membership functions centered at x_i^j:

$$\mu_{ij}(x_i) = \exp(-(\frac{x_i - \overline{x}_i^j}{\sigma})^2) \tag{1}$$

where $\overline{x}_i^j$ and σ are the center and width of the membership function, respectively. The width of the Gaussian membership function is taken as about 20% of the length of the input interval of x_i. A "fuzzy curve" can be produced using defuzzification method:

$$C_i(x_i) = \frac{\sum_{j=1}^{m} y^j \mu_{ij}(x_i)}{\sum_{j=1}^{m} \mu_{ij}(x_i)} \tag{2}$$

The fuzzy curve stands for the $x_i - y$ relationship. It can tell us if the output is changing when x_i is changing. The importance of the input variables are ranked according to the ratio of the range of y covered by the curve to the whole range of y, which is defined as Influence Rate R. The Influence Rate for variable x_i can be written as

$$R_{x_i} = \frac{C_i(x_i^u) - C_i(x_i^l)}{a} \tag{3}$$

where $C_i(x_i^u)$ is the highest point on the curve and $C_i(x_i^l)$ is the lowest point on the curve. a is the whole range of y.

This method is easy to understand and to calculate. The result obtained is straightforward. However the method can not always work well. The distribution of the sampling data set will affect the result. In other words, the influence of the input variables obtained from this method may vary from sampling to sampling. The EMG and force prediction systems are complicated nonlinear systems. It is hard to control the distribution of the sampling data. Thus we can not apply the fuzzy curves method directly to the model.

In [10] the limitation of the fuzzy curve method is discussed and improved with Fuzzy Average with Fuzzy Cluster Distribution (FAFCD). To find out the $x_i - y$ relationship using fuzzy average method, each of the input variable (except x_i) should have roughly the same distribution along the axis of x_i, respectively. But for many practical applications, this requirement is normally hard to meet. The sampling data need to be preprocessed to become a representative data set before being used to determine the influence of input variables.

To transform the sampling data of force prediction system into the required form, we use the fuzzy clustering method in [10] to change the distribution of the data set. First the data points are divided into groups using fuzzy clustering method. The number of data points in each group (fuzzy cluster) will be different since the distribution of the original data is uneven. Then one data point (for instance, the fuzzy cluster center) is used to represent each group to obtain a new data set with the distribution of fuzzy clusters. Since different number of sampling data in small regions will be replaced by the same number of cluster center, we will obtain a new data set with better distribution.

Change the Distribution of the Data Set

Each of the sampling data point (vector) represents a point in the n-dimensional Euclidean space (n is the input dimension). The purpose of clustering is to partition the data set into clusters in such a way that data points in each cluster are highly similar to each other, while data points assigned to different clusters have low degrees of similarity.

To generate even cluster distribution, we partition the input space using fuzzy rules. We build a fuzzy rule base for the nonlinear system. Those data points that can excite a particular fuzzy rule with high firing strength are grouped to the same partition. The fuzzy rule base is in the form of

IF x_1 is A_{11} and x_2 is A_{21} and ... and x_n is A_{n1} THEN y is y^1
IF x_1 is A_{12} and x_2 is A_{22} and ... and x_n is A_{n2} THEN y is y^2
...
IF x_1 is A_{1m} and x_2 is A_{2m} and ... and x_n is A_{nm} THEN y is y^m

where $A_{ij}(i = 1, 2, ..., n; j = 1, 2, ..., m)$ and y^j are fuzzy sets of x_i and y, respectively.

If the width σ of Gaussian membership functions are the same for all the fuzzy sets, the fuzzy partition generated by the fuzzy rules is an even partition.

The method is implemented as follows: the first sampling data point is taken as center of a cluster and a corresponding fuzzy rule is built. The center of the Gaussian membership function is $\overline{x}_i^j = x_i^j$; the width σ' is $1/30$ of the normalized range of the input variable.

For every sampling data point, the firing strength of each existing rule is calculated:

$$G_j = \prod_{i=1}^{n}(\mu_{ij}(x_i)) = \prod_{i=1}^{n} \exp(-(\frac{x_i - \overline{x}_{ij}}{\sigma'})^2) \tag{4}$$

AND operation is used in (4).

If the firing strength $G_j \geq \beta$, then the sampling data point is similar to the data points in the partition. Thus it belongs to this partition. β is a predefined threshold as the least acceptable degree. It decides to what extent the similarity should be in order to be classified into the partition. If the firing strength is less than the threshold β, then a new fuzzy rule (a new partition) should be created.

After all the data are partitioned, Fuzzy c-means (FCM) algorithm is used to cluster data points in each small partition. FCM allows one piece of data to belong to two or more clusters [11]. It provides a method that group data points in multidimensional space into a specific number of clusters. The same number of clusters are set for each small partition so that the distribution can be more even. Or, if the partition is small enough, we can set only one cluster for each partition and find its center by FCM. We would like to use the centers of the clusters to represent the clusters. But for real world systems, the corresponding output of the system to the centers are not available, if the centers are not coincident to the existing data points. So we use the sampling data point closest to the center of a cluster to represent the cluster. The closest data point is decided by its Euclidean distance to the center.

There is a loss of information during this process, but we can control the number of partitions to make sure only redundant data points are removed while keeping enough data points to represent all the sampling data in the input space. This is done by adjusting σ'. If $\sigma' \to 0$, then each sampling data point is a partition; if $\sigma' \to \infty$, there is only one partition.

The procedure of FAFCD is listed as follows:

1. Normalize the original data set.
2. Cluster the original data.
3. Find cluster centers and use them to form a new data set.
4. Calculate the fuzzy curves of y in each input-output space on the new data set.

5. Identify key input variables according to their Influence Rate.

Using FAFCD method, the importance of input variables of the force prediction system can be identified.

2.2 The Key Variables Identified using FAFCD

All the input variables and output variables in the original data set are normalized to the range of $[0, 1]$. Then using the fuzzy clustering method described earlier, a new data set with a different distribution to the original data set was obtained. On this new data set, the fuzzy average of y_j in each $x_i - y_j$ space was calculated. The importance of the input variables are indicated by their Influence Rate R. Based on the Influence Rate, we can identify the key variables.

According to the average Influence Rate to all the muscles, the importance of the kinematic variables and subject variables are ranked as shown in Table 1 and Table 2 respectively. It is clear that kinematic variables have more influence to forces than subject variables. Thus these twelve kinematic variables should all be selected as inputs in modelling. As for subject variables, five variables (standing height, shoulder height, lower arm length, spine length, lower leg length) have bigger influence than the others. These variables should also be taken as inputs in modelling. While examining the Influence Rates of two variables "standing height" and "shoulder height", we found that the Influence Rates of these two variables are very similar, for all forces. In other words, these two variables are correlated. They are dependent variables to each other. Therefore one of them can be removed. So at last twelve kinematic variables and four subject variables were kept. The input dimension is decreased from 27 to 16.

Table 1. Rank kinematic variables by their Average Influence Rate

Rank	Variable Name	Average Influence Rate
1	Sagittal Trunk Moment	0.171
2	Lateral Trunk Moment	0.150
3	Axis Trunk Angle	0.144
4	Sagittal Trunk Velocity	0.134
5	Axis Trunk Moment	0.120
6	Sagittal Trunk Angle	0.118
7	Axis Trunk Acceleration	0.117
8	Sagittal Trunk Acceleration	0.105
9	Lateral Trunk Velocity	0.100
10	Axis Trunk Velocity	0.097
11	Lateral Trunk Angle	0.090
12	Lateral Trunk Acceleration	0.088

Table 2. Rank subject variables by their Average Influence Rate

Rank	Variable Name	Average Influence Rate
1	Standing Height	0.091
2	Shoulder Height	0.090
3	Lower Arm Length	0.080
4	Spine Length	0.078
5	Lower Leg Length	0.066
6	body weight	0.052
7	Trunk Breadth (xyphoid)	0.050
8	Trunk Circumference	0.041
9	Trunk Depth (xyphoid)	0.033
10	Trunk Breadth (pelvis)	0.032
11	Upper Arm Length	0.030
12	Elbow Height	0.027
13	Upper Leg Length	0.017
14	Trunk Depth (pelvis)	0.012
15	Age	0.002

Knowing the influence of each variable, we can use the identified variables as inputs, instead of using hypothetically selected variables. This can greatly reduce the complexity of the model and time of modelling. In building the fuzzy neural model, less input variables means less free parameters and shorter training time.

3 Prediction of EMG Signals of Trunk Muscles using a Neural Network Model

Since there is strong relationship between kinematics and EMG activity, we can build models to simulate such a relationship. Many studies concentrated on predicting torque or kinematics from EMG [12, 13, 14], whereas predicting EMG from kinematics variables has seldom been done. Here an EMG signal prediction model is built using neural network. Kinematics variables and subject variables are selected as inputs of this model. By adding regional connections between the input and the output, the novel architecture of the neural network can have both global features and regional features extracted from the input. The global connections put more emphasis on the whole picture and determine the global trend of the predicted curve, while the regional connections concentrate on each point and modify the prediction locally. Back-propagation algorithm is used in the modeling. A basic structure of neural network designed for this problem is discussed. Then to overcome its drawbacks, a new structure is proposed.

The objective is to predict EMG magnitude of ten trunk muscles during manual lifting tasks. All the kinematics variables and the first four subject variables are selected as inputs (see Section 2). In addition, timing of the motion should also be considered as one of the input variables [15]. Without timing, the system is modeling the static states, instead of the process of the dynamic lifting motion.

3.1 A Basic Model

For the problem described above, a basic feedforward neural network model with one hidden layer can be built. At first, this model has seventeen input variables including twelve kinematics variables, four subject variables, and one timing variable. The outputs are normalized EMG magnitudes of ten trunk muscles (Right Latissimus Dorsi, Left Latissimus Dorsi, Right Erector Spine, Left Erector Spine, Right Rectus abdominus, Left Rectus Abdominus, Right External Oblique, Left External Oblique, Right Internal Oblique, and Left Internal Oblique).

As stated before, timing is used as an input variable in order to represent the process of lifting motion. But if the available measurements of motions are not synchronized, that is to say, the measurements did not capture the motions with the same starting point and the same ending point, then introducing inaccurate timing into the model would make the prediction doubtful in view of the fact that most data have not been synchronized.

In this basic model, we are predicting the EMG signals point by point. Each input vector consists of twelve kinematics variables of one sampling point of one subject, as well as the corresponding four subject variables. The timing variable determines the sampling point of the current input. The kinematics variables are time series, while the subject variables of each subject are constants. All sampling points of all subjects in a same motion were used to train the network one by one. As we can see that, the number of the training examples (training vectors) could be very big. If we have fifty subjects doing a particular motion and we got 100 sampling points for each subject, then the number of training examples will be 5000. The network has been found to often suffer from overtraining. Decreasing the learning rate can be helpful, but this will make the learning process very slow, and the prediction quality is also not good.

3.2 The Improved Model

The unsatisfactory performance of the conventional network model stated above shows that predicting point by point may not be a good idea. After all, we are modeling the whole motion. It might be better for us to predict the entire span of motion at one time. Therefore, another network with all the sampling points of a subject as one whole input vector is built. The outputs are EMG magnitudes of ten muscles of all the 100 sampling points. Thus the input space is composed of 12 (kinematics variables) * 100 (samplings) + 4 (subject variables) elements. The output space is composed of 10 (EMGs) * 100 (samplings) elements.

In this model, each training example is the whole motion of a subject. And the outputs are the EMG signals of the whole motion. This makes the problem very clear and easy to deal with. The global connections form a fully connected feedforward neural network with two hidden layers. To increase the importance of subject variables, the subject variables are connected directly to the second hidden layer [16]. Additional "regional connections" are added to connect the input neurons and output neurons which belong to the same sampling data point. The subject variables are also connected to them.

The simulations show that if proper hidden layers and nodes are selected, the fully-connected neural network without regional connections is able to capture the kinematics-EMG characteristics. However, this model has two drawbacks. Firstly, although no explicit timing variable exists, this model also suffers from asynchronization of the motion. That is because the sampling data points are arranged in time sequence in the input space. Secondly, this big network is insensitive to subject variables. The importance of an individual input is decreased because too many inputs exist. To overcome these limitations, the regional connections are added to the model. Since they only connect the input neurons and output neurons which belong to the same sampling point, this model has a better "locality". When the values in a small area of the input space changed, it will only influence the output of the corresponding small area, without interfering values outside this small area.

3.3 Advantages

The first advantage of this model is that it takes the interactions between muscles into account. The muscle activities are complex in the motion. It is known that the interactions between muscles will influence the EMG signals. By learning the whole motion, the new model can take this into account. This is a global feature that can not be extracted from isolated sampling points. Although the timing is used as one of the inputs in the previous basic model, the input-output pairs are still independent points. When the data of one sampling point are fed into the network, the behavior of the muscles before and after this point is unavailable to the network. But when the data of the whole motion are fed into the network, such information is included. The second advantage is that the training time of this model is much shorter than the training time of the previous one. That is because we are predicting one sampling point at one time in the previous model. But in this model, we are predicting the whole motion of one subject.

The improved model has better locality. For the model without regional connections, although the MAE of the prediction is not bad, the prediction doesn't fit the curve very well in the small regions. By extracting the local features and modifying the output regionally, the model with regional connections can produce a better prediction.

The improved model does not suffer from incorrect timing. As mentioned before, the network without regional connections suffers from incorrect timing. If the data used to train the model captured the whole process of the motion, their EMG pattern normally will be first going up, then going down, ending with a low level of EMG value. This is typical since during the lifting, the muscle will first contract, and then begin to relax. If in a lifting motion, the rear part of the motion is missing (which means the recorded data start from the beginning of the motion, but end before the motion is finished), the incomplete motion will not follow such trend. However, in the basic model, the neural network will take it as complete. This makes the prediction not so satisfactory. After the regional connections are added, the timing is no longer a problem. Since these connections are connected "regionally", local features of each sampling point are extracted by them.

3.4 Results

The developed model gives good prediction quality in most situations. The MAEs of females are normally larger than that of males. For different muscles, we found that four muscles (Right Rectus Abdominus, Left Rectus Abdominus, Right External Oblique, and Left External Oblique) have smaller MAE than others. That is because for the manual lifting motion, the EMG signals of these four muscles are more or less static. Therefore they are easier to predict than others. We also found that the MAEs of the left muscles were larger than MAEs of the corresponding right muscles.

Simulations also indicate that the more complex the motion is, the more difficult to predict its EMG signals. In a trial that the weight of the object to be lifted is 15 lbs, the original height is floor, and the destination height is waist, then the overall MEA of the motion is 7.5%. But if the subject is also requested to turn his body for 60 degrees, the overall MEA increases to more than 10 percent. That is because the motion is not symmetric.

4 Prediction of Spinal Forces using a Recurrent Fuzzy Neural Network Model

Neural and fuzzy approaches have been used to improve or replace the biomechanics model. In [12], Lin et al. predicted the muscle activations from EMG signals using a four-layer feed-forward neural network model trained by back-propagation learning algorithm. Luh et al. built a neural network to model the relationship between the EMG activity and elbow joint torque [17]. Liu et al. useed a neural network to predict dynamic muscle forces from EMG signals [18]. In [19] and [20], neuro-fuzzy models were developed for EMG signal classification and prosthesis control. These findings focus on building the relationship between the EMG signals of muscles and the forces on the joint. They all require the EMG signals be measured in the laboratory, which is time consuming and often impractical in industry.

To predict the dynamic forces on lumbar spine without the measuring of EMG signals and the use of biomechanics model, a recurrent fuzzy neural network (RFNN) model is built. The feedback makes it possible to take past information into account. The output of the model is computed by the current data as well as the preceding data. Time delay is incorporated in the feedback connections. It serves to preserve the past information so that the RFNN is able to handle the dynamics. A learning algorithm is used to modify the RFNN's both premise parameters and consequent parameters in order to correctly identify the nonlinear relationship between the input and output.

In the spinal load estimation model, EMG signals are used as intermediate output and are fed back to the input layer. By doing that, more information (EMG) was provided to the model and the feedback of the intermediate output has a physical meaning (the direct relationship of EMG-force). This reflects the dynamics of the system in a clear and straightforward way. At the same time, the advantages of recurrent property is utilized. The rules generated from the model can be easily interpreted and can help us understand the muscle activities better. This solves the problem that the input and output of the system have no direct and explicit physical connection. At the same time, the advantages of recurrent neural network are utilized.

4.1 Model Construction

The EMG signals of ten trunk muscles are scaled and delayed before they are fed back to the input layer. The delay of EMG is used to represent the muscle activation dynamic properties. The interaction between muscles influences the EMG and the forces on the spine. By presenting the previous EMG to the input, we hope the modle can take such interaction into account. Direct physical relationships (kinematics data-EMG and EMG-force) reside in the model. The identified kinematic variables and subject variables are inputs of the model. Forces on the spine (lateral shear force, A-P shear force and spinal compression) are outputs of the model. They are not the forces measured from the experiments since they can not be measured directly. They are actually the forces obtained from the biomechanics model. After the direct prediction model is built, the biomechanics model will be no longer needed in future.

The function of each layer is described as follows.

Layer 1 is the input fuzzification layer, which represents linguistic sets in antecedent fuzzy membership functions. Each neuron describes a membership function and encodes the center and width of membership functions. The output of this layer is the degree of membership of each input:

$$y_j^1 = \mu_{ij}(x_i) \tag{5}$$

For external input, the following Gaussian membership function is used:

$$\mu_{ij}(x_i) = exp(-(\frac{x_i - \bar{x}_{ij}}{\sigma_{ij}})^2) \tag{6}$$

For the feedback input, the following sigmoid membership function is used:

$$\mu_{ij}(x_i) = \frac{1}{1 + exp(-x_i)} \tag{7}$$

Layer 2 computes the firing strength of each fuzzy rule. Nodes in this layer perform the product operation. Those links establish the antecedent relation which is an "AND" association for each fuzzy set combination (both the external input and the feedback). The output of this layer is the firing strength of each fuzzy rule:

$$y_j^2 = \Pi_{i=1}^{n} \mu_{ij}(x_i) \tag{8}$$

Layer 3 normalizes the firing strength of each fuzzy rule. The output of the third layer is the normalized firing strength of each fuzzy rule:

$$y_j^3 = \frac{\Pi_{i=1}^{n}\mu_{ij}(x_i)}{\sum_{j=1}^{m}\Pi_{i=1}^{n}\mu_{ij}(x_i)} \tag{9}$$

Layer 4 is the defuzzification layer. Center Average defuzzificaiton is used here. The output of this layer is the overall output using Center Average defuzzification:

$$y_j^4 = \sum_{j=1}^{m} y_j^3 W_{jk} \tag{10}$$

During the training process, both the consequent and the premise parameters are tuned simultaneously. The fuzzy rules are discovered from and refined by the given input/Output data. The forces predicted for time t depend on not only the inputs at time t, but also the predicted EMG at time $t-1$, which again depend on the previous inputs. This is a dynamic approach that can represent the dynamic properties of the forces better than a feedforward network.

4.2 Simulations and Results

The performance of the proposed recurrent fuzzy neural network is evaluated with two kinds of data. One is the sagittal symmetric lifting motions, while the other one includes nonsymmetrical lifting motions. To make the results comparable, similar task variables and subject variables are selected for these two motions. In both motions, the weight of the object is 30 lbs, lift height is 30 cm, lift style is stoop, and both-handed.

In a sagittal symmetric lifting motion, the subject does not turn his body. The motion is done sagittally. This kind of motion is simpler and easier to model, comparing to the nonsymmetrical motions. Simulations show that the recurrent fuzzy neural network can model the kinematics-EMG-force relationship and give an satisfactory prediction.

If we are predicting the asymmetrical motions, we could expect that the prediction errors will be bigger compared to the sagittal symmetric motions. That is because the motion is nonsymmetrical, thus more complex than the symmetric motions. The subjects were required to turn their bodies during the lifting task.

Statistical results are used to evaluate the system performance on different types of tasks. The overall Mean Absolute Errors (MAEs) of different tasks are shown in Table 3. The variations of lateral shear force, A-P shear force and the spinal compression are around 300 Newtons, 800 Newtons and 2500 Newtons, respectively. The MAEs are out of such ranges. From the results we can see that the MAEs of the predicted sagittal symmetric tasks are much smaller than those of the prediction for nonsymmetrical tasks. This is reasonable since the muscle activities are much more complicated in the nonsymmetrical tasks.

Table 3. Overall MAEs of different types of motions. The values are forces in Newtons (percentage errors are in brackets).

Force Names	Unsym. Motion	Sag. Sym. Motions
LSF	25.5 (8.22%)	20.5 (6.61%)
ASF	80.0 (9.52%)	68.5 (8.15%)
CMP	192.5 (7.86%)	167.0 (6.82%)

5 Conclusions

This chapter discussed the EMG and spinal force evaluation models using neural and fuzzy approaches. Input variables of the models were identified using a fuzzy approach, which greatly reduce the dimension of the input space of the models. A neural network model and a recurrent fuzzy neural network model were built for EMG evaluation and spinal force evaluation, respectively.

In the neural network model, the global connections provide the model's basic prediction reference, while the additional connections enable the model to extract the relationships among regional inputs. The additional connections can reduce the adverse influence of the problem of incorrect timing.

In the recurrent fuzzy neural network model, EMG was fed back to the input, acting as a bridge between the input and the output. The delayed EMG feedback allows for better representation of the muscle activation dynamics. At the same time, the advantages of recurrent neural network can be utilized. The model predicts forces directly from kinematics data, avoiding EMG measurements and the use of biomechanical model.

References

1. W. Karwowski, W.S. Marras (Editors), "The Occupational Ergonomics Handbook", 1999, CRC Press, Boca Raton.
2. P.A. Crosby, "Use of surface electromyogram as a measure of dynamic force in human limb muscles", *Med. and Biol. Eng. and Comput.*, vol. 16, 1978, pp. 519–524.
3. D.G. Lloyd, Besier, T.F., "An EMG-driven Musculoskeletal Model to Estimate Muscle Forces and Knee Joint Moments in Vivo", *Journal of Biomechanics*, 36 (2003), pp. 765–776.
4. W. Lee, W. Karwowski, W.S. Marras, D. Rodrick, "A neuro-fuzzy model for estimating electromyographical activity of trunk muscles due to manual lifting", *Ergonomics* 46 (1-3), JAN 15 2003, pp. 285–309
5. Y. Hou, J.M. Zurada, W. Karwowski, "Prediction of EMG Signals of Trunk Muscles in Manual Lifting Using a Neural Network Model", *Proc. of the Int. Joint Conf. on Neural Networks*, Jul. 25-29, 2004, pp. 1935–1940
6. Y. Hou, J.M. Zurada, W. Karwowski, W.S. Marras, "Prediction of Dynamic Forces on Lumbar Joint Using a Recurrent Neural Network Model", *Proc. of the 2004 Int. Conf. on Machine Learning and Applications (ICMLA'04)*, Dec. 16-18, 2004, pp. 360–365

7. Y. Hou, J.M. Zurada, W. Karwowski, "A Hybrid Neuro-fuzzy Approach for Spinal Force Evaluation in Manual Materials Handling Tasks", 2005 International Conference on Fuzzy Systems and Knowledge Discovery (FSKD 2005), submitted.
8. Y. Lin, G.A. III Cunningham, "A new approach to fuzzy-neural system modeling"; *IEEE Trans. on Fuzzy Systems*, vol. 3, no. 2, May 1995 pp. 190–198
9. Y. Lin; G.A. Cunningham, S.V. Coggeshall, R.D. Jones, "Nonlinear system input structure identification: two stage fuzzy curves and surfaces"; *IEEE Trans. on Systems, Man and Cybernetics*, vol. 28, no. 5, Sept. 1998 pp. 678 -684
10. Y. Hou, J.M. Zurada, W. Karwowski, W.S. Marras, "Identification of Input Variables using Fuzzy Average with Fuzzy Cluster Distribution"; *IEEE Trans. on Fuzzy Systems*, submitted.
11. S. Auephanwiriyakul, J.M. Keller, "Analysis and efficient implementation of a linguistic fuzzy c-means"; *IEEE Trans. on Fuzzy Systems*, vol. 10, no. 5, Oct. 2002 pp. 563–582
12. L. Wang, T.S. Buchanan, "Prediction of Joint Moments Using a Neural Network Model of Muscle Activations from EMG Signals," *IEEE Transactions on Systems and Rehabilitation Engineering*, 10(1), March 2002 pp. 30–37.
13. T.C. Arthur, F.K. Robert, "EMG-based Prediction of Shoulder and Elbow Kinematics in Able-bodied and Spinal Cord Injured Individuals", *IEEE Transactions on Rehabilitation Engineering*, 8(4), Dec.2000 pp. 471–480.
14. C.S. Pattichis, C.N. Schizas, "Genetics-based Machine Learning for the Assessment of Certain Neuromuscular Disorders", *IEEE Transactions on Neural Networks*, 7(2), March 1996 pp. 427–439.
15. B. Verma, C. Lane, "Vertical Jump Height Prediction Using EMG Characteristics and Neural Networks",*Journal of Cognitive System Research* 1920000 pp. 135–141.
16. S.T. Chen; D.C. Yu, and A.R. Moghaddamjo, "Weather Sensitive Short-Term Load Forecasting Using Nonfully Connected Artificial Neural Network", *IEEE Transactions on Power Systems* 7(3), pp. 1098–1105.
17. J.J. Luh, G.C. Chang, C.K. Cheng, J.S. Lai, T.S. Kuo,, "Isokinetic elbow joint torques estimation form surface EMG and joint kinematic data: Using an artificial neural network model", *J. Electromyogr. Kinesiol.*, vol. 1, no. 9, 1999, pp. 173–183.
18. M.M. Liu, W. Herzog, H. H. Savelberg, "Dynamic muscle force predictions from EMG: An artificial neural network approach", *J. Electromyogr. Kinesiol.*, vol. 9, 1999, pp. 391–400.
19. S.E. Hussein, M.H., Granat, "Intention detection using a neuro-fuzzy EMG classifier", *Engineering in Medicine and Biology Magazine, IEEE*, vol. 21, no. 6, Nov.-Dec. pp. 123–129
20. K. Kiguchi, T. Tanaka, T. Fukuda, "Neuro-fuzzy control of a robotic exoskeleton with EMG signals", *IEEE Transactions on Fuzzy Systems*, vol. 12, no. 4, Aug. 2004 pp. 481–490
21. S. Wu and M. J. Er, "Dynamic fuzzy neural networks-a novel approach to function approximation," *IEEE Trans. on Systems, Man and Cybernetics* B, vol. 30, Apr. 2000, pp. 358–364.
22. S. Wu, M. J. Er, and Y. Gao, "A fast approach for automatic generation of fuzzy rules by generalized dynamic fuzzy neural networks," *IEEE Trans. on Fuzzy Systems*, vol. 9, Aug. 2001, pp. 578–594.
23. C.F. Juang and C.T. Lin, "An on-line self-constructing neural fuzzy inference network and its applications", *IEEE Trans. on Fuzzy Systems*, vol. 6, pp. 12–32, Feb. 1998.
24. C.H. Lee and C.C. Teng, "Identification and control of dynamic systems using recurrent fuzzy neural networks", *IEEE Trans. on Fuzzy Systems*, vol. 8, no. 4, Aug. 2000 pp. 349–366.

25. C.M. Lin and C.F. Hsu, "Supervisory recurrent fuzzy neural network control of wing rock for slender delta wings", *IEEE Trans. on Fuzzy Systems*, vol. 12, no. 5, Oct. 2004 pp. 733–742.
26. C.F. Juang, "A TSK-type recurrent fuzzy network for dynamic systems processing by neural network and genetic algorithms", *IEEE Trans. on Fuzzy Systems*, vol. 10, no. 2, Apr. 2002 pp. 155–170
27. F.J. Lin, R.J. Wai, "Hybrid control using recurrent fuzzy neural network for linear-induction motor servo drive", *IEEE Trans. on Fuzzy Systems*, vol. 9, no. 1, Jan. 2001 pp. 68–90.
28. F.J. Lin, R.J. Wai, C.M. Hong, "Hybrid supervisory control using recurrent fuzzy neural network for tracking periodic inputs", *IEEE Trans. on Neural Networks*, vol. 12, no. 1, Feb. 2001 pp. 102–115.
29. Y.C. Wang, D. Zipser, "A learning algorithm for continually running recurrent neural networks", *Neural Comput.*, vol. 1, no. 2, 1989 pp. 270–280.
30. R.J. Williams, C.J Chien, C.C. Teng, "Direct adaptive iterative learning control of non-linear systems using an output-recurrent fuzzy neural network", *IEEE Trans. on Systems, Man, and Sybernetics*, vol. 34, no. 3, Jun. 2004 pp. 1348–1359.
31. J. Zurada, W. Karwowski, and W. S. Marras, "Classification of Jobs With Risk of Low Back Disorders by Applying Data Mining Techniques", *Occupational Ergonomics*, 2005, vol. 5 no. 1, pp. 1–15.
32. W. Karwowski, "Cognitive Ergonomics: Requisite Compatibility, Fuzziness And Nonlinear Dynamics", Proceedings of *the 14 th Triennial Congress of International Ergonomics Association and the 35th Annual Meeting of the Human Factors Society*, San Diego, CA, 2000, pp. 1–580 * 1–583.
33. W. Karwowski, "The Human World of Fuzziness, Human Entropy, and the Need for General Fuzzy Systems Theory", *Journal of Japan Society for Fuzzy Theory and Systems*, 1992, vol. 4 no. 5, pp. 591–609
34. W. Karwowski, J. Grobelny, Y. Yang, and W.-G. Lee, "Applications of Fuzzy Systems in Human Factors", in: H. Zimmermman (Ed.), *Handbook of Fuzzy Sets and Possibility Theory*, Kluwer Academic Publishers, Boston, 1999, pp. 589–621.
35. W. Karwowski, and A. Mital, (Editors), "Applications of Fuzzy Set Theory in Human Factors", Elsevier Science Publishers, Amsterdam 1986.
36. G. W. Evans, W. Karwowski, and M. R. Wilhelm, (Editors), "Applications of Fuzzy Set Methodologies in Industrial Engineering", Elsevier Science Publishers, Amsterdam. 1989.

Intelligent Interfaces Based on Fuzzy Logic: Example with a Human-Error-Tolerant Interface Approach

C. Kolski and N. Malvache

LAMIH, University of Valenciennes and Hainaut-Cambrésis,
Le Mont-Houy, F-59316 Valenciennes Cedex 9, FRANCE,
christophe.kolski@univ-valenciennes.fr

Summary. The intelligent human-machine interaction domain is huge and rich in concepts, methods, models and tools. Fuzzy logic can be exploited for designing current approaches contributing to intelligent interfaces. As an illustration, this chapter describes the development of a intelligent human-machine interface which is tolerant of human error during the control of a simple industrial process. Human-error-tolerant interfaces (HETI) should be applied to industrial processes in order to keep the human operators sufficiently vigilant to enable them to handle unexpected events. With this goal, a global architecture is proposed; it integrates a human operator model (concerned with possible human actions and potential errors). For the design of this model, preliminary human behaviours and errors during the control of a simulated process have been analysed. This enables to devise general rules, to be used when programming such an interface, using fuzzy logic. The Human-error-tolerant interface design and evaluation are described.

Key words: Human-machine interface; intelligent interface; human error tolerance; fuzzy logic; human behaviour; human operator model.

1 Introduction

Today's increasingly complex industrial systems require highly skilled operators, who need to control several parameters at once in control rooms (Rasmussen, 1986; Gilmore et al., 1989; Kolski, 1997; Moray, 1997). These human operators have often to perform complex cognitive tasks, in various situations, that the automatic devices are not able to realize (Sheridan, 1988; Hoc, 1996). This implies that human reliability should be ensured (Swain and Guttman, 1983; Hollnagel, 1994; Laprie et al., 1995). Certain circumstances may bring about grave errors, even with reliable operators (Rasmussen and Vicente, 1989; Reason, 1990; Senders and Moray, 1991).

C. Kolski and N. Malvache: *Intelligent Interfaces Based on Fuzzy Logic: Example with a Human-Error-Tolerant Interface Approach*, StudFuzz **201**, 339–366 (2006)
www.springerlink.com

One way of avoiding such errors is to develop specialised, intelligent (or adaptive) help systems. In fact, the research domain concerning intelligent interfaces is important at the international level and very rich in concepts, methods, models and tools (Hancock and Chignell, 1989; Schneider-Hufschmidt et al., 1993; Kolski et al., 1992; Avouris et al., 1993; Roth et al., 1997; Kolski and Le Strugeon, 1998, Höök, 2000).

The Human-Error-Tolerant Interface (HETI) corresponds to a special kind of intelligent interface; this concept was proposed during the 1980s (Rouse and Morris, 1985; Coonan, 1986); one that is aimed at minimizing the consequences of certain human errors by keeping human operators alert in the face of an unexpected event. In order to be truly efficient, the HETI has to understand the human actions, and correct them in cases of error. It is why the preliminary analysis and modelling of the human errors is a very important step in the design of the so-called "human error tolerant interfaces". The model must be coherent with what the human operator has to do in summing the application.

Based on (Beka Be Nguema et al., 2000), this chapter is composed of three main parts. In the first part, intelligent interface approaches are classified and examples of approaches using fuzzy logic are given. In the second one, the global principles of the HETI are defined. This part explains akso preliminary experiments aimed at studying and modelling human errors that would be tolerated by the HETI to the greatest degree possible. Based on the data obtained from the preliminary experiments, a HETI is described. Of course, this HETI must be considered as a laboratory prototype, aimed at proving the feasability of such an approach. This HETI was designed using fuzzy logic, which is the practical artificial intelligence method for operator-activity modelling (Rouse and Rouse, 1983; Cacciabue et al., 1990; Shaw, 1993). The main appeal of fuzzy-logic models is that they take into account the imprecisions and uncertainty of human judgement (Zadeh, 1965; Kaufmann, 1972; Pedrycz, 1989; Yager and Filev, 1994; Cox, 1998). The evaluation of the HETI, tested within a laboratory (controlled) environment, is explained in the last part of this chapter.

2 Intelligent Interface Approaches

A major role of HMIs is to bridge the gaps which exist between humans and machines (Card, 1989). In this perspective, research on so-called "intelligent" interfaces appeared at the beginning of the 1980s. A common definition of an intelligent interface is one which provides tools to help minimize the cognitive distance between the mental model which the user has of the task, and the way in which the task is presented to the user by the computer when the task is performed (Hancock and Chignell (1989)). According to Chignell et al. (1989), an "intelligent" HMI is an "intelligent" entity which mediates between two or more interactive agents, each of which has either imperfect understanding of the way in which the others act, or an imperfect understanding of the way in which the others communicate. A global classification about intelligent HMI will be presented. Then examples of intelligent HMI based on fuzzy logic will be given.

2.1 Global Classification

In what follows, we define an "intelligent interface" as any human-machine interface that contains components, which make use of the properties of Artificial Intelligence. Thus we would apply the term to any interface which makes use of, or includes, a knowledge base, a planning mechanism, or heuristics. Equally, we include in our definition interfaces which make use of concepts relevant to Distributed Artificial Intelligence, including functions embodied as agents, including intelligent agents, autonomous agents, intentional agents, etc. (Ferber, 1995; Logan, 1998). As explained in (Kolski and Le Strugeon, 1998), most interfaces have an important characteristic in common, namely adaptability. They differ, however, in how adaptability is achieved: Figure 1 classifies five main types of systems by their degree of intelligence:

- flexible (or adaptable) interface (which we do not here consider to be inherently intelligent) allows adaptation to the preferences of the user, and according to the system in which it is used (Waern, 1989). Note also the very interesting proposition concerning « co-evolution » of interacting systems in which the user contribute directly to the evolution of the tools at his/her disposal by taking into account the acquired experience (Bourguin et al., 2001).
- The human-error-tolerant interface takes account of the behaviour of the user (Rouse and Morris, 1985). This chapter will be focussed on this type of intelligent interface: an approach based on fuzzy logic will be described.
- An adaptive HMI, in itself, should take into account the two previous approaches, but generalise them, and adapt itself to the cognitive behaviour and the tasks of the user (Edmonds, 1981; Kolski et al., 1992, 1993; Schneider-Hufschmidt et al., 1993). New concepts have appeared in the literature, such as context-aware applications (Dey et al., 2001) or HMI plasticity (Thévenin and Coutaz, 1999).

- An Operator Assistant, while having in principle the same abilities as an intelligent interface, has further levels of autonomy, and behaves almost like another human assistant (or co-pilot) (Boy, 1991; 1997). Guy Boy, one of the leaders in this area as a result of his work for NASA on "Intelligent Operator Assistant" projects, gives the following example: *"In an aircraft cockpit, a human co-pilot shares the work with the captain, but does not have the final responsibility: the captain is the captain! The captain can consult his co-pilot at any time during the flight, but the former has the ultimate responsibility. If the captain delegates part of his responsibilities to the co-pilot, then that responsibility becomes a task for the co-pilot to perform. Furthermore, the captain can interrupt a co-pilot's task at any time if he thinks it necessary. However, a co-pilot can take personal initiatives, for example to test parameters, keep himself up to date with the development of a situation, predict which tasks can be foreseen, etc. A co-pilot can make use of the instructions in an operating procedures manual to the request of the pilot. He must be able to explain, at an appropriate level of detail, the results of any such use."* Note that an operator assistant can also be modelled as an agent (see below).
- An intelligent agent has in principle all the above characteristics, but in our opinion represents the arrival of a truly "Intelligent" interface because of its ability to model cooperative human-machine systems, or even socio-technical systems (Wooldridge and Jennings, 1995; Grislin-Le Strugeon et al., 2001; Mandiau et al., 2002): the notion of an intelligent interface using the concepts of intelligent agent(s) arises from the possibility of decomposing the human-machine system into a set of agents. These agents would work in parallel or would cooperate, with the goal of solving their relevant problems in the light of the task to be performed. The results of their activities would be transmitted to the users, but at the same time they would perform a large number of other operations, for example to control the system itself. This domain is the subject of numerous current researches; see for instance (Keeble and Macredie, 2000; Klusch, 2001; Ezzedine et al., 2005).

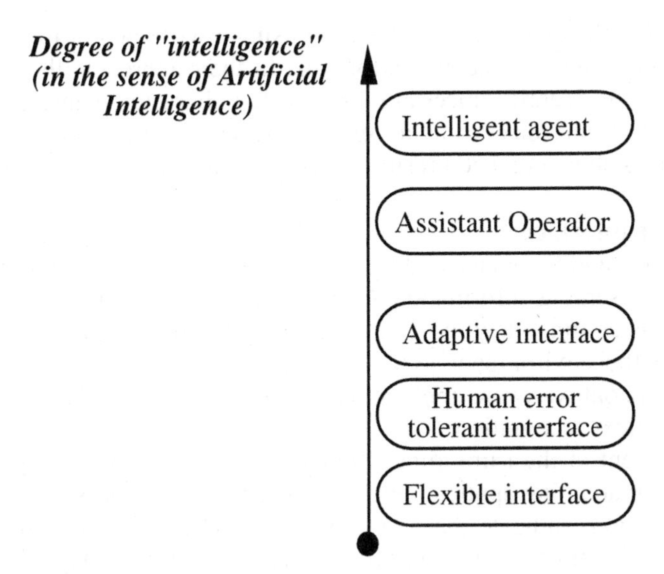

Fig. 1. Types of intelligent interfaces (Kolski and Le Strugeon, 1998)

2.2 Examples of Intelligent Interfaces Using Fuzzy Logic

Modelling technics and tools used in the current intelligent interface approaches are various: rule-based approaches, neural networks, bayesian networks, and so on. More and more approaches are based on fuzzy logic. Five representative examples (one for each type of intelligent interface) are given below.

Flexible interface

Ribeiro and Moreira (2003) describe a flexible query interface built for a relational detabase of the 500 biggest non-financial Portuguese companies. The interface is based on fuzzy logic in which queries in natural languages with pre-defined syntactical structures are performed (for instance, "Has company X a high financial health?"), and the system uses a fuzzy natural language process to provide answers (for instance, "Financial health is average (43 %), because cash flow is above_average (61%) [...] solvency is very small (14%) financial autonomy is high [...]").

Human-error tolerant interface

Pornpanomchai et al. (2001) are interested in situations in which the user do not use a keyboard to interact with a computer. They propose a non-keyboard computer interaction by using a write-pen or mouse to write Thai handwritten characters and words. In their approach, the fuzzy logic set is used to identify uncertain handwritten character shapes (in such approach, we can consider as an error a badly written character or word). There tests show precision results equal to 97.82%.

Adaptive interface

Mäntyjärvi and Seppänen (2003) are focused on the adaptation of applications representing information in handled devices: in these applications, the user is continuously moving in several simultneous fuzzy contexts (for instance the environment loudness and illumination). These authors explain that context-aware applications must be able to operate sensibly even if the context recognition is not 100% reliable and there are multiple contexts present at the same time. Mäntyjärvi and Seppänen propose an approach for adapting applications according to fuzzy context representation. User reactions indicate that (1) they accept adaptation while insisting on retaining the most control over their device, (2) abrupt adaptations and instability should be avoided in the application control.

Operator assistant

During the MESSAGE Project of analysis and evaluation of air-craft cockpits (Boy, 1983; Boy and Tessier, 1985), an operator assistant (copilot assistant) has been designed and evaluated. It is able to generate and execute tasks either in parallel (automatisms), or in sequence (controlled acts). With the aim to reason like a (simplified) copilot, such an assistant is characterized by a cognitive architecture; in its long term memory, so-called situational and analytical representations are implemented and accessible. Fuzzy logic has been used to model different types of situations. For instance, at a given time the *perceived* situation is a particular image of the local environment and is characterized by incomplete, uncertain and imprecise components; the *desired* situation is composed with a set of (fuzzy) goals which the operator intends to reach.

Interface using intelligent agents

Agah and Tanie (2000) propose intelligent graphical user interface design utilizing so-called fuzzy agents. The objective of these agents is to understand the intents of the user, and to transform the deduced intentions into system actions. For instance the motions of the mouse cursor can be interpreted by the agents and the mouse cursor can be moved according to the conveyed intentions; in these conditions, the amount of work required by the user can be reduced. The agents are specialized for different system states and/or situations (environment characteristics, task features...). Each agent is implemented using fuzzy logic control and uses a set of fuzzy rules making possible the identification of user intentions and the proposition of system actions.

3 Illustration: A Human-Error-Tolerant Interface (HETI) Based on Fuzzy Logic (Beka et al., 2000)

3.1 HETI: Global Principles

The development of interfaces that are tolerant of human errors is, in practice, based on preliminary studies of the kinds of errors that humans make in simulated and/or real conditions. In these studies, errors are identified by recording actions that result in the behaviour of the human-machine system failing to meet well-defined criteria of productivity or safety. The idea is to use such studies to develop ways which, in the real world, will make it possible to replace, improve or negate inappropriate human actions (Rouse and Morris, 1985; Hollnagel, 1989, 1994; Beka Be Nguema et al., 1993; Masson and De Keyser, 1992; Masson, 1994). There is no unique or unified architecture for a HETI to be found in the research literature.

A possible architecture of such an interface could consist of three major modules (Figure 2). A decoding module translates the human actions (i.e., the input commands of the human operator) into data that the HETI can use. A second module first identifies the human actions in all control situations. It is based on: (1) a *human actions model*, which describes what the human operator can do in all possible control situations, and (2) a model of the industrial application, which describes what should be done by the human operator in all possible control situations. This second module can then correct the actions in the event of human error. In the research literature, the *human actions model* and the *application model* can be combined into a so-called *human operator model*. This chapter uses this terminology (*human operator model*). A third module is concerned with presentation of information on a graphical screen. It has two roles: it presents the state of the process variables, according to different presentation modes, and it explains to the human operator the problems that the HETI has diagnosed and the advantages to be gained from its proposed intervention (feedback from the module #2).

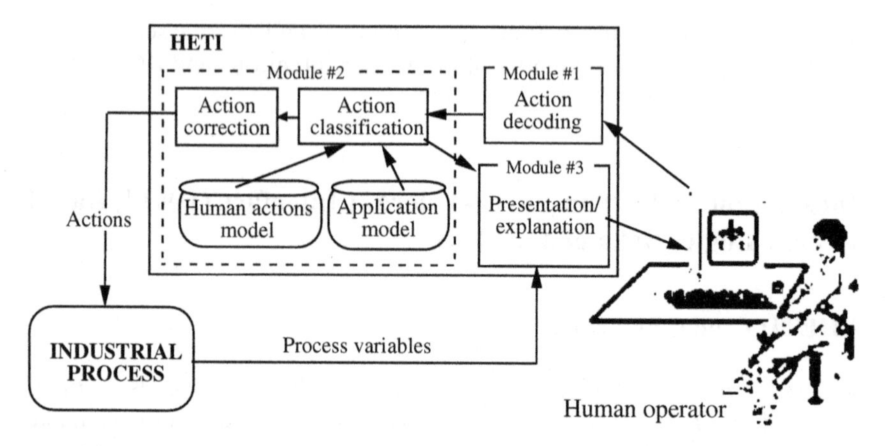

Fig. 2. Global architecture of a HETI (Beka Be Nguema et al., 2000)

3.2 Preliminary Experiments Aimed at Studying and Modelling Human Errors

One of the aims of the HETI is to identify the human operator's action. In the event of human error in context of the application, that action must be corrected. Whatever the application, it is necessary to be aware of what errors the human operator is likely to commit. This is made possible by carrying out preliminary experiments, in a real context or by simulation, with operators, and by observing the errors that they commit during the performance of their process-control tasks. Without knowledge of the possible errors, it is impossible to design the HETI. The goal of the preliminary experiments was to define the specification for the human operator model, to be integrated into the HETI. To achieve this, a study of human operator behaviour during the course of a simple simulated process was conducted, under various task configurations. Analysis of the experimental data allowed classification of the various kinds of behaviour, as well as the kinds of errors encountered in each task configuration. The task configurations used in the study are: presence of thermal inertia, presence of graphic deterioration, and double tasking (with two different tasks):

- A manual task of temperature adjustment, in which the simulated industrial process is a quadruple heat exchanger. This process consists of a cooling system that takes hot water at a temperature (T_{1e}), and flow rate (Q_{1e}), and then cools it using cold water at a temperature (T_{2e}) and flow rate (Q_{2e}). The system is made up of four heat exchangers: e_1, e_2, e_3 and e_4. These are connected in series on the hot-water circuit, and are fed cold water in parallel (see Fig. 3). Each exchanger is controlled by an up-flow dispenser, respectively named d_1, d_2, d_3 and d_4, which sends cold water into the exchanger and redirects it into a secondary pipe when switched off. A similar dispenser, called d_0, controls the hot water input in the cooling system. This redirects hot water into a secondary pipe when switched off, as would be the case in an emergency shutdown.

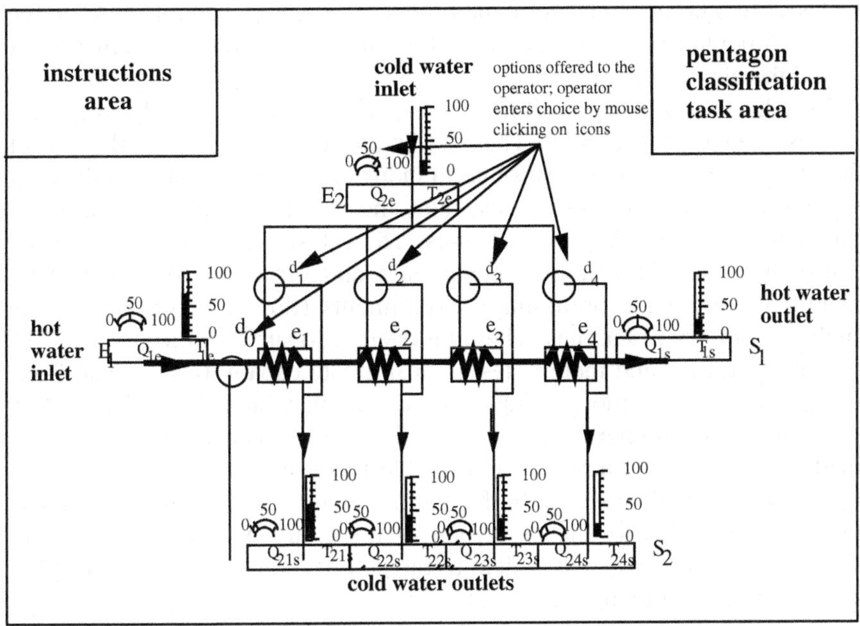

Fig. 3. Diagram illustrating the industrial process.

Figure 3 appears on a graphic screen in front of the subject. The upper left-hand part is where instructions are given to the subject. Temperatures are represented by bar graphs; flow rates are represented by dials. The upper left-hand area is used for a pentagon classification task.

- A second task consists of classifying a series of pentagons. In this classification task, 36 randomly selected pentagons, of any size, are displayed, one by one on the screen. From these, eleven pentagons belong to the "very large" category, eight to the "large" one, five to the "medium" category, four belong to the "very small" one. Each display comes with a multiple-choice question and a space where the operator enters a self-evaluation of the certainty on a scale of 0 to 1, where 0 indicates null certainty and 1 indicates complete certainty about this classification (this gives an evaluation of the degree of confidence of the human operator in performing the task). This pentagon-classification task is an often-used and well-documented study in the authors' laboratory (See for instance Desombre et al., 1995; Louas et al., 1998). It was selected for the present study to increase the complexity of the human task. Moreover, it uses very different assessment skills from those used in the first task. The human operator influences the process manually by clicking icons, and enters answers for the pentagon-classification task in the same way.

- These two tasks can be combined (under a so-called "double task" in the experimental tradition), including both the temperature adjustment task and the classification task.

In order to prove the feasibility of the HETI design, two relatively simple tasks have been chosen. These tasks are not an accurate reflection of the many complex situations found in industry, and particularly in the control rooms of dynamic processes; thus, the results cannot be directly extrapolated to such complex processes. These tasks have been chosen because they allow the human behaviours and errors to be exhaustively identified during the preliminary experiments (this is very important in such exploratory researches); these tasks are also sufficient to overload the human operators, and thereby test their ultimate capabilities as regards error generation. For more complex processes in which the situations can prove to be too numerous to be studied in an exhaustive manner, it is a matter of studying whether it is possible to decompose the process into several simpler sub-systems. In that case, it then becomes possible to apply the same approach to one or several of these sub-systems. This is a research line in its own right which, to the authors' knowledge, has not been studied at international level.

3.2.1 First Experiment (Single Task)

In the first experiment, conducted with 44 subjects (also called "human operators" in chapter section, even though the subjects are not real operators, but university students), the main human task consists of keeping the outlet temperature constant. First, each human operator (i.e., each subject) is instructed to aim for a temperature of between 20°C and 30°C in the outgoing hot water (T_{1s}). To achieve this, the operator may adjust the cold-water flow from Q_{2e} in increments of 10 m^3/s. The operator also has control over the on/off switches of the main hot water dispenser (d_0) and the individual heat exchangers. The operator is provided with continuous temperature and flow-rate readings from the hot water and cold water circuits, as shown in Fig. 4(a).

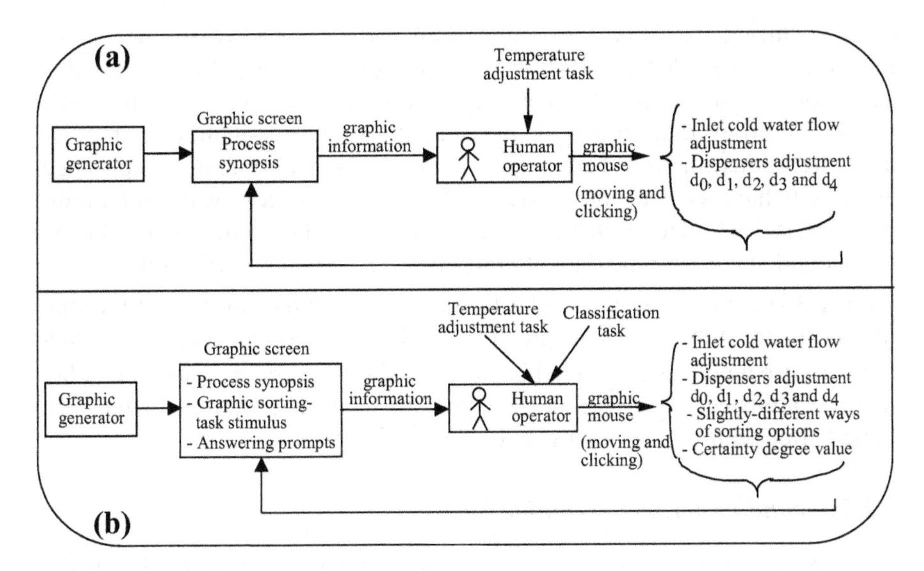

Fig. 4. Experimental system (a) Single task; (b) double task

The test consists of twenty iterations, each lasting twenty seconds. The number of temperature adjustment sequences has been fixed at twenty to provide more easily exploitable scaled results assessment. Operator-performance evaluation is accomplished by counting the number of acceptable temperature adjustments achieved by the operator over the course of the 20 sequences that the operator undergoes. An acceptable temperature adjustment is one where the desired final temperature (20°C-30°C) is achieved in less than 20 seconds. Any sequence where the desired temperature range cannot be reached within 20 seconds, or where the d_0 dispenser is used to stop the temperature-adjustment sequence, is deemed unacceptable. The human operator is unaware of the 20 s time limit. This limit was decided upon following test trials, done to validate the protocol, where 20 seconds was sufficient time for any operator to perform the task under normal operating conditions (to be defined later). However, the operator was asked, at the beginning of the test, to achieve the very best possible results.

This first experiment is divided into four stages: (1) a training stage, which enables the human operator to get familiar with the process; (2) a stage during which the temperature of the cold water inlet (T_{2e}) does not change; during this time the temperature and flow-rate of the hot water inlet vary between 10 and 100; these changes occur every 20 seconds (this stage corresponds to normal operating conditions); (3) a stage where the above conditions deteriorate due to the addition of error-inducing, graphic-data alterations; the aim here was to bring the human operator to produce an error behaviour. During the stages (2) and (3), the cold-water inlet temperature is 15°C. Hot-water inlet parameters are shown in Fig. 6. During the second stage, graphic data alterations P_1, P_2, P_3 and P_4 are introduced. Finally, (4) there is a stage similar to stage (2), with the addition of thermal inertia in the outgoing hot water (T_{2e}). This inertia was selected so that temperature would seem to change slowly. The temperature variation delay may be adjusted according to the intended goal. The optimal value, obtained after preliminary testing, is 0.25 s/°C.

3.2.2 Second Experiment (Double Task)

For this second experiment, 28 of the 44 subjects were available. In the second experiment, each operator is to undertake the following tasks, illustrated in Fig. 4(b): one temperature-adjustment task, as described above, one pentagon-classification task which involves classifying 36 pentagons (appearing one by one on the graphic screen) according to pre-existing templates, then self-evaluating the certainty of this classification on a scale of 0 to 1. This second experiment is divided into three stages: (1) a training stage for the pentagon-classification task, (2) the pentagon-classification task, (3) double tasking, induced by the addition of the pentagon-classification task to the temperature-adjustment task. In every classification task, 36 pentagons are displayed, one by one, on the screen. This number was selected so that the two different tasks would take the same time. The test sequences are shown in Figure 5.

Pentagon appearances are synchronised with the beginning of the sequences; one new pentagon for every two sequences at first, then one pentagon per sequence, then two, then four (the pentagon display rate regularly increases so as to further complicate the task)

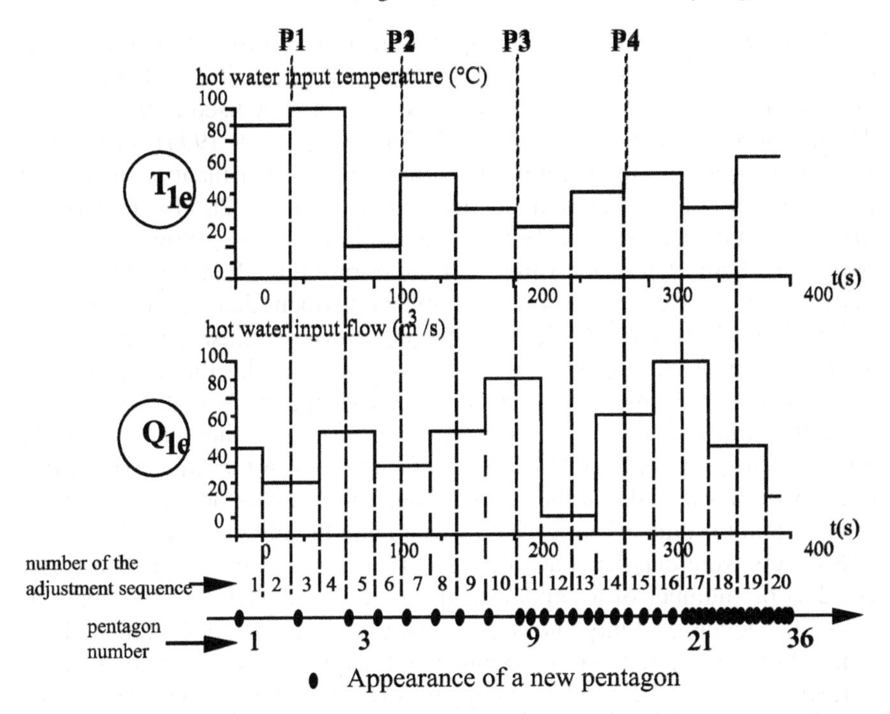

Fig. 5. Test sequences, showing changes in temperature and flow rate on the incoming hot water line. P1, P2, P3 and P4 mark the points at which the graphical display undergoes progressive deterioration (during the second of the two test phases only). P1: all thermometer outlines disappear, and outgoing hot water thermometer starts to behave erratically, P2: outgoing hot water thermometer disappears altogether, P3: outgoing cold water thermometers start to behave erratically, P4: outgoing cold water thermometers disappear altogether.

3.2.3 Results

Each experiment starts after the subject has completed an anthropometric identification questionnaire. This lasts from 40 to 60 minutes, according to the time needed by each operator to become familiar with the process. During the temperature-adjustment stage, the data collected are the variations in temperature, flow-rate and dispenser status parameters over time. At the end of the experiment, subjects are required to fill out another questionnaire; this time concerning the operator's perceptions about the experiment, data deterioration, and whether any available means were left either unused or little used by the operator during the experiment.

The results were processed using classical descriptive statistical methods (Snedecor and Cochran, 1989; Hardy and Bryman, 2004). Cumulative curves, histograms and hierarchical classification were used in the process. These results are fully described in (Beka Be Nguema, 1994). Forty subjects underwent the temperature-adjustment task experiment without data deterioration but with thermal inertia added; of these, 28 also underwent the double-task experiment. Some facts could be noted following the experiments: subject reaction times and the duration of the adjustment were both longer in the case of an unacceptable adjustment than in the case of an acceptable adjustment; subjects either used every adjustment parameter available, or used only the cold-water flow-rate, in the temperature-adjustment task; some errors were due to the operator's inability to estimate the limits of the acceptable temperature range, thus causing slight 'oversteering'; an analysis of the subjects' answers in the after-experiment questionnaire showed that the subjects took into account only the outlet parameter and the adjustment variables when conducting the task.

Subjects were classified according to their performance, which was defined as the number of acceptable adjustments achieved over the total adjustment sequences. Only one subject had a performance of less than 10/20 when doing every temperature-adjustment sequence. Most subjects had a performance over 12/20. Two subjects' strategies gave good results. The first one, used by all but one of the subjects, was to use every available parameter: only the cold-water inlet at first, then the dispensers as needed. Another strategy, used by the remaining subject (who was the exception), was to use only the cold-water inlet, even if two sequences were then impossible to achieve.

Three kinds of behaviours were encountered among the subjects: (1) the "high-risk" takers: these continued the task even when insufficient information was available, or when they did not use "upstream" information; (2) the "measured-risk" takers: these continued the task until a certain critical point (varying from one subject to another) was reached, and then preferred to stop the process; (3) the "no-risk" takers: these stopped the process by activating the emergency d_0 dispenser as soon as something was amiss, especially during the data-deterioration stage.

Four main kinds of errors were observed. These are, from the most frequent to the least frequent, as follows: (1) errors caused by *lack of attention* (Reason, 1990); when the operator used the emergency shutdown during the temperature-adjustment task without thermal inertia and without data deterioration; (2) intended "errors" due to the operator's *lack of motivation* which can be seen during non-critical sequences of the first stage (the subjects concerned do not admit to these errors, which are therefore difficult to analyse); errors caused by *lack of understanding* (Reason,

1990), which are typical of the beginning of the temperature-adjustment task without thermal inertia and without data deterioration; behaviour is hesitant; these could also be delayed lack-of-attention errors; errors due to *poor estimation of the results* (Leplat, 1985); these occur when the operators poorly estimate the outlet hot water temperature or the size of the pentagon. These errors have been considered in the HETI design.

3.3 HETI Design Based on the Data Obtained from the Preliminary Experiments

3.3.1 From Strictly Manual to Automatic Functioning Modes

The system can work using any of the five modes seen in Fig. 6. In the "strictly automatic" mode, an automatic process-control system is implemented by the HETI when requested by the human operator. Actually, the process-control system is a fuzzy controller. The human operator has no further direct control over the process when using this mode. The strictly automatic mode could be useful to an inexperienced operator, by indicating the appropriate method of handling the process. The "strictly manual" mode can only be activated on a request from the human operator. It gives the human operator total control over the process. When this mode is activated, the HETI is prevented from interfering with the process. The "temporarily manual" or "normal" mode is the default functioning mode of the system. In this mode the system is controlled by the human operator, but the HETI is active. The "temporarily automatic" mode can only be activated by the HETI, following a human error. The HETI leaves this mode as soon as the process reaches a non-critical state. It uses the same fuzzy controller as the "strictly automatic" mode. The "transitory" modes are temporarily activated during the transition from the automatic to the manual mode, or vice versa.

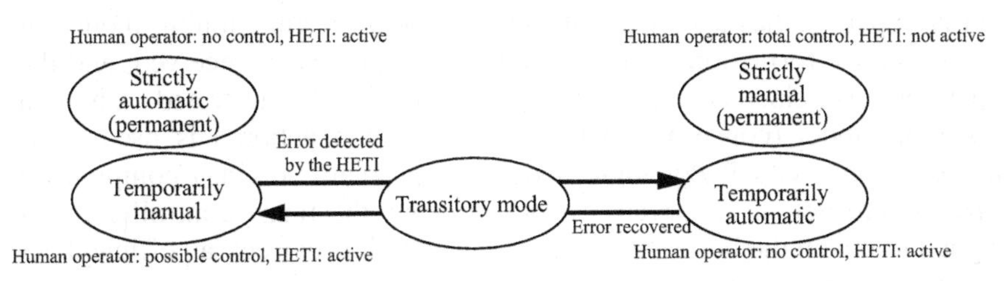

Fig. 6. HETI functioning modes

In addition to the above modes, a help (advisory) function was defined. The purposes of this function are (1) to warn the human operator that an error has probably been made and (2) to give advice on the correct course of action. These actions are the same as those that would be taken by the human-operator model if the HETI were active.

The "strict" modes are permanent modes, where the HETI has a passive role towards the operator, and cannot initiate any change of modes. The "normal" and "temporarily" modes allow the HETI to take an active role in the process.

A three-button menu, related to the functioning modes, was defined. This is accessible via the graphic screen by the human operator. The three buttons are called respectively: AUTO, MANU and HELP (Fig. 7). The AUTO and MANU buttons are mutually exclusive, i.e. selection of the AUTO button deactivates the MANU button, and vice versa. The HELP button works independently of the other buttons.

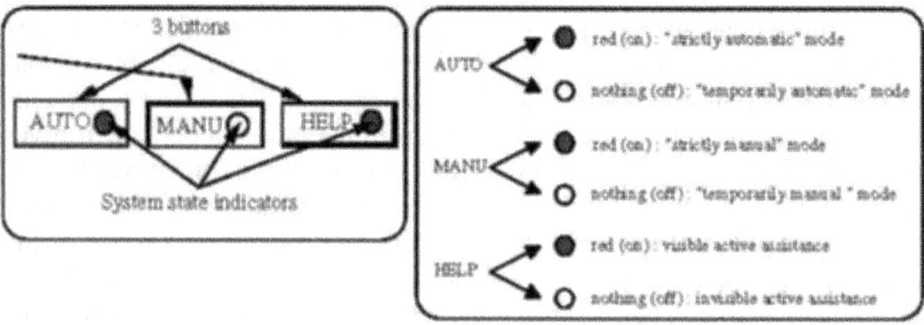

Fig. 7. Mode selection (directly on the graphic screen in a specific zone)

3.3.2 Structure of the HETI

The structure of the HETI is shown in Fig. 8. Throughout each task, the human operator has a number of options about the functioning modes of the system. Information about the state of the process is received, and the operator gets help, as needed, when the "help" mode is activated. A human operator model (concerned with possible human actions) is used. This model (along with fuzzy logic and fuzzy problem solving) comprises a fuzzy controller. In the event of human error, the fuzzy controller is designed to match the best operator strategy, which is correct: an efficient action is then applied to the (simulated) process.

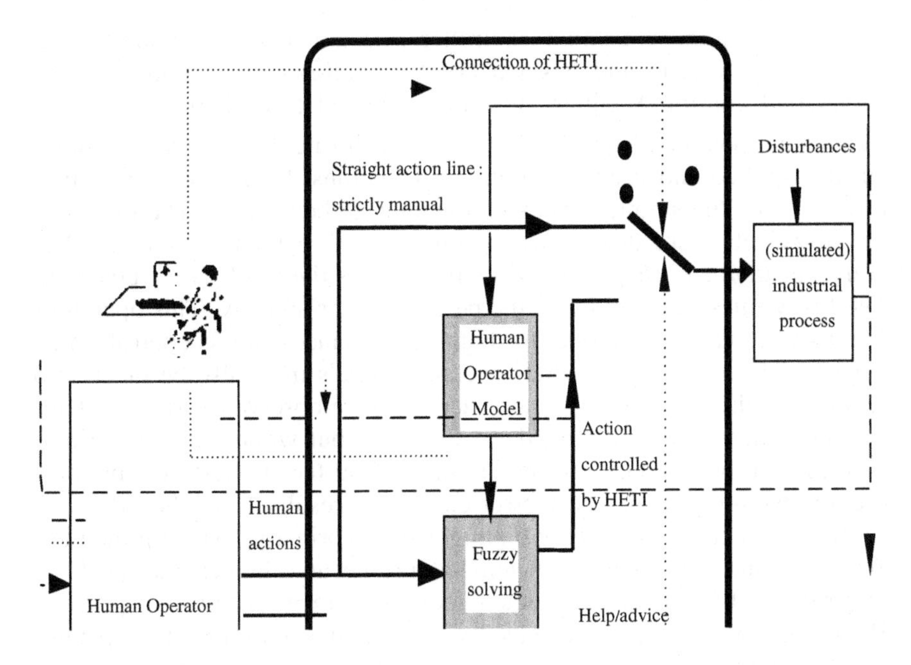

Fig. 8. HETI internal structure. The grey lines show the functioning mode options that the human operator may take. The dotted lines indicate the information output that can be used by the human operator. The bold lines show the input and the main outputs of the HETI. Finally, the fine lines show the processes mode within the HETI.

3.3.3 Description of the Human Operator Model

The temperature-adjusting operator performs the role of a temperature-control device which is responsible for keeping the hot water outlet temperature within a given range. Similarly, the fuzzy model of this human operator is the equivalent of a fuzzy controller. Fuzzy logic was selected for this model because it takes into account human imprecision and uncertainty. Moreover, it allows for descriptive modelling of knowledge and behaviour. The model's role in the HETI is: (1) to provide training to inexperienced human operators (during training, the right actions are shown to the human operator by the model); (2) to provide assistance to human operators in overload situations; in this case the model calculates the preferred course of action, which is then indicated to the human operator; (3) to assume control of the process when the operator is overwhelmed.

The fuzzy-logic reasoning controller selected here is similar to that designed by Sugeno and Nishida (1985); it allows direct output of the defuzzified control. Weights (W_i) are attributed to each rule (i). These weights are obtained from the premise of each rule. Every rule is systematically applied and used for control calculations. Fuzzification was performed using the simpler trapezoid function, to begin with. The rules and the fuzzy sets were determined according to five linguistic values: VN (very negative), N (negative), Z (zero), P (positive), VP (very positive). The fuzzy rules were set using the best operator's strategy. This operator's actions were used as a model for high-performance process control. In an ideal HETI, other (non-optimal) operator models must also be taken into account. In this case this operator's actions were observed during the temperature-adjustment task with thermal inertia, but without data deterioration. Indeed, preliminary testing has shown that the shortest possible procedures would give the best error-correction results from the HETI. A study of the operator's strategy highlighted two primary, logical principles. Whenever the hot water outlet temperature became higher than 30°C or lower than 20°C, the operator acted upon the number of in-use dispensers. However, if the temperature stayed within the desired range, the operator acted upon the cold-water inlet flow rate, which allows easier temperature control. This operator's strategy led to the design of five fuzzy rules, to be described in detail later.

The controller is composed of two fuzzy motors and one "strategy-choice device" (OCS) (Fig. 9), so as to use both temperature-adjustment strategies: (1) acting upon the dispensers, and (2) acting upon the cold-water inlet flow rate. The strategy-choice device compares the outgoing water temperature with a set value of 25°C, which corresponds to a mid-range temperature. This 25°C value was used for stabilising and optimising the temperature control.

Each fuzzy motor receives the $\Delta\tilde{E}$ fuzzy variables (variation of the error between the outgoing hot water temperature and the mid-range value of 25°C, over time) and $\tilde{E}$ (error between the mid-range value of 25°C and the outgoing hot water temperature of the process). However, only one of the motors selected by the OCS, does the controlling calculations. Motor #1 generates a flow-variation command, whereas Motor #2 generates a command to either add or remove a heat-exchanger.

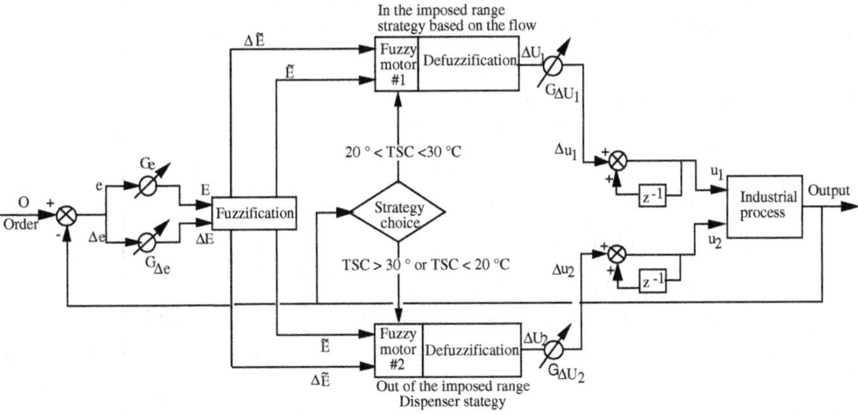

Fig. 9. Principle used for the fuzzy regulation with strategy choice (Beka Be Guema, 1994)

Fuzzification

The fuzzification of error and error variation was found using the best operator's strategy. This was done according to the five linguistic values introduced earlier (VN, N, Z, P and VP) (Fig. 10). The Z linguistic value corresponds to a membership function where a 0°C gap between 25°C and the outgoing water temperature gives an ordinate of 1. In the case of a 2.5 °C/s thermal inertia, for instance, error variation can really take only three values: -2.5°C/s, 0°C/s or +2.5°C/s, Fig. 10(b); these three values correspond to the possible rates of temperature variation within the hot-water outlet.

The five fuzzy rules can be placed in a matrix form (Fig. 11). For example, the rule yielding a very positive Δu command is the following: (if e is VN AND Δe is Z) OR (if e is VN AND Δe is N) THEN (Δu is VP).

A W_i weight, which is independent of the AND and OR fuzzy operators, is given to each "number i" rule (from 1 to 5). Weight calculation allows an estimation of the ratios in which the commands of each rule must be applied. The relative importance of each weight is related to the state of the parameters within the process to be regulated. W_i weight values are given by Guerra (1991):

$$W_i = \underset{\ell}{OR}(AND[\mu_{E_j}(e_0), \mu_{\Delta E_k}(\Delta e_0)]) \quad (1)$$

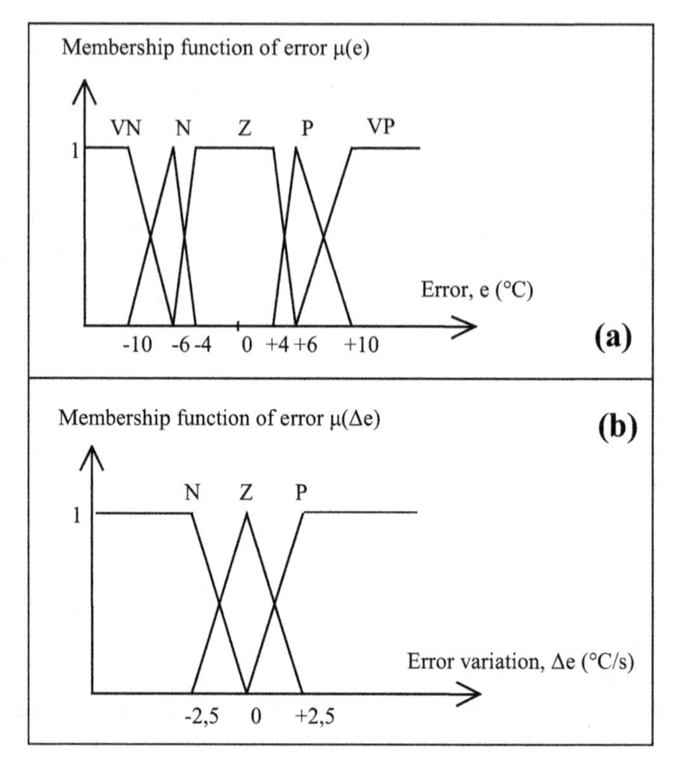

Fig. 10. (a) Membership functions of (a) error (the "no error" category (Z) was widened in order to avoid wobbling within the imposed range) and (b) error variation

Δe \ e	VP	P	Z	N	**VN**
P	VN	Z	P	P	Z
Z	VN	N	Z	P	**VP**
N	N	N	N	-	**VP**

Fig. 11. Regulation rules matrix after adaptation.

where λ is an index that takes into account the number of entry combinations yielding the same Δu command (a Δu command is a command that is acceptable to the operative part, from the fuzzy command); k are indices for the linguistic variables that are taken into account, and $\mu_X(x_0)$ is the membership function of the fuzzy value to the X fuzzy set. The Min/Max logical functions are associated to the AND/OR functions:

$$\begin{aligned} \mathrm{AND}(\tilde{A}, \tilde{B}) &= \mathrm{Min}(\tilde{A}, \tilde{B}) \\ \mathrm{OR}(\tilde{A}, \tilde{B}) &= \mathrm{Max}(\tilde{A}, \tilde{B}) \end{aligned} \tag{2}$$

The weighting formula (1) comes down to a "maximum of minima" calculation, and becomes:

$$W_i = \mathrm{Max}\left(\mathrm{Min}\left[\mu_{E_j}(e_0), \mu_{\Delta E_k}(\Delta e_0)\right]\right) \tag{3}$$

Defuzzification

The controller output is obtained after calculating the weights of each rule. This can be done in many ways. If command variables Δu_i are set at the maximum of their linguistic values, two defuzzifications are possible (Buckley and Ying, 1991): linear and non-linear defuzzifications. Non-linear defuzzification was used here:

$$\Delta u = \sum_{i=1}^{n} W_i . \Delta u_i \Big/ \sum_{i=1}^{n} W_i \tag{4}$$

where n is the number of rules (five in this case) and Δu_i are the maximum values of Δu for the flow rate and dispenser commands (Fig. 11).

Evaluation

The evaluation has been made in two stages.

During the first stage, several preliminary trials (without a human operator interacting with the HETI) have been performed in ways that validate the model technically. For the 20 temperature-adjustment sequences of the experimental protocol with thermal inertia, the model achieved the following performances: (1) for a 20/20 regulation performance, 20 acceptable adjustments were made over the 20 adjustments that had to be done; (2) during a change of input variables in the simulated process, the controller

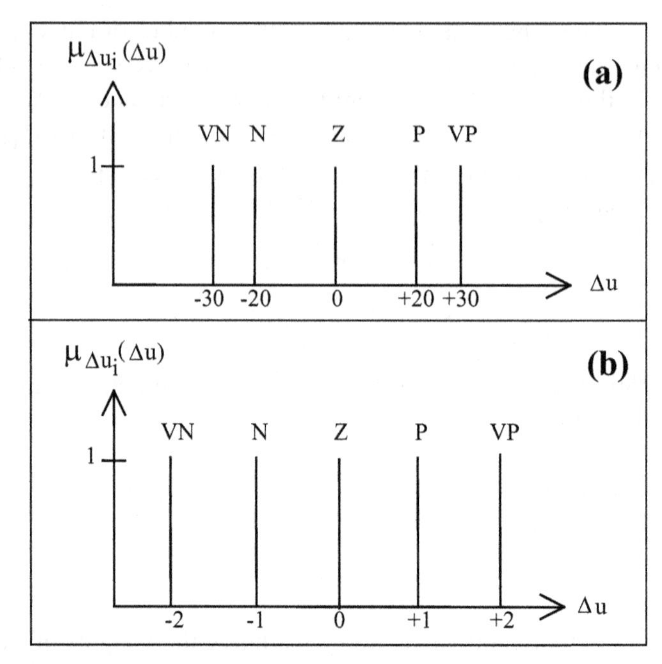

Fig. 12. Membership functions of Δu (a) for fuzzy motor #1: this fuzzy motor is used to control the inlet cold water flow. Control variables Δu_i are set at their maximum linguistic values. The arbitrarily set breakpoints are +20 and +30 for a flow-rate increase, and -20 and -30 for a flow-rate decrease. (b) for fuzzy motor #2: this motor is used to control dispensers d_1, d_2, d_3 and d4. The arbitrarily set breakpoints are +1 and +2 for an increase in the number of available dispensers, and -1 and -2 for a drop in that number.

reacts with a less-than-one second delay; (3) on average, a controller needs 3 s to find the next adjustment during each sequence change; this corresponds to the controlling program's execution time (Note that the very best human operator's execution time was 6 seconds on average, whereas the overall mean, including all subjects, was 12 seconds - the gap between the best operator's time and the controller execution time is due to the human operator's delayed reaction: 4 seconds on average); (4) the controller is stable throughout the 20 sequences; (5) a compromise was found so as to let the "strategy choice" device (Fig. 9) use the best operator's temperature-adjustment strategy while keeping the system stable.

During the second stage, an evaluation was done by five experts in human-machine systems (between 8 and 15 years of experience each in designing and evaluating such systems); they were all familiar with the research work being performed and the means being implemented. First, the experts were considered as human operators (subjects) interacting with the HETI during experiments in laboratory; in this case the aims were: (1) to check that, in situations when the HETI was in fact used, the expert's performance improved (even if it was *a priori* already proved during the preliminary trials); (2) to study their behaviour in relation to the HETI. During these experiments, the experts had to select the HETI option on the menu (Fig. 7) only if they considered it necessary; more the HETI was automatically activated when no temperature adjustment could be achieved within a predefinate period. The evaluation was performed using the double task (described in 3.2.2): after a trainig stage and simple tasks (see 3.2.1), the double task was done first without, and then with the HETI. In the double-task stages, the experts were required to complete the pentagon classification as a priority. The results obtained by these five experts are detailed in (Beka et al., 2000). They were very promising, showing the potential efficiency on such an intelligent interface approach.

These five experts were also considered as evaluators; in this case the aim was: to collect remarks and criticisms before and after the experiment, according to technical and ergonomic criteria (the principles of such classical evaluations are described by many authors, such as Nielsen (1993) or Wilson and Corlett (1996)).

4 Conclusion

Several different approaches are proposed in the intelligent human-computer interface domain, such as: flexible (or adaptable) interfaces, human-error-tolerant interfaces (HETI), adaptive interfaces, operator assistants (or intelligent operator assistants) or intelligent agents. Often, they combine technics and methods issued from artificial intelligence and human-computer interaction domains. In such approaches the modelling of the users and their objectives and tasks is very important: when the human tasks are complex, it is a difficult work for the designers, and a source of many imprecisions or uncertainties. In these conditions, fuzzy logic can be potentially very useful.

As an illustration, a human-error-tolerant interface based on fuzzy logic has been described in this chapter. The fuzzy-logic operator model was designed using an analysis of the best operator's actions after preliminary experiments. For the moment, this particular intelligent interface has been evaluated only in laboratory (with complex and representative human tasks): the first results are promising. A research and development perspective consists in adapting and evaluating this system in real situations in industry.

Acknowledgements

Particular thanks are due to Beka Be Nguema and Denis Waroux for their participation to this research, and to Gaz de France for its financial and scientific support.

References

1. Agah, A., Tanie, K. (2000). Intelligent graphical user interface design utilizing multiple fuzzy agents. Interacting with Computers, 12, 529-542.
2. Avouris, N.M., Van Liedekerke, H. (1993). User interface design for cooperating agents in industrial process supervision and control applications. *International Journal of Man-machine Studies*, 38, 873-890.
3. Beka Be Nguema, M., Waroux, D. and Malvache, N. (1993). Contributing to the design of a human error tolerant interface during the control of a simple simulated industrial process. *Proceedings 12Th European Annual Conference on Human Decision Making and Manual Control*, Kassel, Germany, June 22-24.
4. Beka Be Nguema, M. (1994). *Comportement de l'opérateur humain face à une situation dégradée et imprévue: contribution à la réalisation d'une interface homme-machine tolérante aux erreurs humaines*. Ph.D Thesis, University of Valenciennes, France, December.
5. Beka Be Nguema M., Kolski C., Malvache N., Waroux D. (2000). Design of human error tolerant interface using fuzzy logic. *Engineering Applications of Artificial Intelligence*, volume 13, issue 3, 279-292.
6. Bourguin, G., Derycke, A., Tarby, J.C. (2001). Beyond the Interface : Co-evolution inside Interactive Systems - A proposal founded on Activity Theory, *Proceedings of IHM-HCI 2001* (Lille, France, 10-14 September 2001), Springer Verlag, 297-310.
7. Boy, G. (1983). Knowledge acquisition in dynamic systems: how can logicism and situadness go together? Knowledge acquisition for knowledge-based systems. Proceedings of the 7th European Workshop, EKAW'93, Lecture Notes in Artificial Intelligence 723, Springer-Verlag, Berlin, 23-44.
8. Boy, G. (1991). Intelligent Assistant Systems. New York: Academic Press.
9. Boy, G. (1997). Knowledge elicitation for the design of software agents. In Helander M., Landauer T.K., Prabhu P. (Eds.), Handbook of Human-Computer Interaction, Elsevier Science B.V., 1997, 1203-1233.
10. Boy, G., Tessier, C. (1985). Cockpit analysis and assessment by the MESSAGE methodology. 2nd IFAC/IFIP/IFORS/IEA Conference on Analysis, Design and Evaluation of man-machine systems, Varese, Italy, September, 46-56.
11. Buckley, J.J., Ying, H. (1991). Expert fuzzy controller. Fuzzy sets and systems, 44, 373-390.

12. Cacciabue, P.C., Mancini, G. and Bersini, U. (1990). A model of operator behaviour for man-machine system simulation. Automatica, 26 (6), 1025-1034.
13. Card, S.K. (1989). Human factors and artificial intelligence. In P.A. Hancock and M.H. Chignell (Eds.), Intelligent interfaces : theory, research and design, Amsterdam, North Holland.
14. Chignell, M.H., Hancock, P.A., Loewenthal A. (1989). An introduction to intelligent interfaces. In P.A. Hancock and M.H. Chignell (Eds.), Intelligent interfaces : theroy, research and design, North Holland.
15. Coonan, T.A. (1986). An error tolerant interface using operator and system models. IEEE-CH2364-8186.
16. Cox, E. (1998). The fuzzy systems handbook: a practitioners's guide to building, using and maintaining fuzzy systems. Second edition, AP Professional, New-York.
17. Desombre, L., Malvache, N. and Vautier, J.F. (1995). Graphic communication and human errors in a vibratory environment. Proceedings 6th IFAC/IFIP/IFORS/IEA Symposium on Analysis, Design and Evaluation of Man-Machine Systems, Cambridge, MA USA, 27-29 June.
18. Dey, A.K., Salber, D., Abowd, G.D. (2001). A conceptual framework and a toolkit for supporting the rapid prototyping of context-aware applications, anchor article of a special issue on context aware computing. Human-Computer Interaction Journal, 16 (2-4), 97-166.
19. Edmonds, E.A. (1981). Adaptative Man-Computer Interfaces. In : M.J. Coombs and J.L. Alty (Eds.), Computing skills and the user interface, London, Academic press.
20. Ezzedine, H., Kolski, C., Péninou, A. (2005). Agent oriented design of human-computer interface, application to supervision of an urban transport network. Engineering Applications of Artificial Intelligence, volume 18, 255-270.
21. Ferber, J. (1995). Multi-Agent Systems: Introduction to Distributed Artificial Intelligence. Addison-Wesley.
22. Gilmore, W.E., Gertman, D.I., Blackman, H.S. (1989). User-computer interface in process control, a Human Factors Engineering Handbook. Academic Press.
23. Grislin-Le Strugeon, E., Adam, E., Kolski, C. (2001). Agents intelligents en interaction homme-machine dans les systèmes d'information. In Kolski C. (Ed.), Environnements évolués et évaluation de l'IHM. Interaction Homme Machine pour les SI 2, 207-248. Paris : Éditions Hermes.
24. Guerra, T.M. (1991). Analyse de données objectivo-subjectives: approche par la théorie des sous-ensembles flous. Ph.D Thesis, University of Valenciennes, July.
25. Hancock, P.A., Chignell, M.H. (Eds). (1989). Intelligent interfaces: theory, research and design. North Holland, Amsterdam.
26. Hardy, A.M., Bryman, A. (2004). Handbook of Data Analysis. Sage Publications Ltd.
27. Hoc J.M. (1996). Supervision et contrôle de processus, la cognition en situation dynamique. Grenoble, Presses Universitaires de Grenoble, Grenoble.

28. Hollnagel, E. (1989). The design of fault tolerant systems: prevention is better than cure. Proceedings 2nd European Meeting on Cognitive Science Approaches to Process Control, October 24-27, Siena, Italy.
29. Hollnagel, E. (1994). Human reliability analysis, context and control. Academic Press, London.
30. Höök, K. (2000). Steps to take before intelligent user interfaces become real. Interacting with computers, 409-426, 12.
31. Kaufmann, A. (1972). Introduction à la théorie des sous-ensembles flous - Tome I: éléments théoriques de base. Masson, Paris.
32. Keeble, R.D., Macredie, R.D. (2000). Assistant agents for the world wide web intelligent interface design challenges. Interacting With Computers, 12, 357-381.
33. Klusch, M. (2001). Information agent technology for the internet: a survey. Data and Knowledge Engineering, 36 (3), 337-372.
34. Kolski, C. (1997). Interfaces homme-machine, application aux systèmes industriels complexes. Editions Hermes, Paris.
35. Kolski, C., Le Strugeon, E. and Tendjaoui, M. (1993). Implementation of AI techniques for "intelligent" interface development. Engineering Applications of Artificial Intelligence, 6 (4), 295-305.
36. Kolski, C., Tendjaoui, M., & Millot, P. (1992). A process method for the design of "intelligent" man-machine interfaces: case study: "the Decisional Module of Imagery". International Journal of Human Factors in Manufacturing, vol. 2 (2), 155-175.
37. Kolski, C., Le Strugeon, E. (1998). A Review of "intelligent" human-machine interfaces in the light of the ARCH model. International Journal of Human-Computer Interaction, 10 (3), 193-231.
38. Laprie, J.C. et al. (1995). Guide de la sûreté de fonctionnement. Cépaduès Editions, Toulouse.
39. Leplat, J. (1985). Erreur humaine, fiabilité humaine dans le travail. Armand Colin, Paris.
40. Logan, B. (1998). Classifying Agent Systems. In Software Tools for Developing Agents: papers from the 1998 Workshop, J. Baxter and B. Logan (Eds.), Technical Report WS-98-10, AAAI Press, 11-21.
41. Louas, Y., Desombre, L., Malvache, N. (1998). Human behavior assessment method for a detection task performed in a stressful environment. Proceedings 17th European Annual Conference on Human Decision Making and Manual Control, Valenciennes, France, December.
42. Mandiau, R., Grislin-Le Strugeon, E. Péninou, A. (Eds.) (2002). Organisation et applications des SMA. Paris: Hermes.
43. Mäntyjärvi, J., Seppänen, T. (2003). Adapting applications in handheld devices using fuzzy context information. Interacting with Computers, 15, 521-538.
44. Masson, M., De Keyser, V. (1992). Human errors learned from field study for the specification of an intelligent error prevention system. In S. Humar (Ed.), Advances in Industrial Ergonomics and Safety IV, 1085-1092, Taylor and Francis.

45. Masson, M. (1994). Prévention automatique des erreurs de routine. Ph.D. Thesis in Psychology, University of Liège, Belgium, July.
46. Moray, N. (1997). Human factors in process control. In Handbook of human factors and ergonomics, G. Salvendy (Ed.),_John Wiley & Sons, INC., 1944-1971.
47. Nielsen, J. (1993). Usability engineering. Academic Press.
48. Pedrycz, W. (1989). Fuzzy control and fuzzy systems. Research Studies Press, John Wiley & Sons Inc.
49. Pornpanomchai, C., Batanov, D.N., Dimmitt, N. (2001). Recognizing thai handwritten characters and words for human-computer interaction. International Journal of Human-Computer Studies, 55, 259-279.
50. Rasmussen, J. (1986). Information processing and human-machine interaction, an approach to cognitive engineering. Elsevier Science Publishing.
51. Rasmussen, J., Vicente, K.J. (1989). Coping with human errors through system design: implications for ecological interface design. International Journal of Man-Machine Studies, 31, 517-534.
52. Reason, J. (1990). Human Error. Cambridge University Press, Cambridge.
53. Ribeiro, R.A., Moreira, A.M. (2003). Fuzzy query interface for a business database. International Journal of Human-Computer Studies, 58, 363-391.
54. Roth, E.M., Malin, J.T., Schrecekghost, D.L. (1997). Paradigms for intelligent interface design. In Helander M., Landauer T.K., Prabhu P. (Eds.), Handbook of Human-Computer Interaction, Elsevier Science B.V., 1997.
55. Rouse, W.B., Rouse, S.H. (1983). Analysis and classification of human error. IEEE Transactions on Systems, Man and Cybernetics, SMC-13, 539-549.
56. Rouse, W.B., Morris, N. (1985). Conceptual design of human error tolerant interface for complex engineering systems. Proceedings 2nd IFAC/IFIP/IFORS/IEA Conference on Analysis Design and Evaluation of Man Machine Systems, Varese, Italy, September.
57. Schneider-Hufschmidt, M., Kühme, T., Malinkowski, U. (Eds) (1993). Adaptive User Interfaces. North Holland.
58. Senders, J.W., Moray, N. (1991). Human error, cause, prediction and reduction. Lawrence Erlbaum Associates Publishers.
59. Shaw, I.S. (1993). Fuzzy model of a human control operator in a compensatory tracking loop. International Journal of Man-Machine Studies, 38, 305-332.
60. Sheridan, T.B. (1988). Task allocation and supervisory control. In Handbook of human-computer Interaction, M. Helander (Ed.), Elsevier Science Publishers B.V., North Holland, 1988.
61. Snedecor, G.W., Cochran, W.G. (1989). Statistical Methods. 8th edition, Iowa State Press.
62. Sugeno, M., Nishida, M. (1985). Fuzzy control of a model car. Fuzzy sets and systems, 16, 103-113.
63. Swain, A.D., Guttman, H.G. (1983). Handbook of human reliability analysis with emphasis on nuclear power plant applications. US Nuclear regulatory commission technical report Nureg/Cr 1278.

64. Thévenin, D., Coutaz, J. (1999). Plasticity of user interfaces: framework and research agenda. Proceedings of Interact'99 seventh IFIP Conference on Human-Computer Interaction, Edinburgh, Scotland.
65. Yager, R.R., Filev, D.P. (1994). Essentials of fuzzy modeling and control. Wiley, Chichester.
66. Waern, Y. (1989). Cognitive aspects of computer supported tasks. John Wiley & Sons Ltd.
67. Wilson, J.R., Corlett, E.N. (Eds) (1996). Evaluation of human works, 2nd edition. Taylor and Francis.
68. Wooldridge, M., Jennings N.R. (1995). Intelligent Agents: Theory and Practice. The Knowledge Engineering Review, 10 (2), 115-152.
69. Zadeh, L.A. (1965). Fuzzy sets. Information and Control, 8, 338-353.

Estimation of Ease Allowance of a Garment using Fuzzy Logic

Y. Chen[1], X. zeng[1], M. Happiette[1], P. Bruniaux[1], R. Ng[2], W. Yu[2]

[1] Ecole Nationale Supérieure des Arts & Industries Textiles, 9 rue de l'Ermitage, Roubaix 59100, France
[2] The Hong Kong Polytechnic University, Hong Kong, China

Summary. The ease allowance is an important criterion in garment sales. It is often taken into account in the process of construction of garment patterns. However, the existing pattern generation methods can not provide a suitable estimation of ease allowance, which is strongly related to wearer's body shapes and movements and used fabrics. They can only produce 2D patterns for a fixed standard value of ease allowance. In this chapter, we propose a new method of estimating ease allowance of a garment using fuzzy logic and sensory evaluation. Based on these values of ease allowance, we develop a new method of automatic pattern generation, permitting to improve the wearer's fitting perception of a garment. The effectiveness of our method has been validated in the design of trousers of jean type. It can also be applied for designing other types of garment.

Key words: garment design, ease allowance, fuzzy logic, sensory evaluation

1 Introduction

A garment is assembled from different cut fabric elements fitting human bodies. Each of these cut fabric elements is reproduced according to a pattern made on paper or card, which constitutes a rigid 2D geometric surface. For example, a classical trouser is composed of cut fabrics corresponding to four patterns: front left pattern, behind left pattern, front right pattern and behind right pattern. A pattern contains some reference lines characterized by dominant points which can be modified.

Of all the classical methods of garment design, draping method is used in the garment design of high level [1]. Using this method, pattern makers drape the fabric directly on the mannequin, fold and pin the fabric onto the mannequin, and trace out the fabric patterns. This method leads to the direct creation of clothing with high accuracy but needs a very long trying time and sophisticated techniques related to personalized experience of

Y. Chen et al.: *Estimation of Ease Allowance of a Garment using Fuzzy Logic*, StudFuzz **201**, 367–379 (2006)
www.springerlink.com

operators. Therefore, it can not be applied in a massive garment production. Direct drafting method is more quick and more systematic but often less precise [2]. It is generally applied in classical garment industry. Using this method, pattern makers directly draw patterns on paper using a patter construction procedure, implement in a Garment CAD system. This construction procedure does not determine the amount of ease allowance, but instead generates flat patterns for any given value of ease allowance. In practice, it is necessary to find a compromise between these two garment construction methods so that their complementarity can be taken into account in the design of new products.

To each individual corresponds a pattern whose parameters should include his body size and the amount of ease allowance of the garment. In fact, most of fabrics are extensible and can not be well deformed. Moreover, the amount of ease allowance of a garment, defined as the difference in space between the garment and the body, can be taken into account in the pattern by increasing the area along its outline.

In practice, there exist three types of ease allowance: (1) standard ease, (2) dynamic ease and (3) fabric ease.

Standard ease allowance is the difference between maximal and minimal perimeters of wear's body. It is obtained from standard human body shape for the gesture of standing or sitting still. This amount can be easily calculated using a classical drafting method [2], [3].

Dynamic ease allowance provides sufficient spaces to wearers having non standard body shapes (fat, thin, big hip, strong leg, …) and for their movements (walking, jumping, running, etc.).

Fabric ease allowance takes into account the influence of mechanical properties of fabrics of the garment. It is a very important concept for garment fitting.

Existing automatic systems of pattern generation or garment CAD systems can not determine suitable amounts of ease allowance because only standard ease allowance is taken into account. In this case, 2D patterns are generated according to the predefined standard values of ease allowance for any body shapes and any types of fabric.

In this chapter, we propose a new method for improving the system of garment pattern generation by defining the concept of fuzzy ease allowance capable of taking into account two aspects: standard ease and dynamic ease.

This method permits to generate new values of ease allowance using a Fuzzy Logic Controller (FLC), adapted to body measurements and movements of each individual. The corresponding scheme is given in Figure 1.

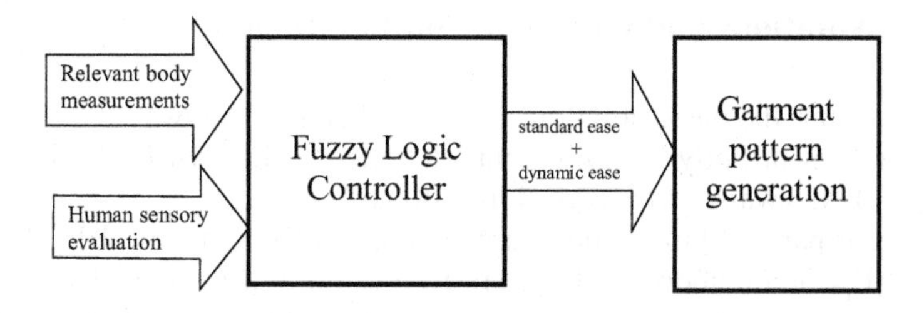

Fig. 1. General scheme of the FLC for generating fuzzy ease allowance

For simplicity, we only study the ease allowance for trousers of jean type and the influence of fabric ease related to physical properties of garment materials are not taken into account.

The FLC used for generating fuzzy ease allowance includes an interface of fuzzification, a base of fuzzy rules, an inference mechanism and an interface of defuzzification. It permits to produce the fuzzy ease allowance at each key body position, i.e. the combination of the dynamic ease and the standard ease, from a number of selected relevant measures on wearer's body shapes and comfort sensation of wearers. The amount of fuzzy ease allowance will be further used for generating more suitable patterns.

The construction of this FLC is based on a learning base built from an adjustable garment sample generating different trouser sizes and a group of representative evaluators or wearers (sensory panel). The c-means fuzzy clustering algorithm is used for optimizing the parameters of each fuzzy variable and the method of antecedent validity adaptation (AVA) is used for extracting fuzzy rules.

This chapter is organized as follows. Section 2 gives the basic notations for the elements used in the following sections. These elements include garment samples, garment sensory evaluations of wearers related to garment comfort, body measurements and body parts. Also in this section, we describe the concept of comfort degree obtained from sensory evaluation of wearers. Section 3 presents the method for selecting the most relevant body measurements in order to constitute the input variables of the FLC. In Section 4, we give details of the procedure of fuzzy modeling for estimating values of ease allowance of garments. In Section 5, experimental results and corresponding analysis are given in order to show the effectiveness of our method. A conclusion is included in Section 6.

2 Basic Notations and Garment Comfort Degree

In this chapter, the basic notations are formalized as follows.

Given a set of body measurements denoted as BM_1, BM_2, …, BM_r. For example, BM_1=waist girth, BM_2=hip girth and so on.

Different parts of human body are denoted as BP_1, BP_2, …, BP_m. For example, BP_1=lateral abdominal region, BP_2=femoral triangle and so on. The comfort degree at each body part can be evaluated subjectively by wearers.

We have produced a special sampling jean whose key body parts can be adjustable in order to generate different values of ease allowance. This sample can be used to simulate jeans of different sizes and different styles. In our project, only the normal size is studied. The corresponding ease allowance at different body parts vary from –1 to 8.

We select a group of n evaluators having different body shapes. These evaluator or wearers are denoted as WS_1, WS_2, …. , WS_n. The values of the comfort degree at each body part is evaluated by these wearers. For each wearer WS_i, the body measurements are denoted as $BM_1(WS_i)$, $BM_2(WS_i)$, …, $BM_r(WS_i)$. In this case, the body measurements for all wearers constitute a (n×r)-dimensional matrix.

In order to take into account the dynamic ease allowance, we ask each wearer to do a series of movements and evaluate the comfort degree at each body part for each movement. These movements include bending leg, bend waist, open legs at sitting and so on and they are denoted as M_1, M_2, …, M_h.

According to the above definitions, the comfort degree can be formalized by $CD(WS_i, BP_j, M_k)$

It is a function of three variables: wearer, body part and movement. It represents the sensory evaluation provided by the wearer WS_i at the body part BP_j (j=1, …, m) when he/she does the movement M_k.

In our experiments, we select 20 (n=20) wearers for evaluating comfort degrees at different body parts of the garment sample of normal size. The total number of body movements is 14 (h=14). The values of the comfort degrees given by wearers vary between 0 and 8, in which 0 represents extremely uncomfortable, 2 very uncomfortable, 4 normal, 6 very comfortable and 8 extremely comfortable.

The general comfort degree of the wearer WS_i at the body part BP_j for all movements can be calculated by

$$GCD(WS_i, BP_j)=Min_{k=1...h}\{ CD(WS_i, BP_j, M_k)\}$$

It is the comfort degree corresponding to the movement in which the wearer WS_i feels the least comfortable at the body part BP_j of the trouser sample of normal size. For example, the comfort degree of the wearer n°2 for all movements at the gluteal region can be calculated by

GCD(WS_2, gluteal region)=
Min{CD(WS_2, Gluteal region, bending leg), CD(WS_2, Gluteal region, Bend Waist) , …}

3 Selection of the Most Relevant Body Measurements

In garment design, there exist a great number of possibilities for taking body measurements. However, for a specific garment, only a very small set of measurements is relevant to the corresponding comfort degree. Then, we should only take this set of body measurements as input variables. The relevant body measurements can be selected by garment designers using their professional experience. In practice, this personalized knowledge is not normalized and each designer selects his/her own relevant body measurements, different from others. Moreover, for a specific garment, it is possible that some important body measurements are neglected by designers because they have no complete knowledge on all concerned human body parts related to their movements.

In our fuzzy model, the relevant body measurements are first selected using the data sensitivity criterion. Then, these selected variables are validated using the general knowledge of garment design. The principle of the data sensitivity criterion is given as follows.

IF a small variation of a body measurement corresponds to a large variation of the garment comfort, THEN this body measurement is considered as a sensitive variable.

IF a large variation of a body measurement corresponds to a small variation of the garment comfort, THEN this body measurement is considered as an insensitive variable.

Moreover, in practice, body measurements related to uncomfortable feeling of wearers seem to be more important than those related to comfortable feeling. According to this principle, we define, for a specific body part BP_k, an importance coefficient P_{ij} in our sensitivity criterion. We have

$$P_{ij} = \rho / (GCD(WS_i, BP_k) + GCD(WS_j, BP_k))$$

where ρ is a constant so that $\sum_{i \neq j} P_{ij} = 1$.

The value of P_{ij} is big if the comfort degrees of the two wearers WS_i and WS_j are low values. In this case, both wearers have uncomfortable feeling at the body part BP_k of the sample of normal size. The value of P_{ij} is small if the comfort degrees of WS_i and WS_j are high values. In this case, both wearers have comfortable feeling at the body part BP_k. In any cases, the value of P_{ij} is inversely proportional to the sum of the comfort degrees of WS_i and WS_j.

For a specific body part BP_k of the sample of normal size, the sensitivity criterion for selecting the most relevant body measurements is denoted as $S(BM_i, BP_k)$. It is defined by

$$S(BM_i, BP_k) = \sum_{s \neq t} \left(\frac{P_{st} \left| GCD(WS_s, BP_k) - GCD(WS_t, BP_k) \right|}{\left| BM_i(WS_s) - BM_i(WS_t) \right|} \right)$$

where $i \in \{1, \ldots, r\}$, $k \in \{1, \ldots, m\}$, $l \in \{1, 2, 3\}$ and $s,t \in \{1, \ldots, n\}$.

From this definition, we can see that the value of the sensitivity S is big if a small variation of BM_i for different wearers causes a big variation of their comfort feelings (from an uncomfortable level to a comfortable level or from a comfortable level to an uncomfortable level). For any specific body part of the studied trouser sample, all body measurements BM_1, BM_2, ..., BM_r can be ranked according to this criterion and the elements having the highest ranks are considered as the most relevant body measurements, which will be taken as input variables in the fuzzy model.

Using this criterion, for the gluteal region and the trouser of normal size, we obtain 8 body measurements with the highest ranks. These parameters can be classified into two classes as follows.

Vertical type: Waist to hip, Out leg, Curved front body rise,

Girth type : Thigh girth, Waist girth, Half back waist, Hip size, Half back hip

These body measurements are shown in Figure 2.

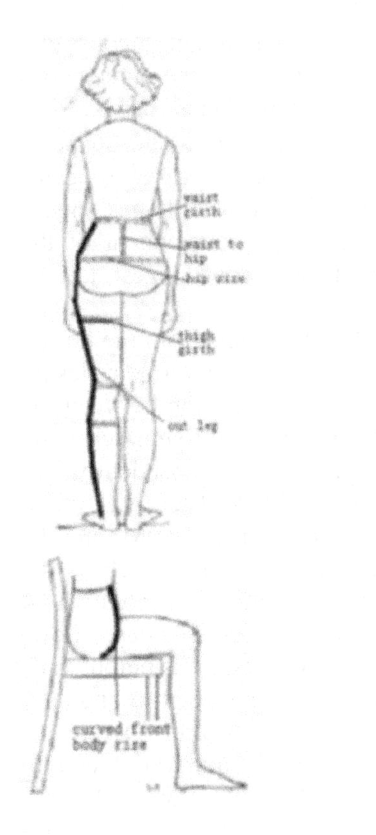

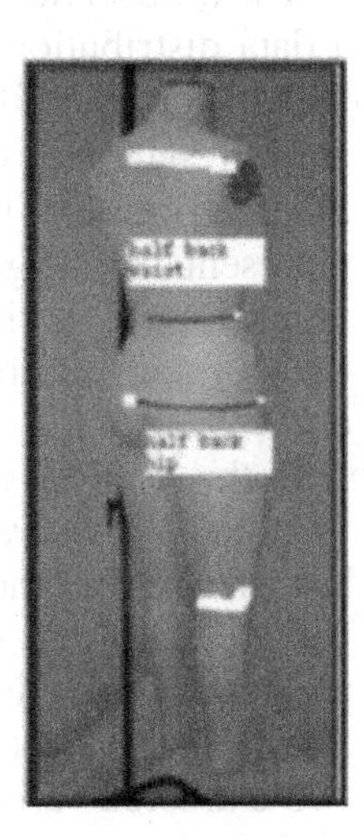

Fig. 2. Relevant body measurements related to the gluteal region and the trouser of normal size

4 Construction of the Fuzzy Model

In Section 3, we select the 8 most relevant body measurements according to the criterion of data sensitivity. However, in the fuzzy modeling procedure, 8 input variables are still too large related to the number of learning data obtained from 20 wearers. In this case, we apply the Principal Component Analysis (PCA) [4] before the procedure of fuzzy rules extraction in order to further reduce the input space. Its principle is described as follows.

PCA performs a linear transformation of an input variable vector for representing all original data in a lower-dimensional space with minimal information lost.

The q observations in $\Re^n$, corresponding to n input variables, constitute a data distribution characterized by the eigenvectors and the eigenvalues which can be easily calculated from the variable covariance matrix. PCA aims at searching for the smallest subspace in $\Re^n$ maintaining the shape of this distribution. The first component of the transformed variable vector represents the original variable vector in the direction of its largest eigenvector of the variable covariance matrix, the second component of the transformed variable vector in the direction of the second largest, and so on.

In this section, we wish to extract two first components from the 8 relevant body measurements selected in Section 3. From the garment design knowledge, we know that there exist a very weak correlation between the parameters of the vertical type and those of girth type. For simplicity, the parameters of each class (vertical type or girth type) are independently projected into two one-dimensional subspaces using PCA. Then, we obtain two extracted variables: the vertical body measurement (x_1: VBM) and the girth body measurement (x_2: GBM). These two variables as well as the general comfort degree (x_3: GCD) obtained from sensory evaluation of wearers are taken as input variables of the fuzzy model. The ease allowance for the corresponding body part, denoted as y, is taken as output variable of the model. Its values are real numbers varying between –1 and 8.

In order to extract significant fuzzy rules, we have to transform measured or evaluated numerical values of the input and output variables into linguistic values. The linguistic values of x_1 (VBM), x_2 (GBM) and y are: {very small (VS), small (S), normal (N), big (B), very big (VB)}. The linguistic values of GCD are {very uncomfortable (VUC), uncomfortable (UC), normal (N), comfortable (C), very comfortable (VC)}.

The corresponding learning input/output data, measured and evaluated on n different wearers, are denoted by $\{(x_{11}, x_{12}, x_{13}; y_1), \ldots, (x_{n1}, x_{n2}, x_{n3}; y_n)\}$. In our experiments, we have n=20. Mamdani method is used for defuzzification [5]. The fuzzy rules are extracted from these input/output learning. For each input variable x_i (i=1, 2, 3), the parameters of its membership functions are obtained using the fuzzy c-means clustering method [6]. This method permits to classify the learning data $\{x_{1i}, \ldots, x_{ni}\}$ into 5 classes, corresponding to the five fuzzy values of x_i. For each learning data x_{ki}, we obtain the membership degrees for these five fuzzy values as follows: $\mu_1(x_{ki})$, …, $\mu_5(x_{ki})$. Assuming that the corresponding membership functions take a triangular shape characterized by $Tr(a_{1i}, b_{1i}, c_{1i})$, …, $Tr(a_{5i}, b_{5i}, c_{5i})$, the 15 parameters a_{1i}, …, c_{5i} are obtained by minimizing the following criterion:

$$\sum_{j=1}^{5}\sum_{k=1}^{n}\left(Tr(a_{ji},b_{ji},c_{ji})-\mu_j(x_{ki})\right)^2$$

An example of the membership functions optimized by fuzzy c-means method is given in Figure 3. In practice, the fuzzy values obtained by this method lead to more precise results than uniformly partitioned fuzzy values because each fuzzy value generally correspond to one aggregation of learning data.

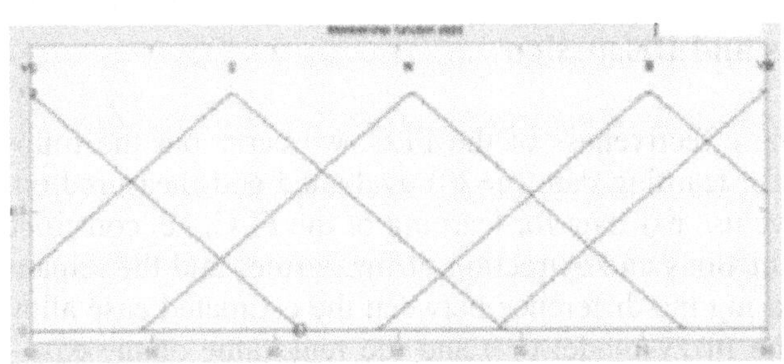

Fig. 3. The membership functions for x_1 (vertical body measurement) for lumbar

The fuzzy rules of the FLC for estimation of ease allowance are extracted from the learning data $\{(x_{11}, x_{12}, x_{13}; y_1), \ldots, (x_{n1}, x_{n2}, x_{n3}; y_n)\}$ using the method of antecedent validity adaptation (AVA) [7]. It is based on the method of Wang-Mendel [8] but some improvements have been done in it. Its principle is essentially a process by which the antecedent validity of each data, with respect to fuzzy values and fuzzy rules, is evaluated in order to adjust the output consequent. Compared with the other fuzzy extraction methods, the AVA method can effectively resolve the conflicts between different rules and then decrease the information lost by selecting only the most influential rules. In our project, the basic idea of applying the AVA method is briefly presented as follows.

According to the previous discussion, the input variable x_i ($i\in\{1, 2, 3\}$) is partitioned into 5 fuzzy values: $FV_i=\{VS, S, N, B, VB\}$. For each learning data $(x_{k1}, x_{k2}, x_{k3}; y_k)$, we set up the following fuzzy rules by combining all fuzzy values of these three input variables:

Rule j: IF (x_1 is A_1^j) AND (x_2 is A_2^j) AND (x_3 is A_3^j), THEN (y is y_k) with $A_i^j \in FV_i$ and the validity degree of the rule

$$D(rule\ j) = \sum_{i=1}^{3} \mu_{A_i^j}(x_{ki})$$

Given a predefined threshold σ, the rule j is removed from the rule base if the following condition holds: D(rule j)<σ. The value of σ should be carefully selected by taking into account the criteria of precision, complexity, robustness and significance. For big values of σ, the remaining fuzzy rules after this elimination procedure are significant, less complex but probably lead to imprecise results. For small values of σ, the number of fuzzy rules is important and some rules are not significant but lead to more precise results.

5 Results and Discussion

To test the effectiveness of the FLC, we carry out the following experiments. Of n learning data (n=20) evaluated and measured on the garment sample, we use n-1 data for learning of the FLC, i.e. construction of membership functions and extraction of fuzzy rules and the remaining one data for comparing the difference between the estimated ease allowance generated by the fuzzy model (y_m) and the real value of the ease y. Next, this procedure is repeated by taking another n-1 data for learning. Finally, we obtain the results for 20 permutations (see Figure 4). For three key body parts: lumbar, gluteal and thigh, the averaged errors between y_m and y for all the permutations are 0.62, 0.72 and 0.75 respectively. This means that the difference between the estimated ease allowance and the real ease allowance is very small (<1). The precision condition of the FLC can be satisfied.

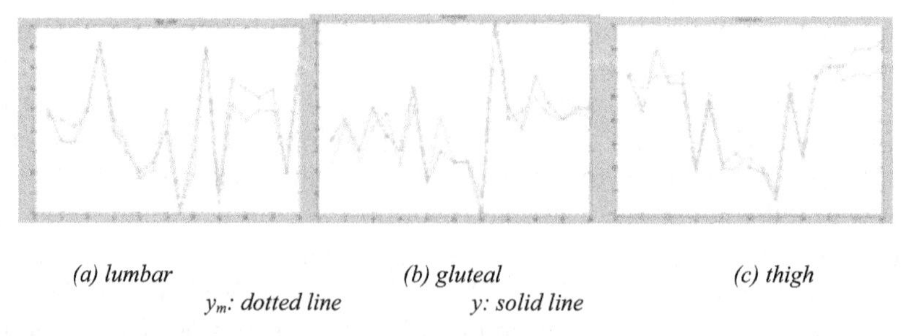

(a) lumbar *(b) gluteal* *(c) thigh*

y_m: dotted line *y: solid line*

Fig. 4. Comparison between the estimated ease allowance and its real value

Figure 4 shows that the estimated ease allowance for lumbar, gluteal and thigh generated from the fuzzy model can generally track the evolution of the real ease. y_m varies more smoothly than y because there exist an averaging effect in the output of the FLC. The difference between y_m and y is bigger for isolated test data because their behaviors can not be taken into account in the learning data.

Moreover, we obtain 16 fuzzy rules for each key body part. For lumbar, the 2 most important rules are given as follows.

1) *IF GBM=big AND VBM=normal AND GCD=very uncomfortable THEN ease=small(0.68).*
2) *IF GBM=big AND VBM=normal AND GCD=very comfortable THEN ease=normal (0.62).*

For gluteal, the 2 most important rules are given as follows.

3) *IF GBM=very big AND VBM=small AND GCD=very uncomfortable THEN ease=very small (D=0.85).*
4) *IF GBM=normal AND VBM=normal AND GCD=normal THEN ease=normal (D=0.79).*

For thigh, the 2 most important rules are given as follows.

5) *IF GBM=big AND VBM=normal AND GCD=very uncomfortable THEN ease=very small (D=0.85)*
6) *IF GBM=normal AND VBM=very small AND GCD=normal THEN ease=small (D=0.71).*

For a specific wearer, a fuzzy pattern of his trouser can be generated using the values of ease allowance at three key body parts: lumbar, gluteal and thigh. Next, we compare the patterns obtained using the classical direct drafting method with standard ease allowance and the fuzzy method proposed in this chapter. The results of this comparison are shown in Figure 5.

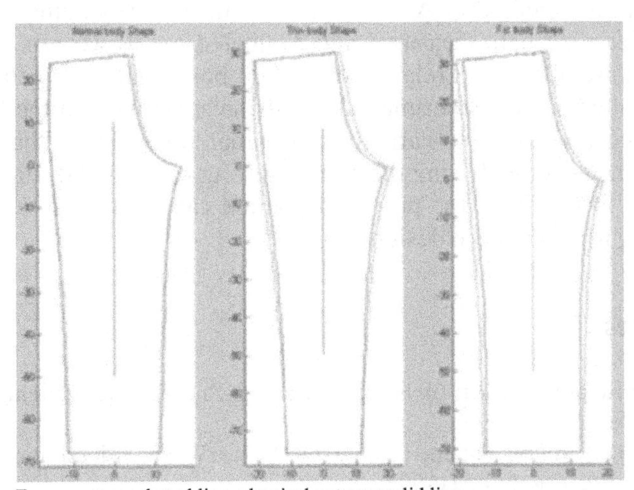

Fuzzy pattern: dotted line, classical pattern: solid line

Fig. 5. Comparison between classical pattern and fuzzy pattern

From Figure 5, we can see that the fuzzy pattern generated using the fuzzy method is very close to that of the classical method for normal body shapes. So, standard ease allowance can be generally taken into account in the proposed fuzzy model. However, there exists some difference between the fuzzy pattern and the classical pattern when human body shapes are abnormal such as fat, thin, strong legs, big hip. The fuzzy pattern is generally more sensitive to variations of human body shapes than the classical method because movements and comfort feeling of wearers have been taken into account in the values of ease allowance generated by the fuzzy model.

6 Conclusion

The proposed method combines the experimental data measured on wearer's body shapes and the wearers' sensory perception on garment samples in the construction of the FLC. The FLC has been used for estimating fuzzy ease allowance at three body parts: lumbar, gluteal and thigh. A data sensitivity based criterion has been proposed for selecting the most relevant body measurements. Then, these selected body measurements are separately projected into two one-dimensional subspaces using PCA in order to generate two features: vertical body measurement and girth body measurement. These two features as well as the comfort degree evaluated using sensory evaluation of wearers constitute the input variables of the FLC. Using the method of AVA, we extract the corresponding fuzzy rules from the learning data measured and evaluated on 20 wearers. Using the values of ease allowance related to the key body parts, we generate new garment patterns. The experimental results have shown that the fuzzy patterns are more sensitive to variations of human body shapes than the classical patterns generated using the direct grafting method with standard ease allowance. In this way, the proposed method can effectively improve the quality of garment pattern design.

References

1. C.A.Crawford, The Art of Fashion Draping, 2nd Edition, Fairchild Publications, New York (1996).
2. W.Aldrich, Metric Pattern Cutting for Men's Wear, 3rd Edition, Blackwell Science Ltd, Cambridge (1997).
3. R.Ng, Computer Modeling for Garment Pattern Design, Ph.D. Thesis, The Hong Kong Polytechnic University (1998).
4. K.Fukunaga, Introduction to Statistical Pattern Recognition, 2nd ed. San Diego, CA: Academic (1990).
5. E.H.Mamdani and S.Assilian, An experiment in linguistic synthesis with a fuzzy logic controller, *Int. J. of Man-Machine Studies*, 7, 1-13 (1975).
6. J.C.Bezdek, Pattern recognition with fuzzy objective function algorithms, Plenum Press (1981).

7. P.T.Chan and A.B.Rad, Antecedent validity adaptation principle for fuzzy systems tuning, *Fuzzy Sets and Systems*, **131**, pp.153-163 (2002).
8. L.X.Wang and J.M.Mendel, Generating fuzzy rules by learning from examples, *IEEE Trans. on SMC*, vol.22, no.6, pp.1414-1427 (1992).

Part V

Fuzzy Techniques in Manufacturing Systems and Technology Management

Intelligent Manufacturing Management

H.-J. Zimmermann

Aachen Institute of Technology Operations Research, Templergraben 64, D-52062 Aachen

Summary. This chapter is on scope and problems of manufacturing management (MM), contexts and uncertainty in MM, intelligent approaches in project management, inventory management, and production logistics. The other subject included is intelligent production planning and control.

Key words. Manufacturing management, intelligent project management, intelligent inventory management, intelligent production planning and control

1 Scope and Problems of Manufacturing Management

Manufacturing Management (MM) is as old as manufacturing is. The main problems have stayed the same, some new problems have emerged, particularly during the last decades, with growing complexity and dynamics and new tools and approaches have emerged due to new and advanced technologies, particularly in information processing. It will be useful to look at the different areas of MM on the strategic, tactical and operational level. These levels are here understood as well in terms of range and scope (importance) of decisions as in terms of the time span that they include. Depending on the context the scales can, of course, be very different. Strategic decisions can, for instance, cover a period of many decades, depending on the type of industry and the type of decision. They might, however, also only reach a few years into the future. Planning decisions may consider years, but also focus on shorter horizons of month and operational decisions may cover month but also go down to (immediate or continuous) online control.

On the strategic level of MM one will generally find decisions concerning basic technologies to be used, physical capacities, lay-outs etc. These can, of course, already concern the type of production technology (flow shop, job shop, group technology, flexible manufacturing systems, etc) but also the physical production logistic system or the type of storage facilities.

H.-J. Zimmermann: *Intelligent Manufacturing Management*, StudFuzz **201**, 383–400 (2006)
www.springerlink.com

On the tactical (planning) level of MM, which might look between weeks and years ahead, one finds the planning of the production program (as well as possible balanced to the structure of the physical capacity), one may find project management (to construct the physical capacities as well as for production itself) and one will also establish material management systems (inventory control). Policies and tools for maintenance and quality control will also have to be considered and the qualitative and quantitative structure of the facilities for production logistics. If the company is part of a supply chain coordination, actions particularly for production planning will have to be performed.

In control, eventually, the time grid may be anywhere between online and days (thinking of project management). On this level one finds most of the areas mentioned above, such as, production control (scheduling and dispatching, line balancing, tooling, process optimization in continuous production), maintenance, qua,lity control, inventory control, logistic control etc.

One may wonder why over the last 5 or 6 decades no computer based system has be designed, that performs most of the functions mentioned above satisfactorily. One rather finds hundreds or even thousands of publications in the area of MM that always concentrate on specific aspects rather than on general systems for MM. The main reason for this might be, that there is no system of this kind, because there are very many types of production and each of them requires a planning and control system that is well adapted to the special situation of a production.

2 Contexts and Uncertainty in MM

Different contexts and different types of products and productions are certainly one reason for the high diversity of different systems of MM. (This makes this area very attractive for publications, but less attractive for practitioners). Another reason is uncertainty occurring in the production area. This might be due to high complexity of the systems to be managed, to the need to plan far ahead (in strategic planning) into an unknown future, the character of the knowledge which is available about the system (which might, for instance, be linguistic in the form of existing experience), or to other reasons. The problem becomes particularly nasty if – as in production planning and control - the planning phase has to fit to the control phase.

Modeling uncertainty is obviously a genuine modeling decision, i.e. the modeler has to decide in a specific environment, whether he wants to use any of the existing uncertainty theories or whether he wants to adopt a "wait-and-see" approach and stick to deterministic, crisp models. On higher levels (strategic and planning) the use of uncertainty models makes generally more sense than on the operational level, partly because the effort is better justifies and partly the available time between the modeling and the action is longer and allows for more computations. In this case, however, one has to decide, which of the more than 25 uncertainty theories (Zimmermann, 2000) is suited in the specific context. In quality control or maintenance nowadays this will generally be probability theory or statistics. In other areas – such as production planning and control – this is still an open question. Since the adequate uncertainty theory also depends on the factor time, the situation becomes particularly difficult if , for instance, production planning (being on a rougher time grid) shall fit to production control (being on a very fine time grid). Some of the existing problems have been attacked by using "intelligent methods", which we will consider in the next section. We are, however, still far from satisfactory solutions to most of our problems of this type.

3 Intelligent Approaches in MM

The term "intelligent" is very attractive and has, therefore, be used often in many different contexts and with different meanings. A book published in 1988 (Wright and Bourne, 1988), for instance, has the title "Manufacturing Intelligence" and there are certainly some justifications for the authors to choose this title. In this chapter, however, the term "intelligent" shall only be used in the sense it is used in "Computational Intelligence" since the beginning of the 90's, i.e., as the union of Fuzzy Sets Theory (and also possibility theory, rough set theory, intuitionistic set theory, grey set theory etc.), Artificial Neural Nets and Evolutionary Computation (Zimmermann, 2000).

Intelligent approaches in this sense have been used in MM since the 80's mainly in three ways: 1. To model quantitatively information, that is only available in linguistic or approximate form; 2. For knowledge based systems (expert systems), and 3.for either fuzzified mathematical models (fuzzy mathematical programming, fuzzy net planning, fuzzy graphs, fuzzy cluster methods, etc.) or neural nets. Hybrid models that either combine fuzzy sets with neural nets or which use evolutionary programming to adjust or optimize neural nets or fuzzy models can also be found. In the following we shall sketch recent applications of that type in the different areas of MM and in section 4 we shall describe in more detail an approach in production planning and control, which uses the last two approaches and seems to be more general in its applications.

3.1 Project Management

One of the most frequently used methods in project management, which can be applied in the more strategic decision making or also in production planning-or even control if the product to be manufactured are large projects (large special machines, buildings, power plants etc. is net planning of the CPM or PERT type. In CPM die durations of the activities are deterministic, in PERT they are assumed to be stochastic, normally modeled as β-distributions. In both cases the durations are assumed to be independent of each other. Under these assumptions longest paths through the networks are determined, which are then denoted "critical". Critical paths (consisting only of critical activities) determine the length of a net plan or project. If the above mentioned assumption do not hold (which they do rarely in practice), the computed critical paths do not reflect the real length of the project.

A number of authors (Okada, 2004; Zielinski, 2005; Zhang et al., 2005) have suggested to model the durations of the activities as fuzzy numbers and to introduce a measure of dependence between the activities. If that is done, deterministic unequivocal critical paths can no longer be determined. Generally possibility theory is used to determine a kind of "criticality measure" and come to more differentiated results than in classical theory. In a similar direction go approaches that suggest labeling algorithms for the fuzzy assignment problem (Lin and Wen, 2004) and for fuzzy location problems on networks (Pérez et al., 2004). Theoretically these approaches are certainly of interest even they are of any real use in practice and they have not yet been proven.

An area with very many applications of fuzzy control and of hybrid models has been the area of robotics. It can either be considered as a part of production control or as a decision about the technology that is to be used in manufacturing (or logistics). A good survey is presented by (Saffiotti et al., 1999) and an interesting newer application by (Hearnes and Esogbue, 2004). We will not deal with this topic in more detail because it is actually more an engineering application than a (manufacturing) management one.

3.2 Inventory Management

There exist a number of different views on inventory management: One can consider (aggregate) inventories as means of smoothing (aggregate) production. Let us call this problem "aggregate inventory management". Most of the publications in inventory theory consider single items and aim at optimal lot sizes under different conditions (let us call this the "lot sizing problem") and even others consider either of these problems for very specific environments (i.e., multi item discounts, perishable goods, inventories in supply chains etc.). In all these problems costs, capacities, demands etc. are considered to be given numerically. This is often unrealistic in practice and, hence, one has tried to circumvent problems of these hidden uncertainties by using fuzzy set theory. We will give an example each for the above mentioned three classes:

Aggregate Inventory Management: The "HMMS-model" or "Management Coefficient Approach" (Holt et al., 1960) was one of the best known classical models for aggregate production and inventory planning. It assumes that the main objective of the production planner is to minimize total cost, which is assumed to consist of costs of regular payroll, overtime and layoffs, inventory, stock-outs, and machine setups. The model assumes quadratic cost functions and then derives linear decision rules linking production level, work-force level and inventory level. This model was tested empirically and performed quite well. Nevertheless it was hardly used in practice assumedly because managers considered it to contain too much mathematics. Rinks (1982) tried to avoid this lack of acceptance by managers by transforming the classical (mathematical) HMMS model into a knowledge based systems in which the algorithms consisted of a series of relational assignment statements (rules) of the form

> If workforce at time t is ...an inventory at time t-1 is... and workforce at time t-1 is....,then production level in period t is...
> Else ...

For the different state variables workforce, inventory etc. he used terms of linguistic variables. The results of this model with forty decision rules were again tested empirically and showed a quite good performance. For more details see Zimmermann (2001a).

A similar problem is considered by Sommer (1981): The management of a company wants to close down a certain plant within a definite time interval. Therefore production levels should decrease to zero, inventory levels should be zero at the end and total cost should be minimized. Sommer (1981) formulates this problem as a kind of dynamic programming model in which the constraint "production should decrease as evenly as possible" and the goal "have the stock level as close to zero as possible at the end" are modeled as fuzzy sets. The model is then solved with fuzzy dynamic programming, where the computational effort can be further decreased by heuristic bounding.

Lot sizing problem: A very interesting – though not too well suited for practical use – model was presented by Mandal et al. (2005): They consider the normal multi item economic order quantity problem with backlogging, but with constraints on storage space, the number of orders and production cost restrictions. Production rate is instantaneous and the production costs as a function of demand are so-called posynomials (positive exponential functions). They aim at minimizing holding cost, shortage cost, set-up cost and production cost. This type of model has been solved in the past by geometric programming, which is particularly well suited for this type of nonlinear programming models. Mandal now assumes the cost parameters, the objective function and the constraints to be fuzzy and modeled as triangular fuzzy numbers. After some kind of defuzzification into a crisp equivalent (nonlinear) model they use the dual form of geometric programming again to solve the problem. An extensive numerical example for sensitivity analysis demonstrates the strength and plausibility of this approach. Naturally it uses rather demanding mathematics and might not be too well suited for practical use in this form. It could, however, probably be transformed into a more user friendly type of rule-based system (similar to Rinks) and then be very attractive even for real use.

Special situations: A very up-to-date application is that of inventory decisions in supply chains. A few publications of the recent past have considered inventories in supply chains: Petrovic et al. (1998,1999) focused on the uncertain demand and on supplier's reliability and developed a fuzzy inventory model to determine the order point (order-up-to-level) for each individual site in the supply chain. It is not possible, however, to evaluate the performance of the entire supply chain as a function of the inventory management. Giannoccaro et al. (2003) developed a periodic review inventory system for supply chains based on the concept of fuzzy echelon stock. The fuzzy quantities in their model are the market demand and the inventory holding cost. Material lead times, supplier's reliability etc. were not considered and, hence, their model can be considered as a model that considers partially the uncertainties in supply chains. Since a supply chain is, however, a very multi facetted system most or even all of the models in this area will be partial.

The so far probably most advanced model that focuses on inventories in supply chains is that by Wang and Shu (2005). Their goal is to minimize inventory cost in a supply chain subject to the constraint of satisfying the target fill rate of the finished product. Fuzzy sets are used to model the fluctuating customer demand, uncertain processing time, and unreliable supply delivery. A decentralized supply chain model, based on possibility theory is developed to evaluate the performance of the supply chain. Based on this model a genetic algorithm is developed to determine order-up-to stock levels of all stock keeping units in the chain that minimize the supply chain inventory cost and satisfy the pre-specified target fill rate of the finished product. Extensive simulations are performed that can serve as a sensitivity analysis. This is also a model that can certainly be extended. The practical acceptance will depend on the willingness of the decision makers to accept mathematical models that are not too transparent and plausible.

3.3 Production Logistics

Production logistics is that part of logistics that links the different parts of production together. Its importance varies widely with the type of production. Sometimes it becomes even part of production capacity (in production cells, flexible manufacturing systems etc.). Often, however, it becomes an independent function into which millions of Dollars are invested. This does not only depend on the type of production but also on the type production is organized: If a piece part production is, for instance, organized such that each department (cutting, lathes, drilling etc.) has their own transportation equipment, then assumedly half of the transportation capacity will be vacant most of the time (because it normally goes empty to get something and returns empty after is has brought something to another place. In this case the management of logistics is done in a decentralized way and is not complicated. If, however, all forklift trucks, hangers, towing vehicles etc. are controlled centrally the utilization of the transport capacity can be considerably, but the optimal control of the transportation devices becomes a combinatorial problem of high complexity, which can no longer be solved by a human expert. The same is the case if the production area includes a large container terminal in which containers and other units are stored for production or between different production stages. In the building industry or in supply chains transportation may become the critical factor between different parts of production. The critical factor in these cases is normally time (Zimmermann, 2004). The systems which control logistics in these cases are very complex. They contain combinatorial algorithms but they also contain modules with approximate reasoning which are interlinked with many other modules and not suited for publication (Zimmermann, 2001b).

In other logistical problems, which may occur inside or outside the production area fuzzy models have been published. Chanas (1984), for instance, suggested a fuzzy model for transportation problems in which the supplies and the demands of the receivers are fuzzy. He used different membership functions and finally solved his problem by using fuzzy linear programming.

Ernst (1982) solved in his dissertation a real problem by suggesting a fuzzy model for the determination of time schedules for containerships which could be solve by branch and bound and for the scheduling of containers on ships, which resulted in a linear program. The model contained in a realistic setting 2,000 constraints and originally 21,000 variables. The system is the core of a decision support system that schedules the inventory, movement, and availability of containers, especially empty containers, in and between 15 harbors, on 10 big container ships worldwide on 40 routes. The problem was formulated as a large LP model with an objective function that maximized profits (from shipping full containers) minus cost of moving empty containers minus inventory cost of empty containers. When comparing the results of the computation with data of earlier periods it turned out that ships had in the past transported many more containers than they officially could. In investigations it was found out that the official capacities of the ship were not really considered to be crisp (as in the model), but that the captains were very good in putting more containers aboard if attractive containers were waiting in a harbor. Hence, the constraints had to be fuzzified. Rather than using the classical way of fuzzifying a linear program (Zimmermann, 2001a), here a penalty was attached to violating "soft constraints" and these penalties were added to the objective function. This enlarged the model by the number of "negative slack variables" for the soft constraints, but kept it a crisp linear program. The success of this fuzzification was impressive: neither management nor captains complained any more about the schedules!

4 Intelligent Production Planning and Control

This is certainly considered to be the most central part of MM. There are computer systems on the transaction level (see e.g. SAP)[1] that claim –and probably do- cover all kinds of production types with their system. This is,

[1] www.sap.com

however, not true for approaches on the intelligent level, i.e. models that try to optimize production.

Generally, one distinguishes between aggregate production planning and production control.

4.1 Production Planning

Production planning as far as it is considered to be part of MM consists of roughly planning production (and inventory) according to the capacities and the sales figures that come from marketing. On this level it is not very relevant what type of product or production organization is involved. Very often it is really only the attempt to fit the forecasts of marketing as well as possible to the existing or expected capacities or the availability of specific raw materials or other resources. It should be mentioned here, that production planning is intimately linked with inventory planning and that approaches such as the already mentioned by Rinks (1982) or by Türksen and Zhong (1990) belong to production planning as well as to inventory planning.

Alternatively linear programming is used to optimize the expected result (cost or profit) by respective production and inventory levels and subject to a number of constraints. We will consider this approach again further down.

Another methodology that has proven quite successful in the 80's and 90's, particularly in Germany, for production planning as well as for production control is that of Fuzzy Petri Nets (Zimmermann, 1993). Crisp Petri Nets are frequently used in computer science to combine several processes that work in parallel and asynchronously to each other, into one scheme. They are also suited to map industrial processes of many types. Crisp Petri Nets, which are essentially 7-tuples, are in a way restricted. They have, therefore, been fuzzified in different ways. Some use fuzzy places, fuzzy transitions and fuzzy switching functions (Lipp et al., 1989) others focus on the time the transactions have to be fired. A recent, more detailed, description of Fuzzy Petri Nets can be found in (Pedrycz and Camargo, 2003). Petri Nets can, in principle, be used for simulations and/or for visualization. There are versions, however, that include also a knowledge based "optimizer" which heuristically improves the parameter of the Petri net (Zimmermann, 1993). In so far production planning of the traditional kind and by linear programming on the one hand and the use of Fuzzy Petri Nets for production planning on the other hand can be considered as analogues.

Practical applications of Fuzzy Petri Nets in different kinds of industries have been reported in the literature (Lipp, 1992, 1993; Khansa et al., 1996; Zimmermann, 1993).

4.2 Production Control (Scheduling)

Production control is the short term, often on-line, activity of assigning activities to available resources or to control the "speed" of resources primarily in continuous production processes. Uncertainty of the future is certainly smaller than in the planning process; the uncertainty of the data depends on the source of the data and the most important source of uncertainty is here probably the human experience that is used for scheduling or control. From the point of view of formal decision making structures there is a significant difference between flow shops and/or continuous production on one hand and job shop production on the other. The former is the domain of continuous control, of line balancing etc. in which nonlinearities play a prominent role, and the latter is that of scheduling, dispatching etc. in which combinatorial problems pose the big challenge. Real production systems are normally located between these extremes or they contain elements of both.

Flow shops and continuous control: The "classical" fuzzy tool for rather simple or embedded control problems is Mamdani's Fuzzy Controller (King and Mamdani, 1977; Mamdani, 1977). It was the first successful application of fuzzy knowledge based systems to (nonlinear) technical control problems.

The classical structure – fuzzification-approximate reasoning-aggregation- defuzzification – has found hundreds of successful applications in production control, in consumer products and in many other areas. To describe them in detail would be carrying coal to Newcastle. There exist quite a number of CASE tools for that type of controllers, generally on the basis of fuzzy min/max theory. These fuzzy controllers have been used on a stand-alone basis as well as embedded controllers in more complex production control systems, where they have been inserted or embedded into rather complex classical control systems.

A different type of fuzzy or neuro-fuzzy models has been used for a macroscopic type of optimization of production systems. Good examples for this type of application are (Khoei et al., 2005) who describe the fuzzy level control of a tank. The goal is to minimize the movements of a valve (reducing cost). The present in their paper interesting practical details and they have implemented their controller in 0.35 μm CMOS technology, such that the chip has the size of 0.052 mm^2.

Another theoretical but interesting model presents Tao and Taur (2005). They consider a fuzzy linear model in which the only uncertain information is that of linguistic information obtained from human experts, which is modeled as interval parameters in the linear control model. They focus their attention on designing a "robust fuzzy controller" which is defined to be a fuzzy controller that for a family of plants provides stable control systems. Discussions on the stability and robustness of systems on the basis of simulations will certainly contribute to the design of macro control systems.

Interesting real applications of a kind of macro control are described by Andújar and Bravo (2005) and by Alvarez-Lòpez et al. (2005). The former model is a waste water plant in the sense of a Takagi-Sugeno controller (Takagi and Sugeno, 1985). Three parameters have to be regulated: Chemical need of oxygen, solids in suspension, and acidity or alkalinity degree of the water. The process model of the plant consists of six functions, which correspond to the decomposition of the nonlinear waste water plant into six fuzzy systems which are necessary to apply the control signal to allow the following of a reference value. The fuzzy control is based on the so-called feedback linearization control method (Guardabassi and Savaresi, 2001), which allows a back propagation similar to neural nets. Hence the model obtains learning capabilities. Simulation shows that the model is very effective.

Alvarez-López et al. (2005) consider the drying process of tobacco leaves in Cuba and develop a knowledge based model on the bases of the expert curers taking part in the natural drying process. In the process two variables are controlled: temperature and relative humidity. The model, also based on a Sugeno controller was developed using the OMRON case tool C200 HG for multiple inputs and a single output and the experience so far again seem to be very promising.

There exist also systems that do not have the "control philosophy" that has been described so far. An example is Ishibuchi et al. (1994), who present a model for flow shop scheduling with fuzzy due dates and also Türksen and Zarandi (1998), who analyze scheduling systems rather than optimizing them.

Normally maintenance and quality control is also part of manufacturing management. In this book an extra chapter is dedicated to it and it shall, therefore, not be considered in detail in this chapter. Nevertheless it may be useful to point to the very large importance of these areas and also to the considerable potential of fuzzy applications, particularly in quality control. This is particularly true for the area of acoustic quality control. When the human ear is substituted by sensors or microphones and fuzzy models are used to analyze noises, remarkable successes can be achieved in practice. The methods used are, however, normally not fuzzy control but fuzzy clustering. It could, for instance, be shown, that the breakdown of the main shaft in a helicopter could be detected hour before an experts ear would recognize it by such a fuzzy system. (Joentgen et al., 1999)

Job shops and scheduling: It was already mentioned that the job shop model is an extreme and that most of the real systems are somewhere between the theoretical flow shop and job shop model. Still, the core of this section is assignment and sequencing under various constraints. What makes it even more difficult to find a more generally valid model is, that real production systems differ from each other and that the complexity of many systems becomes quickly very high, due to the combinatorial character of scheduling problems. It is, therefore, not surprising that – as in classical scheduling models – published scheduling models generally focus on very special low dimensional problems and that in practice heuristics are used, if one tries to optimize at all.

A good survey on single machine scheduling under fuzziness is presented by Vlach (2000). Also the other contributions in the book edited by Slowinski and Hapke (2000) give good surveys of different types of approaches for scheduling under fuzziness. A more general survey on crisp and fuzzy approaches for production planning and scheduling is presented by Türksen and Zarandi (1999). As one older approach which has become a "classic" one should probably mention Bensana et al.(1988) who started from another AI-approach that is used in scheduling very often, constrained based programming, and extended it by fuzzification to the job shop scheduling decision support system OPAL. Rather typical contributions of the recent past are:

Sung et.al. (2003) consider the problem of sequencing a finite number of jobs for processing on a single machine with the objective of minimizing the number of jobs that are not completed by their due dates. Their model is based on an algebraic approach. By specifying suitable binary relations between due dates, between processing times, and between due dates and job completion times they arrive at a general model from which results for fuzzy environments can be obtained as special cases. They show that optimal sequences can be obtained by modifications of Moore's algorithm (Moore, 1968). This is a very formal contribution and, even though valuable, already due to the single machine focus, not readily usable in practice. Also Chanas and Kasperski (2004) are concerned with the single machine scheduling problem. They consider the case in which some or all of the parameters (processing times, due dates, weights etc) are fuzzy numbers. A schedule is represented by the sequence of jobs and it is considered to be optimal if the value of the cost function (which depends on the sequence of the jobs) is minimal. If parameters on which the cost function depends are fuzzy, then also the value of the cost function will be fuzzy. In this case –to determine the optimality of the sequence- an order of fuzzy sets has to be determined. The authors use for their evaluation possibility theory and determine the possibility and the necessity that a schedule is optimal. This is certainly an approach that can be extended and may be useful even in practice in the future.

Kasperski (2005) offers an extension to last paper. Essentially the same problem is considered, but the objective function is varied: He considers 5 cases in which either the weighted max of the possibility or necessity of the due date violation, the sum of the necessities of the latenesses or the max of not exactly meeting the due dates (just in time) is minimized. He then investigates the computational complexity of these 5 cases and finds, that two of those problems can be solved in polynomial time while the rest is NP-hard. For those problems obviously heuristics are the only way to solve them.

5 Linking Production Planning and Production Control

So far production planning and production control has been considered separately, as it is normally done in the literature. In practice, however, the link between production planning and production control is the critical point: Production planning has a relatively long planning horizon and is normally done without explicitly considering uncertainty. Production control is actually supposed to be based on the data of production planning. But when the transition from planning to control is about to come, it turns out, that –due to the neglected uncertainty in production planning- the data from production planning are no longer valid. Usually, even to day, the control is then performed in a kind of experience based way, with a very limited time-and functional information horizon. Hence the results are very sub-optimal, inventories high, service level low and goal-oriented utilization of capacity also low. We will sketch a theoretical and a practical system which successfully avoid these weaknesses:

A method to control Flexible Manufacturing Systems

Flexible Manufacturing Systems (FMS) emerged in the 80'sas a central element of integration and automation of manufacturing. By an FMS we mean an integrated manufacturing system consisting of work stations linked by a computerized material handling system capable of making it possible for jobs to follow diverse routes through the system. Such a system is usually controlled by a PPC (Production Planning and Control System), which consists of a central part for aggregate planning and a decentralized PPC-system for each FMS which coordinates and locally optimizes. If the central system provides the FMS with a schedule of complete orders, the decentralized system is responsible for meeting the due dates and for the internal optimization of the FMS, i.e. to minimize flow times and due date deviations and to maximize machine utilization, goals which are normally conflicting. Hintz and Zimmermann (1989) describe a system that controls such a system, which consists of Master Scheduling, Tool Loading, Release Scheduling and Machine Scheduling.

Master Scheduling was performed by fuzzy linear programming and the remaining 3 (operating) functions by fuzzy knowledge based systems, using the knowledge about priority rules as they had been used in production control in the past already. The main difference was that using single priority rules makes scheduling decisions extremely local and thus globally suboptimal. Figure 1 shows the rules (criteria) which were used for machine scheduling. They were combined by the γ-operator and yielded an overall rating for each of the parts waiting to be assigned to a machine.

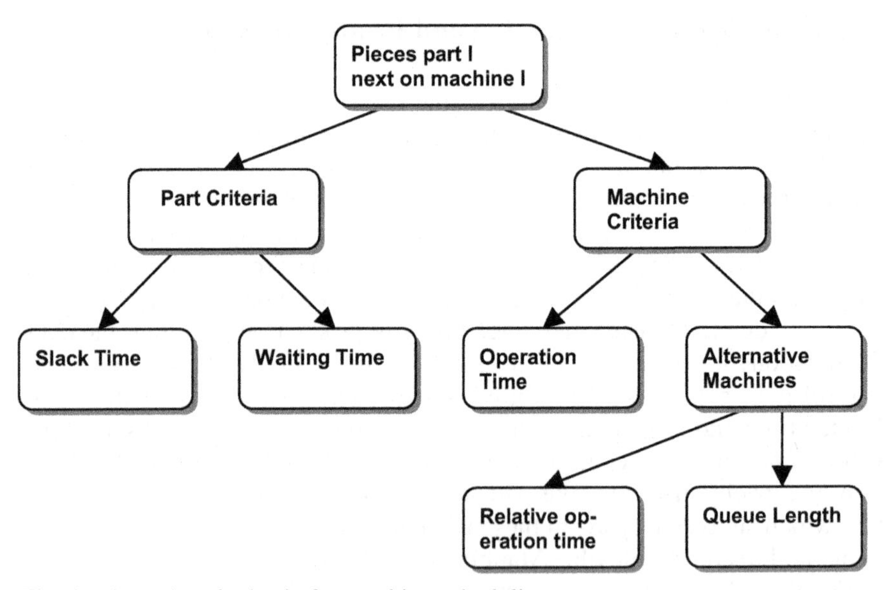

Fig. 1. Hierarchy of criteria for machine scheduling

The fuzzy sets for the rules (criteria) were automatically generated from the data contained in the system (e.g. mean values of operating times etc. of parts in the system.). By simulation the system was then compared to the best priority systems which were available at that time (COVERT) and it turned out , that the above system was superior to this systems in all criteria, i.e. mean in-process waiting time, fraction of parts that met their due dates , and mean machine utilization. (see the details in Hintz and Zimmermann, 1989).

The production system that was modeled was certainly very special and all the decisions were made automatically by the computer. The ideas, however, were transformed into a software system (FELIOS)[2], which has been installed in practice several hundred times and which has shown a very good performance:

An "Add-on" system to control production in small-batch-size production: This system is again aimed at a special type of production: multi stage job shop type production in which the lot sizes or batches are very small (in the extreme only one peace of a type may be produced). This is a typical situation for either the production of large special machines or of products which are offered to the customers in many variations. Aggregate production planning is done in standard PPS-Systems considerable time before the actual production starts. At that time the data in the PPS system are normally no longer correct, due to the uncertainty which is not considered in aggregate production planning. Production control is now very difficult due to rather complex product nets that run over several stages of the production process and make local decisions particularly sub optimal.

FELIOS retrieves from the PPS-system the actual data, for instance, for the next two days, checks the availability of all resources and materials and optimizes this part schedule over all stages and products. This optimization is done by using a combination of classical combinatorial and heuristic methods and approximate reasoning similar to the one described above. It does, however, not optimize the schedule down to the lowest level (i.e. assignment of operations to machines), but leaves the detailed decisions within a shift or a day to the supervisors, which at the end of the day report what has been done during the day. This is again entered into the optimization run of FELIOS for the next day. Hence the expert performs the most detailed decisions himself (because he knows best). Thus the sub optimality due to the limited information horizons of the decision makers is decreased and the complexity of the computer system is also very much diminished. So far the system has been superior to competing systems in term of inventory levels, through put times, service levels (due dates) and

---

[2] www.felios.de

capacity utilization. It should be stressed, however, that this system is only successful for the type of production described above. Hopefully other systems for other types of production will be developed, using classical scheduling theory and also different parts of Computational Intelligence.

References

Alvarez-Lopez I, Llanes-Santiago O, Verdegey JL (2005) Drying Process of tobacco leaves by using a fuzzy controller. FSS 150:493-506

Andújar JM, Bravo JM (2005) Multivariable fuzzy control applied to the physical-chemical treatment facility of a Cellulose factory. FSS 150:475-492

Bensana E, Bel G, Dubois D (1988) OPAL: A multi knowledge based system for industrial job-shop scheduling. Int. J. Prod. Res. 26 795-819

Chanas S, Kolodziejczyk W, Machaj A (1984) A fuzzy approach to the transportation prolem. FSS 13:211-221

Chanas S, Kasperski A (2004) Posssible and necessary optimality of Solutions in the single machine scheduling problem with fuzzy Parameters. FSS 142:359-371

Ernst, E. (1982) Fahrplanerstellung und Umlaufdisposition im Container Schiffsverkehr (Disertation) Aachen

Giannocaro, I, Pontrandolfo P, Scozzi B (2003) A fuzzy echelon approach for inventory management in supply chains.EJOR149: 185-196

Guradabassi GO, Savaresi SM (2001) Approximate linearization via Feedback-an overview. Automatica 37:1-15

Gupta MM, Saridis GN, Gaines BR (edtrs) (1977) Fuzzy Automata and Decision Processes. North-Holland , NewYork, Amsterdam

Hearnes WE, Esogbue AO (2003) Application of a near-optimal Reinforcement learning controller to a robotics problem in Manufacturing. Fuzzy Optim. Decision Making 2:229-242

Holt CC, Modogliani F, Muth JF, Simon HA (1960) Planning production, inventrories, and workforce. Prentice-Hall, Englewood Cliffs

Ishibuchi H, Yamamoto N, Misaki S, Tanaka H (1994) Local search algorithms for flow shop scheduling with fuzzy due dates. Int. J. of Production Economics 33:53-66

Joentgen A, Mikenina L, Weber R, Zimmermann HJ (1999) Automatic fault detection in gearboxes by dynamic fuzzy data analysis. Fuzzy Sets and Systems 105:123-132

Kasperski A (2005) A possibilistic approach to sequencing problems with fuzzy parameters. FSS 150:77-86

Khansa W, Denat JP, Collart-Dutilleul (2003) P-time Petri-Nets for Manufacturing Systems. Proc. of the WODES'96, Edinburgh 94-102

Kim YT, Cho HC, Seo YJ, Jeon HAT, Klir GJ (2002) Intelligent path Planning of two cooperating robots based on fuzzy logic. Int. J. of Gen. Systems 31:359-376

King PJ, Mamdani EH (1977) The Application of Fuzzy Control Systems to Industrial Processes. In:Gupta et.al. 1977:321-330

Lin CJ (2004) A labeling algorithm for the fuzzy assignment problem FSS 142:373-392

Lipp H.-P, Günter R, Sonntag P (1989) Unscharfe Petri Netze- Ein Basiskonzept für computerunterstützte Entscheidungssysteme in Komplexen Systemen. Wissenschaftliche Schriftenreihe der TU Chemnitz 7

Lipp H.-P (1992) Einsatz von Fuzzy Konzepten für das operative Produktionsmanagement. Atp-Automatisierungstechnische Praxis 34:668-675

Lipp H.-P (1993) Fuzzy Technologien in der Industrie-Automatisierung. Superelectronics Jahrbuch, Vogel Verlag Würzburg

Mamdani EH (1977) Application of Fuzzy Set Theory to Control Systems. In: Gupta et. al. 1977:77-88

Mandal NK, Roy TK, Maiti M (2005) Multi-objective fuzzy inventory model with three constraints: a geometric programming approach. FSS 150:87-106

Moore JM (1968) An n job, one machine sequencing algorithm for minimizing the number of late jobs. Mgt. Sc.15:102-109

Moreno Pérez JA, Moreno Vega JM, Verdegay JL (2004) Fuzzy location problems on networks. Fuzzy Sets and Systems 142:393-406

Okada S (2004) Fuzzy shortest path problems incorporating interactivity among paths. Fuzzy Sets and Systems 142:335-359

Pedrycz W (1992) Fuzzy Control and Fuzzy Systems. 2nd Ed. Wiley New York, Chichester, Toronto, Bisbane, Singapore

Pedrycz W, Camargo H (2003) Fuzzy Timed Petri Nets. FSS 140: 301-330

Petrovic D, Roy R, Petrovic R (1998) Modeling and simulation of a supply chain in an uncertain environment. EJOR 109:200-2009

Petrovic D, Roy R, Petrovic R (1999) Supply chain modeling using Fuzzy sets. Int. J. Prod. Econom.59:443-453

Rinks DB (1982) A heuristic approach to aggregate production scheduling using linguistic variables. In: Zimmermann (2001)

Saffiotti A, Ruspini EH, Konolige K (1999) Using fuzzy logic for mobile robot control. in: Zimmermann HJ (1999):185-204

Slowinski R, Hapke M (eds) (2000) Scheduling under Fuzziness. Physica Verlag, Heidelberg New York

Sommer G (1981) Fuzzy inventory scheduling. in: Zimmermann(2001)

Sung SC, Vlach M (2003) Single machine scheduling to minimize the number of late jobs under uncertainty. FSS 139:421-430

Takagi T, Sugeno M(1985) Fuzzy identification of systems and its applications to modeling and control. IEEE Trans. SMC 15:116-132

Tao CW, Taur JS (2005) Robust Control for a Plant with Fuzzy Linear Model. IEEE Transactions on Fuzzy Systems 13:30-411

Türksen IB, Zhong Z (1998) An Approvimate Analogical Reasoning Schema based on Similarity Measures and Interval Valued Fuzzy Sets. FSS 34:323-346

Türksen IB, Facel Zarandi MH (1997) Fuzzy system models for aggregate scheduling analysis. Int. J. of Appr. Reas.19:119-143

Türksen IB, Facel Zarandi MH (1999) Production Planning and Scheduling: Fuzzy and Crisp Approaches. in: Zimmermann (1999) 479-530

Vlach M (2000) Single Machine Scheduling under Fuzziness. in: Slowinski, Hapke (2000) 223-246

Wang J, Shu Y-F (2005) Fuzzy decision modeling for supply chain management. FSS 150:107-127

Wright PK, Bourne DA (1988) Manufacturing Intelligence. Addison-Wesley Publ. Co. Reading, Menlo Park, New York, Amsterdam

Zhang H, Tam CM, Li H (2005) Modeling uncertain activity duration by fuzzy number and discrete-event simulation. Eur. J.f. OR 164:715-729

Zielinski P (2005) On computing the latest starting times and floats of activities in a network with imprecise durations. FSS 150:53-77

Zimmermann HJ (ed) (1993) Fuzzy Technologien. VDI Verlag Düsseldorf

Zimmerman HJ (ed) (1999) Practical Applications of Fuzzy Technologies, Kluwer Academic Publishers Boston, Dodrecht, London

Zimmermann HJ (2000) An application-oriented view of modeling uncertainty. Europ. J. for Operations Research 122:190-198

Zimmermann HJ (2000) Concept origins, present developments and the future of Computational Intelligence. J. of Chaos Theory 5: 3-10

Zimmermann HJ (2001a) Fuzzy Set Theory and its Applications.4th edition, Kluwer Academic Publishers Boston, Dodrecht, London

Zimmermann HJ (2001b) Approximate Reasoning in Logistics Proc. Joint 9th IFSA World Congress, Vancouver pp 134-140

Zimmermann HJ (2004) Online Optimization-From Logistics to Business Intelligence. OR/MS 31:32-37

Metaheuristic Techniques for Job Shop Scheduling Problem and a Fuzzy Ant Colony Optimization Algorithm

Sezgin Kılıç[1] and Cengiz Kahraman[2]

[1] Industrial Engineering Department, Air Force Academy, 34807, Yeşilyurt, İstanbul, Turkey

[2] Industrial Engineering Department, Istanbul Technical University, Maçka, İstanbul, Turkey

Summary: Job shop scheduling (JSS) problem is NP-hard in its simplest case and we generally need to add new constraints when we want to solve a JSS in any practical application area. Therefore, as its complexity increases we need algorithms that can solve the problem in a reasonable time period and can be modified easily for new constraints. In the literature, there are many metaheuristic methods to solve JSS problem. In this chapter, the proposed Ant algorithm can solve JSS problems in reasonable time and it is very easy to modify the artificial ants for new constraints. In addition, it is very easy to modify artificial ants for multi-objective cases.

Key words: Job shop scheduling, ant colony optimization, metaheuristic methods, tabu search, simulated annealing.

1 Job Shop Scheduling Problem

Scheduling is the allocation of resources over time to perform a collection of tasks. It has been examined in the operation research literature since the early fifties (Conway *et al.,* 1967). Variations in problem types are mostly illustrated using the manufacturing domain. For example, a job is a term used to designate a single item or batch of items that require processing on the machines and the processing of a particular job through a particular machine is called an operation. The Job Shop Scheduling (JSS) problem is a well known NP-hard problem (Lawler *et al.,* 1982), meaning that there is no algorithm that can solve such problems in polynomial time with respect to problem size. JSS is the most general case of the scheduling problems

S. Kiliç and C. Kahraman: *Metaheuristic Techniques for Job Shop Scheduling Problem and a Fuzzy Ant Colony Optimization Algorithm*, StudFuzz **201**, 401–425 (2006)
www.springerlink.com

and consists of a finite set of jobs to be processed on a finite set of machines. Each job is characterized by a fixed order of operations, each of which is to be processed on a specific machine for a specific duration. Each machine can handle only one job and each job can be processed by only one machine at a time. Each operation needs to be processed during an uninterrupted period of time on a given machine. A schedule is an assignment of operations to time slots on the machines. We can formulate a JSS problem with n jobs to be scheduled on m machines as an integer programming model.

Let c_{ik} denote the completion time of job i on machine k as the decision variable of the model and t_{ik} denotes the processing time of job i on machine k. If job i must be processed on machine h before machine k, we need the following constraint:

$$c_{ik} - t_{ik} \geq c_{ih} \tag{1.a}$$

On the other hand, if the processing on machine k comes first, the constraint becomes

$$c_{ih} - t_{ik} \geq c_{ik} \tag{1.b}$$

Thus, we need to define an indicator variable x_{ihk} as follows:

$$x_{ihk} = \begin{cases} 1, \text{if processing on machine } h \text{ precedes that on machine k for job } i \\ 0, \text{otherwise} \end{cases}$$

We can then rewrite the aforementioned constraints as follows;

$$c_{ik} - t_{ik} + L(1 - x_{ihk}) \geq c_{ih}, \qquad i = 1,2,\dots,n \quad h,k = 1,2,\dots,m \quad h \neq k \tag{1}$$

where, L is a large positive number. Consider two jobs, i and j, that are to be processed on machine k. If job i comes before job j, we need the following constraint:

$$c_{jk} - c_{ik} \geq t_{jk} \tag{2.a}$$

Otherwise, if job j comes first, the constraint becomes

$$c_{ik} - c_{jk} \geq t_{ik} \tag{2.b}$$

Therefore, we also need to define another indicator variable y_{ijk} as follows:

$$y_{ijk} = \begin{cases} 1, \text{if job } i \text{ precedes job } j \text{ on machine } k, \\ 0, \text{otherwise} \end{cases}$$

We can then rewrite the aforementioned constraints as follows;

$$c_{jk} - c_{ik} + L(1 - y_{ijk}) \geq t_{jk,} \quad i, j = 1,2,...,n, \quad k = 1,2,...,m \quad i \neq j \tag{2}$$

The JSS problem with a makespan objective can be formulated as follows:

$$\min \max_{1 \leq k \leq m} \left[\max_{1 \leq i \leq n} [c_{ik}] \right]$$

$$s.t.$$

$$c_{ik} - t_{ik} + L(1 - x_{ihk}) \geq c_{ih}$$

$$c_{jk} - c_{ik} + L(1 - y_{ijk}) \geq t_{jk}$$

$$c_{ik} \geq 0$$

$$x_{ihk} = 0 \; or \; 1$$

$$y_{ijk} = 0 \; or \; 1$$

$$h, k = 1,2,...,m \quad i, j = 1,2,...,n \quad h \neq k \quad i \neq j$$

The model mentioned above is a general case for JSS problems. It can be specified by changing the objective function and adding specific constraints for desired purposes.

2 Solution Methodologies for JSS Problem

The JSS is amongst the hardest combinatorial optimization problems. One of the earliest works on scheduling theory is Johnson's (1954). Up to time many researchers tried to generate optimum or near optimum schedules to the JSS problem by various techniques. An $n \times m$ size JSS has an upper bound of $(n!)^m$ possible solutions, thus a 15×15 problem may have at most 5.591×10^{181} different solutions (with minimized idle times). Due to this factorial explosion only small instances can be solved with exact methods in acceptable time periods and problems of dimensions greater than 15×15 are usually considered to be beyond the reach of the exact methods.

2.1 Main Methodologies Used to Solve JSS Problem

There exists a rapid and extensively growing body of literature for JSS. Basically we can cluster the techniques used to solve JSS into two groups; optimization algorithms and approximation algorithms. Various techniques applied to solve JSS are given below and a comprehensive survey of JSS techniques can be found in Jain and Meeran (1999).

- **Optimization algorithms**
  - Efficient methods – solvable in polynomial time
  - Enumerative methods
    - Mathematical Formulations
      - *Integer Linear Programming*
      - *Mixed Integer Linear Programming*
      - *Langrangian Relaxation*
      - *Surrogate (constraint) Duality*
      - *Decomposition Techniques*
    - Branch and Bound
- **Approximation Algorithms**
  - Constructive Methods
    - *Priority Dispatch Rules*
    - *Insertion Algorithms*
    - *Bottleneck Based Heuristics*
  - Iterative Methods (General Algorithms)
    - Artificial Intelligence (AI)
      - *Constraint Satisfaction*
      - *Neural Networks*
      - *Expert Systems*
      - *Ant Optimization*
  - Local Search
  - Problem Space Methods
    - *GRASP*
    - *Genetic Algorithms*
    - *Reinsertion Algorithms*
    - *Threshold Algorithms*
      - *Simulated Annealing*
      - *Threshold Accepting*
      - *Iterative Improvement*
  - Tabu Search
  - Large Step Optimization

Most of the recent research approaches for JSS are dispatching rules, decomposition methods, and metaheuristic search techniques. Simple heuristic rules are dominantly used in JSS research area. Thus, there are a considerable number of dispatching rules that have been developed across many industries (Zoghby, 2004). With simple dispatch heuristics we schedule a job when it arrives at an empty machine and when the machine is available; we simply schedule the highest priority job currently available at the machine.

2.2 Metaheuristic Methods for JSS Problem

Metaheuristics are semi-stochastic approaches for solving a variety of hard optimization problems. The main differences between traditional mathematical methods and metaheuristics can be determined based on solution optimality and CPU time. The beauty of traditional mathematical methods is that they provide an optimal solution to the stated problem. However, when large size instances are treated, the CPU time required to achieve an optimal solution becomes an issue. In addition, traditional approaches most often require simplifications of the original problem formulation. Some times due to oversimplifications, computed solution does not solve the original problem. Furthermore, one of the fundamental advantages of the metaheuristics approach over traditional mathematical methods is its capability to adapt to the considered problem (Back *et al.*, 1997). All metaheuristics should not be considered as ready to use algorithms but rather as a general concept that can be applied to most of the real world applications. In many cases metaheuristics are augmented with additional approaches to enhance performance for specific problems. Genetic algorithms, Tabu Search, Simulated Annealing and Ant Colony Optimization are among the most successful ones. The success of these methods depends on their ability to find good (not necessarily optimum) solutions to the hard combinatorial optimization problems in short time periods. They are generally easy to implement and they can handle specific constraints in practical applications easily. In addition, developing models for solving difficult combinatorial optimization problems characterized with uncertainty is a very important and challenging research task. Metaheuristic techniques need to be combined with fuzzy sets theory techniques for solving complex problems characterized with uncertainty.

2.2.1 Genetic Algorithms (GA)

Genetic Algorithms (GAs) are adaptive heuristic search algorithm premised on the evolutionary ideas of natural selection and genetic. The basic concept of GAs is designed to simulate processes in natural system necessary for evolution, specifically those that follow the principles first laid down by Charles Darwin of survival of the fittest. They represent an intelligent exploitation of a random search within a defined search space to solve a problem. First pioneered by John Holland (1962), Genetic Algorithms have been widely studied, experimented and applied in many fields in engineering worlds. One of the first attempts to approach a simple JSS problem through the application of GAs can be seen in the research of Davis (1985). Since then, a significant number of GAs to JSS problems have been appearing. The first applications of GAs for fuzzy JSS problem were made by Ishii (1995), Tsujimura (1995), and Sakawa (1995). In these works, the values of processing times and duedates were considered as fuzzy numbers due to man-made factors.

2.2.2 Tabu Search (TS)

The basic concept of Tabu Search as described by Glover (1986) is "a meta-heuristic superimposed on another heuristic". The overall approach is to avoid entrainment in cycles by forbidding or penalizing moves which take the solution, in the next iteration, to points in the solution space previously visited (hence "tabu"). The Tabu search begins by marching to a local minima. To avoid retracing the steps used, the method records recent moves in one or more Tabu lists. The original intent of the list was not to prevent a previous move from being repeated, but rather to insure it was not reversed. At initialization the goal is make a coarse examination of the solution space, known as 'diversification', but as candidate locations are identified the search is more focused to produce local optimal solutions in a process of 'intensification'. The Tabu search has traditionally been used on combinatorial optimization problems. Many of the applications in the literature involve integer programming problems, scheduling, routing, traveling salesman and related problems. Dell'Amico and Trubian (1991) have solved the makespan job shop problem to obtain higher accuracy results at high computation times than earlier tabu search results.

2.2.3 Simulated Annealing (SA)

As its name implies, the Simulated Annealing (SA) exploits an analogy between the way in which a metal cools and freezes into a minimum energy crystalline structure (the annealing process) and the search for a minimum in a more general system. The algorithm is based upon that of Metropolis *et al.* (1953), which was originally proposed as a means of finding the equilibrium configuration of a collection of atoms at a given temperature. The connection between this algorithm and mathematical minimization was first noted by Pincus (1970), but it was Kirkpatrick *et al.* (1983) who proposed that it form the basis of an optimization technique for combinatorial (and other) problems. SA's major advantage over other methods is an ability to avoid becoming trapped at local minima. The algorithm employs a random search which not only accepts changes that decrease objective function, but also some changes that increase it. Van Laarhoven *et al.* (1992) used SA for finding the minimum makespan in JSS problem. They found that SA is not effective and efficient as some other heuristics they used.

2.2.4 Ant Colony Optimization (ACO)

Real ants leave on the ground a deposit called *pheromone* as they move about, and they use pheromone trail to communicate with each other for the purpose of finding shortest path between food resource and their nest. Ants tend to follow the way in which there is more pheromone trail. When an obstacle is placed on their existing path, which is between food and nest, some ants will prefer to go around it by the left side and the others by the right side. Those that have chosen the shortest path will rejoin the previous path more quickly, and this will result a more pheromone trail on the shorter path, and more ant will be attracted to the shorter path. In ACO algorithms, artificial ants with the above described characteristics and the additional features stated below (Dorigo, 1999) collectively search for good quality solutions to the optimization problem.

- Artificial ants live in a discrete world and their moves consist of transitions from discrete states to discrete states.
- Artificial ants have an internal state. This private state contains the records of its past actions.
- Artificial ants deposit an amount of pheromone, which is a function of the quality of the solution found.
- Artificial ants timing in pheromone laying is problem dependent and often does not reflect real ants behavior. For example, in many cases artificial ants update pheromone trails only after having generated a solution.
- To improve overall system efficiency, ACO algorithms can be enriched with extra capabilities like look ahead, local optimization, backtracking, and so on, that can not be found in real ants.

Based on the generic framework of ACO, numerous ant algorithms have been developed. They are named according to the different procedures used within them. Ant System (AS) (Dorigo *et al.*, 1991, Dorigo, 1992) is the first algorithm to fall into the framework of ACO and it was introduced using the Traveling Salesman Problem (TSP) as an example application. Although it was found to be inferior to state-of-the-art algorithms for TSP, its importance mainly lies in the inspiration it provided for a number of extensions that significantly improved performance and are currently among the most successful ACO algorithms (Dorigo and Stützle, 2004).

In this section we present the AS to be a basis for the proposed algorithm in Sect. 4. Artificial ants construct solutions for Travelling Salesman Problem (TSP) by moving on the related graph from one city to another according to the AS rules. In each iteration, each ant moves from one city to another. If the problem is consisting of N cities, after N iterations every ant will make a complete tour. During each iteration, each ant applies a probabilistic decision rule to select the next city that will be added to the ants' tour (3)

$$P_{ij}^{r}(t) = \begin{cases} \dfrac{[\tau_{ij}(t)]^{\gamma} \times [\eta_{ij}]^{\theta}}{\sum_{j \notin tabu_k} [\tau_{ij}(t)]^{\gamma} \times [\eta_{ij}]^{\theta}} & if\ j \notin tabu_k \\ 0 & if\ j \in tabu_k \end{cases} \tag{3}$$

$P_{ij}^{r}(t)$: the transition probability from town i to town j for the r_{th} ant at time t.

$\tau_{ij}(t)$: the intensity of the trail on edge (i,j) at time t.

$\eta_{ij} = 1/d_{ij}$: visibility of town j from town i

d_{ij} : distance between town i and town j

$tabu_r$: dynamically growing set which contains the tabu list of the r_{th} ant

γ, θ : parameters that control the relative importance of trail versus visibility

Tabu list saves the towns visited for each ant in order to forbid the ant to visit them again. When a tour is completed, it shows the solution generated by each ant. At time t, each ant chooses the next town, where it will be at time $t + 1$. If there are m ants, m moves will be carried out in the interval $(t, t + 1)$, and when the iteration number is N, a cycle will be completed. When a cycle is completed, every ant will have generated a complete tour and the trail intensities at the edges will be updated according to the equation (4)

$$\tau_{ij}(t+N) = \rho.\tau_{ij}(t) + \Delta\tau_{ij}(t) \tag{4}$$

ρ : the coefficient of evaporation $(0 < \rho < 1)$

$\Delta\tau_{ij} = \sum_{r=1}^{R} \Delta\tau_{ij}^{r}$

$\Delta\tau_{ij}^{r}$: the quantity of trail substance laid on edge (i,j) by the r_{th} ant at the end of the each cycle.

$$\Delta\tau_{ij}^{r} = \begin{cases} \dfrac{Q}{L_r}, & \text{If } r_{th} \text{ ant used edge } (i,j) \text{ in its tour} \\ 0, & \text{otherwise} \end{cases} \tag{5}$$

Q is a constant and L_r is the tour length of the r_{th} ant.
Formally the AS algorithm is (Dorigo *et al.*, 1996):

Step-1. Initialize:
Set t: = 0 (t is the time counter)
Set NC: = 0 (NC is the cycle counter)
For every edge (i,j) set an initial value $\tau_{ij}(t)$ = c for trail intensity and $\Delta\tau_{ij} = 0$

Step-2. Set s: = 1 (s is the tabu list index)
For r: = 1 to R do
Place the starting town of the r_{th} ant in $tabu_r(s)$

Step-3. Repeat until tabu list is full
Set s: = s+1
For r: = 1 to R do
Choose the town j to move to, with probability $P_{ij}^{r}(t)$ given by equation (3)
Move the r_{th} ant to the town j
Insert town j in $tabu_r(s)$

Step-4. For r: = 1 to R do
Move the r_{th} ant from $tabu_r(N)$ to $tabu_r(1)$
Compute the length L_r of the tour described by r-th ant
Update and save the shortest tour found
For r: = 1 to R do
Compute $\Delta\tau_{ij}^{r}$ by equation (5)
$\Delta\tau_{ij} = \Delta\tau_{ij} + \Delta\tau_{ij}^{r}$

Step-5. For every edge (i,j) compute $\tau_{ij}(t+N)$ according to equation (4)
Set t: = t+N
Set NC : = NC+1
For every edge (i,j) set $\Delta\tau_{ij} = 0$

Step-6. If ($NC < NC_{max}$) and there is not stagnation behaviour
then
Empty all tabu lists
Goto Step-2
else
Print the shortest tour
Stop

Numerous successful implementations of the ACO metaheuristics have been applied to many different combinatorial optimization problems like routing, assignment, scheduling, subset and network routing (Dorigo and Stützle, 2004). The applications for scheduling problems are summarized in Table 1.

Table 1. Applications of ACO for scheduling (Dorigo and Stützle, 2004)

Problem Name	Main References
Job Shop	Colorni, Dorigo, Maniezzo, Trubian (1994)
Open Shop	Pfahringer (1996)
Flow Shop	Stützle (1998)
Total Tardiness	Bauer, Bullnheimer, Hartl, Strauss (2000)
Total Weighted Tardiness	den Besten, Stützle, Dorigo (2000)
	Merkle, Middendorf (2000, 2003)
	Gagne, Price and Gravel (2002)
Project Scheduling	Merkle, Middendorf and Schemeck (2000, 2002)
Group Shop	Blum (2002, 2003)

Many researchers used ACO algorithms to solve various types of the scheduling problems and it can be seen that ACO is among the best-performing approaches for some types of scheduling problems such as single-machine total weighted tardiness problem, the open shop problem, and the resource constrained project scheduling problem. However, ACO research results are far behind the state of the art for classic scheduling problems, like the permutation flow shop problem and the job shop problem, which will be in our interest in this work.

3 Fuzzy Job Shop Scheduling Problem

3.1 Fuzzy Sets and Fuzzy Logic

Fuzzy sets were introduced by Zadeh in 1965 to represent/manipulate data and information possessing nonstatistical uncertainties. Fuzzy logic provides an inference morphology that enables approximate human reasoning capabilities to be applied to knowledge-based systems. The theory of fuzzy logic provides a mathematical strength to capture the uncertainties associated with human cognitive processes, such as thinking and reasoning.

There are two main characteristics of fuzzy systems that give them better performance for specific applications;

- Fuzzy systems are suitable for uncertain or approximate reasoning, especially for the system with a mathematical model that is difficult to derive.
- Fuzzy logic allows decision making with estimated values under incomplete or uncertain information.

3.1.1 Operations with Fuzzy Numbers

If we denote a triangular fuzzy number $\tilde{A}$ by a triplet (a^1, a^2, a^3) then addition of two positive triangular fuzzy numbers $\tilde{A}$ and $\tilde{B}$ is calculated by the following formula:

$$\tilde{A} + \tilde{B} = (a_1, a_2, a_3) + (b_1, b_2, b_3) = (a_1 + b_1, a_2 + b_2, a_3 + b_3)$$

The addition of two triangular fuzzy numbers will also be a triangular fuzzy number.

According to the extension principle of Zadeh, the membership function $\mu_{\tilde{A}\vee\tilde{B}}(z)$ or $\tilde{A} \vee \tilde{B}$ through the $\vee$ (max) operation becomes as follows:

$$\mu_{\tilde{A}\vee\tilde{B}}(z) = \sup_{z=x\vee y} \min(\mu_{\tilde{A}}(x), \mu_{\tilde{B}}(y))$$

The result of the $\vee$ (max) operation does not always become a triangular number. For simplicity, we approximate the $\vee$ (max) operation with the following formula as in Sakawa and Kubota (2000).

$$\tilde{A} \vee \tilde{B} = (a_1, a_2, a_3) \vee (b_1, b_2, b_3) \approx (a_1 \vee b_1, a_2 \vee b_2, a_3 \vee b_3) \tag{6}$$

The ranking method used in this work involves three ordered criteria (Kaufmann and Gupta, 1988);

Criterion 1. The greatest associate ordinary number, $C_1(\tilde{A}) = \dfrac{a^1 + 2a^2 + a^3}{4}$, is used as a first criterion for ranking.

Criterion 2. If C_1 does not rank the fuzzy numbers, those which have the best maximal presumption $C_2(\tilde{A}) = a^2$ will be chosen as a second criterion.

Criterion 3. If C_1 and C_2 do not rank the fuzzy numbers, the difference of the spreads $C_3(\tilde{A}) = a^3 - a^1$ will be used as a third criterion.

3.2 Fuzzy Job Shop Scheduling Problem

Although the JSS problem has often been investigated, very little of this research is concerned with the uncertainty characterized by the impression in problem variables. In most of the work about JSS it is assumed that the problem data are known precisely at the advance or the prevalent approach to the treatment of the uncertainties in the JSS problem is use of probabilistic models. It is obvious that, if there exists a human interaction in the system there will be uncertainties in processing times. However, the evaluation and optimization of probabilistic models is computationally expensive and the use of probabilistic models is realistic only when description of the uncertain parameters is available from the historical data (Balasubramanian, 2003).

The main application area of fuzzy set theory to scheduling is in the systematic framework it provides for the representation, application, propagation and relaxation of imprecise constraints, and the evaluation of schedules with respect to vaguely defined goals. The desirability of a particular schedule is given by the degree to which it simultaneously satisfies all the goals and constraints, which may be interpreted as the schedule's degree of membership of the intersection of fuzzy constraint/goal sets (Slany, 1994).

McCahon and Lee (1992) were the first to illustrate the application of fuzzy set theory as a means of analyzing performance characteristics for a flow shop problem. Ishii *et al.* (1992) investigated two machine open shop problems with maximum lateness criteria and an identical machine scheduling problem with fuzzy due dates. Fortemps (1997) used simulated annealing in order to minimize the makespan of JSS with fuzzy durations. That was the first significant application which considers the uncertainty in time parameters, six-point fuzzy numbers were used to represent fuzzy durations. Flexible solutions which can cope with all possible durations were generated for benchmark fuzzified problems. Balasubramanian and Grossmann (2003) present a mixed integer linear programming (MILP) for flow shop scheduling with fuzzy task durations and compute optimistic and pessimistic values of makespan. The model was computationally tractable only for reasonably sized problems. Sakawa and Kubota (2000) introduced fuzzy job shop scheduling problems (FJJS) by incorporating the fuzzy processing time and fuzzy duedate. On the basis of the agreement index of fuzzy duedate and fuzzy completion time, multi objective FJJS problems have been formulated as three-objective ones which not only maximizes the minimum agreement index but also maximize the average agreement index and minimize the maximum fuzzy completion time. They proposed a genetic algorithm for generating solutions for multi objective FJSS problem. Lin (2002) considered a FJSS problem with imprecise processing time. He used fuzzy numbers and level $(\lambda,1)$ interval valued fuzzy numbers for the representation of vague processing times. He used Johnson's constructive algorithm for JSS problems. Ghrayeb (2003) presented a bi-criteria genetic algorithm approach to solve FJSS problem, in which the integral value and the uncertainty of the fuzzy makespan, which are conflicting objectives, are minimized. Fuzzified benchmark problems were used to show the effectiveness of the proposed approach.

Using fuzzy numbers to represent the uncertainty in processing times is very plausible for real world applications. If a decision maker estimates the processing time (t_{jk}) of the job j on machine k as an interval rather than a crisp value then the interval can be represented as a fuzzy number. The use of interval $\left[t_{jk} - \Delta_{jk1}, t_{jk} + \Delta_{jk2}\right]$ is more appropriate than the crisp t_{jk} value. The decision maker should carefully determine the parameters Δ_{jk1} and Δ_{jk2}, which satisfy $0 < \Delta t_{jk1} < t_{jk}$ and $0 < \Delta t_{jk2}$. As shown in Fig. 1 the fuzzy processing time is represented by a triangular fuzzy number.

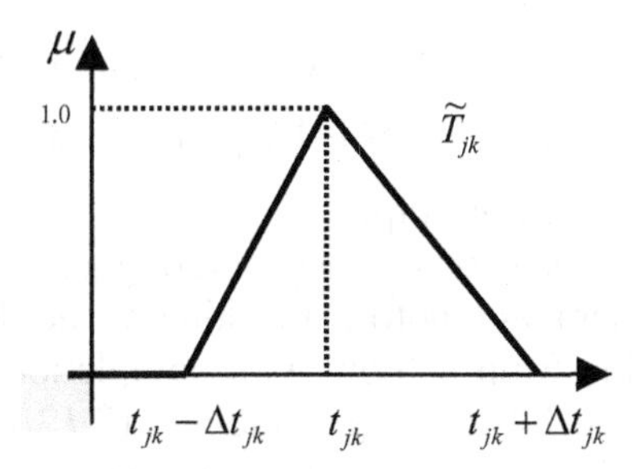

Fig. 1. Fuzzy processing time

In addition, as in many real-world situations, if a certain amount of delay (Δd_j) may be tolerated we can also use fuzzy numbers for due dates, Fig. 2. The fuzzy due date of job j is represented by the degree of satisfaction with respect to the job completion time.

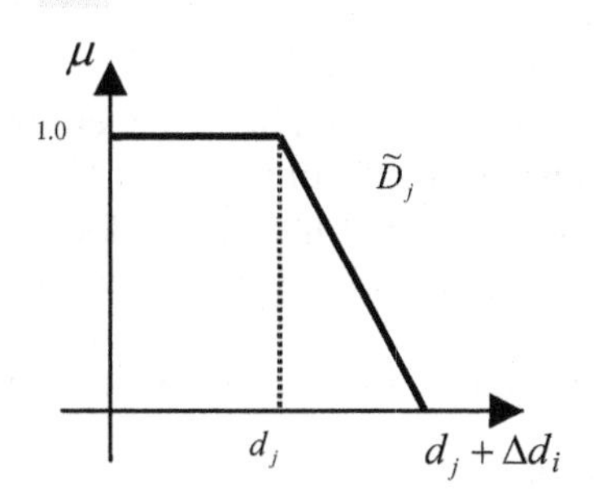

Fig. 2. Fuzzy due date

As we use fuzzy processing times for jobs, the completion time of the each job will also be a triangular fuzzy number (TFN). The fuzzy completion time for each job is denoted by $\tilde{C}_i$. Now we can model the FJSS similar to the crisp JSS model but with fuzzy processing times and fuzzy due dates;

$$\tilde{C}_{ik} - \tilde{T}_{ik} + L(1 - x_{ihk}) \geq \tilde{C}_{ih} \qquad i = 1,2,...,n \qquad h,k = 1,2,...,m \tag{8}$$

$$\tilde{C}_{jk} - \tilde{T}_{jk} + L(1 - y_{ijk}) \geq \tilde{C}_{ik} \qquad i,j = 1,2,...,n \quad k = 1,2,...,m \tag{9}$$

$$\tilde{C}_{ik} \geq 0 \qquad i = 1,2,...,n \qquad k = 1,2,...,m$$

$$x_{ihk} = 0 \; or \; 1 \qquad i = 1,2,...,n \qquad h,k = 1,2,...,m$$

$$y_{ijk} = 0 \; or \; 1 \qquad i,j = 1,2,...,n \quad k = 1,2,...,m$$

It can be shown that if $x_{jhk} = 1$ and $y_{ijk} = 1$, equations (7) and (8) implies

$$\tilde{C}_{jk} \geq (\tilde{C}_{jh} \vee \tilde{C}_{ik}) + \tilde{T}_{jk} \qquad i,j = 1,2,...,n \quad h,k = 1,2,...,n$$

according to the extension principle.

As an example assume that we are generating a schedule and Job-1 must be processed on machine-4 before machine-5 and Job-2 is the last job scheduled on machine 5 up to now with completion time (8,10,14). The completion time of the Job-1 on machine-4 is (7,12,13). If we want to assign Job-1 on machine-5, the minimum completion time of Job-1 on machine-5 can be calculated as;

$$\tilde{C}_{15} = (\tilde{C}_{14} \vee \tilde{C}_{25}) + \tilde{T}_{15}$$

$$\tilde{C}_{15} = [(7,12,13) \vee (8,10,14)] + \tilde{T}_{15} = (8,12,14) + \tilde{T}_{15}$$

For an objective function which minimizes the makespan of the schedule we can use the following one;

$$\text{Minimize } z = \max \tilde{C}_i \qquad i = 1,...,n$$

However the aim of the schedule might be generating a schedule in order to minimize the lateness of maximum late job from its due date. If this is the case, we can use the agreement indices (Sakawa and Kubota, 2000). The agreement index (AI) expresses how much the $\tilde{C}_i$ is inside the boundaries of the fuzzy due date ($\tilde{D}_i$). As shown in Fig. 3.;

$$\text{AI} = [\text{area}\,(\tilde{C}_i \cap \tilde{D}_i)]/[\text{area}\,(\tilde{C}_i)]$$

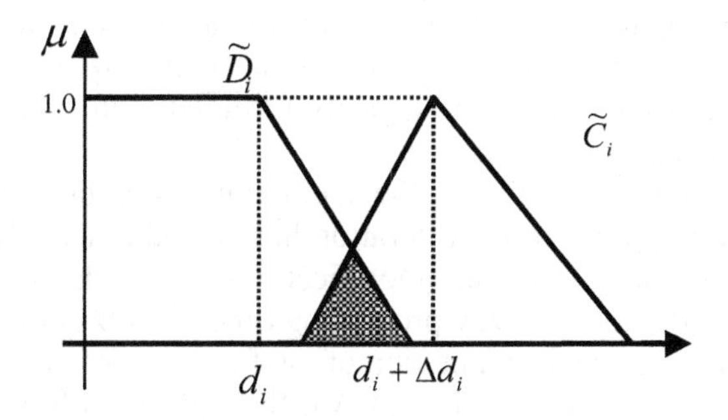

Fig. 3. Agreement index

And, the objective function will be;

$$\text{Maximize } z = \min AI_i \qquad i = 1,\ldots,n \tag{9}$$

4 Fuzzy Ant Colony Optimization Algorithm for Fuzzy JSS Problem

In this section we propose an implementation of fuzzy logic in ACO framework for JSS problems. ACO algorithms show good performance in solving problems that are combinatorial in nature. However, some of the real-life problems are characterized both by uncertainty and by combinatorial in nature. There may be four main types of imprecision in the JSS problems. They are fuzzy processing time, fuzzy due date, fuzzy preferences and fuzzy objective functions with fuzzy criteria (Lin, 2002). We tried to handle the fuzziness in JSS problem in Sect. 3 for fuzzy processing times and fuzzy due dates with a objective function which maximizes the minimum agreement index between fuzzy due dates and fuzzy completion times.

Although some work has been done on combining fuzzy rules with ant-based algorithms for optimization problems, to our knowledge until now fuzzy rules have not yet been used to control the behavior of the artificial ants in a scheduling algorithm. In addition there exists only a little work which combines fuzzy logic with ant-based algorithms. Schockaert *et al.* (2004) proposed a clustering algorithm with fuzzy ants. Lucic (2002) proposed a fuzzy ant system (FAS) for transportation problems.

Human operators use subjective knowledge or linguistic information on a daily basis when making decisions. The environment in which a human expert (human controller) makes decisions is often complex, making it difficult to formulate a suitable mathematical model. Thus, the development of fuzzy logic systems seems justified in such situations. The control strategies of artificial ants can be easily formulated in terms of numerous descriptive rules (Lucic, 2002).

Through many publications, the methodology of how ants make the choice where to go in the next iteration has stayed practically unmodified for an entire decade. We make modifications to the classical Ant System algorithm in order to use fuzzy processing time and fuzzy due date. But the basic modification was in adaptation of the fuzzy rules instead of transition probability function in the classical Ant System algorithm. Therefore the proposed algorithm is not only an ant algorithm which can operate with fuzzy variables but also an algorithm which uses fuzzy rules for generating solutions.

In principle, there is an infinite number of feasible schedules for any JSS problem, because arbitrary amounts of idle time can be inserted. However, this is not useful for a given known sequence and regular objectives. By definition a local shift occurs in a feasible schedule if one operation can be moved left to start earlier and keep feasibility. A schedule is semi-active if no local left shifts are available. It is possible to further improve semi-active schedules in many cases. Although an activity may not be moved immediately to the left, it may be possible to "jump" over obstructions to the left and obtain a better schedule. A global left shift occurs if one operation can be shifted into a hole earlier in the schedule and preserve feasibility. The set of all schedules in which no global left shift can be made is called the set of active schedules. The number of active schedules is still much too large for all but the smallest problems when using search. Thus attention often focuses on an even smaller subset called dispatch schedules or nondelay schedules. A dispatch or nondelay schedule is one in which no machine is kept idle at a time when it could begin processing some operation. The set of all feasible schedules may be classified as follows. A set of all nondelay schedules is part of the set of all active schedules, which is part of the set of all schedules (Fig. 4). There is always an active schedule that is optimal for a regular objective. The best nondelay schedule may not be optimal (Morton and Pentico, 1993).

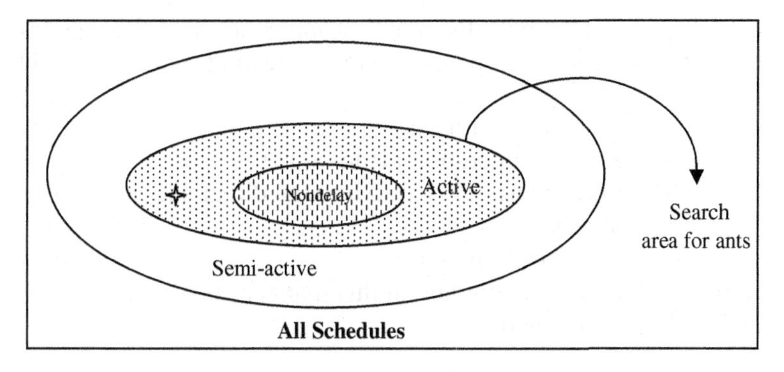

Fig. 4. Venn diagram of schedule relationships

Schedule generation procedures treat operations in an order that is consistent with the precedence relations of the problem. In other words, no operation is considered until all of its predecessors have been scheduled. Once we schedule all the predecessors of an operation, that operation becomes schedulable. Generation procedures operate with a set of schedulable operations at each stage. At each stage, the operations that have already been assigned starting time make up a partial schedule. Let

PS(t) : a partial schedule containing t scheduled operations.

SO(t) : the set of schedulable operations at stage t, corresponding to a given PS(t).

s_j : the earliest time at which operation $j \in SO(t)$ could be started.

f_j : the earliest time at which operation $j \in SO(t)$ could be finished.

For a given active partial schedule, the potential start time for schedulable operation j, denoted s_j, is determined by the completion time of the direct predecessor of operation j and the latest completion time on the machine required by operation j. The larger of these two quantities is s_j. The potential finish time f_j is simply s_j+t_j, where t_j is the processing time of operation j. A systematic approach to generating active schedules works as follows (Baker, 1997).

Active Schedule Generation Algorithm

Step 1. Let t = 0 and begin with PS(t) as the null partial schedule. Initially, SO(t) includes all operations with no predecessors.

Step 2. Determine $f^* = \min_{j \in SO(t)}\{f_j\}$ and the machine k^* on which f^* could be realized.

Step 3. For each operation $j \in SO(t)$ that requires machine k^* and for which $s_j < f^*$, calculate a priority index according to a specific priority rule.

Find the operation with the smallest index and add this operation to PS(t) as early as possible, thus creating partial schedule, PS(t + 1).

Step 4. Update the data set as follows:

(a) Remove operation j from SO(t).
(b) Form SO(t + 1) by adding the direct successor of job *j* to SO(t).
(c) Increment t by one.

Step 5. Return to Step 2 and continue in this manner until an active schedule has been generated.

Proposed Model

The proposed model in this paper searches the best active schedule with artificial ants. Basically, the main steps of the proposed algorithm is as follows,

Step 1: Set parameters, initialize the pheromone trails.
Step 2: While (termination condition is not met)
do the following:
construct solutions;
improve the solutions by local search (optional);
update the pheromone trail or trail intensities;
Step 3: Return the best solution found
Check the termination criterion:
Go to *Step 2* or Stop.

Step 1. Setting parameters and initializing the pheromone trails
Values of the parameters are determined in accordance with our experiences gained in the numerical experimentations and the recommendations based on the previous studies. Number of ants (m) constructing complete solutions at each iteration is equal to the number of jobs (J) to be scheduled. The initial trail intensities are chosen as $\tau_{jn}^{k}=1$, where τ_{jn}^{k} denotes the trail intensity for the j_{th} job for the n_{th} position on machine *k*. The persistence of trail (ρ) is set to 0.95, where ($1-\rho$) denotes the evaporation rate.

Step 2. Construction of schedules
An ant starts constructing a schedule by determining f^* and k^* in order to narrow the search space. For each candidate job, $j \in SO(t)$, that requires machine k^* and for which $ş_j < f^*$, the ant will have greater or lesser perceived utility towards it depending on the total time of operations remaining

to be scheduled for job j, and its trail intensity for the position. Remaining times for jobs and trail intensities may be represented by fuzzy values (Fig. 5 and Fig. 6).

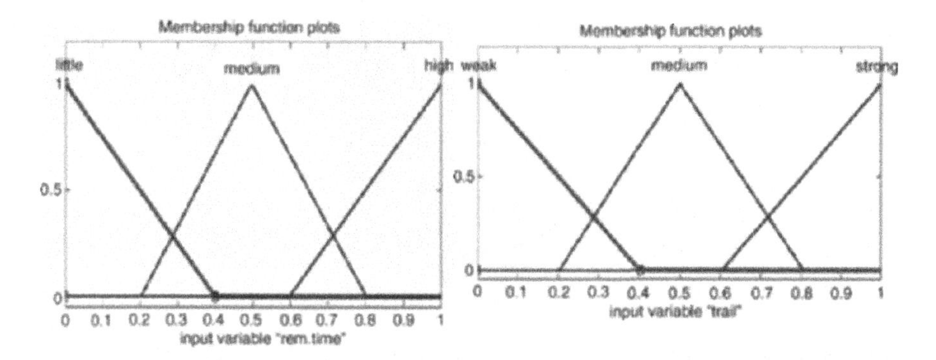

Fig. 5. Fuzzy remaining time

Fig. 6. Fuzzy trail intensity

The approximate reasoning algorithm for calculating the ant's perceived utility of choosing an operation consists of the following 9 rules in Table 2:

Table 2. Fuzzy rule base for artificial ants

Rule	Trail	Remaining Time	Utility
1	WEAK	LITTLE	VERY LOW
2	WEAK	MEDIUM	LOW
3	WEAK	HIGH	MEDIUM
4	MEDIUM	LITTLE	MEDIUM
5	MEDIUM	MEDIUM	HIGH
6	MEDIUM	HIGH	VERY HIGH
7	HIGH	LITTLE	VERY HIGH
8	HIGH	MEDIUM	VERY HIGH
9	HIGH	HIGH	VERY VERY HIGH

As an example, Rule-6 is:

If trail intensity is **MEDIUM** and remaining time is **HIGH**
Then utility is **VERY HIGH**

Graphical representation of the surface of the fuzzy rule base is shown in Fig. 7. It gives the utility value of selecting of a job for a position, according to the rules in Table. 2. This is the main differences between crisp ACO and the fuzzy ACO algorithm. In Sect. 2.2.4 the utility values for candidates were calculated by equation (1). We use the fuzzy rule base instead as represented in Table 2.

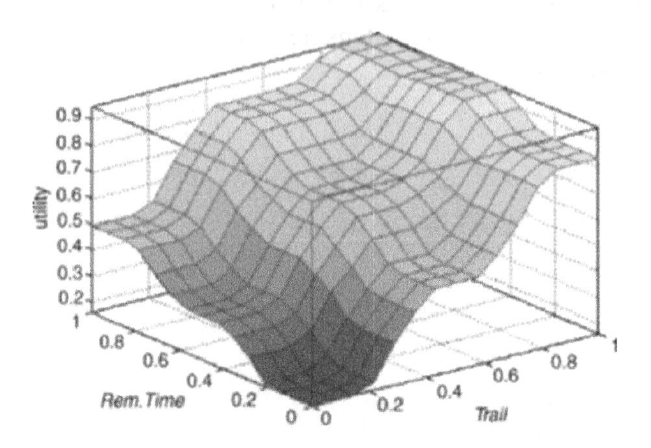

Fig. 7. Utility value of selecting of a job for a position, according to the fuzzy rules

After obtaining utilities for each candidate operation with fuzzy rule base, it is possible to calculate probabilities associated with each operation. The probability of each operation will represent its probability of being selected by the ant.

u_{jn}^{k} : utility of job j being scheduled for the n_{th} position of machine k.

$$p_{jn}^{k} = \frac{u_{jn}^{k}}{\sum_{j \in m^{*}} u_{jn}^{k}} \tag{10}$$

Final choice would be obtained in a proportional manner as in crisp ACO algorithm known as the roulette wheel selection. The selected job for machine m^{*} will be added on the Gantt chart of ant r. In this way one ant can generate a complete schedule. Artificial ants live in discrete time, time index (t) is zero at the beginning of the algorithm and it is increased one by one as the jobs are added to the Gantt chart and when t reaches the value of total number of operations a complete schedule is generated. Since there are R ants, total of R schedules will have been generated in a cycle. After finishing a cycle trail intensities will be updated.

The original equation (1) used in the crisp Ant System is replaced by approximate reasoning. In this way, it will also be possible to calculate transition probabilities even if some of the input data were only approximately known.

Updating Trail Intensities

Trail intensities made the jobs more desirable for the specific positions if they had been frequently chosen and the objective function value of the schedule for these positions had been greater at the earlier cycles. Furthermore, evaporation of pheromone lessen the desirability of the jobs for the specific positions if they had been seldom chosen and the objective function value of the schedule for these positions had been smaller at the earlier cycles. In this work we wanted to minimize the lateness of maximum late job from its due date so the agreement indices (AI), defined in section 3.2, are used for determining the quantity of the trails to be left by ants;

Let, $\Delta\tau_{jn}^{kr}$ is the quantity of trail substance to be left, on position n for job j on the k-th machine by ant r, at the end of a cycle. Each ant left trail substance for each job for its position in the schedule as much as the minimum AI value of the jobs of the schedule generated by the considered ant, the quantity of the trail to be left is denoted by;

$$\Delta\tau_{jn}^{kr} = \min \mathrm{AI}_{j} \qquad j = 1,2,\ldots,n \tag{11}$$

After each cycle, trail intensities are updated by the equation below;
R: total number of ants ($r = 1,2,\ldots,R$)

$$\tau_{jn}^{k,new} = \rho \times \tau_{jn}^{k,old} + \sum_{r=1}^{R} \Delta\tau_{jn}^{k,r} \tag{12}$$

Step 3.
We should set some termination conditions to stop the search. The algorithm may be stopped if it finds a solution which is known to be a optimal or a near optimal solution. It can also be stopped after it makes a predefined number of iterations. In addition, we can also end the search if it is in a stagnation behavior. If all ants start to construct same solutions because of the high trail intensification on some positions for specific jobs we do not need to wait for algorithm to escape from this behavior because it is almost impossible.

5 An Application for JSS Problem Using Fuzzy ACO

Sakawa and Mori (1999) applied Genetic Algorithm (GA) and Simulated Annealing (SA) to the FJSS problem as shown in Table 3;

Table 3. Fuzzy JSS problem (6 jobs, 6 machines)

	Processing Machines (Fuzzy Processing Time)					
Job 1	1(5,6,13)	5(3,4,5)	2(1,2,3)	6(3,4,5)	4(2,3,4)	3(2,3,4)
Job 2	1(3,4,5)	2(2,4,5)	3(1,3,5)	6(4,5,6)	4(5,6,7)	5(6,7,8)
Job 3	3(1,2,3)	6(5,6,7)	5(4,5,6)	4(3,4,5)	2(1,2,3)	1(1,2,3)
Job 4	6(2,3,4)	5(1,2,3)	4(2,3,4)	2(2,3,5)	1(3,4,6)	3(3,4,5)
Job 5	6(3,4,5)	5(2,3,4)	4(1,2,3)	3(2,3,4)	2(4,5,6)	1(2,3,4)
Job 6	5(6,7,8)	6(4,5,6)	1(2,3,4)	2(3,4,5)	3(2,3,4)	4(1,2,3)
	Job 1	Job 2	Job 3	Job 4	Job 5	Job 6
Fuzzy due date	30,40	35,40	20,28	32,40	30,35	40,45

We also tested the proposed fuzzy ACO for FJSS problem on the same problem and made 10 trails as they did. Results by proposed fuzzy ACO are represented on the Table 4 with the Sakawa and Mori (1999)'s solutions. Results represent the minimum agreement indices for the solutions obtained from each trail. The second term in parenthesis in ACO column represents the first cycle in which the max (min AI_i) is captured.

Table 4. Results for the FJSS problem

Trail	GA	SA	ACO	Trail	GA	SA	ACO
1	0,69	0,59	0,69 (84)	6	0,69	0,38	0,42 (220)
2	0,69	0,39	0,69 (99)	7	0,69	0,38	0,69 (242)
3	0,69	0,36	0,69 (188)	8	0,69	0,38	0,69 (120)
4	0,69	0,36	0,39 (124)	9	0,69	0,36	0,69 (87)
5	0,69	0,39	0,69 (285)	10	0,69	0,36	0,69 (162)

As a termination criteria they use 50 generations with population size 30 for GA (almost 30*50 = 1500 iterations) and 3000 iterations for SA. We run the proposed algorithm for 250 cycles and as there exist 6 ants each generating a solution in each cycle, it makes about 1500 iterations.

It should be also noted that we did not use a local search algorithm to improve solutions generated by artificial ants in order to show the explicit force of the Ant algorithm. It is possible to reach good solutions in shorter time periods with local search algorithms.

6 Conclusion

The JSS problem is NP-hard in its simplest case and we generally need to add new constraints when we want to solve a JSS in any practical application area. Therefore, as its complexity increases we need algorithms that can solve the problem in a reasonable time period and can be modified easily for new constraints. As it is mentioned, exact algorithms can be used in small sized problems due to very long solution times and it can be very hard to modify an exact solution algorithm only for a new constraint. However, the proposed Ant algorithm can solve JSS problems in reasonable time and it is very easy to modify the artificial ants for new constraints. In addition, it is very easy to modify artificial ants for multi-objective cases.

Determining parameter values (γ, θ, ρ, m) for an Ant algorithm is a very difficult and time consuming task for researchers. Generally, they are found after many trails or same values are used from previous works. Parameter values have great effect on solution time and it is also possible to reach optimal solution for specific values for them but after a slight differences in these values it can be impossible to reach near-optimal solutions. In the proposed fuzzy Ant algorithm model, γ and θ parameters, which control the relative importance of trail versus visibility, are not used as crisp values. We stand in fuzzy rule base instead of equation (1) in order to generate transition probabilities. And it became more easier to find appreciate rules than determining crisp parameter values.

JSS problems are generally handled with precise and complete data in theoretical studies. There are various exact and approximate solution techniques for these problems. However, it is a usual case to work with insufficient data in practical applications. As the uncertainty and the problem size increases, such techniques became inadequate to generate applicable solutions. Nevertheless, specialized human experts often use their knowledge and experience when they have to deal with such kind of uncertainties. Therefore, they can usually able to generate near optimum and applicable solutions. Actually, the urgent attribute of the Fuzzy Ant System is its ability to represent specialized human experts' knowledge by using a fuzzy rule base. Accordingly, it can deal with uncertain and incomplete data and generate near optimum solutions in shorter time periods.

References

Adams, J., Balas, E. Zawack, D. (1988), The Shifting Bottleneck Algorithm for Job-Shop Scheduling. Management Science 34, 391–401.

Back, T., Hammel, U., Schwefel, H.P. (1997), Evolutionary Computation: Comments on the History and Current State, IEEE Transactions on Evolutionary Computation 1,1, 3–17.

Balasubramanian, J., Grossmann, I.E. (2003), Scheduling optimization under uncertainty – an alternative approach. Computers and Chemical Engineering 27, 469–490.

Baker, K.R. (1997), Elements of sequencing and scheduling. Kenneth R. Baker.

Conway, R.W., Maxwell, W.L., and Miller, L.W., (1967), Theory of Scheduling. Addison Wesley.

Dorigo, M., Stützle, T. (2004), Ant Colony Optimization. MIT.

Fortemps, P. (1997). Jobshop scheduling with imprecise durations: a fuzzy approach. IEEE Transactions on Fuzzy Systems 5 (4), 557.

Gen, M., Cheng, R. (1997), Genetic Algorithms & Engineering Design. New York:Wiley.

Ghrayeb, O.A. (2003), A bi-criteria optimization: minimizing the integral value and spread of the fuzzy makespan of job shop scheduling problems. Applied Soft Computing 2/3F, 197–210.

Glover, F. (1986). Future paths for Integer Programming and Links to Artificial Intelligence. Computers and Operations Research 5, 533–549.

Holland, J.H. (1962), Outline for a logic theory of adaptive systems. Journal of the ACM 3, 297.

Johnson, S.M. (1954), Optimal two-and three-stage production schedules with setup times included. Naval Research Logistics Quartely 1, 61–68.

Jain, A.S., Meeran, S. (1999). Deterministic job-shop scheduling: Past, present and future. European Journal of Operational Research 113, 390–434.

Kaufmann, A., Gupta, M. (1988), Fuzzy Mathematical Models in Engineering and Management Science, North-Holland, Amsterdam.

Kirkpatrick, S., Gerlatt, C. D. Jr., Vecchi, M.P. (1983), Optimization by Simulated Annealing. Science 220, 671–680.

Lin, F. (2002), Fuzzy Job Shop Scheduling Based on Ranking Level (λ,1) Interval Valued Fuzzy Numbers. IEEE Transactions on Fuzzy Systems Vol. 10, No.4.

Lawler, E.L., Lenstra, J.K. Rinnooy Kan, H.G. (1982), "Recent developments in deterministic sequencing and scheduling: A survey ," in Deterministic and Stochastic Scheduling, Dempster, M., Lenstra, J., and Rinnooy Kan, H. Eds. Dordrecht, The Netherlands: Reidel.

Lucic, P. (2002), Modeling Transportation Systems using Concepts of Swarm Intelligence and Soft Computing. PhD thesis, Virginia Tech.

McCahon, C.S., Lee, E.S. (1992), Fuzzy job sequencing for a flow shop. European Journal of Operational Research 62, 294.

Metropolis, N., Rosenbluth, A.W., Rosenbluth, M. N., Teller, A.H. Teller, E., (1953). Equations of State Calculations by Fast Computing Machines, J. Chem. Phys. 21, 1087–1092.

Morton, T.E., Pentico, D.W. (1993), Heuristic scheduling systems with applications to production systems and project management, Wiley Series in Engineering and Technology, John Wiley and Sons, Inc.

Pincus, M. (1970), A Monte Carlo Method for the Approximate Solution of Certain Types of Constrained Optimization Problems, Operations Research 18, 1225-1228.

Roy, B., Sussmann, B. (1964), Les problémes d'ordonnancement avec contraintes disjonctives, SEMA, Paris, Note DS 9 bis.

Sakawa, M., Kubota, R. (2000), Fuzzy programming for multiobjective job shop scheduling with fuzzy processing time and fuzzy duedate through genetic algorithms. European Journal of Operational Research 120, 393-407.

Van Laarhoven, P.J.M., Aarts, E. Lenstra, J.K. (1992), Job shop scheduling by simulated annealing. Operations Research 40, 113-125.

Zadeh, L.A. (1965), Fuzzy Sets, Information and Control 8, 338-353.

Zoghby, J., Batnes, J.W., Hasenbein J.J. (2004), Modeling the reentrant job shop scheduling problem with setups for metaheuristic searches. European Journal of Operational Research.

Fuzzy Techniques in Scheduling of Manufacturing Systems

Stratos Ioannidis[1] and Nikos Tsourveloudis[2]

[1] Department of Mathematics, University of the Aegean,
83200 Karlovasi, Samos, Greece.
efioan@aegean.gr
[2] Department of Production Engineering and Management,
Technical University of Crete, University Campus,
73100 Chania, Crete, Greece
nikost@dpem.tuc.gr

Summary. A *distributed* and a *supervisory* scheduling architecture for manufacturing systems and their realization in MATLAB are presented. The distributed architecture uses a set of lower level fuzzy control modules that reduce Work-In-Process (WIP) and synchronize the production system's operation. The production rate in each stage is controlled so as to satisfy demand, avoid overloading and eliminate machine starvation or blocking. Performance tuning of the distributed controllers has been assigned to a supervisory control architecture. The scheduling objective is to keep the WIP as low as possible maintaining, at the same time, quality of service by keeping backlog into acceptable levels. It is also shown, in this chapter, how MATLAB's SIMULINK may be used to construct effective production systems simulators. SIMULINK has become very popular in academia and industry and provides a number of powerfull tools, that is almost impossible to find in a dedicated tool for discrete event systems simulation.

Key words: Production systems; Fuzzy Scheduling; Work-in-process; Backlog; SIMULINK models

1 Introduction

A production system can be viewed as a network of machines and buffers. Items receive operations at each machine and wait for the next set of operations in a buffer with finite capacity. The machines can break down in a random order and sometimes can be incapable of producing more parts because of starvation and/or blocking phenomena. Due to a failed machine with operational neighbors, the level of the downstream buffer decreases, while the upstream increases. If the repair time is big enough, then the

S. Ioannidis and N. Tsourveloudis: *Fuzzy Techniques in Scheduling of Manufacturing Systems*, StudFuzz **201**, 427–452 (2006)
www.springerlink.com

broken machine will either block the next station or starve the previous one. This effect will propagate throughout the system.

From the methodology point of view, production control policies include, among others, research on simulation studies of specific systems; queuing theory based performance analysis, stability and optimal control, as well as fluid approximations of discrete systems. According to Gershwin [1] production control policies may be classified as token-based, time-based and surplus-based [4]. Token-based systems, including Kanban, Production Authorization Card [2] and Extended Kanban Control Systems [3], involve token movement in the manufacturing system to trigger events. When an operation is performed or a demand arrives, a token is either created or taken from a specific location. Only when a token exists at the first location (and only if space for tokens exists at the second location) the operation takes place.

When considering simple manufacturing systems, analytical results produced thus far have demonstrated the superiority of surplus-based systems [5, 6]. More specifically, hedging point policies have been proven optimal in minimizing production cost in single-stage, single-part-type system scheduling [5, 6]. Generalizations to more than one-part-type or production stages have proven to be difficult [7, 8], but obtained solutions [9, 10], may be successfully applied in real manufacturing systems [11, 12].

In summary, it is a common belief among many researchers that for complex production systems the problem of scheduling, in order to minimize costs due to inventory and non-satisfaction of demand, cannot be solved analytically. Since neither analytical nor computational solutions are achievable, heuristic policies are suggested to control job flow within production systems [13, 14, 15, 16, 17, 18].

This chapter presents theoretical background of two fuzzy scheduling architectures along with their practical implementation in MATLAB's simulation environment, known as SIMULINK. The distributed scheduling scheme uses a set of lower level fuzzy control modules regulating production rate in each production stage in a way that demand is satisfied and machine overloading is avoided. The second architecture uses a fuzzy supervisor controller in order to tune the performance of the distributed fuzzy controllers. The overall production control system is viewed as a two-level surplus-based system with the overall control objective to keep the WIP as low as possible maintaining at the same time quality of service by keeping backlog into acceptable levels.

The rest of the paper is organized as follows. Section 2 describes the distributed fuzzy scheduling architecture. In Section 3 the supervisory fuzzy scheduling approach is presented. Section 4 describes how MATLAB's SIMULINK [21] may be used in order to construct effective production systems simulators. Issues for discussion and suggestions for further development are presented in Section 5.

2 Distributed Fuzzy Scheduling

Production scheduling of realistic manufacturing plants must satisfy multiple conflicting criteria and also cope with the dynamic nature of such environments. Fuzzy logic offers the mathematical framework that allows for simple knowledge representations of the production control/scheduling principles in terms of IF-THEN rules.
In Fuzzy Logic Controllers (FLCs), the control policy is described by linguistic IF-THEN rules, which model the relationship between control inputs and outputs with appropriate mathematical representation. A rule antecedent (IF-part) describes conditions under which the rule is applicable and forms the composition of the inputs. The consequent (THEN-part) gives the response or conclusion that should be taken under these conditions.

The advantage of the distributed FLCs is that they are computationally simple and therefore facilitate application to real time control/scheduling. In the distributed fuzzy scheduling system presented in [16], have been introduced three basic subsystems, namely *transfer line*, *assembly* and *disassembly* module. The majority of the real production networks can be decomposed into these basic subsystems. Each subsystem can be seen as a distributed fuzzy logic controller with input and output variables summarized in Table 1.

The control objective of the distributed scheduling approach, is to satisfy the demand and, at the same time, to keep WIP as low as possible. This is attempted by regulating the processing rate r_i at every time instant. The expert knowledge that describes the control objective can be summarized in the following statements:

- If the surplus level is satisfactory then try to prevent starving or blocking by increasing or decreasing the production rate accordingly.
- If the surplus is not satisfactory that is either too low or too high then produce at maximum or zero rate respectively.

The above knowledge may be more formally represented, for each one of the control modules of Table 1, by fuzzy rules of the following form:

Transfer Line Rule: IF $b_{j,i}$ is $LB^{(k)}$ AND $b_{i,l}$ is $LB^{(k)}$ AND s_i is $LS_i^{(k)}$ AND x_i is $LX^{(k)}$ THEN r_i is $LR_i^{(k)}$,

Assembly node Rule: IF $b_{j,i}$ is $LB^{(k)}$ AND … AND $b_{k,i}$ is $LB^{(k)}$AND $b_{i,l}$ is $LB^{(k)}$ AND s_i is $LS_i^{(k)}$ AND x_i is $LX^{(k)}$, THEN r_i is $LR_i^{(k)}$,

Table 1. Control modules

Module	Schema	Input	Output
Line	$\cdots \to B_{j,i} \to M_i \to B_{i,l} \to \cdots$	b_{ji} buffer level of buffer $B_{j,i}$ b_{il} buffer level of buffer $B_{i,l}$ s_i state of machine M_i x_i production surplus of M_i	
Assembly	$B_{j,i}$, $B_{k,i} \to M_i \to B_{i,l} \to \cdots$	b_{ji} buffer level of buffer $B_{j,i}$ b_{ki} buffer level of buffer $B_{k,i}$ b_{il} buffer level of buffer $B_{i,l}$ s_i state of machine M_i x_i production surplus of M_i	r_i processing rate of M_i
Disassembly	$\cdots \to B_{j,i} \to M_i \to B_{i,k}$, $B_{i,l} \to \cdots$	b_{ji} buffer level of buffer $B_{j,i}$ b_{ik} buffer level of buffer $B_{i,k}$ b_{il} buffer level of buffer $B_{i,l}$ s_i state of machine M_i x_i production surplus of M_i	

Disassembly node Rule: IF $b_{j,i}$ is $LB^{(k)}$ AND $b_{i,k}$ is $LB^{(k)}$ AND … AND $b_{i,l}$ is $LB^{(k)}$ AND s_i is $LS_i^{(k)}$ AND x_i is $LX^{(k)}$, THEN r_i is $LR_i^{(k)}$,
where k is the rule number, i is the number of machine or workstation, LB is a linguistic value of the variable *buffer level b* with term set B= {*Empty*, *Almost Empty*, *OK*, *Almost Full*, *Full*}, s_i denotes the state of machine i, which can be either 1 (operative) or 0 (stopped); consequently S= {0, 1}. LX represents the value that surplus x takes and it is chosen from the term set X= {*Negative*, *OK*, *Positive*}. The *production rate r* takes linguistic values LR from the term set R= {*Zero*, *Low*, *Almost Low*, *Normal*, *Almost High*, *High*}. The processing rate r_i of each machine at every time instant is

$$r_i' = \mathrm{f}_{\mathrm{IS}}(b_{j,i}, b_{i,l}, x_i, s_i) = \begin{cases} 0 & \text{if } s_i = 0 \\ \dfrac{\sum r_i \mu_R^*(r_i)}{\sum \mu_R^*(r_i)} & \text{if } s_i = 1 \end{cases} \tag{1}$$

where, $\mathrm{f}_{\mathrm{IS}}(b_{j,i}, b_{i,l}, x_i, s_i)$ represents a fuzzy inference system [19] that takes as inputs the level $b_{j,i}$ of the upstream buffer, the downstream buffer level $b_{i,l}$, x_i is the surplus (cumulative production minus demand) and s_i is a non fuzzy variable denoting the state of the machine, which can be either 1 (operative) or 0 (stopped). In Eq. (1), $\mu_R^*(r_i)$ is the membership function of the aggregated production rate, which is given by

$$\mu^*_R(r_i) = \max_{b_{j,i}, b_{i,l}, x_i} \min\,[\mu^*_{AND}(b_{j,i}, b_{i,l}, x_i),\ \mu_{FR^{(k)}}(b_{j,i}, b_{i,l}, x_i, r_i)] \tag{2}$$

where $\mu^*_{AND}(b_{j,i}, b_{i,l}, x_i)$ is the membership function of the conjunction of the inputs and $\mu_{FR^{(k)}}(b_{j,i}, b_{i,l}, x_i, r_i)$ is the membership function of the k-th activated rule. That is

$$\mu^*_{AND}(b_{j,i}, b_{i,l}, x_i) = \mu^*_B(b_{j,i}) \wedge \mu^*_B(b_{i,l}) \wedge \mu^*_X(x_i) \tag{3}$$

and

$$\mu_{FR^{(k)}}(b_{j,i}, b_{i,l}, x_i, r_i) = f_{\rightarrow}[\mu_{LB^{(k)}}(b_{j,i}),\ \mu_{LB^{(k)}}(b_{i,l}), \mu_{LX^{(k)}}(x_i), \mu_{LR^{(k)}}(r_i)] \tag{4}$$

In Eqs. 3 and 4, $\mu^*_B(b_{j,i})$ and $\mu^*_B(b_{i,l})$ are the membership functions (MFs) of the actual upstream and downstream buffer levels and $\mu^*_X(x_i)$ is the membership function of production surplus.

2.1 Simulation Results and Comparisons

The assumptions made for all simulations are the following:

- Machines fail randomly with a failure rate p_i.
- Machines are repaired randomly with rate rr_i.
- Time to failure and time to repair are exponentially distributed.
- Demand is continuous with rate d.
- All machines operate at known, but not necessarily equal rates.
- The initial buffers are infinite sources of raw material and consequently the initial machines are never starved.
- Buffers between adjacent machines M_i, M_j have finite capacities.
- Set-up times or transportation times are negligible or are included in the processing times.

The loading times for each machine are determined by a heuristic policy known as the staircase strategy [14]: if the machine is available, load into the machine the part having the maximum positive difference between the integral of production rate and the actual cumulative production.

MATLAB's Fuzzy Logic Toolbox [20] and Simulink [21] have been the software tools for building and testing all simulations. The performance of the distributed fuzzy scheduling approach is evaluated through a series of test cases.

2.1.1 Test Case 1: Single-Part-Type Transfer Lines

The developed fuzzy controller is first tested for the case of the single product transfer line presented in Figure 1. The production line consists of five machines and four intermediate buffers. The first buffer, denoted B_I, is an infinite source while the last buffer B_F has infinite storage capacity. The system is balanced. All machines have the same processing time, r_i =2 (i=1,...,5), and same failure and repair probabilities, p_i =0.1 and rr_i =0.5, respectively.

The transfer line of Figure 1 is identical to one presented by Bai and Gershwin (test case 5 in [14]), and it was selected to facilitate comparisons. The performance of our controller is compared to a) The classical produce-at-capacity approach, according to which the machines produce in their maximum rate when they are operational (up, not blocked, not starved), and b) The approach presented by Bai and Gershwin in [13], [14], [23] and elsewhere. Their method is based on the determination of a desirable production surplus value, the hedging point.

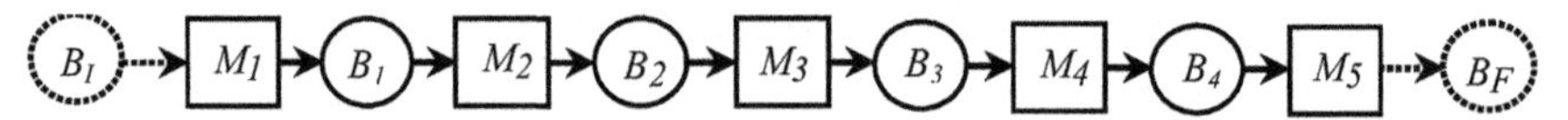

Fig. 1. The transfer line of test case 1

When the machine is up, the control law is summarized in the following:

- If the actual surplus is less than the hedging point, then the machine should produce at its maximum rate.
- If surplus is equal to the hedging point, the production rate should be equal to demand.
- If surplus is greater than the hedging point, stop producing.

The control methods are examined for five different demand values. Buffer capacities are given and presented in Table 2. It can be seen that the proposed approach keeps the in-process inventories significantly lower than the other two methods for all demands. Numerical results are presented in Table 3.

Table 2. Buffer sizes and demand levels for test case 1

Demand *d*	**Buffer Capacity**			
	BC_1	BC_2	BC_3	BC_4
1.6	6	6	8	5
1.4	3	3	10	1
1.2	1	4	5	1
1.0	1	1	1	1
0.6	1	1	1	1

Table 3. Average buffer level and WIP inventory for test case 1

Demand	Average Buffer levels for Fuzzy WIP Control				Work-in-process inventory		
	B_1	B_2	B_3	B_4	Fuzzy WIP control	Hedging point method	Produce at capacity
1.6	0.97	1.42	2.77	0.77	9.93	16.5	13.67
1.4	0.74	0.67	3.22	0.35	8.48	10.7	12.36
1.2	0.78	1.36	1.02	0.21	6.37	7.5	11.96
1.0	0.83	0.43	0.26	0.21	4.23	5	7.71
0.6	0.52	0.33	0.19	0.26	2.8	3.9	20.6

2.1.1 Test Case 2: Multiple-Part-Type Transfer Lines

In this section the performance of the transfer line controller introduced in this chapter is demonstrated, for the case of multiple-part-type production lines. The production system under consideration consists of three machines and produces three product types. Buffers with finite storage capacity are located between machines. The first buffer for each product is assumed to be an infinite source while the last is an infinite sink. All machines are subject to random failures and repairs with known rates.

Each machine is "virtually" divided in as many sub-machines as the number of part types to be processed. Consequently, the 3-part-type system under study, illustrated in Figure 2, can be approximated by three single-type systems similar to one analyzed in the previous section. Partial machines are presented in Figure 2 with dotted line squares. The processing times τ_{i_j} are given in Table 4.

Each machine i performs operations on parts of j type and it is divided into m_{ij} partial machines (here $i = j = 1,\ldots,3$). Each m_{ij} does one operation on the type j part, which then waits in the b_{ij} buffer for an operation at the $m_{i+1,j}$ partial machine. The demand for parts of type j is d_j. The failure and repair rate of machine i is p_i and rr_i, respectively.

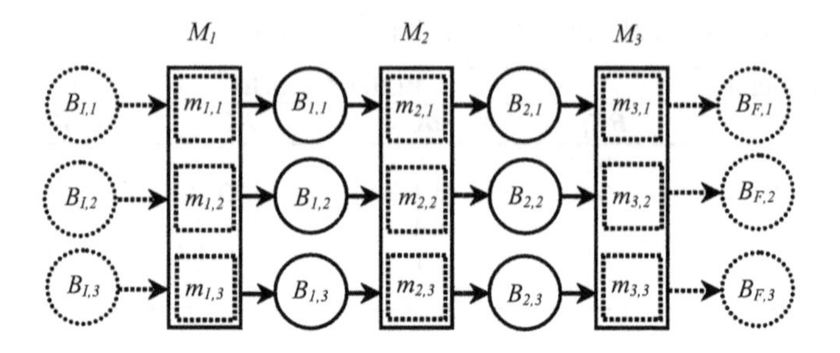

Fig. 2. Test case 2: 3-machine, 3-part-type production line

Table 4. Machines Processing Times.

	Machine		
Part type	1	2	3
1	0.5	0.3	0.4
2	0.3	0.2	0.3
3	0.4	0.4	0.5

Failure rates are $p_1 = 0.1$, $p_2 = 0.01$ and $p_3 = 0.2$, and repair rates are $rr_1 = 0.5$, $rr_2 = 0.8$, $rr_3 = 0.6$. Buffer sizes are all equal to 1 and demand is assumed to be constant over time for each part type ($d_1 = 0.8$, $d_2 = 0.6$, $d_3 = 0.3$). The WIP was calculated for each product after multiple simulation runs using different seeds. These results are shown below:

$WIP_1 = 1.369$ $WIP_2 = 0.988$ $WIP_3 = .0.667$

The fuzzy WIP controller satisfied the demand (which is selected low anyway) and kept the total WIP 20 % less than in [13].

3 Supervised Fuzzy Scheduling

In control systems literature a supervisory controller utilizes macroscopic data of higher hierarchies to adjust the overall system's behavior. Potentially, this may happen by modifying the lower level controllers in a way to ultimately achieve desired specifications. The supervisory controller's task, introduced in [17] and its optimization discussed in the next paragraph, is the tuning of the distributed fuzzy controllers presented in the previous paragraph, in a way that improves certain performance measures without causing a dramatic change in the control architecture. Therefore, the overall scheduling approach remains modular since the production control modules are not modified but simply tuned by the additional supervisory controller.

In the supervisory scheduling scheme it is assumed that the demand and the cumulative production are known. This is important for the production surplus monitoring and control and, consequently, for scheduling decisions based on production surplus. The expert knowledge that describes the supervisory control objective builds on the assumption that adaptive surplus bounds may improve the production systems performance and can be summarized in the following statements:

- If the upper surplus bound is reduced there is an immediate reduction of WIP.
- If the upper surplus bound is increased there is an increase of WIP and the total production rate leading to a small reduction of backlog.
- If the lower surplus bound is increased a substantial reduction of backlog and an increase in WIP is achieved.
- If there is a reduction of lower surplus bound as a result we have a deterioration of backlog with an improvement of WIP.

Surplus bounds are decided by the output of IF-THEN rules of the following form:

IF mx_e is $LMX^{(k)}$ AND e_x is $LE_x^{(k)}$ AND e_w is $LE_w^{(k)}$
THEN u_c is $LU_c^{(k)}$ AND l_c is $LL_c^{(k)}$,

where, k is the rule number, mx_e is the mean surplus of the end product, *LMX* is a linguistic value of the mx_e with term set *MX*= {*Negative Big, Negative Small, Zero, Positive Small, Positive Big*}, e_x denotes the error of end product surplus (= the difference between surplus x_e and the lower bound of surplus), with linguistic value term set E_x= {*Negative, Zero, Positive*}. The relative deviation of *WIP* is denoted e_w which is:

$$e_w = \frac{WIP(t) - \overline{WIP(t)}}{\overline{WIP(t)}} \tag{5}$$

where $\overline{WIP(t)}$ is the mean WIP and LE_w is the corresponding linguistic value is chosen from the term set E_w= {*Negative, Zero, Positive*}. The upper surplus bound correction factor u_c takes linguistic values LU_c from the term set U_c= {*Negative, Negative Zero, Zero, Positive Zero, Positive*} and the lower surplus bound correction factor l_c takes linguistic values LL_c from the term set L_c= {*Negative, Negative Zero, Zero, Positive Zero, Positive*}.

The crisp arithmetic values, u_c^* and l_c^*, of the corrections of the upper and lower surplus bounds, respectively, are given by the following defuzzification formulas:

$$u_c^* = \frac{\sum u_c \cdot \mu_{U_c}^*(u_c)}{\sum \mu_{U_c}^*(u_c)} \quad \text{(a)} \qquad l_c^* = \frac{\sum l_c \cdot \mu_{L_c}^*(l_c)}{\sum \mu_{l_c}^*(u_c)} \quad \text{(b)} \tag{6}$$

where $\mu_{U_c}^*(u_c)$ and $\mu_{L_c}^*(l_c)$ are the membership functions of the upper and lower surplus bounds, respectively. These membership functions represent the aggregated outcome of the fuzzy inference procedure. The correct selection of input and output membership functions characterizes the performance of the overall scheduling task.

3.1 Simulation Results and Comparisons

The proposed supervisory control approach is tested and compared to the unsupervised/distributed production control approach presented in Section 2, which has given very good results compared with other production control approaches. The assumptions we made for all simulations, are the same as in Section 2.1, except that here demand is stochastic and follows the Poisson distribution.

3.1.1 Test case 3: Single-Part-Type Production Networks

The developed fuzzy supervisory control architecture is tested for the case of a single part production network presented in Figure 3.

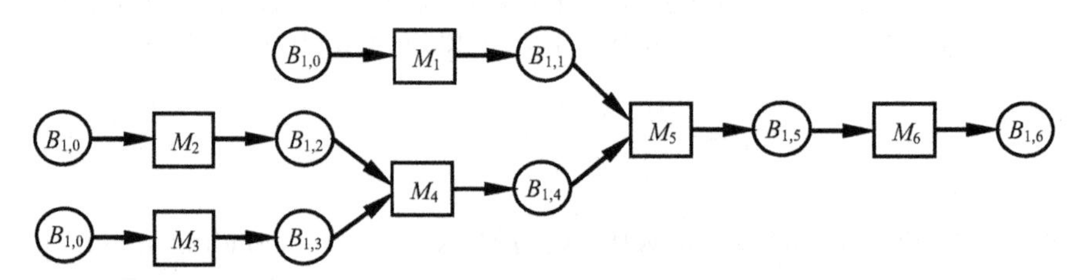

Fig. 3. Test case 3: single part type production network

The production system under consideration consists of six machines producing one part type. The failure and repair rates of all machines are equal. The repair rates are rr_i=0.5. All assembly and disassembly factors are assumed equal to 1. The processing times are $\tau_{i,}$= 0.3 (here i=1,...,6) equal for all machines. Comparative results for the WIP and Backlog in relation to demand for various buffer capacities and failure rates are shown in Figure 4.

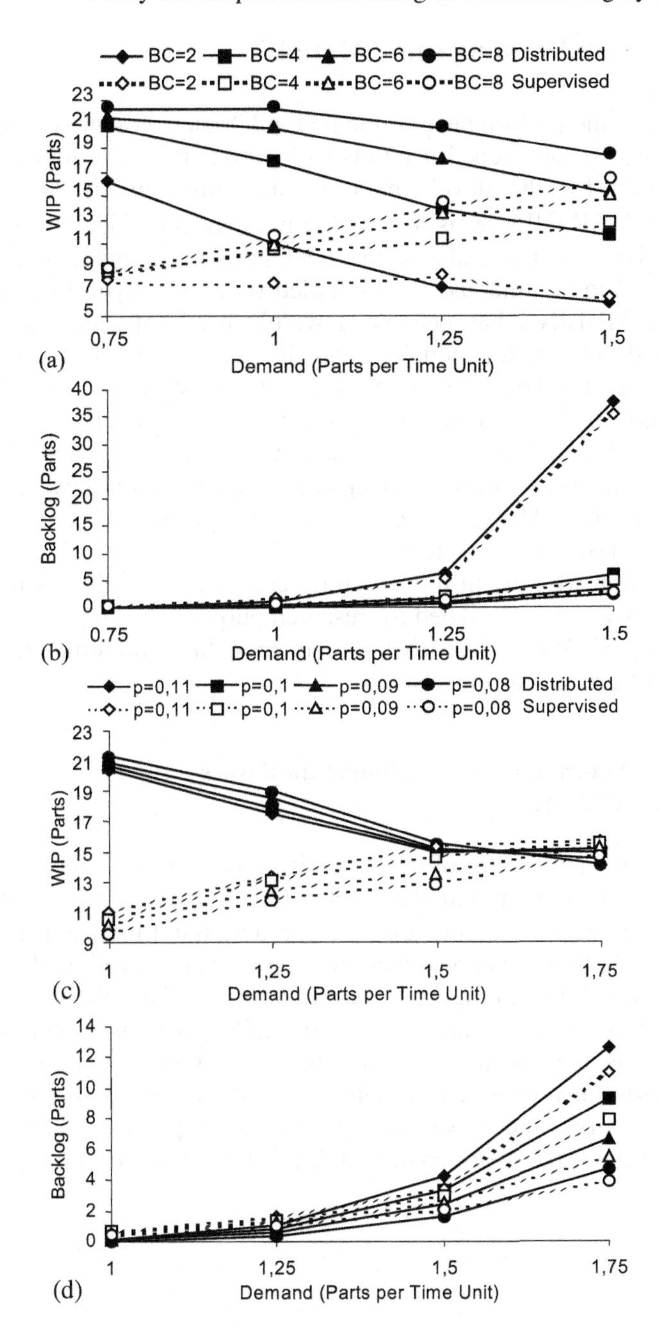

Fig. 4. Test Case 2: (a) WIP versus Demand for various Buffer Capacities (BC); (b) Backlog versus Demand for various BC; (c) WIP versus Demand for various Failure Rates (p); (d) Backlog versus Demand for various p

4 Simulation and Testing using Matlab

In order to test the performance of the methodologies presented in the previous sections we developed a number of production system simulators. The tool used for the development of the simulators is MATLAB's SIMULINK©[1]. SIMULINK is a software package for modeling, simulating, and analyzing dynamical systems. It supports linear and nonlinear systems, modeled in continuous time, sampled time, or a hybrid of the two. Although, SIMULINK has become a widely used software package in academia and industry for modeling and simulating dynamical systems, it hasn't been used for the modeling of manufacturing systems.

SIMULINK provides a Graphical User Interface (GUI) for building models as block diagrams, using click-and-drag mouse operations. Models are hierarchical, so one may build models using both top-down and bottom-up approaches. This approach provides insight into how a model is organized and how its parts interact.

In this section we present the structural components of the manufacturing simulators we have designed for research purposes and reported mainly in [16] and [17]. We explain their use in modeling and simulation of a wide range of manufacturing systems.

4.1 Building Manufacturing Systems Simulators with the SIMULINK

Here we present a simulator of the supervisory control architecture applied to a single product production network. As we have mentioned, SIMULINK models are hierarchical so one may build models using both top-down and bottom-up approaches. Here, however, we follow the hierarchical approach shown in Figure 5 and explained in what follows.

Figure 5 presents the higher level SIMULINK model, which consists of two subsystems: the *Production System subsystem* (shown in Figure 6) and the *Supervisor subsystem*. The Production System subsystem models the production network under examination, and the Supervisor is the implementation of the supervisory control architecture presented in the previous section.

---

[1] MATLAB and SIMULINK are registered trademarks of The MathWorks, Inc.

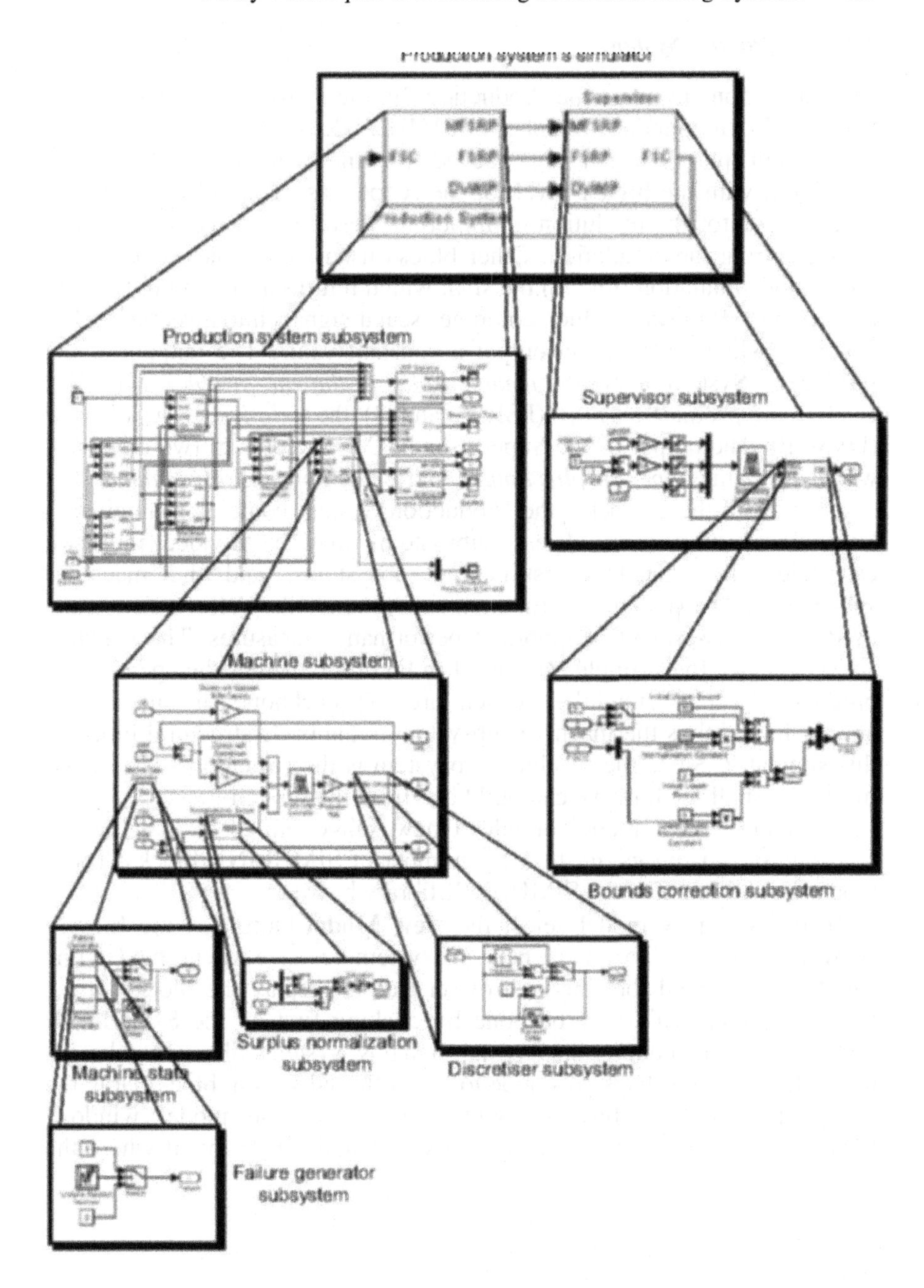

Fig. 5. Hierarchical analysis of a SIMULINK model for production systems

4.1.1 Production System

In Figure 6, one may see the Production System subsystem as designed in SIMULINK. It contains a number of SIMULINK blocks such as the input and output ports, which are needed for the communication of the subsystem with the higher level, or the scopes, as Mean WIP, which are used to monitor the evolution of important performance measures of the system during the simulation. Other blocks used are, 1) the *clock*, which returns the simulation time, 2) the *sum*, which returns as an output the sum of its inputs, 3) the *mux* which combines scalar signals into a vector, and 4) *constant*, which gives as an output a constant value. The outputs of the Production System are 1) the *mean surplus of the end product* (MFSRP), 2) the *end product surplus* (FSRP), and 3) the *relative error* of WIP (DVWIP). The output of the Supervisor FSC is a vector of two elements, which are 1) the *upper surplus bound*, and 2) the *lower surplus bound*.

Apart from these blocks, the Production System contains a number of lower level subsystems. These subsystems may be divided into two categories: the statistical estimation subsystems, and the machines subsystems. The statistical estimation subsystems, like WIP Statistics, are used for the assesment of inportant performance measures. The *machine subsystem* (see for example Machine 1 in Figure 6) models the operation of production control modules, which are the backbone of our control approach, and thus the machine subsystem is the basic structural block of this simulator. Since the machine subsystem is the structural basis of our model we shall see how we can build one from the scratch.

To create a new model or add a new subsystem to a model under construction it is necessary to use a number of SIMULINK blocks. These blocks can be found in the SIMULINK library browser.

To create a new model select the **New Model** button on the Library browser's toolbar. To add a new subsystem to an existent model one should first expand the library browser tree to display the blocks in the Sources library. This may be done by clicking first on the SIMULINK node, then on the node to display the Signal & Systems library blocks. Finally click on the subsystems node to select the subsystem block. Drag the subsystem from the browser, and drop it into the model window. SIMULINK creates a copy of the subsystem node at the point where the node icon is dropped.

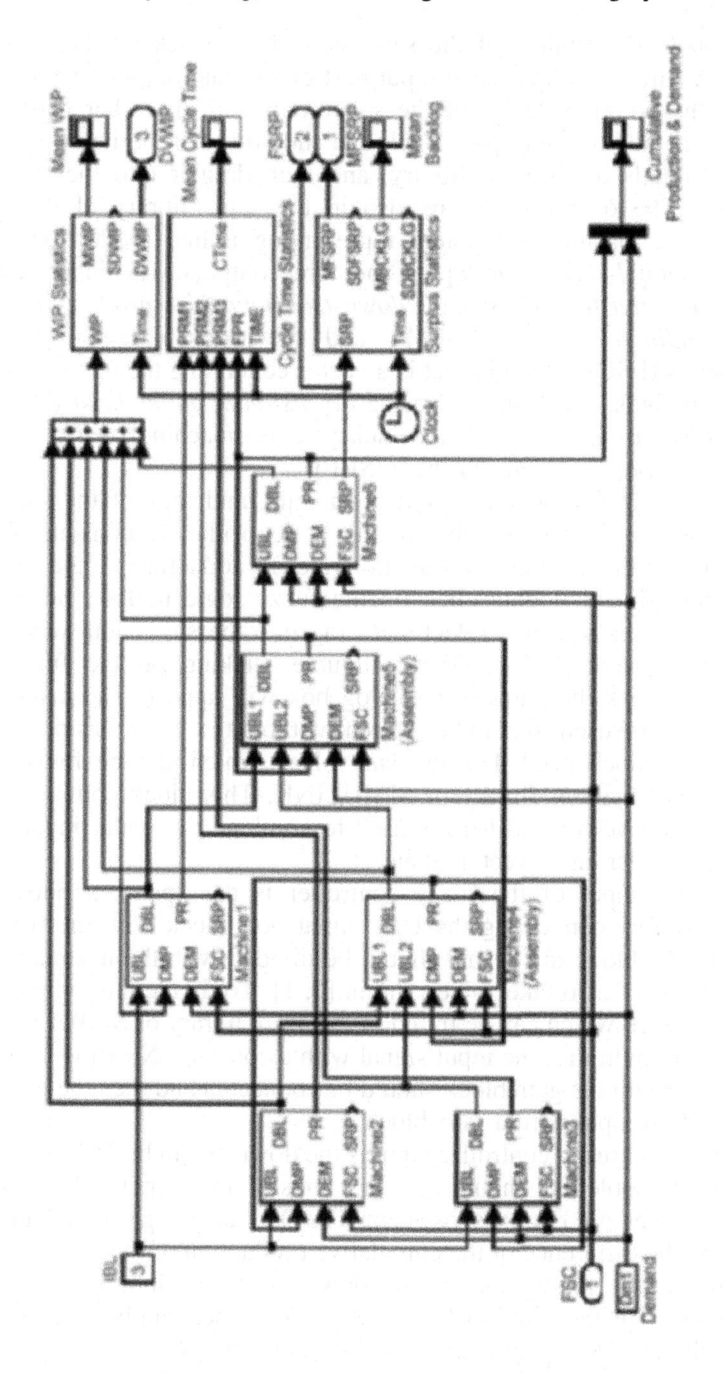

Fig. 6. The Production System subsystem

To open the window of the subsystem doubleclick on the subsystem block. At first one should add input port blocks and output port blocks, to ensure the communication of the subsystem with the other parts of the system. To add an input port block one should select the input port block in the Signals & Systems library, and then drag it into the subsystem window. The output ports are also in the same library of the library browser. A machine subsystem representing a line production control module should have four inputs and three outputs. The inputs are: the *upstream buffer level* (UBL), the *downstream machine production* (DMP), the *cumulative demand* (DEM), and the *fuzzy supervisory control* parameters (FSC), which in fact is a vector containing the upper and lower surplus bounds. The outputs are the *downstream buffer level* (DBL), the *cumulative production* of the subsystems machine (PR), and the *production surplus* of this machine (SRP).

The core of a production control module, and thus of the subsystem modelling it, is the distributed fuzzy controller responsible for the regulation of the machine's production rate. To add a fuzzy logic controller block one should double click on the fuzzy logic toolbox node in the library browser and then select and drag into the subsystem window the fuzzy logic controller block. By double clicking on the fuzzy logic controller block the parameters dialog box will appear. The name of the fuzzy structure used, should be entered (without the .fis extension).

Another block needed is the *mux* block which can be found in the *Signals and Systems* library of SIMULINK. This block combines scalar inputs into a vector and here is used to combine the scalar inputs of the fuzzy controller into a vector signal.

The first input of the fuzzy controller is the upstream buffer level (UBL). Before connecting the UBL input port block with the first input port of mux block this signal should be divided with the upstream buffer capacity in order to take values within [0, 1]. To divide this signal we use the *gain* block which can be found in the *Math* library of SIMULINK. The gain block multiplies the input signal with a constant. Now the UBL input port block with the gain block should be connected and then the gain block with the first input port of mux block.

The second fuzzy controller input is the downstream buffer level (DBL). This is a variable that should be also a subsystems output (UBL) and thus it should be assesed in the subsystem. The DBL at any given time instant is given by the difference of the cumulative production of the machine minus the cumulative production of the downstream machine. To add or to subtract signals the sum block is needed. This block can be taken from the Math library. The specification of the input port where the signal to be subtracted should be connected should be defined in the parameters dialog box. The DMP block must be connected with this port of the sum block. Since the cumulative production of this machine is not yet assesed we will add this signal in a next stage of the subsystem creation. The output of the sum block gives the downstream buffer level and thus it must be connected with the DBL output port block. The downstream buffer level is also necessary for the operation of the fuzzy logic controller block but it should be divided by the buffer capacity before in order to be normalised. This is done with the use of a gain block.

The third input of the fuzzy controller is the machine state (MS). A lower level subsystem is created for the generation of machine state. This new subsystem may be divided into two lower level subsystems: the *failure generator* and the *repair generator* subsystems. The failure generator is used for the generation of failures and the repair generator is used for the generation of repairs when the machine is considered to be out of order. As we have mentioned in the previous chapters the times between failures and repairs are exponentially distributed (see section 3.1). Here the rundom numbers generated, are following the geometrical distribution, which is the discrete equivalent of the exponential distribution. The geometrical distribution is counting the number of Bernoulli trials until an event occurs. Equivalently in this system, the time steps until failure are counted.

To create the failure generator a uniform random number block is needed. This block can be found in the *Sources* library of SIMULINK. This block generates uniformly distributed random numbers. In the parameters dialog box the seed, the upper and lower limits of the the generated numbers and the sample time should be defined. Another block used in the failure generator is the switch block. The switch bolck pass through input 1 when input 2 is greater than or equal to threshold; otherwise, pass through input 3. Finally two constant blocks from the sources library are added to the model. The constant block outputs a constant. The first constant is set equal to one and it's connected to the first input port of the switch. The second constant block is set zero and is connected to the third input port of the switch block. The uniform random number block is connected to the second input port of the switch (see Figure 7). Thus if the number generated by the uniform random number block is greater then the switch threshold the output is one (the machine is in working order otherwise is zero (the machine is under repair). The threshold of the switch block is chosen in such a way that the mean time to failure is the desired.

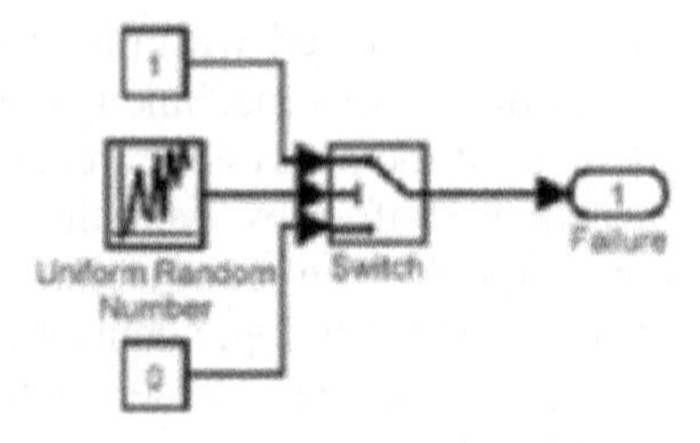

Fig. 7. Failure generator subsystem

The repair generator is constructed similarly.

After building failure and repair generators one should combine them in order to construct the machine state generator. An important thing to remember is the previous state of the machine. A machine can break down only if in previous time step was working. If the machine is out of order then it should be repaired and therefore the repair generator is used. In order to combine the failure and repair subsystems, a switch block and a transport delay block, from the Continous library of SIMULINK are needed. The transport delay block delays the input by a specified amount of time. If the specified amount of time (as it should be here) is the simulation time step, the transport delay block returns the input value at the previous time instant. The failure generator is connected with the first input port of the switch block and the repair generator with the third input port of the switch. The switch output is connected to the transport delay input and the transport delay output is connected to the second input port of the switch. The switch is returning the machine state. If the state in the previous time instant is 1 (operating), which is given by the transport delay then the current state is decided by the failure generator. Otherwise if the machine state in the previous time instant is 0 (under repair) then the current machine state shoulb be decided by the repair generator. The machine state subsystem is shown in Figure 8.

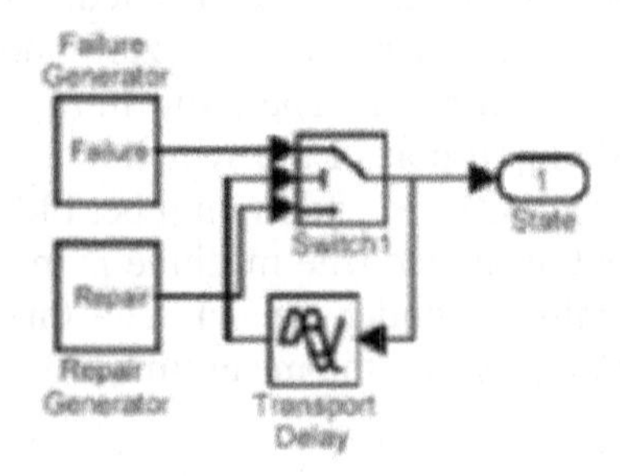

Fig. 8. Machine state subsystem

The last input of the fuzzy controller is the *surplus*. Surplus is given by the difference of the cumulative production minus the cumulative demand. Thus a sum block should be added with an adding input port and a subtracting input port. The DEM input block is connected to the subtracting input port and the production should be the adding input. Surplus should be normalised in order to be used by the fuzzy controller. Normalisation is necessary, because of the continous adaptation of upper and lower surplus bounds from the supervisory controller. The surplus x is normalised according to the following relation:

$$x_n = \frac{x - l_b}{u_b - l_b} \tag{7}$$

where x_n is the normalised surplus and u_b, l_b, are the upper and lower surplus bounds. After the normalisation $x_n \in [0, 1]$ if $l_b \le x \le u_b$, $x_n < 0$ if $x < l_b$, and $x_n > 1$ if $x > u_b$. A new lower level subsystem is used for the normilasation of surplus. It needs to have two inputs the SRP (surplus) and the FSC, which is a vector with two components the upper and lower surplus bounds. At first this vector should be decomposed with the use of a demux block (Signals & Systems library). Then the lower bound is subtracted from the surplus and then from the upper bound. The result of the first subtraction is divided by the result of the second. Finally a saturation block (Nonlinear library) is used to limit the output. The normalisation subsystem should look like this. To get the production rate of the machine at every time instant the output of the fuzzy controller, which is a number between zero and one, should be multiplied with the maximum production rate of the machine. This is done with a gain block. To estimate the cumulative production a new lower level subsystem is added. In Figure 10 one may see how this subsystem should look like.

In this subsystem the production rate is integrated with the use of an integrator block (Continous library). The result of this integration gives us the theoretical cumulative production, which by no means should be greater than the actual cumulative production. The theoretical cumulative production is subtracted by the actual cumulative production at the previous time instant. If the result is negative, meaning that the theoretical production is greater than the actual, then the actual production is increased by one, otherwise the actual production remains the same. The output of theprevious subsystem is the cumulative production.

After the completion of the connections the machine subsystem is ready and it is shown in Figure 11.

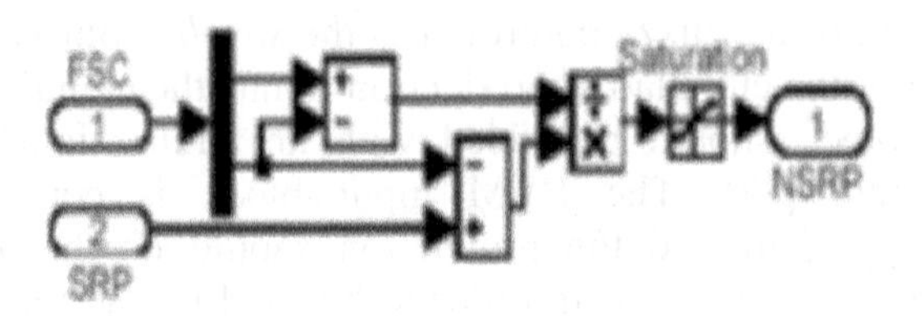

Fig. 9. Surplus normalization subsystem

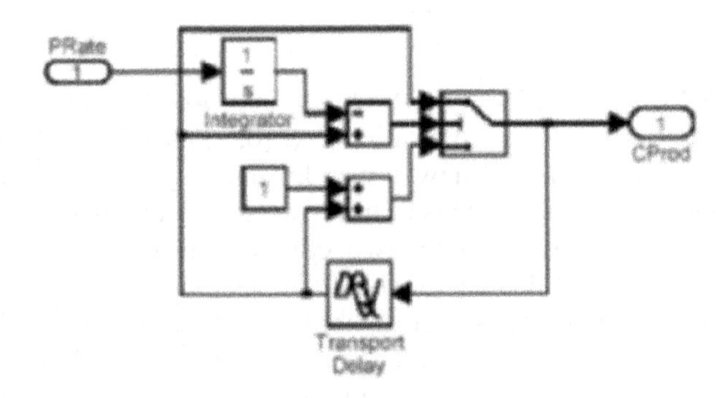

Fig. 10. Discretiser subsystem

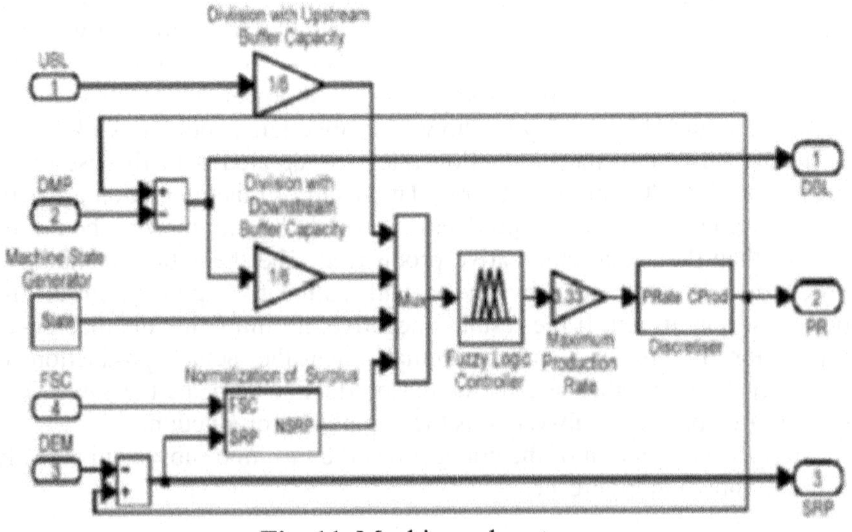

Fig. 11. Machine subsystem

We have allready mentioned that machine subsystem is the structural basis of our simulator. From Figure 6 one may see that there are seven machine subsuystems in the production system subsystem of this simulator.

Apart from the machines subsystems there are some other subsystems used in the production system subsystem. The demand subsystem generates the orders. The orders are assumed to follow the Poisson destribution and thus the times between successive arrivals are exponentially distributed. The demand subsystem is similar to the failure and repair generator subsystems.

Finally one may observe in Figure 6 three subsystems used for the estimation of various performance measures. The WIP statistics subsystem is shown in Figure 12. This subsystem is used for the estimation of mean WIP, standard deviation of WIP, and the relative error of WIP.

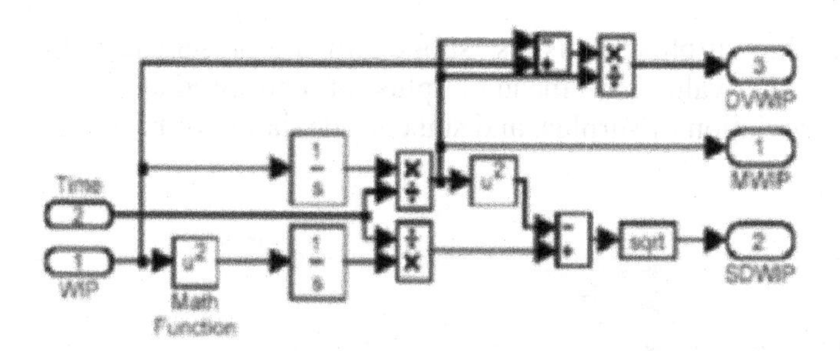

Fig. 12. WIP statistics subsystem

The mean of a signal is assessed by integration in time, and then by division by time. To estimate the standard deviation σ of a signal x one may use the well known result:

$$\sigma^2 = [E(x)]^2 - E(x^2) \tag{8}$$

where $E(x)$ is the mean of a signal x and $E(x)^2$ the mean of x^2. The relative deviation of *WIP* is given by Eq. (5). A new block used in this subsystem is the Math function block (Math library) which performs numerous common mathematical functions such as square, square root, exp, etc.

In the cycle time statistics subsystem the mean cycle time is estimated. Cycle time subsystem is shown in Figure 13.

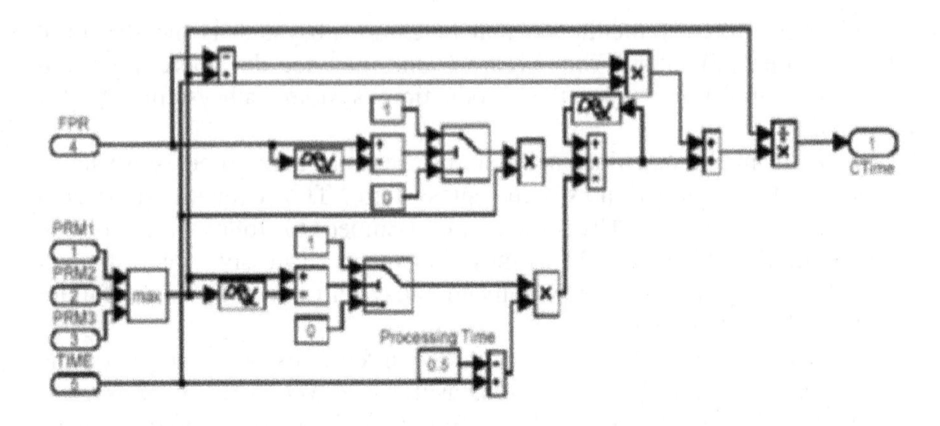

Fig. 13. Cycle time statistics subsystem

In Figure 14 the surplus statistics subsystem is presented. This subsystem returns the values of mean surplus of end product, mean backlog, standard deviation of surplus, and standard deviation of backlog.

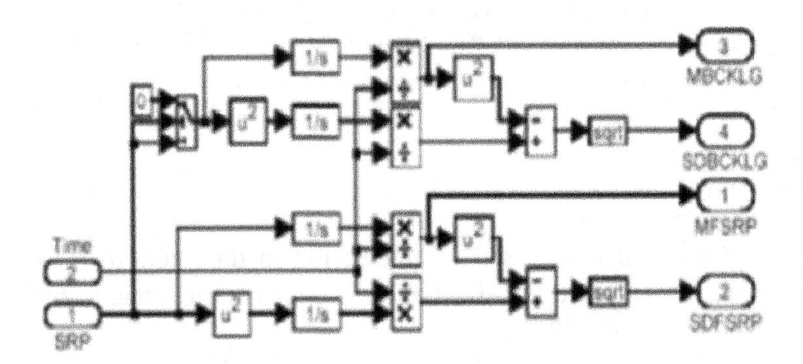

Fig. 14. Surplus statistics subsystem

4.1.2 Supervisor

As we have already seen in the higher level the simulator consists of two subsystems: the production system subsystem, and the supervisor subsystem. The first is described in detail in the previous section. The second is presented in Figure 15. The hart of the subsystem is the fuzzy logic control block which represents the fuzzy supervisory controller described at the beginning of this chapter. The fuzzy controller needs three inputs to operate: the mean surplus of the end product, the error of end product surplus, and the relative WIP error. Apart from the surplus error, the other inputs are also inputs of the supervisor subsystem. To get the surplus error the initial lower bound should be subtracted from the end product surplus, which is an input of the supervisor subsystem. After a normalisation procedure the inputs are feeded to the fuzzy logic controller block.

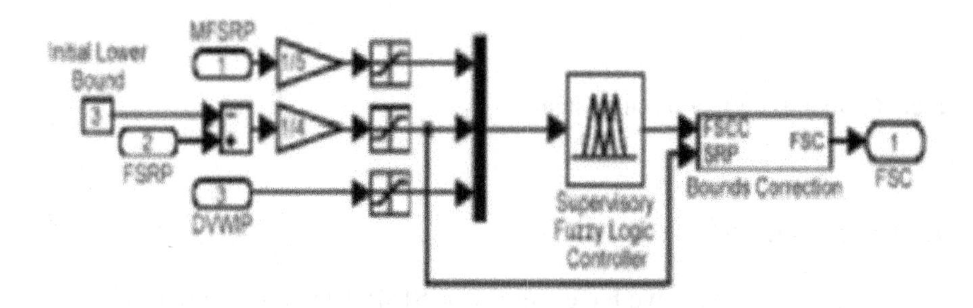

Fig. 15. Supervisor subsystem

The fuzzy logic controller returns as outputs the upper and lower bounds correction factors. Since the supervisor subsystem returns as outputs the upper and lower surplus bounds a bounds correction subsystem, responsible for the adaptation of surplus bounds, is used. The inputs of that subsystem are the surplus correction factors (FSCC), and the end product surplus. The surplus bounds are altered with the use of these correction factors accordingly.

$$
\begin{aligned}
u_b &= I_u + u_c n_u + min(x_e, 0) \\
l_b &= min[(I_l + l_c n_l), u_b]
\end{aligned}
\tag{9}
$$

where u_b, l_b, are the upper and lower surplus bounds respsctively, I_u, I_l, are the initial upper and lower surplus bounds, u_c, l_c, are the upper and lower surplus bounds correction factors, and n_u, n_l, are normalizing constants. The structure of bounds correction subsystem is presented in Figure 16.

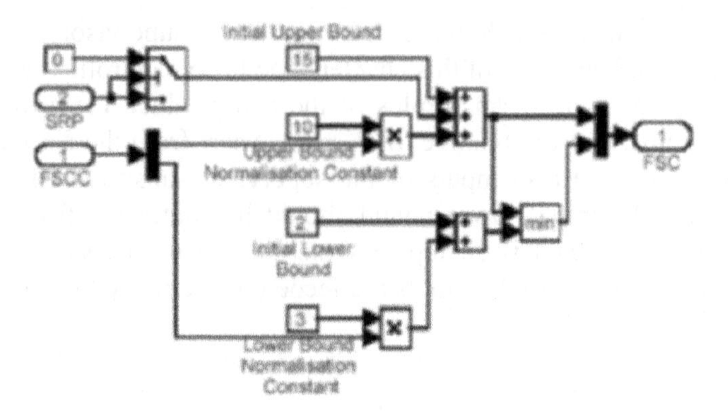

Fig. 16. Bounds correction subsystem

4.2 Discussion

In the previous section we have suggested SIMULINK as the building tool for effective simulation of manufacturing systems. There are many advantages that make SIMULINK appealing for this kind of applications. First of all SIMULINK has become the most widely used software package in academia and industry for modelling and simulating dynamical systems, thus there are many people familiar with this tool. Apart from that the graphical user interface (GUI) makes SIMULINK very easy to use. The easy integration of MATLAB codes and toolboxes provides to SIMULINK users a number of powerfull tools, such as Fuzzy logic toolbox, Neural networks toolbox, etc. that is almost impossible to find in a dedicated tool for discrete event systems simulation. Finally the modularity of the described method reduces the effort needed for the construction of a simulator for another manufacturing system. To change the geometry of the production system to be simulated the only thing one has to do, is to copy and paste or delete a few machine subsystem, and then adjust their parameters and the connections between them.

One disadvantage of SIMULINK for modelling manufacturing systems of realistic size, is that it is a rather slow simulation tool. This is maybe not important for research purposes but it surely can be a major disadvantage in industrial applications.

5 Conclusions

Two fuzzy scheduling approaches have been presented. In the distributed fuzzy scheduling approach a set of distributed fuzzy controllers is used in order to reduce WIP while satisfying demand. This approach was compared with other commonly used production control techniques and has proven to be superior. For the test cases examined it turned out that the distributed fuzzy approach provides lower WIP and smaller product cycle time.

A fuzzy supervisory approach is used in order to tune the distributed fuzzy controllers. This is made in a way that performance measures which are more important in every case are improved. More specifically WIP and Cycle Times are substantially reduced, while backlog is kept in low levels. Simulation results, for a production network with stochastic demand, have shown noticeable improvement of performance. These results are achieved while modularity and distributivity of the control architecture is maintened.

We have also seen how SIMULINK and MATLAB's Fuzzy logic toolbox may be used to construct production systems simulators. The ease and modularity in relation with the numerous capabilities offered by MATLAB and its toolboxes are the most important advantages of the described simulation approach.

In the future it would be very interesting to consider the case of time varying demand and to include production costs due to WIP and Backlog in the supervisory control scheme in an effort to minimize them. Another interesting extension would be the integration of the proposed approach with a design mechanism in order to optimize the selection of important system's parameters which are now selected by tuning and seeking further improvement of supervisors performance. Some of these parameters could be buffer capacities and initial surplus bounds. This could be done with the use of reinforcement learning control methods.

References

1. S. B. Gershwin, "Design and operation of manufacturing systems: the control-point policy", *IIE Transactions*, **32**, 891-906 (2000).
2. J. A. Buzacott and J. G. Shanthikumar, *Stochastic Models of Manufacturing Systems* (Prentice Hall, Englewood Cliffs, NJ, 1993).
3. Y. Dallery and G. Liberopoulos, Extended kanban control system: combining kanban and base stock, *IIE Transactions on Design and Manufacturing*, **32**(4), 369-386 (2000).
4. L. Wein, Scheduling semiconductor wafer fabrication, *IEEE Transactions on Semiconductor Manufacturing*, **1**(3), 115-130 (1988).

5. R. Akella and P. R. Kumar, Optimal Control of Production Rate in a Failure Prone Manufacturing System, *IEEE Transactions on Automatic Control*, **31**2), 116-126 (1986).
6. T. Bielecki and P. R. Kumar, Optimality of zero-inventory policies for unreliable manufacturing systems, *Operations Research*, **36**(4), 532-541 (1988).
7. N. Srivatsan, Synthesis of optimal policies for stochastic manufacturing systems, Ph.D. dissertation, Operations Research Center, Massachusetts Institute of Technology, Cambridge, MA., 1993.
8. G. J. Van Ryzin, S. X. C. Lou and S. B. Gershwin, Production control for a tandem two-machine system, *IIE Transactions*, **25**(5), 5-20 (1993).
9. S. X. C. Lou, S. P. Sethi and Q. Zhang, Optimal feedback production planning in a stochastic two-machine flowshop, *European Journal of Operational Research*, **73,** 33-345 (1994).
10. S. X. C. Lou and G. Van Ryzin, Optimal control rules for scheduling job shops, *Annals of Operations Research*, **17**, 233-248 (1989).
11. S. X. C. Lou and P. Kager, A robust production control policy for VLSI wafer fabrication, *IEEE Transactions on Semiconductor Manufacturing*, **2**, 159-164 (1989).
12. H. Yan, S. X. C. Lou, S. P. Sethi, A. Gardel and P. Deosthali, Testing the robustness of two-boundary control policies in semiconductor manufacturing, *IEEE Transactions on Semiconductor Manufacturing*, **9**(2), 285-288 (1996).
13. S. X. Bai and S. B. Gershwin, Scheduling Manufacturing Systems with Work-in-Process Inventory Control: Multiple-Part-Type Systems, *International Journal of Production Research*, **32**(2), 365-386 (1994).
14. S. X. Bai, S. B. Gershwin, Scheduling Manufacturing Systems with Work-in-Process Inventory Control: Single-Part-Type Systems, *IIE Transactions*, **27**, 599-617 (1995).
15. L. Custodio, J. Senteiro and C. Bispo, Production Planning and Scheduling using a fuzzy decision system, *IEEE Transactions on Robotics and Automation*, **10**(2), 160-168 (1994).
16. N. C. Tsourveloudis, E. Dretoulakis and S. Ioannidis, Fuzzy work-in-process inventory control of unreliable manufacturing systems, *Information Sciences*, **127**, 69-83 (2000).
17. S. Ioannidis, N. Tsourveloudis, K. Valavanis, Fuzzy Supervisory Control of Manufacturing Systems, *IEEE Transactions on Robotics and Automation*, **20**, 379-389 (2004).
18. H. P. Wiendahl, *Load–Oriented Manufacturing Control* (Springer-Verlag Hannover, 1995).
19. D. Driankov, H. Hellendoorn and M. Reinfrank, *An Introduction to Fuzzy Control* (Springer-Verlag, New York, 1993).
20. N. Gulley and J.-S. R. Jang, *Fuzzy Logic Toolbox User's Guide* (The MathWorks, 1995).
21. SIMULINK: Dynamic System Simulation for Matlab, (The MathWorks, 1999).

Part VI

Fuzzy Optimization Techniques

Fuzzy Cognitive Mapping for MIS Decision-Making

Amir Sharif and Zahir Irani

Information Systems Evaluation and Integration Network Group (ISEing)
School of Mathematics, Information Systems & Computing,
Brunel University, UK
[1] *Zahir.Irani@Brunel.ac.uk*

Abstract: Traditional methods for modelling and defining Management Information System (MIS) decision-making tasks, have tended to centre around a systems science view of the world. This is largely in terms of modelling the individual, group, organisation, or system in relation to process and environmental boundaries. Whilst such approaches are excellent for citing the situational context of such decision-making tasks, this and other orthodox Operational Research (OR) techniques do not necessarily highlight or show those causal interdependencies which are dependent upon vague or ill-defined, ambiguous information. Fuzzy Logic, at its core, provides the researcher (be they academic or practitioner) with a multitude of techniques for handling uncertainty in this respect. As such, this article discusses and outlines the development of the application of Fuzzy Cognitive Mapping (FCM) to the MIS decision-making task of Information Systems Evaluation (ISE), in terms of the on-going research interests of the authors. Through defining the nature of ISE, the authors present two models of such investment appraisal techniques, and through the generation and simulation of their respective FCMs, provide further insight into this MIS decision making task. Finally the chapter concludes by discussing and analysing the development of this fuzzy technique as a valuable OR tool.

Key words: Information Systems Evaluation, Fuzzy Cognitive Mapping, OR

1 Introduction: Modelling Investment Justification

Information Technology (IT) is a collective term used to describe the convergence of computers, telecommunications, electronics and resulting technologies (Willcocks, 1994). Industrial applications of IT, such as those within manufacturing organisations are often given generic names such as Computer Integrated Manufacturing (CIM), Computer Aided Design (CAD), Manufacturing Resource Planning (MRPII), Computer Aided Manufacturing (CAM) and so on, whereas in non-manufacturing organisa-

A. Sharif and Z. Irani: *Fuzzy Cognitive Mapping for MIS Decision-Making*, StudFuzz **201**, 455–504 (2006)
www.springerlink.com

tions, such as banks, hospitals, schools and government departments, most computer based systems are simply referred to as IT (Primrose, 1991).

The implementation of new technology is clearly one of the most lengthy, expensive and complex tasks that a firm can undertake (Small and Chen, 1995). In recent years, many sectors of manufacturing, such as aerospace and their related supply chain industries, have been reported as being significant investors in Information Technology (IT) and/or Information Systems (IS) (CEAS 1997 ; Irani *et al.,* 1998). The superconvergence of many forms of on-line, remote and mobile computing devices means that investing in new IT projects has become a significant matter of concern (Farbey *et al.,* 1993; Willcocks, 1994; Butler, 1997).

The level of investment and high degree of uncertainty associated with the adoption of such capital expenditure therefore implies that issues involving project justification should assume great importance (Primrose, 1991). As discussed by Meredith (1987), many companies continue to attain a significant competitive advantage through the adoption of IT. Furthermore, others find that they too must install more computers, implement new technology as a means of gaining or even maintaining a competitive edge. However, in order for senior management to fully commit themselves to this increasing level of expenditure, they need to be convinced of the business justification of such investments. Small and Chen (1995) have identified a variety of concerns expressed by many industries, with regard to the justification of IT. These typically include:

- The fact that many of the achievable benefits are considered to be qualitative and hence difficult to quantify;
- A lack of readily accessible and acceptable techniques for appraising all project costs and benefits;
- The ability to assess the true performance of a system, as it is diminished if all benefits are not quantified during the justification process; and,
- An insufficient level of internal skills (managerial and technical) to appraise proposed systems.

As a result of manufacturing industry's inability to address these concerns, many corporate managers may be forced to adopt one of the following strategies:

- Refuse to undertake IT projects which could be beneficial to the long-term future of the organisation;
- Invest in projects as an 'act of faith';
- Use creative accounting as a means of passing the budgetary process.

The efficient management and operation of business processes are considered closely aligned with the development of a comprehensive IT/IS infrastructure. Industry's innovative development of IT/IS in manufacturing is evident in its evolution, from a limited data processing perspective, to an expanded organisational-wide scope of manufacturing computer-based activities, where information is recognised as a corporate resource, with much potential to improve strategic and operational processes. Therefore, it would appear that during the evaluation process, there is much need for suitable mechanisms that can acknowledge the 'full' implications of an IT/IS deployment. The consideration of such issues; constructs for success, clearly needs developing, as it supports investment decision-making, facilitating a rigorous evaluation process.

This is crucial, as the absence of such a criterion may be affecting the success of many IT/IS deployments. Also, organisations are appreciating the significance of human and organisational factors, and seeking to address these factors, as their contribution is acknowledged as supporting the successful deployment of IT/IS (Meredith, 1987). In addressing the need for structured evaluation tools, many researchers have approached investment decision making from a variety of perspectives. Much of this effort has been focused on developing a 'single' generic appraisal technique, which can deal with all types of projects, in all circumstances. This has resulted in the development and use of the widely known 'traditional' appraisal techniques (Farbey *et al.*, 1993; Irani *et al.*, 1999b). It would appear that more attention has been focused in recent years on prescribing how to carry out investment appraisal, rather than taking a holistic view of the evaluation process, which identifies those factors that support the rigorous evaluation of IT/IS.

Thus, in recent years, the authors have attempted to address this issue by carrying out research into seeking to understand the nuances of those interdependencies and critical success factors (CSFs) which impinge upon the investment justification process in this regard. This has principally been achieved through recognizing that traditional, orthodox Operational Research (OR) techniques do not necessarily provide assistance in defining tacit, and sometimes implicit, people-focussed, knowledge which is used for this decision-making task. The authors have taken the view that the decision-making processes inherent within ISE encompass not only explicit, orthodox, appraisal techniques but also include those intuitive, and some would say tacit, unorthodox factors which are dependent upon human expertise and judgement.

As such, this paper discusses and outlines the development of the application of Fuzzy Cognitive Mapping (FCM) and associated optimisation techniques to Information Systems Evaluation (ISE). As is well known and understood in the Fuzzy Logic community, this technique is a powerful means to encapsulate and represent vague, ill-defined pr ambiguous information in order to model uncertainty better. Through this combination of quantitative and qualitative modelling, the authors have been able to gain further insight into those managerial factors that influence the resulting evaluation of capital and other investments.

In order to ground and provide a basis for the context of the research application of fuzzy logic, Section 2 provides a brief overview of Fuzzy Cognitive Mapping (FCM) concepts, and how such an artificial intelligence approach can be used in the guise of simulation and machine learning. Subsequently, Section 3 provides an outline of the research methodology used in the formation of the development of FCMs for investment justification, within this paper. Section 4 details alternative models for investment justification, based upon the on-going research efforts of the authors. As such, the authors then show the application of the FCM technique in modelling IS evaluation and investment justification processes within an organisation in the UK manufacturing industry. Thus, the development of these cognitive mappings is chiefly based upon the realisation of Strategic, Tactical, Operational and Financial criteria on the one hand; and the representation of a functional view of investment appraisal factors on the other. The resulting developed FCMs for ISE are then analysed and assessed with respect to their effectiveness in modelling example numerical simulation (so-called FCM machine learning scenarios), which attempt to mimic real-world example of an investment justification situations. The article concludes in Section 5 by discussing and analysing the development of this fuzzy technique as a valuable tool, which can be used alongside other OR techniques in this context, alongside proposed avenues for further research.

2 Fuzzy Cognitive Maps: A Brief Introduction

2.1 Fuzzy Cognitive Mapping (FCM): the Fuzzy Logic view

Fuzzy logic dictates that everything is a matter of degree. Instead of variables/answers in a system being either 'Yes' OR 'No' to some user-specified question, variables can be 'Yes' AND 'No' to some degree. The principles that form the genesis of fuzzy logic are built on the notion of variable(s) existing/belonging to a set of numerical values to some degree or not. Membership of variables to a certain set can be both associative and distributive: the whole can also be a part (Kosko, 1990). By extending this view further, fuzzy logic allows the membership of more than 1 set of concepts and consequently allows sets of statements overlap and merges with one another. Thus, there is no definitive 'Yes' and 'No' answer (black OR white) to a question but more of a MAYBE answer (black AND white = grey). It is not within the scope of this paper to present an overview of fuzzy logic and the reader is directed to the seminal work on the subject by Zadeh (1965) and in the more recent non-mathematical text by Kosko (1990). Cognitive and causal rules model the system and thus allow some of the inherent qualitative objectives to be seen more clearly. It should be noted that this is not the same idea as proposed by Zadeh (1996), with respect to 'computing with words' (CW) were words and semantic structure are used instead of numbers to achieve a better modelling of reality. An FCM is essentially an Artificial Neural System (ANS) which seeks to mimic how the human brain associates and deals with different inputs and events, and is best summarised by (Kosko, 1990: p. 222):

"An FCM draws a causal picture. It ties facts and things and processes to values and policies and objectives... it lets you predict how complex events interact and play out".

An FCM is essentially a non-hierarchic digraph from which the effect of changes in local nodal parameter values have an affect on other graph nodal values. Thus, each parameter is a statement or concept that can be linked to another such statement or concept to produce the nodes of the FCM, i.e. the nodes of the digraph. This can be achieved via some direct but usually indirect and vague association that the analyst of the system understands but cannot readily quantify in numerical terms. Changes to each statement, hence the fuzzy concept, are thus governed by a series of causal increases or decreases in fuzzy weight values.

The advantage of modelling dynamic multi-parametric systems with an FCM, is that even if the initial mapping of the problem concepts is incomplete or incorrect, further additions to the map can be included, and the effects of new parameters can be quickly seen. This has the advantage over many quantitative methods in that no laborious 'accounting' of each parameter needs to be done (in the sense of ensuring and maintaining error bounds for each dataset). Thus, within an FCM each concept which makes up the digraph can be judged not only on its own merits, but can also be related to and put in context with the other nodes and parameters that make up the mapping via the defined causal fuzzy weights. Furthermore, and most importantly, the analysis of a particular problem mapped using this fuzzy approach, allows the researcher to view a holistic picture of the scenario being modelled, as the system parameters evolve, therefore allowing the incorporation of the wider strategic perspective.

Hence, such mappings have proved useful in analysing interrelationships within complex adaptive systems which cannot normally be described via traditional 'flow-graph' methods – problems such as those in political science, control engineering and macro economics (Kosko, 1990; Simpson, 1990). Traditional approaches to modelling such problems as geo-political conflict, engineering and process control, rely upon 'orthodox' notions of input and output states for a prescribed set of discrete conditions (Axelrod, 1976; Mentezemi and Conrath, 1986). Instead, by applying the associative and commutative laws of fuzzy logic within an FCM, provides the researcher to attribute specific parameters of the system using fuzzy / vague quantifiers in the form of words or numerical weights. The positive (+) and negative (-) signs which connect each fuzzy concept, denote causal relationships in terms of descriptors. Typically, such causal notation usually denotes meanings such as 'has greater effect on' and 'has lesser effect on' respectively (although many other bespoke causal terms can be used, these are the most frequently presented within the literature and research applications).

Fuzzy terms are additionally used to delimit the meaning of causal relationships. For example, '+ often' would be read as 'often has greater effect on', etc. The inclusion of additional parameters into the mapping is simple and re-appraisal of interrelationships can be carried out in a straightforward manner. As such, an FCM can provide a compact holistic view of a given adaptive system. Since no hierarchical relationship exists between each fuzzy concept / parameter, this type of mapping can be read in an arbitrary fashion. However, in order to highlight a particular interrelationship within the map, a starting or root concept should be chosen from which other fuzzy concepts can be related via the given causal relationship between them. However, certain aspects of the problem being modelled may appear to be unimportant in the light of so-called 'hidden patterns of inference', within the causal links relating to each statement. As such, an FCM is a dynamic system model, which thrives on feedback from each concept (i.e. intercommunication). Once again, this is a key difference between the FCM and other cognitive maps that have been used frequently in psychology. Because of this structured approach to relating and mapping concepts using weightings, FCMs are highly amenable for enumeration and simulation, and can be programmed relatively easily without too much difficulty. From an artificial intelligence perspective, an FCM is a supervised learning neural system, where as more and more data becomes available to model the problem, the FCM becomes better at adapting itself and developing a solution (Simpson, 1990). Hence FCM's (and their Neural Network counterparts, Neuro-Fuzzy systems) are very good at producing responses from a given set of initial conditions and are also thus a good example of a parallel expert system.

2.2 Machine Learning using an FCM: causal simulation

One important and attractive feature of FCMs is that these forms of graph representations can be used not only to outline and define connectivity and causal relationships between disparate real-world concepts, but can also be subsequently "run" as computational artificial intelligence systems, which describe and highlight system or machine learning characteristics. This is primarily achieved by noting that the connectivity of any digraph so constructed, along with a set of input concept states, i.e. training data, realises and / or approximates an abstract memory construct, much akin to a neural network. As an aside, it is also for this reason that there is vibrant and continuing research into using fuzzy logic techniques for the tuning and calibration of neural networks in order to optimise and produce performant learning intelligent systems. As such, the logistics of the learning behaviour for an FCM, in terms of computational steps, is now given and is relatively simplistic in nature.

Given an FCM with a number of nodes, Ci where $i = 1,\ldots, n$ exists, the value of each node in any iteration, can be computed from the values of the nodes in the preceeding state, using the following equation:

$$C_i^{t+1} = f\left(\sum_{j=1}^{n} W_{ij}\, C_i^t\right) \qquad (1)$$

where C_i^{t+1} is the value of the node at the $t + 1$ iteration, f is a given threshold or transformation function, W_{ij} is a corresponding fuzzy weight between two given nodes, i and j, and C_i^t the value of the interconnected fuzzy node at step t. The threshold function, f, can be constructed as the following responses:

Bivalent,

$$f(x) = \begin{cases} 1, \; x \geq 1 \\ 0, x \leq 0 \end{cases} \qquad (2)$$

Trivalent,

$$f(x) = \begin{cases} -1, & x \leq -0.5 \\ 0, & -0.5 < x < 0.5 \\ 1, & x \geq 0.5 \end{cases} \qquad (3)$$

Hyperbolic,

$$f(x) = \tanh(x) \tag{4}$$

Logistical,

$$f(x) = \frac{1}{1+e^{-cx}} \tag{5}$$

where c is a constant. This function is a continuous-output response, where c is critical in determining the degree of fuzzification of the function. In order to simulate the dynamic behaviour of the FCM, therefore requires the additional definition of the fuzzy weights, W_{ij}, within a connection matrix, W, and the initial or starting input vector at time t, C_i^t. As such, the latter is a 1 x n rowvector with the values of all concepts, $C_1, C_2, \ldots, C_n$ for n concepts or nodes in the FCM, whilst the former is a n x n matrix of weights between any two fuzzy nodes, w_{ij}. If there is no direct relationship between the ith and jth nodes, then the value of the connection strength is zero. As such, the connection, adjacency, influence matrix, W, can be written as:

$$W = \begin{bmatrix} \ldots & \ldots & \ldots \\ \ldots & w_{ij} & \ldots \\ \ldots & \ldots & \ldots \end{bmatrix} \tag{6}$$

Whilst the initial row vector can be represented as:

$$C^0 = \left(w_{i,j}^1, \ldots, w_{i+1,j+1}^n\right) \tag{7}$$

for n nodes in the FCM. The simulation therefore proceeds by computing C_i^{t+1} based upon this initial starting vector, and the given threshold function in f, as well as the causal connection strengths in the n x n matrix, W. Each subsequent $t + 1$ iteration then uses the values of the preceeding $t - 1$ row vector in S. Therefore by calculating each subsequent value of equation (1), the FCM subsequently simulates the dynamical system being modelled. Each corresponding causally-linked node within the mapping reacts and responds to its respective inputs – the state of each, in a cumulative sense, presage any underlying modality or hidden pattern of inference, which belies the implicit system dynamic of the FCM. As such, if the input adjacency or influence matrix in equation (6) is likened to a set of training data, in a sense, the application of the iterative process belies a learning process. This can be seen further if the behaviour of the computed rowvector, C, is also analysed in terms of limit cycle, fixed point or chaotic attractor behaviour.

An FCM can be said to approach a limit cycle, when the values of the states of the constructed FCM are effectively "trapped" in a repeating or alternating sequence of states, the divergence from which is not ascertainable with respect to any increases in the number of iterations carried out. For n iterations, a limit cycle can be said to exist if for each rowvector C_i^{t+1} computed,

$$C^t \rightarrow C^{t+1} \rightarrow ... \rightarrow C^{t+n} \rightarrow C^t \rightarrow ... \quad (8)$$

In (8), C^t is the initial phase of the rowvector C when the limit cycle begins, and C^{t+n} is the last face, after which the initial or preceding rowvector is then repeated, ad infinitum (regardless of any change in iterations). In contrast, a fixed point can be said to be the convergent case of a limit cycle, in that once a particular value for the computed rowvector is achieved, any increase in the number of iterations carried out to compute the $t + 1$ solution, does not change the resulting state of the FCM any further. Thus, the following provides the definition of a fixed point with respect to a rowvector, C^t.

$$C^t \rightarrow C^t \rightarrow C^t \rightarrow ... \quad (9)$$

Finally, an FCM can be said to deviate from either a limit or fixed point cycle towards a chaotic (i.e. non-linear, transitive dynamic) response, if for any subsequent iteration, for each node weighting in W,

$$C^t \neq C^{t+1} \rightarrow ... \neq C^{t+n} \neq C^t \neq ... \quad (10)$$

That is the result of each proceeding rowvector, is different from the preceeding one and no limit or fixed point cycle is reached even for a modest to large increase in the number of computed cycles for the FCM. By carrying out a simulation of the FCM and assessing the behaviour of the computed rowvectors in the limit in this manner, the cognitive mapping can be assessed in terms of its realistic accuracy and / or connections. Typically, FCMs with a small number of interconnections, say with $n = 4$ or 6 nodes, and w_{ij} values which are close or equal to zero (for which the adjacency matrix, W, can be said to approximate tri-diagonal form), limit cycle behaviour can frequently occur. Likewise, graphs with a similar number of connections which are dominated by bivalent responses of the form of equation (2), and have a small deviation of w_{ij} values, tend to display fixed point behaviour. Subsequently, those mappings which describe numerous interconnections between each fuzzy node (i.e. where the adjacency matrix, is non-sparse and densely populated), employ sigmoidal or hyperbolic responses such as in (4), (5) and are initiated by diverse starting rowvector formations in (7), divergent and chaotic response behaviour may be seen.

3 Research Methodology

The authors now outline the research methodology used in applying the concepts of Fuzzy Cognitive Mapping to an investment justification process within a given case study organisation. The FCM approach used by the authors was applied in order to model and highlight all of the interconnected aspects which impinge upon this decision-making task, and which traditionally, tend to highlight defective investment justification processes. Briefly, this approach entailed identifying a typical investment justification process within an organisation involved in an information systems evaluation (ISE) task; gathering the relevant case data; generating models of the investment justification process; developing specific FCMs of each model; and finally, generating and analysing causal simulations of each FCM in relation to case organisation decision making criteria (i.e. initial or training set data to be applied to the base FCM).

Therefore, in viewing the investment justification process, the normative information systems evaluation literature suggests that the primary reason why organisations fail to operate a robust evaluation process lies with a lack of understanding and agreement on what constitutes meaningful evaluation from a human, organisational, management and process perspective (Pouloudi and Serafeimidis 1999; Remenyi *et al.*, 2000; Serafeimidis and Smithson 2000; Irani *et al.*, 2001; Irani and Love, 2001). To acquire an understanding of the significance of human and organizational issues involved with IS evaluation, the development of a research methodology that involves and enfranchises organizations and their staff is needed. Considering the originality and contextual surroundings of this research, a case study research strategy was followed as advocated by interpretivist researchers such as Bonoma (1985); Hakim (1987) and Yin, (1994).

3.1 Case Organisation

The case organisation used for the research, herein known as Company A, is a British manufacturing organisation specialising in the discrete manufacture of aerospace, automotive, and other engineering components (Irani *et al.*, 1999; Irani *et al.*, 2001). The development and growth of this company has largely been attributed to previous successful innovations in both business and manufacturing technology investment over the years of its existence. As such, the organisation's approach to evaluating and assessing investments in projects was found to be worthy of further investigation by the authors. In particular, analysing the wholly financially quantitative justification criteria, against human / psycho-sociological qualitative decision criteria in this context was deemed to be suitable for modelling using the FCM approach defined. The specific decision-making scenario which was investigated by the authors within the case organisation, involved analysing the evaluation of investment capital into an integrated Manufacturing Resource Planning (MRPII) system. At the time of investigation, the investment of time and money into such a system would enable the organisation to gain and maintain competitive advantage within its market sector through the innovative use of such an integrated computer integrated manufacturing system.

Company A was not systematically sampled, and as a result, it is not possible to generalize the findings to a wider population. However, the findings are considered appropriate to provide others with a frame of reference when seeking to use apply such modelling and simulation / evaluation approaches to complement existing investment evaluation techniques.

3.2 Data Collection

The data collection procedure followed the major prescriptions of the normative literature for doing fieldwork research in Information Systems (Fielder, 1978; Yin, 1994). Primary data in the form of one-on-one interviews and participant observation, were used to gather the core case data for the generation of the FCMs in this paper. This also included illustrative materials (e.g., newsletters and other publications that form part of the case study organization's history) and archived documentation. These data was collected from senior management and other project team members in the organisation, and involved the observation and analysis of an investment justification decision-making process within the firm. The authors were careful to ensure that the interviewees were fully informed about the purpose of the interviews (taking steps to put the interviewees at ease so that a two-way, open communications climate existed) Shaughnessy and Zechmeister (1994) suggest that interviewer bias needs to be addressed, which often results from the use of probes. These are follow-up questions that are typically used by interviewers to get respondents to elaborate on ambiguous or incomplete answers. Care was taken to reduce bias to a minimum through refraining, as much as possible from asking leading questions. In trying to clarify the respondent's answers, the interviewer was careful not to introduce any ideas that may form part of the respondent's subsequent answer. Furthermore, the interviewer was also mindful of the feedback respondents gained from their verbal and non-verbal responses. The interviewer therefore avoided giving overt signals such as smiling and nodding approvingly. After every interview that was undertaken, notes were given to each person to check to resolve any discrepancies that may have arisen and eliminate any interviewer bias. This approach to interviewing has proved successful in similar research as reported by Irani *et al.*, (2001; 2005).

A variety of secondary data sources were also used to collect data, such as internal reports, budget reports, and filed accounts that were later transcribed. The authors have extensive industrial experience in the carrying out research of this nature and have used this experience, together with a predefined interview protocol to determine the data necessary to explore post implementation evaluation of enterprise technologies.

3.3 Analysis and Synthesis of Results: Generating the FCM

The typical approach used for FCM construction is to:

- Choose a relevant topic which has a number of clearly defined issues or decision points which can be sublimated into FCM nodes;
- Gather a set or a subset of the given concepts to be mapped, by eliciting knowledge about the topic from primary or secondary sources (i.e. subject matter experts, observation, documentary evidence, supporting raw data / information, etc);
- Assign causal weightings (e.g. "a lot", "some", "a little", "none" etc)
- Construct an influence matrix which describes the connection strengths, i.e. causal weightings, between each concept;
- Generate the resulting digraph by connecting each node, i.e. graph edge with a causal weighting, i.e. the connection strength / fuzzy degree of membership.

If a simulation of the FCM is required, then the following steps are also taken:

- Choose a suitable response / threshold function, f.
- Choose a suitable initial input rowvector, C^0;
- Calculate the resulting $t + 1$ output rowvector, C_i^{t+1}, from each preceding $t - 1$ rowvector, using equation (1);
- Evaluate convergence towards a limit cycle, fixed point, or identify non-linear / chaotic instability (divergence from either of the two previous limit focii respectively);

For the purposes of this paper, and following the research methodology defined earlier, the approach used by the authors was as follows:

- Gather data from experts / case situation;
- Arrange / rank data into issue groupings;
- Define and agree on causal weightings and their meaning;
- Draw an initial FCM by arbitrarily drawing causal relationships between each issue, i.e. node concept, applying causal weightings as appropriate;
- Review this initial FCM by performing a concensus walkthrough through all nodes with the expert and the researchers to validate or falsify nodes and causal node weightings
- Construct the influence matrix, *W*, separately from the above initial FCM;
- Create a secondary FCM based upon the defined influence matrix;
- Compare against the initial FCM generated and adjust the weight matrix and causal weightings as required;
- Carry out the simulation steps as defined above;

By applying such an approach, the authors felt it was possible to encapsulate all the various permutations of the mapping by re-evaluating the fuzzy connectivity at least once. Furthermore, this approach helped to support the active involvement of the subject matter expert in the formation of the FCM, and also for the researchers to better understand any nuances inherent in the formation of each causal node in the graph. In addition, such a reflexive approach used was also seen to be a useful method to eradicate, or at least to provide a coarse filter to, any augmented biases that may have crept into the knowledge elicitation and / or connectivity mapping processes outlined above.

In order to generate the results shown in the proceeding sections, the authors used a spreadsheet-based model employing matrix multiplication and graph drawing add-ins in Excel (Volpi, 2004) in order to both generate the directed (di)graph representation and also in order to run cognitive simulation scenarios based upon causal weightings and input vector states. In addition, the authors experimented with using different threshold functions, whereupon in order to elucidate and contextualise the responses to the case data more accurately, the hyperbolic function (4) was used, alongside different variations of input vectors as appropriate.

4 Models for Information Systems Evaluation and Justification

In recent years the changing role of IT/IS in organizations has given new impetus to the problem of IT/IS evaluation. As such, Information Systems Evaluation (ISE) is generally seen to be a complex undertaking, in that the extensive time and money invested in such technologies are frequently not perceived to be delivering the business benefits which were initially intended (Ezingeard *et al.*, 1999; Irani *et al.*, 1999, 2000; Khalifa *et al.*, 2000; Remenyi, 2000). The high expenditure on IT/IS, growing usage that penetrates to the core of organizational functioning, together with disappointed expectations about IT/IS impact, have all served to raise the profile of how IS investments can and should be evaluated. According to Willcocks (1992), 'IS evaluation is not only an under-developed, but also an under-managed area which organizations can increasingly ill-afford to neglect'. The increased complexity of IT/IS combined with the uncertainty associated with IT/IS benefits and costs (Irani *et al.*, 1997; 2000) point to the need for improved evaluation processes.

In the majority of manufacturing companies, a formal justification proposal must be prepared and accepted by decision-makers, prior to any expenditure. Primrose (1991) identifies industry's perception of investment justification as a budgetary process that gives a final 'yes' or 'no' - 'pass' or 'fail' verdict on the success of a project's proposal. As a result, managers may view project justification as a hurdle that has to be overcome, and not as a technique for evaluating the project's worth. This has significant implications, as during the preparation of a project's proposal, managers spend much time and effort investigating its technical aspects and become committed to the belief that the project is essential. Therefore, team members may be easily susceptible to persuasion by vendors and consultants, and be prepared to accept untypical demonstrations that show unrealistically high levels of savings. Hence, project members may focus their efforts on trying to identify and estimate maximum benefits and savings at the expense of overlooking full cost implications.

Traditional project appraisal techniques such as Return on Investment (RoI), Internal Rate of Return (IRR), Net Present Value (NPV) and Payback approaches are often used to assess capital investments (Willcocks 1994). These methodologies are based on conventional accountancy frameworks and often facilitated under the auspices of the finance director. Specifically, they are designed to assess the bottom-line financial impact of an investment, by often setting project costs against quantifiable benefits and savings predicted to be achievable (Farbey *et al.* 1993; Hochstrasser 1992). However, the vast array of traditional and nontraditional appraisal techniques leaves many organisations with the quandary of deciding which approach to use, if any.

Consequently, debate about the types of techniques that constitute meaningful justification have been ubiquitous (Small and Chen 1995). Table 1 categorises many of the available investment appraisal methodologies into appropriate groups. These various approaches have been compressed into four principal classifications; economic approaches, analytic approaches, strategic approaches and integrated approaches. The classification of economic and analytic approaches to project justification has been further divided into two respective groups. Appraisal techniques that fall into each of the four classifications have been identified. These are summarised in Table 1. Farbey *et al.* (1992) states that those companies using traditional approaches to project appraisal, often indicate an uncertainty of how to measure the full impact of their investments in IT. Furthermore, Hochstrasser and Griffiths (1991) suggests that those evaluation techniques exclusively based on standard accounting methods, simply do not work for organizations replying on sophisticated IT environments to conduct their business.

Table 1. Summary of Appraisal Techniques

Researchers	Economic*							Strategic**						
	1	2	3	4	5	6	7	8	9	10	11	12	13	14
Garrett (1986)														
Swamidass and Waller (1991)			x	x	x	x	x	x						
Burstein (1986)														
Huang and Sakurai (1990)	x	x												
Primrose (1991)	x			x	x									
Nelson (1986)														
Meredith & Suresh (1986)								x	x	x				
Gaimon (1986)														
Suresh and Meredith (1985)		x	x	x	x									
Naik and Chakravarty (1992)														
Badiru et al (1991)														
Parker et al (1988)		x			x									
Griffiths and Willcocks (1994)														
Barat (1992)														

*Includes: Payback Period Technique, Return on Investment (RoI) , Benefit/Cost Analysis, Net Present Value (NPV) (Demmel and Askin, 1992), Internal Rate of Return (IRR), Equivalent Uniform Annual Value (EUAV), Future Value (FV)

**Includes: Technical Importance, Competitive Advantage, Research and Development, Management Commitment, Look Long Term (Huber, 1985), Emphasis on Intangibles (Kaplan, 1986), Business Strategy First (Huber, 1985)

Table 1. Summary of Appraisal Techniques (*Cont'd*)

	Analytic*							Integrated**		
Researchers	15	16	17	18	19	20	21	22	23	24
Garrett (1986)									x	
Swamidass and Waller (1991)		x		x						
Burstein (1986)			x							
Huang and Sakurai (1990)										
Primrose (1991)										
Nelson (1986)		x								
Meredith & Suresh (1986)				x	x					
Gaimon (1986)			x							
Suresh and Meredith (1985)	x	x		x						
Naik and Chakravarty (1992)									x	
Badiru et al (1991)		x								
Parker et al (1988)								x		
Griffiths and Willcocks (1994)				x						
Barat (1992)										x

* Includes: Non-numeric Scoring Models, Computer Based Techniques, Risk Analysis, Value Analysis, Analytic Hierarchy Process, Expert Systems (Sullivan and Reeve, 1988)
** Includes: Multi-attribute Utility Theory, Scenario Planning and Screening, Information Economics

Many managers have become too absorbed with financial appraisal, to the extent that practical strategic considerations have been overlooked (Van Blois, 1983) suggests that. In a similar fashion, Hochstrasser and Griffiths (1991) identified industries overwhelming belief that they are faced with outdated and inappropriate procedures for investment appraisal and that all responsible executives can do is cast them aside in a bold 'leap of strategic faith'. However, when the purpose of IT investment is to support an operational efficiency drive, the benefits of reduced costs and headcounts may be easy to quantify in financial terms. Thus, the use of many traditional approaches to project justification are often the natural choice, as they are usually in widespread use appraising other types of capital expenditure, such as the purchase of new machinery (Primrose 1991).

Maskell (1991) and Farbey *et al.* (1993) suggest that traditional approaches to project justification are often unable to capture many of the qualitative benefits that IT brings. They suggest that these techniques ignore the impact that the system may have in human and organisational terms. As a result, many companies are often left questioning how to compare a strategic investment in IT that delivers a wide range of intangibles, with other corporate investments whose benefits are more tangible. Hill (1993) suggests a shift in current justification emphasis, towards a strategic based review process. It is proposed that project focus should be placed on where progress can be measured against its contribution towards the corporate strategy, and not how well it meets the criteria determined by accounting rules and regulations.

Similarly, Hares and Royle (1994) suggest that companies should identify opportunities for making strategic investments in projects pertinent to the objectives of their business and that investment decisions should not be made on the sole basis of monetary return alone. However, Kaplan (1985) explains that, if companies, even for good strategic reasons, consistently invest in projects whose financial returns are below its cost of capital, they will inevitably be on the road to insolvency.

The apparent success of traditional appraisal techniques on non-IT based projects has led many practitioners to search for appropriate evaluation methods that can deal will all IT projects, in all circumstances. Farbey *et al.* (1993) explains that this quest for the 'one best method' is fruitless because the range of circumstances to which the technique would be applied, is so wide and varied that no one technique can cope. Primrose (1991) suggests an alternative perspective, claiming that there is nothing special about IT projects, and that traditional appraisal techniques used for other advanced manufacturing technologies can be used to evaluate IT projects. Clearly, there are few universally excepted guidelines for evaluating IT projects, with current research suggesting that many companies have no formal IT justification process, and lack adequate post-implementation audit techniques against which project objectives can be measured (Hochstrasser 1992; Remenyi, *et al.* 2000).

In view of the developed models for investment justification so far in Section 2, the authors now propose the development of a more systematic approach to justifying IT investments based on the exploration of the limitations of traditional appraisal techniques (Irani *et al.,* 1998), but principally based upon the application of the Fuzzy Logic technique of Fuzzy Cognitive Mapping. It is suggested that this approach is more relevant to understanding the specifics of investment evaluation within real-world appraisal situations, for example as in the case of Manufacturing Resource Planning (MRPII) and / or Enterprise Resource Planning (ERP), and one which reflects the views of Carlson (1992) and Farbey *et al.*, (1993), who suggests that no single technique exists, or will be able to act as a generic approach to justifying all investment projects. This viewpoint is considered the case because there are thought to be too many complex variables to be included in a generic evaluation framework. However, if the problem is approached from an application-specific view point, the phenomenon becomes easier to model and understand. As such the authors now discuss in detail, the generation of two fuzzy cognitive maps on the investment justification problem within Company A.

4.1 A Strategic, Tactical, Operational and Financial Model

Hochstrasser and Griffiths (1991) and Farbey *et al.* (1993) suggest that the major problems associated with the application of traditional appraisal techniques are their inability to take into account qualitative project implications. As a result, the authors propose the de-coupling of the relative dimensions of the project, and the division into strategic, tactical, operational and financial dimensions, for further analysis. This methodology [originally proposed by Irani *et al.,* (1999)], as with Naik and Chakravarty (1992), proposes the analysis of each project on individual merits, unlike Garrett (1986) who begins with the analysis of a number of different proposals, resulting in the selection of the 'best' perceived option. The proposed conceptual model is detailed in Figure 1, and provides a broad range of variables for consideration when evaluating investments in IT.

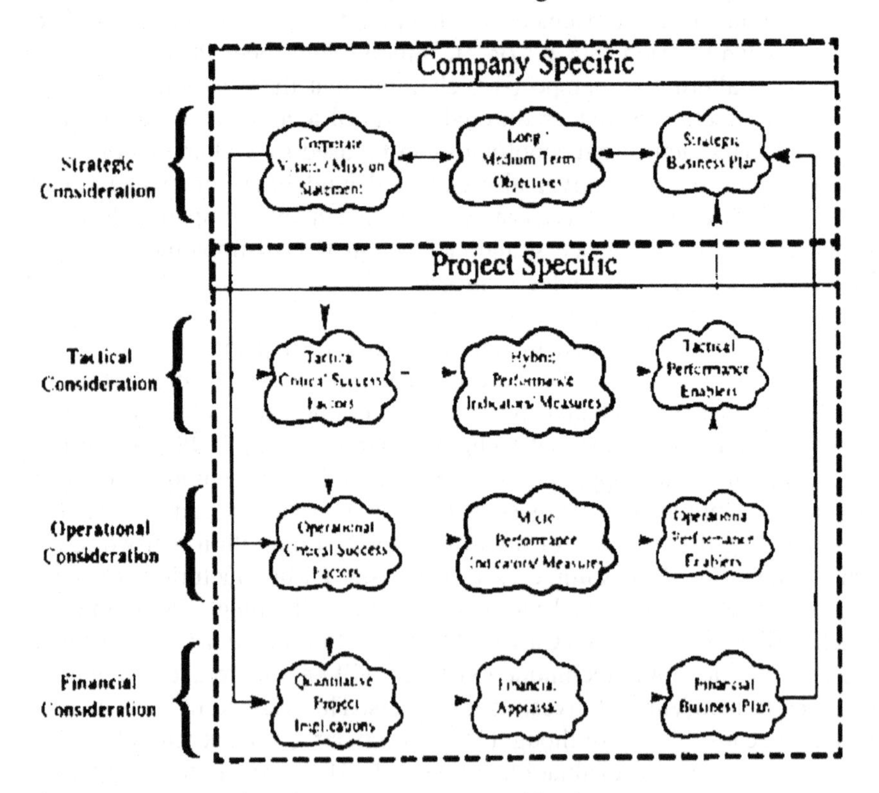

Fig. 1. A conceptual model for Investment Justification

This model is divided into four hierarchical levels of evaluation; strategic, tactical, operational and financial. At the level of *strategic* evaluation, the emphasis is placed on the relative impact of the project in relation to the delivery of a competitive advantage. The framework begins with a consideration of the corporate philosophy, core values and beliefs, which translate into a mission statement.

This provides a vivid description of what the organisation will be like when 'success' has been achieved. Furthermore, these issues are reflected in the organisations strategic business plan, which broadly identifies where the company is relative to its' market place; where it is going in its' market place; and how it is going to get there, with all these issues being broken down into long/medium term objectives. Once the strategic 'game plan' has been 'mapped-out' and resources identified, there is then a need to identify *tactical* project critical success factors (CSF's). These are project requirements that must be fulfilled at a tactical level, by isolating detailed tasks, processes and resources, to ensure short-term project success. If these CSF's are not achieved, they will ultimately become obstacles to corporate progress and will result in a loss of business, and a failure in the achievement of project objectives (Hochstrasser and Griffiths, 1991). It is essential that when tactical CSF's are identified, appropriate 'hybrid' performance indicators are identified.

4.1.1 An FCM of Strategic, Tactical, Operational and Financial Performance Measures

After analysing the responses of the case study company via direct and indirect data gathering techniques (interviews, observation and documentation), the authors generated the FCM in Figure 2, based upon the managerial view of the case study company's ISE approach, using the Strategic, Tactical, Operational and Financial model as detailed in Section 2.1. The FCM given starts with the application of a suitable appraisal technique, from a financial accounting viewpoint (as this was the single most cited reason used for such appraisals within Company A). Practically, this would be in the form of accounting the fiscal benefits available to the company after initiating the project.

Each consideration, hereby a fuzzy concept in the FCM, is related to every other concept (i.e. to each fuzzy node) by linking it with an arrow, which shows where a relationship exists. It should be noted that there is no hierarchy between these fuzzy concepts and the letters (A, B, C, and D) which have been represented in the map for brevity. Further, the '+' and '– signs situated above the lines connecting the encircled variables are not

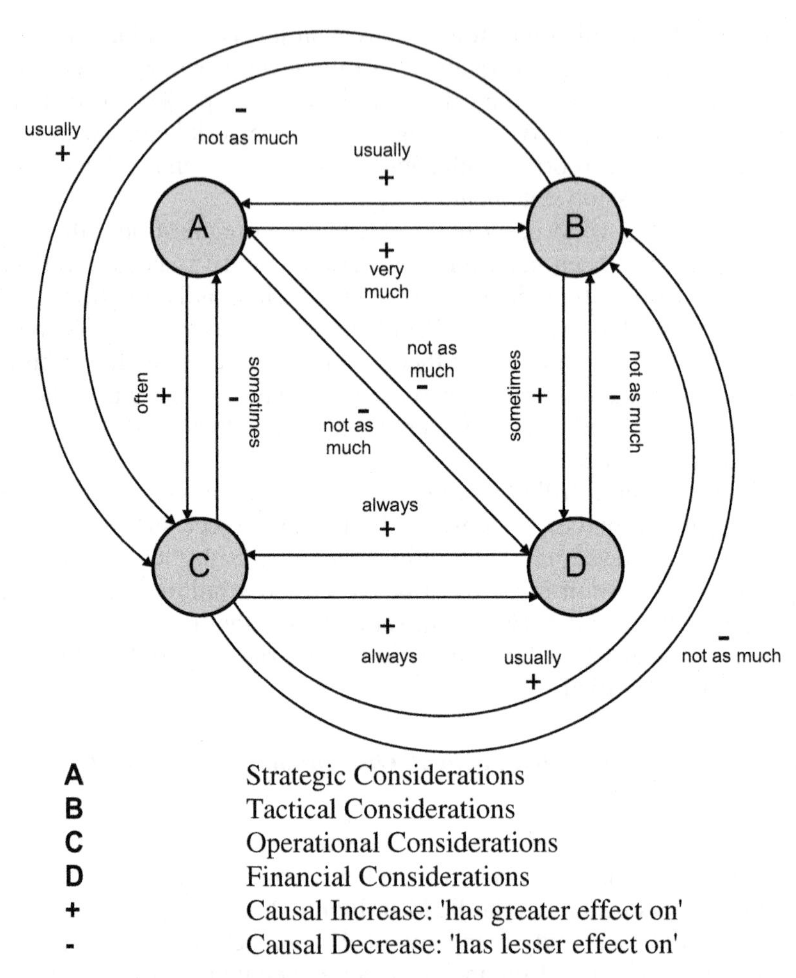

Fig. 2. Conceptual FCM for Investment Appraisal

numerical operators or substantiators, in that they do not show (absolute) scalar quantity increases or decreases between each system concept. Instead these signs denote causal relationships in terms of descriptors, which in this case mean 'has greater effect on', and 'has lesser effect on' respectively. Additional fuzzy terms can also be used to delimit the meaning of these operators e.g. '+ often' would be read as 'often has greater effect on', etc.

The map can be read in any direction and relationships can be viewed in terms of any root concept, as it is a non-hierarchic flow diagram (as stated earlier). However, in order to clarify and highlight pertinent relationships between the key variables in the map, it is often easier to begin from a starting/root concept. The map is read by seeing which concept is linked together with another one, and uses the '+' or '-' signs above each arrowed line to provide a causal relationship between them. For brevity in what follows, we will denote such relationships in the following manner:

'<concept_1> <concept_2> [+ or -]'

For example, 'AB-' would mean an arrowed line connects 'A' to 'B' and would be read as "concept A has little effect on concept B". Taking the finance-orientated viewpoint to project justification ('D'), it was found that by applying the fuzzy mapping, Figure 2 shows that:

- Justifying a project purely on financial terms has little effect on the strategic considerations ('A'). This has been read as the arrow going from ('D') to ('A') and taking the '-' sign above the line to mean 'has less effect on', i.e. 'DA-'. Similarly, strategic considerations have little impact on the financial justification process (i.e. 'AD-') as many of the benefits are largely qualitative and hence not financially quantifiable;
- Justifying a project tactically (i.e. 'BD+') is more quantitative than an a strategic investment but less so than an operational investment;
- Operational considerations can be appraised financially without much difficulty (i.e. 'DC+' and 'DC-') as 'day-to-day' operations can be quantified in terms of current resources and operational CSF's;
- Strategic issues help to justify investments and substantiate tactical considerations/tactical CSF's and vice versa (i.e. 'AB+' and 'AB-'): since tactical and strategic dimension can be viewed as being long/medium term processes. Appraising a project in terms of any of these two would mean that a tactically based justification would be well suited to meeting the strategic goals of the company eschewed in the corporate mission/vision statement;
- Since strategic considerations take account of long-term objectives and goals, appraising a project based on operational factors is best suited to traditional methodologies, largely because of their quantitative nature. If an operational project was to be appraised solely on its operational characteristics, the strategic consideration for this would be weak (i.e. 'CA-'), or rather would not be substantial enough to justify a project by itself; and,
- In order to justify projects solely on operational or tactical grounds, via a financial project appraisal impetus, it can be argued that operational considerations have greater effect justifying tactical considerations and

vice versa (i.e. 'CB+' and 'BC+'). This is due to the fact that operational processes can be accommodated within the slightly longer time scales involved with tactical goals and objectives. However, these relationships are not always applicable to all types of investment, (i.e., 'CB-' and 'BC-') and can be detrimental to the appraisal of a project by any other means (either strategic or financial).

The above causal route through the FCM is but a single pattern that has emerged from the mapping of the conceptual framework. Other patterns can be found by adopting a similar method of beginning a causal route from a starting concept (i.e. from 'A', 'B', 'C' or 'D' respectively) and seeing how each concept can, potentially, be related to any other. The FCM itself shows a low-level representation of the key considerations of the project evaluation model, as opposed to the much higher level conceptual framework given in Figure 1. Further detailing of the exact nature of each consideration ultimately helps to develop a more comprehensive map, which will show causal patterns that cannot ordinarily be seen explicitly in the given mapping. As such, a detailed analysis of this FCM is now given in terms of several machine learning numerical simulations in the following sections.

4.1.1.1 FCM Numerical Simulation Results and Analysis

Using the fuzzy algebra outlined in Section 2, we now present results of executing the FCM in order to elucidate the dynamics of the investment justification model which outlines Strategic, Tactical, Operational and Financial measures (henceforth known as the acronym, STOF). For the FCM given in Figure 2, the causal modifiers are assigned as (to 4 decimal places):

Table 2. Causal Weights for STOF FCM

Causal weight	Weight
Never	0.000
Not as much	0.125
Often	0.250
Usually	0.375
Sometimes	0.475
Very much	0.750
Always	1.000

Likewise, the fuzzy connection matrix, W, is given as:

Table 3. Fuzzy connectivity / adjacency matrix for STOF FCM

Fuzzy Nodes	Investment factors	A Strategic	B Tactical	C Operational	D Financial
A	Strategic	0.000	0.750	0.250	0.125
B	Tactical	0.375	0.000	0.375	0.480
C	Operational	0.475	0.125	0.000	1.000
D	Financial	0.125	0.125	1.000	0.000

In other words,

$$W = \begin{bmatrix} 0.000 & 0.750 & 0.250 & 0.125 \\ 0.375 & 0.000 & 0.375 & 0.475 \\ 0.475 & 0.125 & 0.000 & 1.000 \\ 0.125 & 0.125 & 1.000 & 0.000 \end{bmatrix} \tag{11}$$

As noted earlier, the threshold function, f, was set to be the hyperbolic function in equation (4). Table 4 shows a number of initial rowvectors which were used by the authors for C^0, whereupon f and W were also applied into formula (1) to yield computational results, which are discussed in the sections that follow. The Theoretic response are defined as those vectors which provide a highly simplified view of the FCM, in that only a single causal node is 'switched on'. In contrast, the Realistic response variables can be defined as those where the some assumed or *a-priori* relationship exists between each and any of the variables. This is such that it is assumed that Financial considerations are always inherently a part of any investment justification, in the present of any other variable which may be inherently dominant.

Table 4. Initial starting vectors for the STOF FCM

C^0 vector	Strategic	Tactical	Operational	Financial
		Theoretic response		
Strategic-A	1.00	0.00	0.00	0.00
Tactical-A	0.00	1.00	0.00	0.00
Operational-A	0.00	0.00	1.00	0.00
Financial-A	0.00	0.00	0.00	1.00
		Realistic response		
Strategic-B	1.00	-1.00	-1.00	1.00
Tactical-B	-1.00	1.00	-1.00	1.00
Operational-B	-1.00	-1.00	1.00	1.00
Financial-B	-1.00	-1.00	-1.00	1.00

Theoretic response

Using the Strategic-A initial starting vector, it can be seen that the FCM given in Figure 3 reaches an equilibrium steady-state fixed point within 13 iterations, with some mild non-linearity relating to the Tactical concept (as shown in Figure 3). Initially, we see that with a fully switched on state for Strategic considerations, other non-strategic factors do in fact exist at some level of activation, i.e. 0.500. By iteration 2, the impact of the strategic viewpoint within the FCM on Tactical considerations starts to come into play.

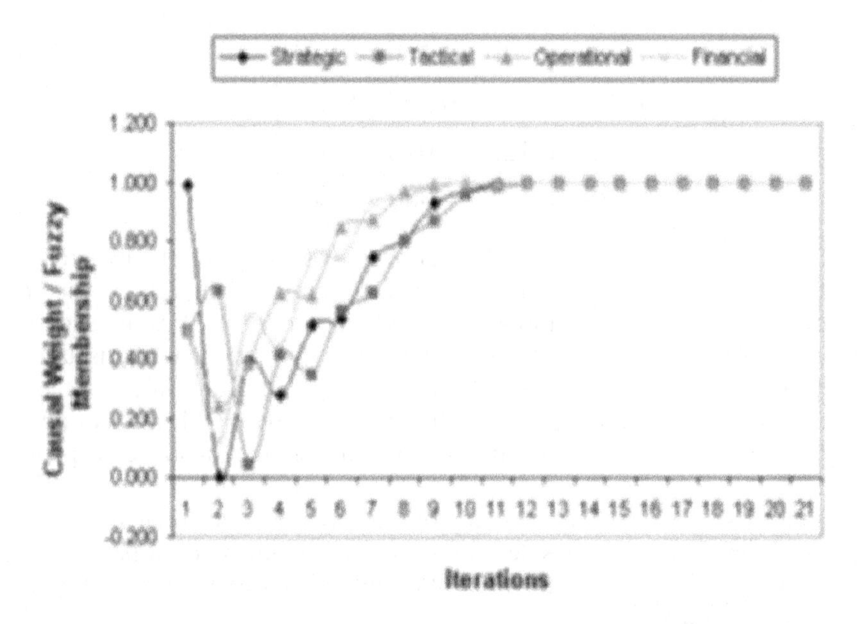

Fig. 3. Strategic-A simulation result

At iteration 5, strategic considerations track financial considerations to the extent that the "ends justify the means", hence up front investment tends to support strategic initiatives in the long term (although this seems to have an inverse impact upon operational factors). Furthermore, by iteration 8, the costs of investments are gradually subsumed within operational considerations (i.e. operational factors track financial considerations).

The application of the Tactical-A vector provides the simulation results as shown in Figure 4. Once again, we see some initial instability across all the factors involved within the FCM, which converges to a fixed point

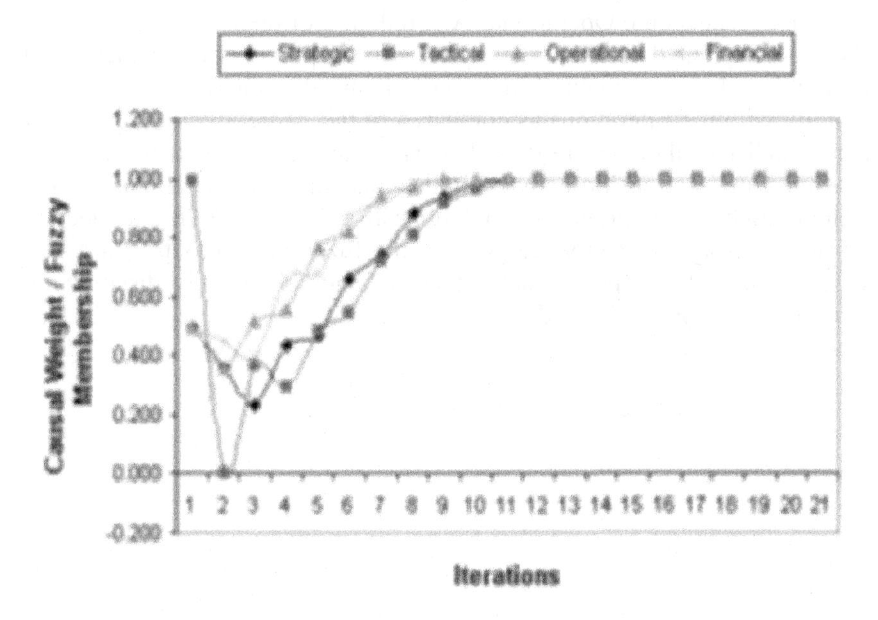

Fig. 4. Tactical-A simulation result

equilibrium within approximately 13 iterations. This is largely similar to the previous Strategic-A model in this respect. Once again, both financial and operational considerations are interlinked in the sense that the convergence of one, has an effect on the convergence of the other (i.e. Operational leads Financial).

Aside from this interrelationship seen here, the Tactical consideration can be seen to be dominant once again and lags all other factors in terms of its dynamic response. One particular aspect of this simulation occurs at iteration 5, where it can be seen that strategic considerations are largely in phase with operational considerations.

This overlapping association highlights the fact that at such a stage in the simulation, the effects of the Tactical FCM concept also has an effect on the other concepts. In such a respect this means that, by and large, the short/medium-term nature of Tactical investments occur in between Strategic and Operational ones. Conversely, it could also be argued that the stabilizing nature of Financial factors denotes that in the medium-term at least, a Tactical approach to investment appraisal is dominated by Operational and Strategic factors. Thus it can be said, that both Strategic and Operational factors underpin and stabilize the Tactical response.

Figure 5 shows the Operational-A simulation of the FCM. These results show a different dynamic as compared to the previous two functions in Figure 3 and 4. The main point to note being that the system converges rapidly, within 10 iterations as compared to the 13 iterations of both Strategic and Tactical considerations. This is because of the principal and deep inherent interrelationship between both Financial and Operational factors, as highlighted in the previous simulations also. Because of this connection, all of the remaining factors respond to the fuzzy connectivities in a likewise manner, the response rates of each factor being only dissimilar in terms of their amplitudes and phases.

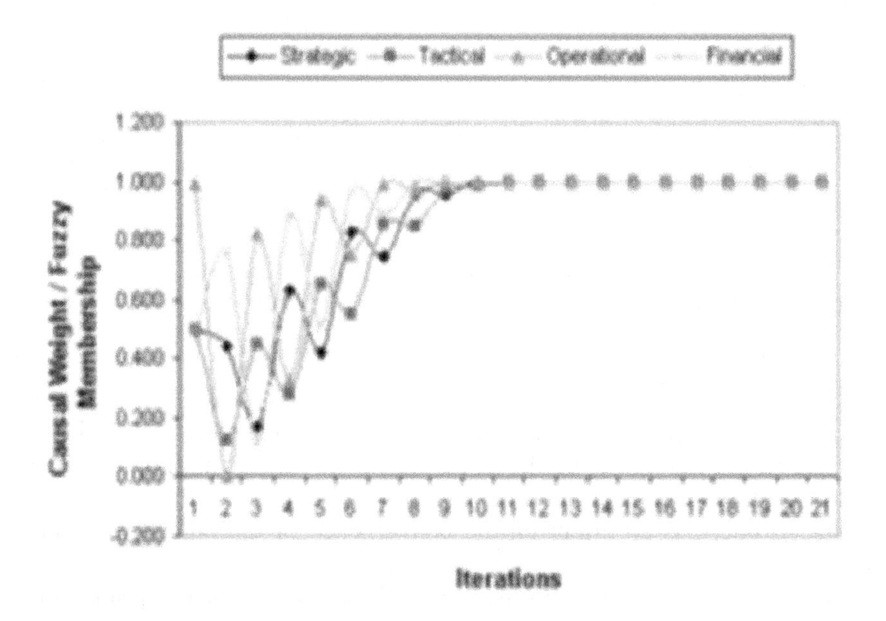

Fig. 5. Operational-A simulation result

A particular point of interest can be seen in iteration 2. Here, the authors suggest that short-term, day-to-day operational factors such as running or maintenance costs, have a negative impact upon strategic factors. Such financial considerations can be justified as having this level of impact in a tactical sense (i.e. the importance and dominance of Operational factors are as a direct result of their existence as a part of any overall lifecycle view of investment justification. It can be further summised that Operational considerations are dominated by Financial considerations although, once more, no one single factor dominates in this model (and subsequently, all factors are considered to be dominant, or “on”). Therefore, and noting the rapid convergence to a fixed point, it can also be concluded that for appraisal of projects in an Operational sense, involves a quicker return on investment as the effects of any such investment are seen sooner rather than later.

Finally, Figure 6 shows the Financial-A simulation results. Once again, the response of the initial starting vector in this case, shows a large correlation and similarity to the Operational-A results. This is so much so, that it can be seen that the Financial and Operational consideration responses are almost reversed when comparing Figure 6 with Figure 5. Here, Strategic considerations lead Tactical considerations, and are in phase with Financial considerations.

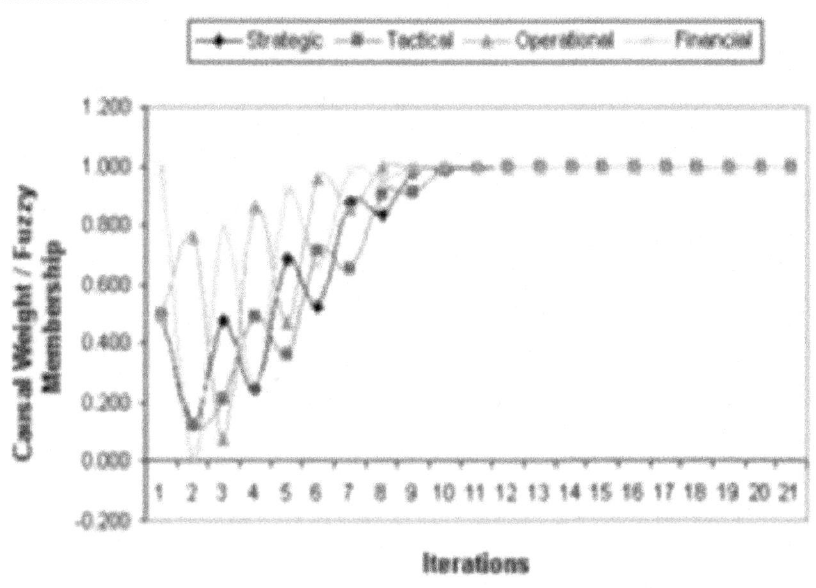

Fig. 6. Financial-A simulation result

Likewise, Operational and Tactical considerations are also in phase with each other. As can be seen from the graph, the responses of all of the variables within this simulation rapidly converge within 10 iterations also. Comparing with the previous Operational-A scenario, there appears to a much more increased level of damping before convergence occurs. This could be possibly due to the fact that Financial considerations drive all other factors, and hence the dynamic interplay between the Strategic, Tactical and Operational aspects are ostensibly linked together.

Realistic response

Figure 7 shows the first of the "realistic" simulations. As can be clearly seen from this and the proceeding results, the dynamic response is markedly different, and instead of all considerations converging to an "on" state, converges to an "off" state instead. As such the Strategic-B simulation shows a high periodicity with both Financial and Strategic and Tactical and Operational factors grouped together, responding out of phase with each other. Here, convergence of the system occurs much later, within approximately 17 iterations.

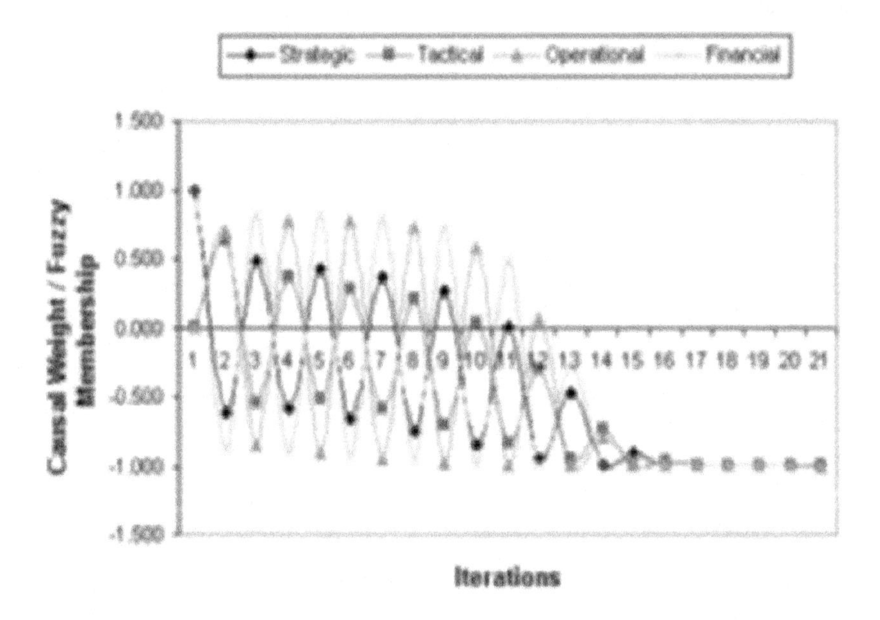

Fig. 7. Strategic-B simulation result

Within these results it can be seen that operational and financial considerations are almost double those of the Strategic and Tactical ones. This possibly highlights the dynamic of taking long term costs and / or investments into account, until Strategic goals can be achieved. As such, it can be seen that overall, the influence of investment costs tend to reduce over time as Tactical and then Operational factors are brought into play. It is also interesting to note that the periodic response of this scenario is primary driven by the reducing effect of Tactical considerations, from iteration 4 onwards. This effect has a direct and immediate effect on firstly Operational then Strategic, then Financial considerations. Indeed between iteration 11 and 12, when Strategic and Financial factors are very closely in phase and intersect one another, the cumulative effect of reducing Tactical considerations, suddenly has a strong dampening effect on the whole system. Thus, during iterations 13 through 15, each aspect within this FCM shows fixed point behaviour with Operational, Financial, Strategic and finally Tactical considerations converging.

The Tactical-B response shown in Figure 8, the inherent result shows a noticeably longer convergence than for any of the other simulations thus far. As such, the system converges within 23 iterations. The high level of periodicity and non-linearity displayed here, can be said to be due the strong correlation between Tactical and Strategic considerations once again.

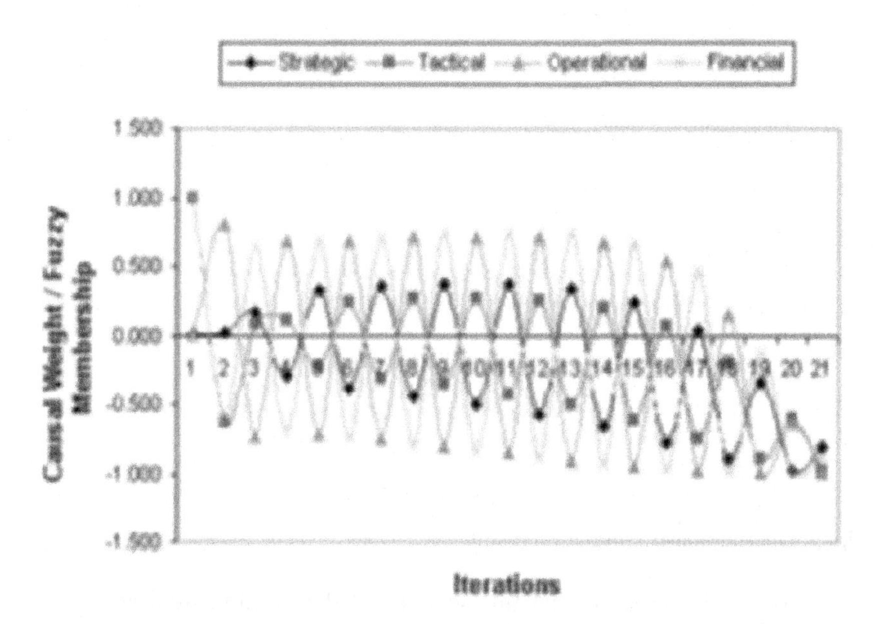

Fig. 8. Tactical-B simulation result

The medium-term response of both of these factors can be quite clearly seen starting from iteration 2 towards iteration 19. As such, both of these drive and dominate the response of the other two. However, the small decrease in Tactical considerations at iteration 14, soon precipitates a general increase in fixed point stability, which ultimately leads to convergence from iteration 19 onwards. This simulation clearly shows that even if a purely Tactical approach to investment is taken, both Operational and Financial considerations still tend to dominate, until such a time as the interplay between Strategic and Tactical aspects stabilises the overall system.

In stark contrast, the Operational-B simulation in Figure 9 shows an altogether different response profile. This shows the clear dominance of Operational and Financial considerations over Strategic and Tactical considerations by some considerable margin. As can be seen, both of these respective grouped vectors are out of phase with one another, but also tend to converge with respect to one another also. Thus for the Operational and Financial considerations, a fixed point is reached within 12 iterations, whilst for Strategic and Tactical considerations, the same occurs but within 14 iterations.

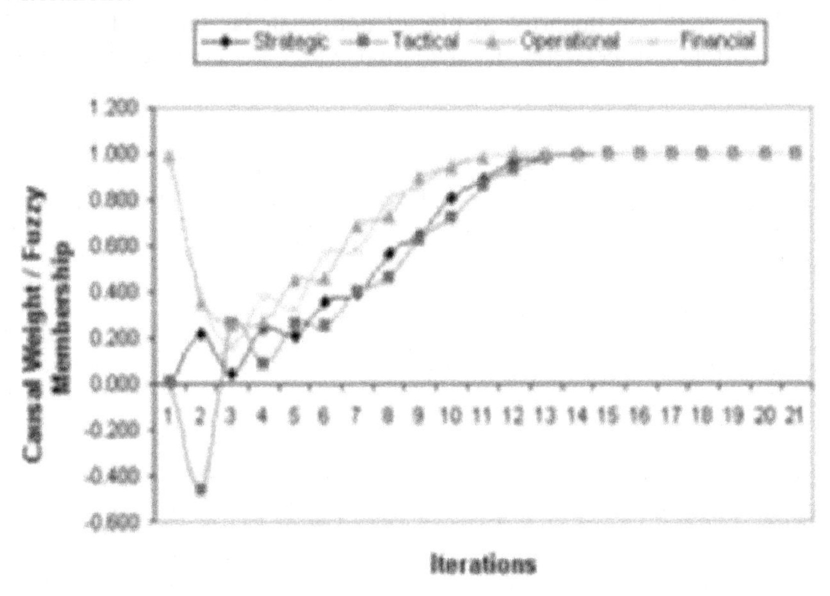

Fig. 9. Opertaional-B simulation result

Of principal interest within this scenario is the rapid switching effect of Tactical factors, from iteration 1 through to iteration 4. Within this circumstance, it appears that such a factor does not have any direct input into an overall Operational point of view taken. However, once again, an interplay exists between Tactical and Strategic aspects, which in some measure can also be seen to be a direct result of an underlying in-phase relationship between Strategic and Financial factors. It can be seen that from iteration 3 towards iteration 12 for both of these conceptual nodes, that they are clearly in some sort of harmony with one another, signifying that such factors are never completely unassociated with the initial starting vector used.

Finally, Figure 10 shows the Financial-B results. Again, the features of this response are unique and different to those that have been discussed so far. The overall effect of the initial Financial starting vector in this case shows a highly damped response, with fixed point convergence occurring within 9 iterations, which is not dissimilar to the Theoretic responses seen earlier. The simulation profile seen here is perhaps the easiest to explain as compared to the others. Quite simply, pure cost-based (i.e. Financial) investment justification approaches are overall quite stable, but there is little or no effect of Strategic, Tactical or Operational ever coming into play.

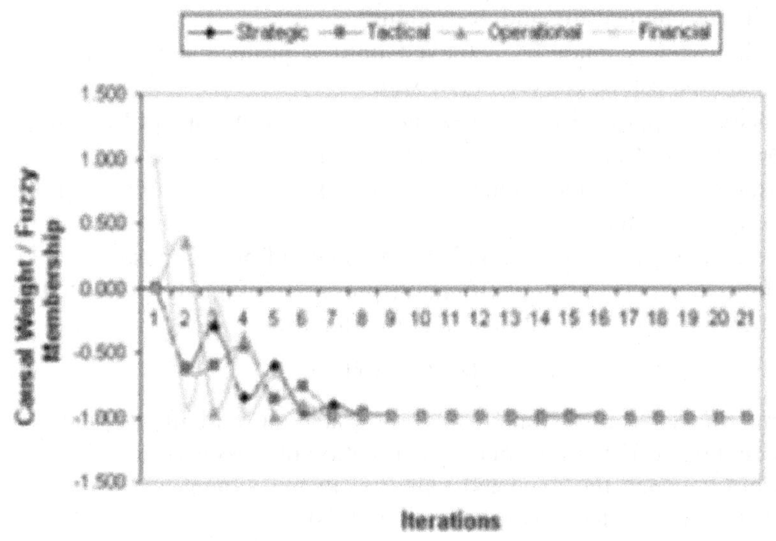

Fig. 10. Financial-B simulation result

That is to say, that in such a case if a project is evaluated purely on such terms, it becomes apparent that the inclusion of the remaining considerations is heavily negated – and therefore, risks attributed to Strategic, Tactical and Operational considerations are not taken into account.

4.2 A Functional Risks and Benefits Model

As an alternative to the former view of assessing the investment justification lifecycle, the authors also now consider a more specific functional model, which identifies the various issues involved in the justification of IT within Company A. The functional model presented below goes some way to conceptualising the phenomena of investment justification, and focuses on a number of key justification criteria; Value, Project Benefits,

Project Costs, Financial Appraisal, and Project Risk Factors. Thus, at a high level, the investment justification process can be succinctly encapsulated within the following expression:

$$JC = f\,[V,\ FA] \tag{12}$$

where *JC* are the justification criteria, *V* is the project value, and *FA* is the financial appraisal of the project. However, the following needs to be observed also:

- Indirect costs need further definition in terms of human and operational costs (re-engineering and re-training);
- Risk review cannot be achieved until a project is implemented and evaluation can be carried out *in-situ*;
- Financial appraisal techniques themselves require quantification within the context of the project being evaluated;
- Strategical and Operational benefits appear to be more tangible (although non-finanical);
- Indirect costs appear as a major component of project costs.

Thus, the investment justification process previously presented in the functional equation in (12), can be rewritten as:

$$JC = f\,[V,\ FA,\ RR] \tag{13}$$

where *RR* is the post-implementation risk review of the project. This is written thus, as the aim of many justification processes is to identify a relationship between the expected value of an investment and a quantitative analysis of the project costs, benefits and risks.

The aim of many justification processes is to identify a relationship between the expected value of an investment and a quantitative analysis of the project costs, benefits and risks. The parameters that are encoded within this model are now discussed in more detail, to obtain more insight into the parameters and their influence in the justification of investment projects. In what follows, an explanation of equation variables relate to those described in equation (12) and hence do not need further repetition. Measuring the perceived value implications of an investment project is a highly subjective process. However, in order to assess the implications impacting on the value of an investment, the authors propose the following relationship, given by the value assessment of the project:

$$V = f\,[(PB/PC)RF] \tag{14}$$

Project benefits are an integral part of any investment justification processes. Until recently, the focus has predominantly been on achievable tangible operational benefits. The reason for this is largely due to the simplicity of quantification, in relation to their values. However, the failure to include strategic benefits in many traditional justification frameworks is largely due to their intangible nature. Since many IT investments now often deliver benefits of a strategic nature, their inclusion in any justification framework is essential. Therefore, accounting for the full implications of the investment. Hence, the holistic implications of project benefits can be denoted for both strategic benefits, *SB*, and operational benefits, *OB*, as:

$$PB = f\,[SB,\ OB] \tag{15}$$

Project costs encompass both the financial and non-financial implications on an investment. Traditionally, much emphasis has been placed on accounting for the direct project costs of an investment, even though much research suggests that these cost factors are largely underestimated (Irani *et al.*, 1997). However, it is the indirect cost implications of an investment which clearly need integrating into a robust justification framework. The reason for their inclusion is emphasised by Hochstrasser (1992), who suggests that indirect cost factors maybe up to four times as high as direct project costs. The holistic project cost implications of an investment can therefore be expressed as:

$$PC = f\,[DC,\ IC] \tag{16}$$

where *DC* are direct project costs. Furthermore, a functional relationship for the indirect costs can be attributed to *HC*, human costs, and *OC*, organisational costs :

$$IC = f\,[HC, OC] \tag{17}$$

Indirect costs are largely difficult to define (Irani *et al.,* 1997). Because of this intangible aspect, *IC* is assumed to have an equal, or indeed greater, relevance than *DC*. Indeed, indirect costs can be up to 4 times greater than direct costs as stated by Hochstrasser (1992).

There is inevitably a risk factor associated with the adoption of any IT project, with Griffiths and Willcocks (1994) suggesting that the degree of risk and uncertainty increases with the size of IT deployment. Therefore, risk management should be considered as an integral part of any holistic justification criteria and must be carried out over the life cycle of the IT project (Hahen and Griffiths, 1996). Using the life cycle process described by Yeate (1991), a projects' risk factor can be represented mathematically as:

$$RF = f\,[RI,\ RA,\ RR] \tag{18}$$

where *RA* is the risk assessment and *RR* is the risk review. With respect to risk identification, *RI,* this can be considered as the initial stage in the process of determining the risk factor. This risk factor is whereby the boundaries and positioning of the project are established. Hence, the functional relationship of risk identification can be represented as:

$$RI = f\,[FR, TR, IR, FUR, SR] \tag{19}$$

where FR are the financial risk implications of the project, TR are the technological risks associated with the project, IR is the corporate specific infrastructural risk, FUR is functional risk of the system and SR is the systemic risk.The second variable in the risk factor equation (17) is that of risk assessment. This is a process where an arbitrary value is assigned to each identified risk along with its significance. This can be done through a number of methods, such as the Analytical Hierarchy Process (AHP) (Saaty, 1980). The third and final variable in the risk factor equation (17), is the risk review process. This is carried out at the end of the projects' life-cycle, through which the effectiveness of a risk assessment exercise can be traced. The risk review process also provides an opportunity to culminate the relevant sources of risk knowledge into a risk file (Hahen and Griffiths 1996).

Many traditional investment decisions are made on the limited basis of financial appraisal. The reason for this is because organisational capital budgeting processes often rely exclusively on conventional appraisal techniques. However, the major limitations in using traditional appraisal techniques are that these methods are unable to accommodate the intangible benefits and indirect costs associated with an IT deployment. Although this may question the predictive value of their use, Kaplan (1986) explains that many companies may be on the road to insolvency, if they consistently invest in projects whose financial returns are below its cost of capital. Therefore, a financial dimension has been integrated into the justification criteria identified in equation (13). This financial perspective can be represented analytically as:

$$FA = f\,[TC\,,\, TB] \tag{20}$$

$$TC\,,\, TB = f\,[RF] \tag{21}$$

$$FA = f\,[TC, TB]\,.\,f\,[RF] \tag{22}$$

where FA is the company preference financial appraisal technique, TC are the tangible cost implications, TB are the tangible benefit implications and RF is the risk factor associated with the project.

4.2.1 An FCM of Costs, Benefits, Risks and Value

In contrast to the FCM shown in Section 4.1, the authors now also propose to generate an FCM based upon those functional factors of costs, benefits, risks and value for Company A. Whilst this approach duely considers specific aspects of investment appraisal rather than generic Strategic, Tactical,

Operational and Financial considerations as discussed so far, the proposed functional representation of the IT justification process consists of a large number of variables, some of which cannot easily be quantified. The subjective aspect of this process, limits the effective optimisation of the given variables. This also restricts the methodical evaluation of justifying these forms of investments. Additionally, the varying nature of IT/IS projects, means that the entire justification process sublimates into a complex adaptive system subject to external as well internal influences. As such, by producing an FCM of this functional representation of the investment appraisal process, it is hoped a much deeper and clearer understanding of the inter-relationships between each functional variable can be found.

Hence, we will now summarise summise the relationship between Project Benefits and the other parameters in the following manner. Project Benefits (*PB*) have increasing effects upon a projects' value (*V*), i.e. '+ highly valued'. *PB* also provides an effective input to the assessment of risk (*RF*), i.e. '+ consistent benefits'. The financial appraisal of project (*FA*) is also greatly enhanced by tangible project benefits, i.e. '+ attractive'. A negative causal relationship exists between project costs (PC) and value (V), i.e. '- high PC', which translates to the rising cost of a project decreasing its overall worth. In such a way, the remaining fuzzy concepts can be related to one another by reading and assessing the fuzzy quantifiers between them. This is shown in Figure 11. In a similar manner to the previous FCM shown, the authors now present modelling and simulation results for this model also.

4.2.1.1 Numerical Simulation Results and Analysis

Once again, using the fuzzy cognitive mapping algebra outlined in Section 4, the results of executing the FCM in Figure 12 are now presented, with the causal modifiers assigned as (to 2 decimal places):

Table 5. Causal weights for Functional FCM

Causal weight	Weight
Attractive	0.75
Increasingly	0.50
Consistent Benefits	0.25
Highly Valued	1.00
Low V	-0.25
High PC	-0.50
Rising Costs	-0.75
Unattractive	-1.00

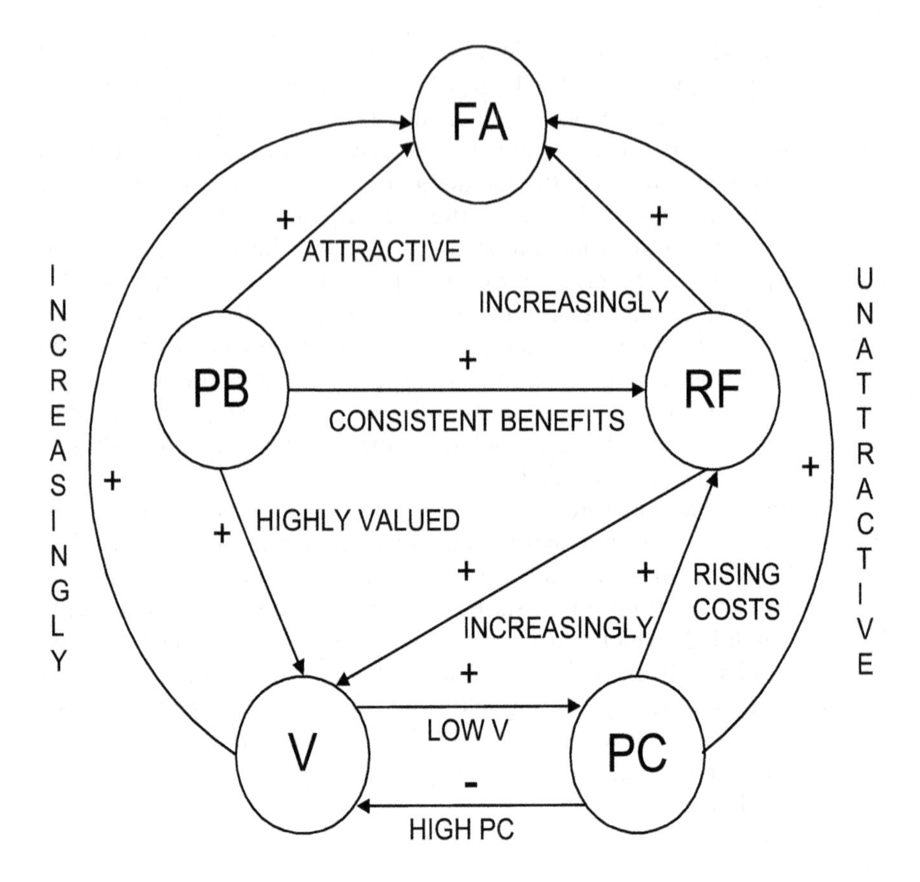

Fig. 11. Fuzzy Cognitive Mapping (FCM) of investment justification criteria

The adjacency matrix, *W*, is given as:

Table 6. Fuzzy connectivity / adjacency matrix for Functional FCM

Fuzzy Nodes	Justification criteria	C1	C2	C3	C4	C5
		FA	PB	RF	V	PC
C1	FA	0.000	0.000	0.000	0.000	0.000
C2	PB	0.750	0.000	0.250	1.000	0.000
C3	RF	0.500	0.000	0.000	0.500	0.000
C4	V	0.500	0.000	0.000	0.000	-0.250
C5	PC	-1.000	0.000	-0.250	-0.500	0.000

In other words,

$$W = \begin{bmatrix} 0.000 & 0.000 & 0.000 & 0.000 & 0.000 \\ 0.750 & 0.000 & 0.250 & 1.000 & 0.000 \\ 0.500 & 0.000 & 0.000 & 0.500 & 0.000 \\ 0.500 & 0.000 & 0.000 & 0.000 & -0.250 \\ -1.000 & 0.000 & -0.250 & -0.500 & 0.000 \end{bmatrix} \tag{23}$$

Again, the threshold function, f, was set to be the hyperbolic function in equation (4). Table 7 shows a number of initial rowvectors used by the authors for C^0, the results for which are discussed in the sections that follow.

Table 7. Initial starting vectors for the Functional FCM

C^0 vector	FA	PB	RF	V	PC
A	1.00	1.00	-1.00	1.00	1.00
B	1.00	-1.00	-1.00	-1.00	1.00
C	1.00	-1.00	1.00	-1.00	1.00
D	1.00	1.00	1.00	-1.00	1.00
E	1.00	1.00	1.00	1.00	1.00

Figure 12 shows the FCM simulation results for simulation A, i.e. where no risk factors for the investment justification are taken into account. Here we can see a rapid convergence to a fixed point, within 8 iterations where no single concept dominates. Although known costs have an adverse effect on risks, these costs cannot be realised until a projects value emerges (in terms of realisable / evident benefits at iteration 2). In other words, once a project's value is defined / realised project risks stabilise. As can be seen from the graph, risk factors RF, tend to stabilise and approach the fixed point equilibrium first, followed by project benefits, PB; project value, V; project costs, PC; and finally financial appraisal techniques, FA. This furthermore denotes that in the absence of any associated project risks, investment evaluations which include at least benefit, value, and cost factors tend to dominate any financial motivations which may be involved or that may drive the initial justification (as shown from iteration 3 to iteration 6).

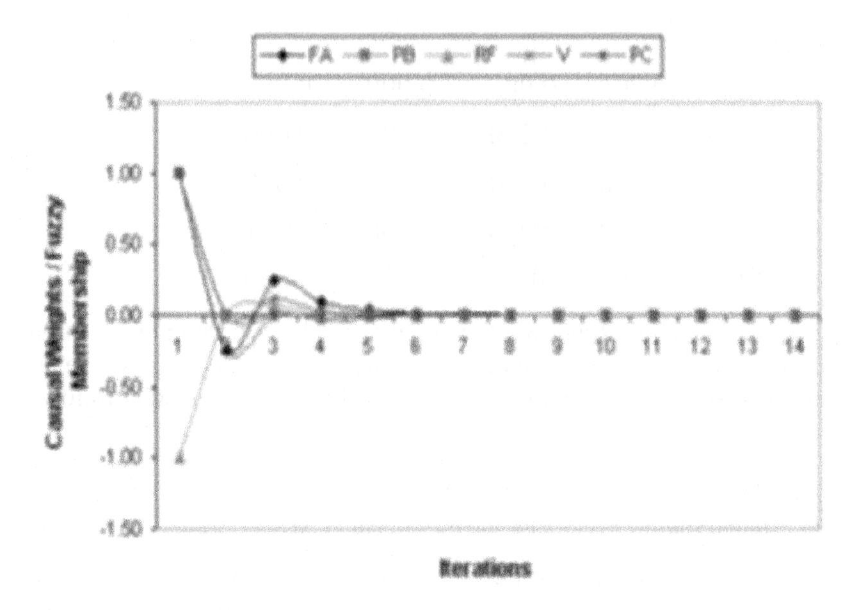

Fig. 12. Simulation A result

Both Figure 13 and 14 show very similar responses for different initial input vectors. In the case of the Simulation B result in Figure 13, this simulation profile was based upon benefits, risks and value not being taken into account, i.e. only financial appraisal techniques and costs being switched on. For Simulation C in Figure 14, risk factors were enabled only (simulating a situation where benefits and value either are not immediately realisable or cannot be defined or measured). As can be seen from both diagrams, the response is quite similar, the only difference being a more damped response in terms of project benefits and value, PB and V in the case where project risk factors, RF are taken into account (i.e. Figure 14).

It can also be seen and summised that in Simulation C, RF are inherently linked to PC, whilst previously they were more explicitly linked to project value factors, V in Simulation B. What is interesting to note is that in both cases, once again, even if purely financial considerations are enabled, i.e. FA, they have little or no direct impact on the resulting response. Rather, in both cases, project costs, PC, dominate more – whilst in the case of Simulation C, the inclusion of project risk factors "up front" tend to smooth the response of the other remaining factors. This dampening effect highlights the fact that the inclusion of project risk stabilises the effect of other investment appraisal factors. However, as can be seen this is only to a limited degree.

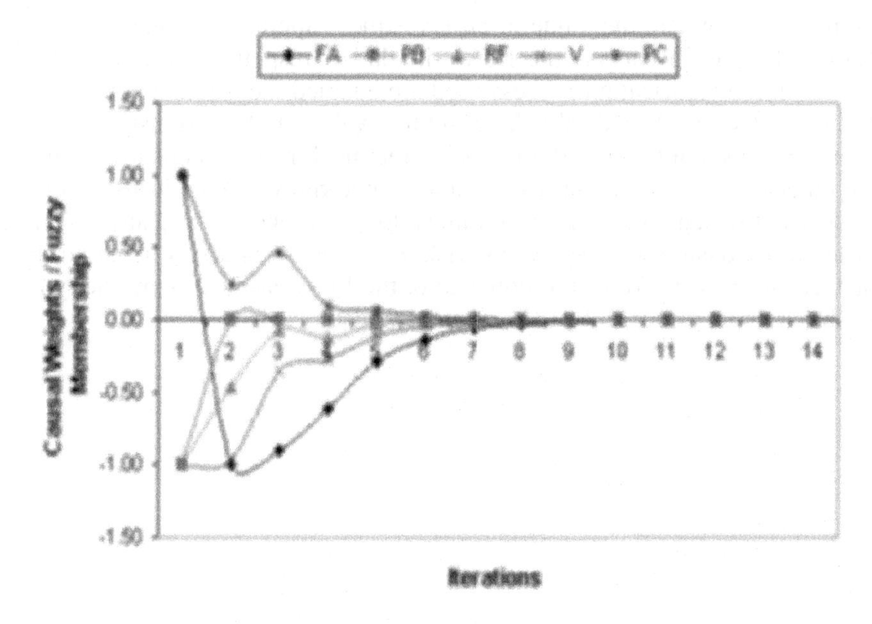

Fig. 13. Simulation B result

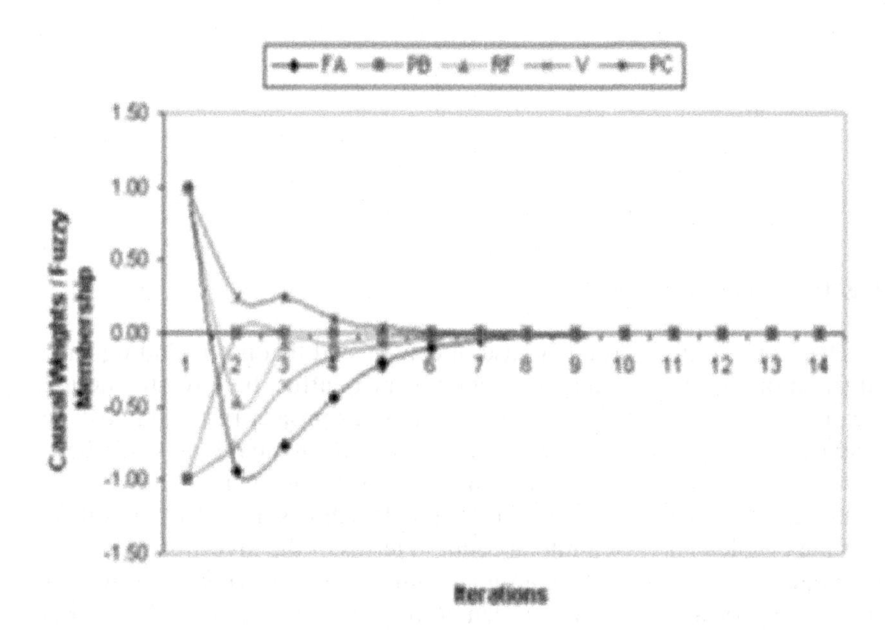

Fig. 14. Simulation C result

In the case where the value of an investment appraisal technique is unknown, Figure 15 shows the FCM simulation response. Here, all other factors in the justification process used are enabled, i.e. set to a value of 1, and project value, V is disabled, i.e. set to a value of -1. Curiously, this response shows that the use of a financial appraisal method tends to stabilise the other factors whilst a project's value is not known. What is also useful to note is the negation of project values does not tend to have any impact or interrelationship with a projects' risk, i.e. V does not affect RF at all. In fact as can be seen from Iteration 3 to 6, the low non-linear periodicity of RF has a direct affect on FA, PB, V and PC also.

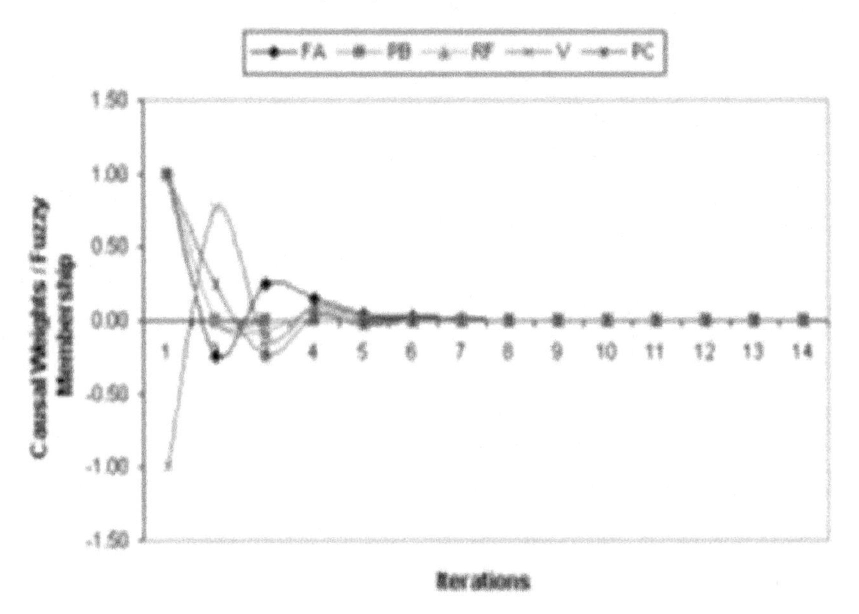

Fig. 15. Simulation D result

Therefore, in this particular case, it can be said that not including an estimation of a project's value does not adversely affect the investment justification process in any way. Rather, as long as project risks and costs are enabled, such a scenario ensures stability within a relatively few number of iterations (although as before, no single factor dominates).

Finally, Simulation E shown in Figure 16, shows a case where all factors are enabled and taken into account in terms of the investment appraisal technique used. This denotes an interesting response, whereby a given financial approach used, tends to denote that risk factors are stable and are subsumed relative to project costs, PC. This is even though the latter

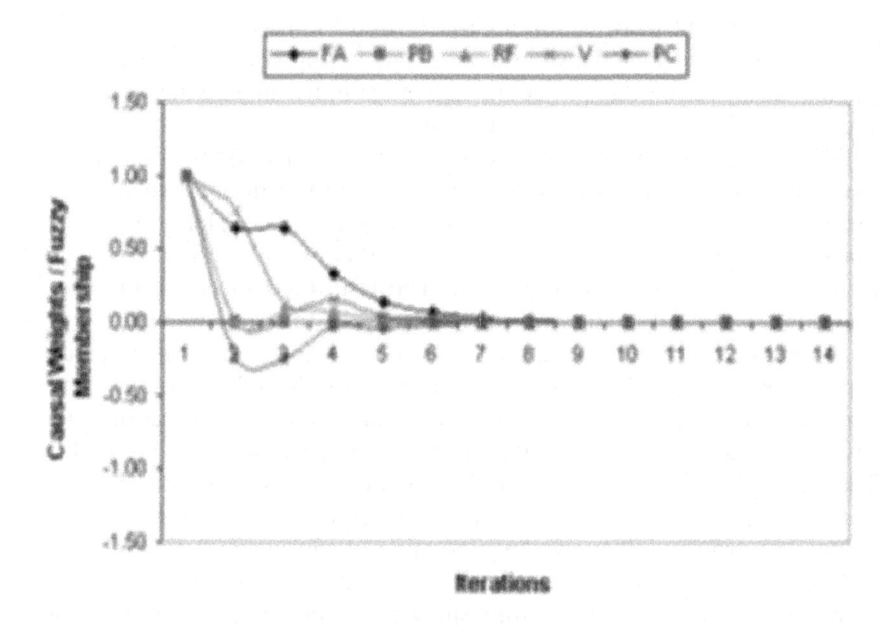

Fig. 16. Simulation E result

project costs, PC, may be negatively impacting the investment justification process (as shown in iteration 2 through to iteration 4). Project benefits, PB, are assumed to arise as a function of the risk profile, RF, also. This can be explained by stating that if there is a better appreciation of risks within any investment appraisal process more benefits can be more effectively realised.

5 Conclusions

Within this paper, the authors have shown the application of fuzzy logic in the guise of Fuzzy Cognitive Mapping (FCM) in order to elucidate key aspects of the investment justification process, within Information Systems Evaluation (ISE). This has allowed for the identification of causal relationships within such a decision-making task, via the analysis of the given case study organisation. These FCMs were derived as a result of two models of investment justification, which involved the inclusion of Strategic, Tactical, Operational and Financial criteria; and of Project benefits, risks, costs and value respectively.

The investment justification approach used by the case study organisation was thus abstracted by observing and noting decision-making tasks carried out by senior and project management individuals. Since traditional appraisal techniques focus on non-strategic, short-term, tangible benefits, with the 'larger picture' often missing from the formal justification process, the nature of many IT projects change from short-term operational investments, to ones with long/medium term strategic focuses. Hence, the limitations of traditional approaches to investment appraisal are barriers to project justification and implementation. The issues associated in the use of traditional appraisal techniques for justifying IT investments were thenceforth identified from the literature and were used to develop the given conceptual models. By modelling and improving knowledge about the interdependencies which exist within such investment justification tasks, several causal maps and simulation results were found and highlighted. Through applying a number of different training data, i.e. initial starting rowvectors, further insight into the investment justification process and model was provided, through the generation of causal simulation results.

In the case of the Strategic, Tactical, Operational, and Financial investment justification FCM, the results showed that no single factor dominates the FCM, with all nodes within the FCM approaching an enabled fixed point value of '1'. Using the theoretic initial rowvectors (i.e. applying only a specific investment justification criterion in singularity), it was seen that direct interrelationships existed between Tactical and Strategic and Operational and Financial considerations. More significantly, it was found that Tactical considerations always tend to stabilise last, and Financial considerations tended to reach a fixed point equilibrium first. These results showed that when this particular FCM was generated based upon the case data the inherent decision-making approach being used within case Company A, was being purely driven from a Tactical viewpoint, with respect to an emphasis on using Financial considerations for the investment evaluation.

When the realistic initial rowvectors were used, i.e. when specific Strategic, Tactical, or Operational considerations were introduced alongside constant Financial considerations, the response of the FCM simulation was markedly different. These results showed greater periodicity in their response and stabilisation to fixed point equilibrium, which again showed no single factor dominated the scenario. In addition, the fixed point stabilisation occurred such that all nodes within the FCM were disabled, or set to a value of '-1' (apart from the Operational-B simulation result, which reached a fixed point of '1' instead). In these results, the interrelationships between each investment criterion were reversed as compared to the theoretical responses. In other words, all the responses showed strong phase relationships between Tactical and Operational and Strategic and Financial factors respectively. Once again, it was found that Financial considerations would stabilise first, and Tactical considerations stabilise last. Overall in looking at these realistic response results, it can be said that taking Financial considerations into account alongside other singular factors, have a direct affect on Operational and Strategic factors. This can be potentially explained as the fact that as Tactical and Financial considerations are taken into account over the lifecycle of the investment justification, Operational benefits cannot be immediately realised. Hence, this prolongs the convergence of the FCM simulation, and once Operational considerations begin to be overtaken by the effects of long-term Strategic considerations (i.e. as time goes on), the benefits of a Strategic view of the investment becomes more apparent. This FCM which encapsulates the underlying four-factor model shows that such a conceptual view of this decision-making task, is inherently biased towards Tactical and Financial considerations, without taking Operational and Strategic views into account. This largely reflects the potential inherent approach used by the case company management team, which traditionally approached such tasks based upon intuitive, 'gut-feel' decisions. In comparison, viewing the results of the functional risks and benefits model shows that this approach to modelling the investment justification task highlights the importance of taking project benefits, costs and risk factors into account. In all of the simulation cases in this respect, it was found that the consideration of project value does not adversely affect the decision-making task in any way. Furthermore, as in the previous FCM, no single factor dominates the simulation, with all factors reaching a fixed-point equilibrium of '0', i.e. neither fully enabled or disabled (hence this can be seen as a null result). Thus, modelling the investment justification task using such a model, shows that this approach provides a more holistic view of the investment justification task by incorporating risks, benefits, costs and financial appraisal techniques in a single manner.

Although this paper has highlighted the application of the FCM approach to the topic of Information Systems Evaluation (ISE) and investment justification, the authors believe multiple avenues of research still exist which may benefit research and development in this area. Of primary concern, is further investigation into methods by which knowledge about the topic being modelled by the FCM in question, can be elicited. At the moment, it is very much up to the researchers and experts in the field, to filter and define the problem area in question, and thenceforth to prescribe the causal weights, nodes and relationships as they see fit. The inclusion of additional expert views and knowledge would obviously alleviate and normalise any causes of bias in the initial formulation of the FCM data. However, further investigation into the use of multiple experts and data sources, as well as knowledge filtering and verification / validation techniques, needs to be carried out to understand these affects. In addition, this research only focussed on produced singular FCMs from the case data so collected – in other words, a single FCM for each of the conceptual models defined was created. These FCMs were in essence generated as a result of simplification and filtering of the qualitative fieldwork data, from a variety of interviews and other sources. It would be useful therefore to investigate further the cumulative effect of creating multiple mappings for each data source, taking each the context of each interviewee, say, into account: therefore, generating a single FCM of the situation from a number of sub-FCMs which would describe specific contexts of each knowledge expert within the scenario.

In terms of the computational simulation / machine learning facet of this cognitive mapping approach, additional investigation into the effects of applying different threshold functions to the fuzzy system would be interesting to note. Ultimately this should only serve to fit the response of the mapped weights to a particular discrete range – however, the full effects of using bivalent, trivalent, hyperbolic and / or sigmoidal functions applied to the same FCM would be of further research interest also. Furthermore, most fuzzy cognitive mappings are concerned with providing an analysis of a particular scenario with respect to a specific moment in time (from whence the fuzzy causal weights and nodes were defined), as opposed to taking temporal effects of varying weights and nodal values as the simulation progresses. By examining the effect of employing discrete time step variations to each set of FCM simulations, fuzzy causal learning scenarios can potentially capture more realistic time-based decision-making scenarios. This may well involve the inclusion of sequential timestep information either within the given threshold function, or as a product of the adjacency matrix itself.

Finally, the authors believe that the application of this mapping approach to investment justification, is but a small insight into using such a technique within the field of Information Systems Evaluation. As such, this research has only focussed on the decision-making process within a single manufacturing organisation. There is obvious scope for extending this analysis technique towards other manufacturing organisations in order to understand and validate the results that have been found in this example. Supplementary usage of this approach to other MIS decision-making situations would also be useful: for example the optimisation of IT/IS resources; business continuity planning; analysis of service level agreement contracts for outsourcing and business process re-engineering projects also.

References

1. Aleksander I., and Morton H., (1990), *An introduction to neural computing*, Chapman and Hall.
2. Axelrod R., (1976), *Structure of design,* Princeton University Press.
3. Badiru A., Foote B. and Chetupuzha J., (1991), A multi-attribute spreadsheet model for manufacturing technology justification'. *Computer & Industrial Engineering*, 21(1-4): 29-33.
4. Baily, M.N. and Chakrabarti, A. (1988) *Innovation and the Productivity Crisis*, Brookings Institution, Washington, DC.
5. Barat J., (1992), Scenario planning for critical success factor analysis, *Journal of Information Technology*, 7, 12-19.
6. Bauer, R.J. 1994. *Genetic Algorithm and Investment Strategies.* New York, NY : John Wiley.
7. Bentley, P.J. & Wakefield, J.P. 1997. Finding Acceptable Solutions in the Pareto-Optimal Range using Multiobjective Genetic Algorithms. (Eds. P.K. Chawdhry, R. Roy, and R.K. Pant). *Soft Computing in Engineering Design and Manufacturing* (2nd On-line World Conference on Soft Computing in Engineering Design and Manufacturing (WSC2), 23-27 June 1997). London : Springer Verlag. Part 5, pp. 231-240.
8. Boaden R.J. and Dale B., (1990), Justification of computer integrated manufacturing: some insights into the practice, *IEEE Transactions on Engineering Management,* 37(4): 291-296.
9. Bonoma, T.V. (1985). Case research in marketing: Opportunities, problems, and a process. *Journal of Marketing Research,* 12:199 - 208.
10. Bougon, M., Weich, K., Binkhorst, D., (1977), Cognition in Organization: Analysis of the Utrecht Jazz Orchestra, Administrative Science Quarterly, Vol.22, pp. 606-639.
11. Brynjolfsson, E. (1993) The Productivity Paradox of Information Technology, *Communications of the ACM, 36, 12*, pp. 67-77.
12. Burstein M.C., (1986), Finding the economical mix of rigid and flexible automation for manufacturing systems, *Proceedings of the Second* ORSA/TIMS *Conference on Flexible Manufacturing Systems,* 69-81.
13. Butler, M. 1997. Future Fantastic. *Information Week*, 19, pp.53.
14. Carlson, W. 1992. Basic principles for measuring IT value. *I/S Analyser.* 30 (10) 1-14.
15. CEAS (Confederation of European Aerospace Societies), 1997. Aeronautical Research and Technology - A Strategic Imperative for Europe, Position Paper June 1997, insert with *Aerospace International,* 24 (6), pp. 1-11.
16. Coats, P.K., (1991), A critical look at expert systems for business information applications, *Journal Information Technology*, 6, 208-216
17. Datta, V., Sambasivarao, K.V., Kodali, R., Deskmukh, S.G., (1992), Multi-attribute decision model using the analytic hierarchy process for the justification of manufacturing systems, *International Journal of Production Economics*, Vol. 28, No.2, 227-234.

18. Davis, L. 1987. *Genetic Algorithms and Simulated Annealing*. San Francisco, CA : Morgan Kaufmann.
19. Demmel, J. G. and Askin, R.G., (1992), Integrating financial, strategic, and tactical factors in advanced manufacturing system technology investment decisions, *Economic and Financial Justification of Advanced Manufacturing Technologies*, H.R.Parsei, W.G., Sullivan, T.R. Hanley (Editors), Elsevier, 209-243.
20. Dilts D.M., and Turowski D.G., (1989), Strategic investment justification of advanced manufacturing technology using a knowledge-based system, *Proc. 3rd Int. Conf. Exp. Sys. and the Leading edge in Prod. and Op. Man., University of South Carolina Press, 193-206.*
21. Ezingeard J.-N., Irani Z. and Race P. 1999. 'Assessing the value and cost implications of manufacturing information and data systems: An empirical study'. *European Journal of Information Systems*, 7(4): 252-260.
22. Farbey B., Land F. and Targett D., (1992), Evaluating investments in IT, *Journal of Information Technology*, 7, 109-122.
23. Farbey B., Land F., Targett D., (1993), *IT investment: A study of methods and practices,* Published in association with Management Today and Butterworth-Heinemann Ltd, U.K.
24. Farbey B., Land F., Targett D., (1995), System - Method - Context: a contribution towards the development of guidelines for choosing a method of evaluating I/S investments, *Proceedings of the Second European Conference on Information Technology Investment Evaluation,* Henley Management College, Henley on Thames, England, 11-12 July.
25. Farbey B., Targett D., Land F., (1994), Matching an IT project with an appropriate method of evaluation: a research note on evaluating investments in IT, *Journal of Information Technology*, 9(3): 239-243.
26. Fiedler, J. (1978). *Field Research: A Manual for Logistics and Management of Scientific Studies in Natural Settings*. Jossey-Bass, San Francisco, USA.
27. Gaimon C., (1986), The strategic decision to acquire flexible technology, Proceedings of the Second ORSA/TIMS Conference on Flexible Manufacturing Systems, 43-54.
28. Gale, D. 1968. A Mathematical Theory of Optimal Economic Development. *Bull. Amer. Math. Soc.* 74 207-23.
29. Garrett S.E, (1986), Strategy first: A case in FMS justification, Proceedings of the Second ORSA/TIMS Conference on Flexible Manufacturing Systems, 17-29.
30. Goldberg, D.E. 1989. *Genetic Algorithms in Search, Optimisation and Machine Learning*. Reading, MA : Addison-Wesley.
31. Goonatilake S., and Khebbal S., (1994), *Intelligent hybrid systems,* John Wiley.
32. Grierson D.E., and Cameron G.E., (1987), A knowledge based expert system for computer automated structural design, The application of artificial intelligence techniques to civil and structural engineering, Proceedings of the 3rd international conference on civil and structural engineering, Civil Comp. Press, 93-97.

33. Griffiths, C. and Willcocks, L. 1994. Are major IT projects worth the risk ?. In A. Brown and D. Remenyi (editors.). *Proc. 1st European Conference on IT Investment Evaluation*, City University, London, UK. pp. 256-259.
34. Hahen, G. and Griffiths, C. 1996. A quantitative model for Technological Risk Assessment in the process of IT Transfer. In A. Brown and D. Remenyi (editors). *Proc. 3rd European Conference on the Evaluation of Information Technology,* Bath University, Bath, UK, 29th November 1996. pp. 35 -42.
35. Hakim, C. (1987). *Research Design: Strategies and Choice in the Design of Social Research.* Allen and Unwin, London, UK.
36. Hares J. and Royle D., (1994), *Measuring the value of information technology,* John Wiley and Sons Ltd, U.K.
37. Hill T., (1993), Manufacturing strategy; The strategic management of the manufacturing function, Second Edition, The Macmillan Press, U.K.
38. Hinton E., (1992), How neural networks learn from experience, *Scientific American* #267, 144-151.
39. Hochstrasser B. and Griffiths C., (1991), *Controlling IT investment; strategy and management,* Chapman and Hall, London, U.K.
40. Hochstrasser B, (1990), Evaluating IT investments - matching techniques to projects, *Journal of Information Technology,* 5, 215-221.
41. Hochstrasser B., (1992), Justifying IT investments, Conference Proceedings: Advanced Information Systems; The new technologies in today's business environment, 17-28.
42. Holland J.H., (1992), *Adaptation in natural and artificial systems: An introductory analysis with applications to Biology, Control and Artificial Intelligence*, MIT Press.
43. Huang P. and Sakurai M., (1990), Factory automation: The Japanese Experience, *IEEE Transactions on Engineering Management*, 37(2): 102-108.
44. Huber R.F., (1986), CIM: inevitable but not easy, *Production Engineer,* April, 52-57.
45. Huber R.F., (1985), Justification- barrier to competitive manufacturing, *Production* , 46-51.
46. Irani Z., Ezingeard J.-N. and Grieve R.J. 1997. 'Integrating the costs of an IT/IS infrastructure into the investment decision making process'. *The International Journal of Technological Innovation, Entrepreneurship and Technology Management (Technovation)*, 17(11/12): 695-706.
47. Irani Z., Ezingeard J.-N., and Grieve R.J. 1999a. The Testing of an information systems investment evaluation framework. *Int. J. Tech. Man.* (forthcoming).
48. Irani Z., Ezingeard J.-N., Grieve R.J. and Race P. 1999. 'Investment justification of information technology in manufacturing'. *The International Journal of Computer Applications in Technology*, 12(2): 90-101.
49. Irani Z., Ezingeard, J.-N., and Grieve, R.J. 1997. Identifying the Cost Implications of Manufacturing IT/IS Investments. *Proc. Portland International Conference on Management of Engineering and Technology (PICMET '97)*, Portland, Oregon, July 27-31st 1997, Portland, OR : Portland State University / IEEE / Informs.

50. Irani Z., Jones S. Love P.E.D., Elliman T. and Themistocleous M. (2005). Evaluating Information System Investments in Local Government: Drawing Lessons from Two Welsh Cases. *Information Systems Journal* – Accepted.
51. Irani Z., Love P.E.D. and Hides M.T. 2000. 'Investment Evaluation of New Technology: Integrating IT/IS Cost Management into a Model'. Association for Information System, 2000 Americas Conference on Information Systems (AMCIS 2000) [CD Proceedings], August 10-13th, 2000, Long Beach, California, USA.
52. Irani, Z., and Love, P.E.D. (2001). The propagation of technology management taxonomies for evaluating information systems. *Journal of Management Information Systems* 17(3) : 161-177.
53. Irani Z. and Sharif A. (1997). '*Genetic Algorithm Optimisation of Investment Justification Theory*'. Proceeding of the Late Breaking Papers at Genetic Programming 1997, 13-16 July, Stanford University, California, USA.
54. Irani, Z., Sharif, A.M. and Love P.E.D. 2001. Transforming failure into success through organizational learning: An analysis of a Manufacturing Information System. *European Journal of Information Systems,* 10(1): 55-66.
55. Jackson P., (1990), *Introduction to expert systems,* Addison-Wesley.
56. Kaplan R.S., (1985), *Financial justification for the factory of the future,* Harvard Business School, U.S.A.
57. Kaplan, R.S. 1986. Must CIM be justified on faith alone. *Harvard Business Review*. 64 (2), pp. 87-97.
58. Kassicieh, S.K., Paez, T.L., and Vora, G. (1998). Data transformation methods for Genetic Algorithm-based investment decisions. *Proc. 31st Ann. Hawaii Int. Conf. Sys. Sci.,* Hawaii, USA. Los Alamitos : CA. IEEE. pp 96-101.
59. Khalifa G, Irani Z and Baldwin L.P. 2000. 'IT Evaluation Methods: Drivers and Consequences'. Association for Information System, 2000 Americas Conference on Information Systems (AMCIS2000) [CD Proceedings], August 10-13th, 2000, Long Beach, California, USA.
60. Kosko, B. 1990. *Fuzzy thinking : The new science of Fuzzy Logic*, London : Flamingo Press / Harper-Collins.
61. Maskell B, (1991), Performance measurement for world class manufacturing: A model for American companies, Productivity Press, U.S.A.
62. Mentazemi A., and Conrath D., (1986), The use of cognitive mapping for information requirement analysis, *Manufacturing Information Systems Quarterly*, March.
63. Meredith J.R. and Suresh N.C., (1986), Justification techniques for advanced technologies, *International Journal of Production Research.* 24(5): 1043-1057.
64. Meredith, J.R. 1987.Implementing the automated factory. *J. Man. Sys*. 6 75-91.
65. Naik B. and Chakravarty A.K., (1992), Strategic acquisition of new manufacturing technology: a review and research framework, *International Journal of Production Research,* 30(7): 1575-1601.
66. Nelson C.A., (1986), A scoring model for flexible manufacturing systems project selection, *European Journal of Operations Research*, 24(3): 346-359.

67. Nozicka, G.J., Bonham, G.M., Shapiro, M.J., (1976), Simulation Techniques, Structure of Decision,Princeton University Press, Princeton, NJ, in R. Axelrod (ed.), pp. 349-559.
68. Parker M., Benson R. and Trainor H., (1988), *Information economics: linking business performance to information technology,* Prentice-Hall International, U.S.A.
69. Pouloudi, A. and Serafeimidis, V. (1999). Stakeholders in information systems evaluation:experience from a case study. In *Proceedings of the 6th European IT Evaluation Conference*, November 4-5th, London, UK, pp. 90-98.
70. Primrose, P.L. 1991. *Investment in manufacturing technology.* London : Chapman and Hall.
71. Remenyi D., Money A., Sherwood-Smith M., Irani Z. 2000. *'The Effective Measurement and Management of IT Costs and Benefits'*. Butterworth Heinemann/Computer Weekly - Professional Information Systems Text Books series, Second Edition, ISBN 0 7506 4420 6, UK.
72. Roach, S.S. (1991) Services Under Siege: The Restructuring Imperative, *Harvard Business Review, September/October*, pp. 83-91.
73. Saaty, T.L. 1980. *The Analytical Hierarchy Process, planning, priority setting, resource allocation.* USA : McGraw-Hill.
74. Sarkis J. and Liles D., (1995), Using IDEFO and QFD to develop an organisational decision support methodology for the strategic justification of computer-integrated technologies, *International Journal of Project Management*, 13(3):177-185.
75. Serafeimidis V and Smithson S. 2000. 'Information Systems Evaluation in Practice: a case study of organisational change', *Journal of Information Technology*, 15(2): 93-105.
76. Sharif A and Irani Z. 1997. *Fuzzy Cognitive Mapping as a Technique for Technology Management. Proc. Portland International Conference on Management of Engineering and Technology (PICMET '97)*, Portland, Oregon, July 27-31st 1997, Portland, OR : Portland State University / IEEE / Informs, pp. 871.
77. Shaughnessy, J.J. and Zechmeister, E.B. (1994). *Research methods in Psychology. 3rd edition.* New York. McGraw Hill.
78. Simpson, P.K, (1990), Artificial neural systems: foundations, paradigms and applications, Mcgraw-Hill.
79. Simpson, P.K. 1990. *Artificial Neural Systems : foundations, paradigms and applications.* San Diego, CA : Mcgraw-Hill. pp. 96-100.
80. Small M, H. and Chen J, (1995), Investment justification of advanced manufacturing technology: An empirical analysis, *Journal of Engineering and Technology Management*, 12(1-2): 27-55.
81. Sullivan, W.G. and Reeve, J.M., (1988), XVENTURE: Expert systems to the rescue- They can help justify investments in new technology, Management Accounting, 51-58.

82. Suresh N.C. and Meredith J.R., (1985), Justifying multi-machine systems: An integrated strategic approach, *Journal of Manufacturing Systems*, 4(2): 117-134.
83. Swamidass P. and Waller M., (1991), A classification of approaches to planning and justifying new manufacturing technologies, *Journal of Manufacturing Systems*, 9(2):181-193.
84. Van Blois J., (1983), *Economic models: The future of robotic justification,* Proceedings of the Thirteenth ISIR/Robots 7 Conference, April 17-21.
85. Vedarajan, G., Chi Chan, L., and Goldberg, D.E. 1997. Investment Portfolio Optimization using Genetic Algorithms. In *Proc. Late Breaking Papers, Genetic Programming 1997,* Stanford University, Stanford, USA, July 13-16th, 1997, Stanford, CA : Stanford University Bookshop, pp. 255-265.
86. Volpi, L., (2004). *Matrix 1.8 – Matrix and Linear Algebra functions for EXCEL.* Available. [on-line]. http://digilander.libero.it/foxes/index.htm
87. Willcocks L., (1994), Information management: the evaluation of information systems investments, Chapman and Hall, London, U.K.
88. Willcocks, L., 1994. Introduction of Capital importance. In L. Willcocks (editor). *Information management: the evaluation of Information Systems Investments.* London : Chapman and Hall. Pages 1-24.
89. Yeates D.A. 1991. *Project Management for Information Systems,* London, Pitman.
90. Yin, R.K. (1994). *Case study research: Design and Methods – 2nd Ed.* Sage Publications : Thousand Oaks, CA, USA.
91. Zadeh L.A, (1965), Fuzzy sets, *Information and Control*, 8, 338-353.
92. Zadeh L.A, (1996) Fuzzy logic = computing with words, *IEEE Transactions on Fuzzy Systems* 4(2): 103-11.

Fuzzy Sets based Cooperative Heuristics for Solving Optimization Problems

Carlos Cruz, Alejandro Sancho-Royo, David Pelta, and José L. Verdegay

Group on Models of Decision and Optimization
Depto. de Ciencias de la Computación
e Inteligencia Artificial
E.T.S. Ingeniería Informática
Universidad de Granada, Spain

Summary. This paper makes a deeper study of a multi-thread based cooperative strategy, previously proposed by us, to solve combinatorial optimization problems. In this strategy, each thread stands for a different optimization algorithm (or the same one with different settings) and they are all controlled by a *coordinator*. Both, the *solvers* threads and the *coordinator* thread have been modeled by soft computing techniques. We evaluate the performance of the strategy according to the number of threads using instances of the knapsack problem.

Key words: Cooperative heuristics; optimization problems; solvers threads; memetic algorithms; parallel metaheuristics; multi agent systems

1 Introduction

Several studies have shown that heuristics and metaheuristics are successful tools to provide reasonable good (excellent in some cases) solutions using a moderate amount of resources. Just a view at several recent books [16, 25, 18] is enough to check the wide variety of problems and methods that appear under the global topic of heuristic optimization.

However, from a practical point of view, there are several difficulties in order to use these methods in optimization. In particular, the problem of choosing the best algorithm for a particular instance is far away from being easily solved due to many issues, and it is being tackled from different points of view [29, 23, 24]. Nowadays, it is accepted that no algorithm outperforms the other in all circumstances [34] and everyone with some experience in the field of optimization may agree that problem instances can be grouped in classes and there exists an algorithm for each class that solves the problems of that class most efficiently [29].

C. Cruz et al.: *Fuzzy Sets based Cooperative Heuristics for Solving Optimization Problems*, StudFuzz **201**, 505–519 (2006)
www.springerlink.com

One way to overcome this problem may arise from a simple idea: let us have a portfolio of solving strategies and use all of them in a parallel and coordinated fashion to solve the problem.

Nowadays, solving combinatorial optimization problems with the help of parallel and distributing processing is a very active field of research. These methods offer the possibility to speed up computations and especially the multi-search metaheuristics, with varying degrees of cooperation, are often used to perform a greater exploration of the search space.

Several fields of research apply this kind of approach directly or in a more subtle way:

- The field of *memetic algorithms*: evolutionary algorithms that apply a local search process to refine solutions to hard problems. One step beyond are multimemetic algorithms where a *set* of local searchers is available to improve solutions [18]. Self adaptation can be used to promote and disseminate those local searchers that provide the best results, thus leading to robust and adaptive search strategies [22, 21, 20, 26]. The interested reader in this field may refer to [18] for recent advances in the topic.
- The field of *parallel metaheuristics*, where several researchers have been applying this kind of techniques to the solution of different problems. From the point of view of algorithmic development, several ideas are being explored: 1) self-guidance by modulating the intensification/diversification tradeoff, 2) combination of population-based and trajectory-based algorithms, 3) cooperation among similar meta-heuristics, and (4) multilevel hybridization of trajectory-based algorithms [1]. Theoretical and practical developments on parallel metaheuristics can be found in [7, 6, 10, 5, 4].
- The field of *multi agent systems*: where a) simpler entities, like "ants" [12, 11] or "particles" [14, 13] form ensembles that shows a cooperative behavior able to solve optimization problems, or b) more sophisticated set of "agents" are used in a multi-population scheme to solve a specific task, like in the so-called *A-Team* [30, 31].

In this work we are interested in the field of parallel metaheuristics.

Following the classification of [7] regarding parallel metaheuristics, we will focus on the so called *Type 3 parallelism or Multi-search metaheuristics* (Other classification is presented in [10]): several concurrent threads search the solution space. Each concurrent thread may or may not execute the same method, they may start from the same or different initial solutions, etc. If the threads communicate during the search, such strategies are called *co-operative multi-thread* methods; if they communicate at the end of the run, they are called *independent search* methods.

The cooperative search threads exchange and share information collected along the trajectories they investigate. This shared information is implemented either as global variables stored in a shared memory, or as a pool in the local memory of a dedicated central processor which can be accessed by all other processors.

In [7], Crainic and Toulouse have pointed out that:

> *Asynchronous, cooperative multi-thread search metaheuristics appear to have been less studied but are being increasingly proposed. In fact, this strategy is probably the strongest current trend in parallel metaheuristics...*

and they go on saying go.

> *The results reported in the literature seem to indicate that asynchronous, cooperative multi-thread parallelization strategies offer better results than synchronous and independent searches. More theoretical and empirical work is still required in this field, however.*

In this context we proposed a cooperative parallel strategy multi-thread based in previous work [9], where each solver agent represented a particular optimization algorithm (with a different parameter setting each) and they were controlled by a *coordinator* agent whose main tasks were: to collect the performance information of the solvers; and to send them orders to alter their behavior.

This strategy uses a memory to keep track of both, the performance of the individual threads and the characteristics of the solutions being obtained. Then this memory is used to define the fuzzy concept of *low* performance or solution quality, and then, such concept allowed us to define a fuzzy rule to determine when a particular thread is not performing fine.

This last idea puts our strategy in the area of fuzzy sets-based heuristics for optimization. It is a novel topic recently addressed in [33], that tries to combine the best of Fuzzy Sets and Systems ideas together with heuristic techniques to obtain robust and adaptive optimization tools.

In this work, we propose to analyze how the number of agents affects the search results, in order to gain knowledge about the global behaviour of the strategy.

To reach the previous goal, we organized the paper as follows: in Section 2 we introduce our proposal in a general manner and then we go into the technical details in section 3. To test our ideas, we use instances of the knapsack problem. The experiment descriptions and the results are shown in Section 4. Finally, Section 5 is devoted to the discussion of the results, its implications and to line out future lines of research.

2 Details of the Proposal

The idea of our proposal can be explained with the help of the scheme shown in Fig. 1. Given some concrete problem to solve, we have a set of *solvers* to deal with it. Each *solver* develops a particular strategy to solve the problem in an independent fashion, and the whole set of *solvers* works simultaneously without direct interaction.

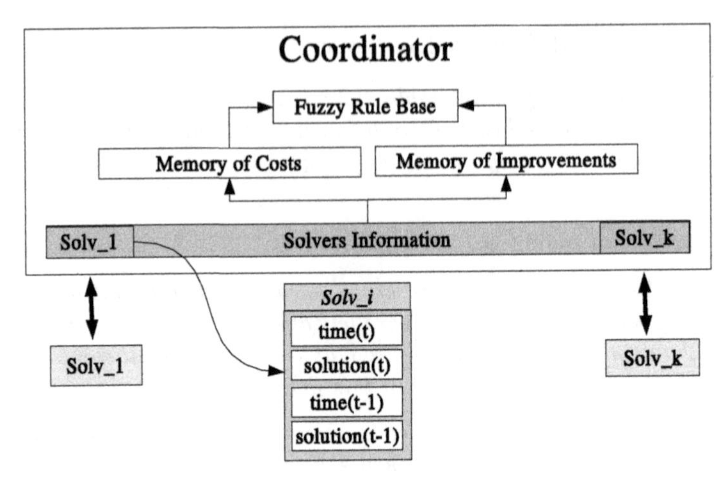

Fig. 1. Scheme of the Proposal

In order to coordinate the *solvers*, there is a *coordinator*, that knows all the general aspects of the concerned problem and the particular features of *solvers*. The *coordinator* receives reports from the *solvers* with the results being obtained and send them back some directives.

The inner workings of the strategy are quite simple. In a first step, the *coordinator* determines the initial behavior (set of parameters) for every *solver*, which models a particular optimization strategy. Such behavior is passed to every *solver* and then, they are executed.

For each *solver*, the *coordinator* keeps the last two reports containing: the time of them, and the corresponding solutions at such times. From the set of *solvers* information, the *coordinator* calculates performance measures used to adapt its internal fuzzy rule base.

This fuzzy rule base is generated manually trying to address an essential principle: *If a* solver *is working fine fine, keep it; but If a* solver *seems to be trapped, do something to alter its behavior.*

As said before, *solvers* execute asynchronously, sending and receiving information. The *coordinator* checks which *solver* provided new information and decides if its behavior needs to be adapted using the fuzzy rule base. If this is the case, it will obtain a new behavior which will be sent to *solvers*.

The operation of the *solvers* is quite simple. They are running alone, alternating the send and reception to obtain adaptation information from the *coordinator* or to inform their performance. Both operations are controlled by conditions.

3 Technical Details

Having in mind the global view provided in the previous section, we describe here how the *solvers* are implemented, what information is exchanged and how the coordination is performed.

3.1 Implementation of the Solvers

The *solver* threads are implemented by means of the Fuzzy Adaptive Neighborhood Search algorithm(*FANS*) [2, 27]. There are three main reasons for this decision:

1. *FANS* is essentially a local search technique so, it is easy to understand, implement and does not need too much computational resources.
2. *FANS* can be used as a *template* of heuristics. Each set of parameters implies a different behavior of the method, thus the qualitative behavior of other local search techniques is easily simulated[2].
3. The previous point allows to build different search schemes and to drive diversification and intensification procedures easily. Thus, we do not need to implement different algorithms but just use *FANS* as a template[1].

A key aspect related to item 2, is the use in *FANS* of a fuzzy valuation representing some (maybe vague and hence fuzzy) property P together with the objective function to obtain a semantic evaluation of the solution. In this way, we may talk about solutions satisfying P in certain degree. Here, the fuzzy valuation is based on a notion of 'acceptability'[2] that is assessed through the cost of a reference solution. In this way, those neighbors solutions that improve the reference cost, are considered as acceptable with a degree of 1. Those solution with cost worst than certain threshold are given zero degree of acceptability. Solution in between, achieve degrees of membership varying in (0,1).

Figure 2 presents a graphical scheme of the label *Acceptability* where α is the reference cost and β is the 'limit' of acceptability.

FANS moves between solutions satisfying the idea of acceptability with at least certain degree λ. Variations in this parameter, enables the algorithm to achieve the qualitative behavior of other classical local search schemes [2]. For example, using the membership function shown in Fig. 2 and setting $\lambda = 1$, *FANS* will achieve a hill climber like or greedy scheme: just cost improving solutions will be considered as acceptable with enough degree. At the other extreme, when $\lambda = 0$, *FANS* will produce random walks. Other values of the parameter will lead to behaviors where the transitions to non-improving solutions will be produced, which in turn may allow to escape from local optima.

[1] A similar justification may be used to implement a set of threads based on Simulated Annealing, each one having a different cooling scheme.

[2] Consequently, we have the fuzzy set of acceptable solutions

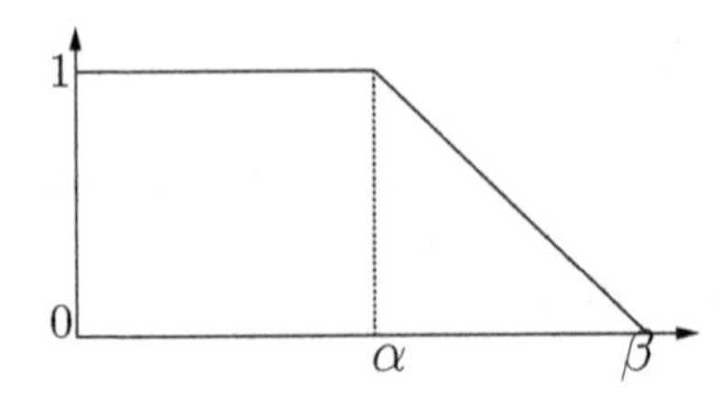

Fig. 2. Membership function of the fuzzy set of 'Acceptable' solutions. Parameter α is the reference cost and β is the 'limit' of acceptability

When no solution from the neighborhood has enough degree of acceptability, we consider the search as being trapped in a local optimum. In this situation, the operator used to construct solutions is changed, so solutions coming from different neighborhoods are explored next (as in variable neighborhood descent search [17]. This process is repeated once for each of a set of available operators until some finalization criterion for the local search is met.

Some readers may found similarities between the acceptance criterion of simulated annealing (at constant temperature) and the one use here. However, the former has a probabilistic nature while the later is deterministic.

3.2 Description of the Cooperation Process

A key aspect of our proposal is the use of "memory" in the *coordinator* to keep information of the behavior of the *solvers* and of the quality of the solutions obtained.

To fully characterize the cooperation process (as suggested in [3]) we need to describe: a) the information management strategy (what kind of information is stored and exchanged); b) how each thread use the information available; and what model of parallelism is implemented.

The Information Management Strategy

The information in our proposal flows in two ways: from the *solvers* to the *coordinator* and vice versa.

From solvers *to* coordinator*:*

Our *solver* agents are local search based heuristics, so we decided that the information sent by the *coordinator* to each *solver* agent is a new solution, i.e. a new point in the search space.

From coordinator *to* solvers*:*

Each $solver_i$ thread sends to the *coordinator*, a sample time or time stamp t, its current solution at that time s^t and the cost of $f(s^t)$. With this information, the *coordinator* calculates an improvement rate as:

$$\Delta_f = \frac{(f(s^t) - f(s^{t-1}))}{(time^t - time^{t-1})} \tag{1}$$

where $time^t - time^{t-1}$ represents the elapsed time between two consecutive reports.

Then, the values Δ_f and $f(s^t)$ are stored in fixed length "memories" M_{Δ_f}, M_f. The memories are kept sorted.

Now to decide if a *solver* agent is not performing fine, the *coordinator* compares the last report values of such *solver* against those stored in M_{Δ_f}, M_f using just the following fuzzy rule:

IF `the quality of the solution reported by` $solver_i$ `is` **low**
AND `the improvement rate of` $solver_i$ `is` **low**
THEN `send to` $solver_i$ `a new solution.`

The label *low* is defined as follows:

$$\mu(x, \alpha, \beta) = \begin{cases} 0 & \text{if } x > \beta \\ (\beta - x)/(\beta - \alpha) & \text{if } \alpha \leq x \leq \beta \\ 1 & \text{if } x < \alpha \end{cases}$$

where x is the percentile rank of a value (an improvement rate or a cost) in the samples stored in M_{Δ_f} or M_f respectively, and the other parameters were fixed to $\alpha = 10$ and $\beta = 20$. In short, what the rule says is: if the reported values by a *solver* are among the worse in the memories, then such *solver* should be changed in some way.

The *new solution* could be a completely new one, for example generated randomly, or could be the best one saw by the *coordinator* until time t. This last option is the one implemented in this work which, in other words, may be understood as a greedy coordination scheme.

How each *solver* use the Information Available?

When a *solver* receives a new solution from the *coordinator*, the local search process is restarted from that new point in the search space. So, as the strategy progresses, the *solvers* will concentrate around the most promising regions of the search space that will be sampled using different schemes (the ones defined by the *solver* threads itself). In this way the chance of finding better and better solutions is increased.

What model of parallelism is implemented?

The parallel strategy has been implemented using the Parallel Virtual Machine 3.4 (PVM) package [15]and C++ language. PVM provides a framework where parallel programs can be developed in an efficient manner; besides, its computing model is simple and accommodates a wide variety of application program structures. We implemented an asynchronous master-slaves model in which a master (the *coordinator*) is responsible for spawning, initialization, collection and display of results, while the slaves (the *solvers*) perform the computation involved (solving the optimization problem). Each slave or *solver* is spawned as a clone with different initialization parameters in a PVM task.

From a hardware point of view, the computational experiments have been performed on a cluster with 8 computers with Dual AMD Athlon processors at 2 GHz, 256 k cache memory, and 2 GBytes of RAM.

4 Computational Experiments

In previous work [9, 8], we showed that the coordination strategy proposed, obtained lower error values than the independent search strategy and sequential algorithms over a set of instances of a combinatorial optimization problem. This conclusion also holds when different types of instances and sizes were considered. Besides, the use of coordination made a better use of the resources available. Also, the results showed the importance of having a set of *solver* threads with balanced roles between intensification and diversification to solve the hardest problem instances.

The influence of number of *solver* threads never was analyzed in those former works, where the number of *solver* threads was fixed; so the goal of this experiment is to evaluate the performance of the strategy according to the number of *solver* threads, using the same adjusting parameters in each *solver* threads.

As in previous works, we perform experiments over instances of the knapsack problem. The mathematical formulation is as follows:

$$Max \sum_{j=1}^{n} p_j \times x_j$$

$$s.t. \sum_{j=1}^{n} w_j \times x_j \leq C,\ x_j \in \{0,1\}, j = 1, \ldots, n$$

where n is the number of items, x_j indicates if the item j is included or not in the knapsack, p_j is the profit associated with the item j, $w_j \in [0, .., r]$ is the weight of item j, and C is the capacity of the knapsack. It is also assumed $w_j < C$, $\forall j$ (every item fits in the knapsack); and $\sum_{j=1}^{n} w_j > C$ (the whole set of items does not fit).

The choice of knapsack as test bed is based on the fact that we can construct test instances of varying hardness following three characteristics: size of the instance, type of correlation between weights and profits, and range of values available for the weights.Although knapsack is believed to be one of the "easier" NP-hard problems, is still possible to construct "hard to solve" instances that are considered a challenge for most state of art exact algorithms [28, 19].

In this work, we deal with a set of instances Weakly Correlated, WC: $w_i = U(1, R), p_i = U(w_i - \frac{R}{10}, w_i + \frac{R}{10})$ with $p_i \geq 1$ generated and provided to us by Dr. David Pisinger.

The weights w_i are uniformly distributed in a given interval with data range $R = 10^4$. We considered three problem sizes, $n = 1000, 1500, 2000$ and we taken the 50 pair instances from a randomly generated a serie of $H = 100$ instances. We execute one times the algorithm for each instance in each size, while the number of *solver* threads was varying between 1 to 6. The capacity of each instance h (of every type and value of R), with $h = 1 \ldots H$ is calculated as $c = \frac{h}{1+H} \sum_{i=1}^{n} w_i$. All *solver* threads have the same "greedy" search behaviour. The optimum of every instance is known, it was also kindly provided and calculated using the best exact algorithm, specifically designed to deal with knapsack problems. As a consequence, to assess the quality of the solutions returned by the strategy, we record an *Error*:

$$Error = 100 * \frac{Optimum - Obtained\,Value}{Optimum}$$

We make two experiments. Firstly, we execute the strategy dividing the running time per number of threads. The running time is fixed to 60 seconds and made use of 6 *solver* threads, all with $\lambda = 0.95$, so the acceptability of solutions being worst than the reference one is low. The reason to use the very same definition for each thread is to confirm if the differences arise as a consequence of the cooperation strategy defined and are not a consequence of other factors like parameter settings for the threads.

In Table 1 we analyze the results in terms of error according to the number of threads in each size. The average is over 50 instances.

Table 1. Average values and standard deviation of error for different sizes and number of *solvers*. The time (60 seconds) is equally distributed between the threads available

Threads	1000	1500	2000
1	3,48 (3,26)	4,12 (4,14)	4,39 (4,51)
2	2,11 (1,96)	3,04 (2,88)	2,90 (2,96)
3	2,32 (2,21)	3,18 (3,09)	3,16 (3,37)
4	2,49 (2,45)	3,43 (3,39)	3,32 (3,37)
5	2,63 (2,60)	3,55 (3,60)	3,46 (3,53)
6	2,72 (2,54)	3,57 (3,60)	3,48 (3,79)

The best values are obtained with 2 *solvers* in every size tested. The difference on the average error between the use one thread and two or more has statistical significance.

It is also possible to verify that the error increases as the number of threads available increase. This is related with the fact that as the number of threads increases, a strategy with a greater exploration capacity is induced. However, the running time available is divided among the *solver* threads. In this way, the possibility of exploitation intensity in the search decreases when the number of *solver* threads grows.

The policy used to determine the running times available for each thread also affects the result. In this case, it is better to use few threads, letting them to run for more time.

It is also interesting to check how many instances were solved with $error \leq 1$. These values are shown in Table 2.

The results present a small variation according to the number of threads. It is clear that the use of more than one thread allowed to achieve a higher number of "well solved" instances. In particular, the use of 2 *solvers* in the smaller instances, obtained extremely good solutions on the 50% of instances tested. The highest number of "successes" was achieved with 2 *solvers* in all sizes.

Table 2. Number of instances solved with $error \leq 1$ .The algorithms were run once per each instance. The time (60 seconds) is equally distributed between the threads available

Clones	1000	1500	2000
1	15	14	14
2	24	18	19
3	21	16	19
4	21	15	17
5	18	16	16
6	17	16	16
Total	116	95	101

In the second experiment we give 15 seconds to each *solvers.* Just one run is made for each instance and we record here the best value obtained at 0, 3, 6, 9, 12, 15 seconds to better understanding the dynamic behavior of the system.

We analyze the results in terms of the number of *solvers* for each size using the values of data gathered every 3 seconds. (see Table 3).

Table 3. Average values and standard deviation of error for each size and different number of *solvers* and time = 15 seconds

Clones	1000	1500	2000
1	3,54 (3,31)	4,18 (4,29)	4,51 (4,63)
2	1,86 (1,80)	2,88 (2,95)	3,45 (3,52)
3	1,79 (1,71)	2,76 (2,78)	3,36 (3,38)
4	1,79 (1,78)	2,73 (2,70)	3,31 (3,41)
5	1,77 (1,72)	2,68 (2,69)	3,30 (3,45)
6	1,70 (1,73)	2,60 (2,72)	3,35 (3,51)

The differences on the average error are similar to last experiment. Again, the differences between the average error obtained with one thread and more than one were statistically signigicant. As in the previous experiment, the differences on the average error using 2 or more threads, are very slightly. However, in this case, as the number of threads increased, the average error decreased. This means that when the set of threads is allowed enough time to run, the coordination strategy starts to be profitable.

In Table 4 we show the number of instances that were solved with $error \leq 1$. Again, the differences are clear between the use of just one thread or more than one. When two or more threads are considered, the number of "well solved" instances does not vary too much. However, these values (obtained with a different time management policy) are slightly higher than those showed in Table 2.

Table 4. Number of instances solved with $error \leq 1$. The algorithms were run once per each instance

Threads	1000	1500	2000
1	14	16	13
2	26	19	15
3	25	19	16
4	27	19	17
5	25	20	16
6	27	19	17
Total	144	112	94

To conclude the analysis, we present in Fig. 3 the evolution of the average *error* during time for each size and different number of threads.

The error shows a continue improvement independent of the number of threads. It decreases quickly during the first 9-10 seconds and decreases slightly until the end of running time in all instances.

At every time step considered, the use of two or more coordinated threads obtained better values of error than those provided by one algorithm.

5 Discussion

We analyzed the behaviour of a cooperative multi-agent based strategy, where each *solver* is modeled as an optimization algorithm, executed asynchronously, and controlled by a *coordinator*.

It is clear that as the number of threads increases, a strategy with a greater exploration capacity is induced. However, the results could be improved, but the same definition (greedy) of the all *solvers* do not permit balanced roles between intensification and diversification in the strategy. The greedy definition

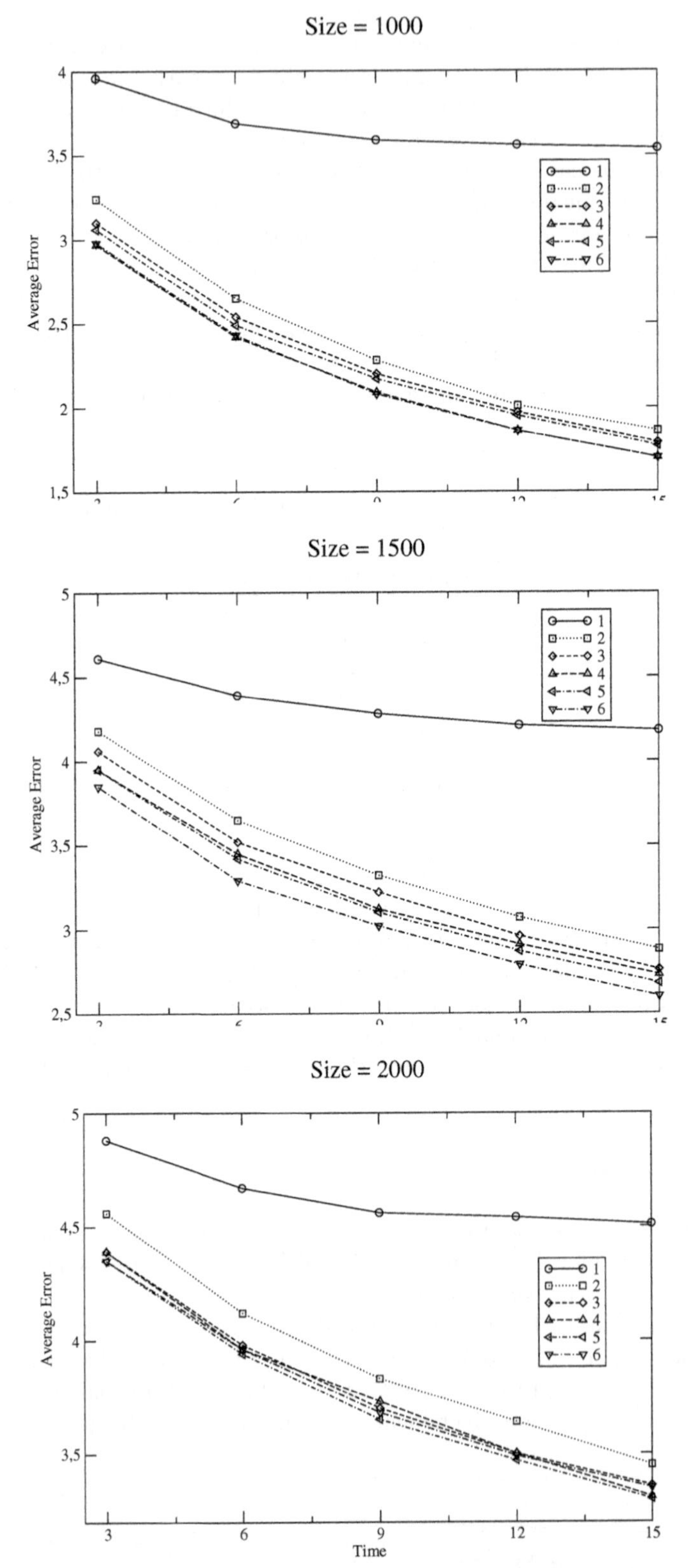

Fig. 3. Evolution of the average *error* during time for each size and number of threads

of the *solver*, (inducing a great intensification capacity), is more important than the diversification achieved with an increase number of *solver*.

More research is needed in order to understand the balance between the quality of the *solvers* and the cooperation mechanism defined. In this context, the use of easy to implement, modify and manipulate *solvers*, like *FANS*, may help to cope with the inherent complexity of the task to be done.

The improvement of the fuzzy rule base is another topic to address. For example, incorporating rules to model fuzzy stopping criteria [32] or finding ways to dynamically adjust the parameters of the base.

The type of information sent from the *coordinator* to the *solvers* should be explored for alternatives. In this work, we changed the reference solution, but the behavior of *solvers* can be also modified through the change of certain parameters that can, in turn, modify the search trajectories of the optimizers.

Acknowledgments

This work was partiall y supported by Projects TIC-2002-04242-C03-02 and SINTA-CC-TIN 2004-01411.

References

1. Enrique Alba. *New Advances on Parallel Meta-Heuristics for Complex Problems*, volume 10 of *Special Issue of the Journal of Heuristics*. Kluwer Academic Publishers, 2004.
2. A. Blanco, D. Pelta, and J.L. Verdegay. A fuzzy valuation-based local search framework for combinatorial problems. *Journal of Fuzzy Optimization and Decision Making*, 1(2):177–193, 2002.
3. Alexandre Le Bouthillier and Teodor Gabriel Crainic. A cooperative parallel meta-heuristic for the vehicle routing problem with time windows. *Computers and Operations Research*, 2003. In press. Available at www.sciencedirect.com.
4. W. Bozejko and M. Wodecki. *Artificial Intelligence and Soft Computing ICAISC 2004*, volume 3070 of *Lecture Notes in Artificial Intelligence*, chapter The New Concepts in Parallel Simulated Annealing Methods, pages 853–859. Springer Verlag, Berlin, 2004.
5. K. Chung, Y. Deng, and J. Reinitz. Parallel simulated annealing algorithms by mixing states. *Journal of Computational Physics*, 148:646–662, 1999.
6. T. Crainic, M. Gendreau, P. Hansen, and N. Mladenovic. Cooperative parallel variable neighborhood search for the p-median. *Journal of Heuristics*, 10(3):293–314, 2004.
7. Teodor G. Crainic and Michel Toulouse. *Parallel Strategies for Metaheuristics*, volume Handbook of Metaheuristics, pages 475 – 513. Kluwer Academic Publisher, 2003.
8. Carlos Cruz, Alejandro Sancho-Royo, and David Pelta Jos L. Verdegay. Analizing the behaviour of a multi-agents based cooperative, parallel strategy for

combinatorial optimization. In Ahmad Lotfi, editor, *Proceedings of the 5th International Conference in Recent Advances in Soft Computing*, pages 512–517, 2004.
9. Carlos Cruz, Alejandro Sancho-Royo, and David Pelta Jos L. Verdegay. A soft computing based cooperative multi-thread strategy. *Information Science*, 2005. Submitted.
10. Van-Dat Cung, Simone L. Martins, Celso Ribeiro, and Catherine Roucairol. *Essays and Surveys in Metaheuristics*, chapter Strategies for the Parallel Implementation of Metaheuristics, pages 263–308. Kluwer Academic Publisher, 2001.
11. M. Dorigo, G. Di Caro, and L. Gambardella. Ant algorithms for discrete optimization. *Artificial Life*, 5:137, 172, 1999.
12. M. Dorigo and T Stützle. *Handbook of Metaheuristics*, chapter Ant Colony Optimization Metaheuristic. Kluwer Academic, 2003.
13. R. Eberhart and J. Kennedy, editors. *Swarm Intelligence.* Academic Press, 2001.
14. R. Eberhart and Y. Shi. Particle swarm optimization: developments, applications and resources. In *Proc. congress on evolutionary computation 2001.* IEEE service center, Piscataway, NJ, 2001.
15. Al Geist, Adam Beguelin, Jack Dongarra, Weicheng Jiang, Robert Manchek, and Vaidy Sunderam. *PVM: Parallel Virtual Machine. A user's guide and Tutorial for Networked Parallel Computing.* The MIT Press, Cambridge, Massachusetts, England, 1994.
16. Fred Glover and Gary A. Kochenberger, editors. *Handbook of Metaheuristics.* Kluwer Academic, 2003.
17. P. Hansen and N. Mladenović. An introduction to variable neighborhood search. In S. Voss, S. Martello, I. Osman, and C. Roucairol, editors, *Metaheuristics: Advances and Trends in Local Search Procedures for Optimization*, pages 433–458. Kluwer, 1999.
18. W. Hart, N. Krasnogor, and J. Smith, editors. *Recent Advances in Memetic Algorithms.* Studies in Fuzziness and Soft Computing. Physica-Verlag, 2004.
19. Hans Kellerer, Ulrich Pferschy, and David Pisinger. *Knapsack Problems.* Springer Verlag, 2004.
20. N. Krasnogor. Self-generating metaheuristics in bioinformatics: The proteins structure comparison case. *Genetic Programming and Evolvable Machines*, 5(2), 2004.
21. N. Krasnogor, B.P Blackburne, E. Burke, and J. Hirst. Multimeme algorithms for protein structure prediction. In *Proceedings of Parallel Problem Solving from Nature, PPSN'02*, 2002.
22. Natalio Krasnogor and David Pelta. *Fuzzy Sets based Heuristics for Optimization*, chapter Fuzzy Memes in Multimeme Algorithms: a Fuzzy-Evolutionary Hybrid. Studies in Fuzziness and Soft Computing. Physica-Verlag, 2003.
23. M.G. Lagoudakis and M.L. Littman. Algorithm selection using reinforcement learning. In *Proc. 17th International Conf. on Machine Learning*, pages 511–518. Morgan Kaufmann, San Francisco, 2000.
24. M.G. Lagoudakis, M.L. Littman, and R. Parr. Selecting the right algorithm. In *Proc. of the 2001 AAAI Fall Symposium Series: Using Uncertainty within Computation*, 2001.
25. Panos Pardalos and Mauricio Resende, editors. *Handbook of Applied Optimization.* Oxford University Press, 2002.

26. D. Pelta and N. Krasnogor. *Recent Advances in Memetic Algorithms*, chapter Multimeme Algorithms using fuzzy logic based memes for protein structure prediction. Studies in Fuzziness and Soft Computing. Physica-Verlag, 2004.
27. David Pelta, Armando Blanco, and Jose L. Verdegay. Applying a fuzzy sets-based heuristic for the protein structure prediction problem. *International Journal of Intelligent Systems*, 17(7):629–643, 2002.
28. David Pisinger. Where are the hard knapsack problems? Technical report, University of Copenhagen, Denmark, 2003.
29. J. Prez, R. Pazos, J. Frausto, G. Rodrguez, D. Romero, and L. Cruz. *WEA 2004*, volume 3059 of *Lecture Notes in Computer Science*, chapter A Statistical Approach for Algorithm Selection, pages 417–431. Springer Verlag, 2004.
30. S. Talukdar. Collaboration rules for autonomous software agents. *International Journal of Decision Support Systems*, 24:269–278, 1999.
31. S. Talukdar, L. Baerentzen, A. Gove, and P. de Souza. Asynchronous teams: cooperation schemes for autonomous agents. *Journal of Heuristics*, 4:295–321, 1998.
32. J.L. Verdegay and E. Vergara-Moreno. Fuzzy termination criteria in knapsack problem algorithms. *MathWare and Soft Computing*, 7(2-3):89–97, 2000.
33. Jose Luis Verdegay, editor. *Fuzzy Sets based Heuristics for Optimization*. Studies in Fuzziness and Soft Computing. Physica-Verlag, 2003.
34. D.H. Wolpert and W.G. Macready. No free lunch theorems for optimization. *IEEE Transactions on Evolutionary Computation*, 1(1):67–82, 1997.

Solving a Fuzzy Nonlinear Optimization Problem by an "ad hoc" Multi-objective Evolutionary Algorithm

F. Jiménez[1], G. Sánchez[1], J.M. Cadenas[1], A. Gómez-Skarmeta[1], and J.L. Verdegay[2]

[1] Dept. Ingeniería de la Información y las Comunicaciones. Facultad de Informática. Universidad de Murcia.
fernan@dif.um.es gracia@um.es jcadenas@um.es skarmeta@um.es
[2] Dept. Ciencias de la Computación e Inteligencia Artificial. E.T.S.I.I.. Universidad de Granada.
verdegay@decsai.ugr.es

Summary. A fuzzy optimization problem arising in some import-export companies in the south of Spain is presented. In Fuzzy Optimization is desirable that fuzzy solutions can be really attained because then the decision maker will be able of making a decision "a posteriori" according to the current decision environment. In this way, no more runs of the optimization technique are needed when decision environment changes or when decision maker requires check out several decisions in order to establish the more appropriates. Multi-objective optimization can obtain the solution of a fuzzy optimization problem, since capturing the Pareto front we are able composing the solution fuzzy. Multi-objective Evolutionary algorithms have been shown in the last few years as powerful techniques to solve multi-objective optimization problems because they can search for multiple Pareto solutions in a single run of the algorithm. In this contribution we first introduce a multi-objective approach for nonlinear constrained optimization problems with fuzzy costs, and then an "ad hoc" multi-objective evolutionary algorithm to solve the former problem. A case-study of a fuzzy optimization problem is analyzed and the proposed solutions from the evolutionary algorithm here considered are shown.

Key words: Nonlinear optimization; multi-objective optimization; evaluationary algorithms; Pareto-based multi-objective evaluationary algorithm

1 Introduction

In exporting-importing companies in the South of Spain [8], optimization problems arise in which managers describe some data with uncertainty. Due to this uncertainty, managers prefer to have not just one solution but a set of

F. Jiménez et al.: *Solving a Fuzzy Nonlinear Optimization Problem by an "ad hoc" Multi-objective Evolutionary Algorithm*, StudFuzz **201**, 521–533 (2006)
www.springerlink.com

them, so that the most suitable solution can be applied according to the state of existing decision of the production process at a given time and without increasing delay. In these situations fuzzy optimization is an ideal methodology, since, on the one hand, it allows us to represent the underlying uncertainty of the optimization problem, and on the other, to find optimal solutions that reflect the uncertainty of the problem and then to apply to its possible instances once the uncertainty has been solved. In such a situation a model of the behavior of the solutions based on the uncertainty of the optimization problem is obtained. A brief description of the pointed problem is the following:

A number n of products for export are to be produced using m different processes. The production of one unit of product x_i $(i = 1, \ldots, n)$ requires, approximately, a_{ij} minutes of processing time in the j department $(j = 1, \ldots, m)$. The total time available for each production process is b_j minutes for j department. When sold abroad, product x_i yields an uncertain profit per unit and an uncertainty discount per unit, here uncertainty is produced by money change from currencies and so profit and discount can be modelled by means of fuzzy costs. The managers want to maximize the benefits.

In terms of Mathematical Programming, the above problem can be stated as follows:

$$\begin{array}{ll} max \displaystyle\sum_{i=1}^{n} \widetilde{cost_i} x_i - \sum_{i=1}^{n} disc(x_i, \widetilde{dcost_i}) & \\ s.t.: & \\ \qquad \displaystyle\sum_{i=1}^{n} appr(a_{ij}, x_i) \le b_j, \; j = 1, \ldots, m & \end{array} \tag{1}$$

where $disc(x_i, \widetilde{dcost_i})$ is an increasing function which gives the required discount according to the amount of production and $appr(a_{ij}, x_i)$ is a decreasing function which gives approximately minutes of processing time in the j department.

Fuzzy optimization problems have been extensively studied since the seventies. In the linear case, the first approaches to solve the so-called fuzzy linear programming problem were made in [15] and [18]. Since then, important contributions solving different linear models have been made and these models have been the subjects of a substantial amount of work. In the nonlinear case, [1, 6, 14], the situation is quite different, as there is a wide variety of specific and both practically and theoretically relevant nonlinear problems with each having a different solution method.

In the following we consider a *Nonlinear Programming* problem with fuzzy costs. From a mathematical point of view the problem can be addressed as:

$$\begin{array}{l} max \; f(\mathbf{x}, \widetilde{\mathbf{c}}) \\ s.t.: \\ \qquad g_j(\mathbf{x}) \le b_j, \; j = 1, \ldots, m \end{array} \tag{2}$$

where $\mathbf{x} = (x_1, \ldots, x_n) \in \mathbb{R}^n$ is a n dimensional real-valued parameter vector, with $x_i \in [l_i, u_i] \subset \mathbb{R}$, $l_i \geq 0$, $i = 1, \ldots, n$, $b_j \in \mathbb{R}$, $g_j(\mathbf{x})$ are arbitrary functions, and $f(\mathbf{x}, \widetilde{\mathbf{c}})$ is an arbitrary function with fuzzy costs $\widetilde{\mathbf{c}} = (\widetilde{c_1}, \ldots, \widetilde{c_n})$. We assume fuzzy costs characterized by membership functions of the following form:

$$\mu_i(v) = \begin{cases} 0 & \text{if } v \leq r_i \text{ } or \text{ } v \geq R_i \\ h_{1i}\left(\frac{v - r_i}{\underline{c}_i - r_i}\right) & \text{if } r_i \leq v \leq \underline{c}_i \\ h_{2i}\left(\frac{R_i - v}{R_i - \overline{c}_i}\right) & \text{if } \overline{c}_i \leq v \leq R_i \\ 1 & \text{if } \underline{c}_i \leq v \leq \overline{c}_i \end{cases} \tag{3}$$

where h_{1i} and h_{2i} are assumed to be strictly increasing and decreasing continuous functions, respectively, such that $h_{1i}(\cdot) = h_{2i}(\cdot) = 1$, $\forall i = 1, \ldots, n$.

The membership function of the fuzzy objective is defined in [5] as follows:

$$\mu(\mathbf{c}) = \inf_i \mu_i(c_i), \mathbf{c} = (c_1, \ldots, c_n) \in \mathbb{R}^n, i = 1, \ldots, n$$

and the following fuzzy relation is induced $\forall \mathbf{x}, \mathbf{y} \in X$:

$$\mu(\mathbf{x}, \mathbf{y}) = sup\{\alpha / f(\mathbf{x}, \widetilde{\mathbf{c}}) \geq f(\mathbf{y}, \widetilde{\mathbf{c}}), \forall c \in \mathbb{R}^n : \mu(\mathbf{c}) \geq 1 - \alpha\}$$

with $X = \{\mathbf{x} \in \mathbb{R}^n / g_j(\mathbf{x}) \leq b_j, x_i \in [l_i, u_i], l_i \geq 0\}$, $\alpha \in [0, 1]$.

This relation is a fuzzy preorder as shown in [5] and the solution of (2) can be obtained by solving the following parametric problem:

$$\begin{aligned} & max \ f(\mathbf{x}, \mathbf{c}) \\ & s.t.: \\ & \qquad g_j(\mathbf{x}) \leq b_j, \ j = 1, \ldots, m \\ & \qquad \mu(\mathbf{c}) \geq 1 - \alpha, \quad \mathbf{c}, \mathbf{x} \in \mathbb{R}^n, \alpha \in [0, 1] \end{aligned} \tag{4}$$

Then by using the results obtained in [5, 17], the problem (4) can be transformed into a parametric interval programming problem as follows:

$$\begin{aligned} & max \ f(\mathbf{x}, \mathbf{I}(\alpha)) \\ & s.t.: \\ & \qquad g_j(\mathbf{x}) \leq b_j, \ j = 1, \ldots, m \end{aligned} \tag{5}$$

with $\mathbf{I}(\alpha) = (I_1(\alpha), \ldots, I_n(\alpha))$, $I_i(\alpha) = [h_{1i}^{-1}(1 - \alpha), h_{2i}^{-1}(1 - \alpha)]$, $i = 1, \ldots, n$, where h_{1i}^{-1} and h_{2i}^{-1} are the inverse functions of h_{1i} and h_{2i} respectively.

Solution of the problem (5), fixed $\alpha \in [0, 1]$, is composed by the set of efficient points.

$\mathbf{x}^* \in X$ is said to be an efficient point of (5) iff $\nexists \mathbf{x} \in X \ / \ f(\mathbf{x}, \mathbf{c}) \geq f(\mathbf{x}^*, \mathbf{c}) \ \forall \mathbf{c} \in \mathbf{I}(\alpha)$, and $\exists \mathbf{c} \in \mathbf{I}(\alpha) \ / \ f(\mathbf{x}, \mathbf{c}) \neq f(\mathbf{x}^*, \mathbf{c})$.

According to the Representation Theorem for fuzzy sets, the fuzzy solution of the problem (2) is

$$\widetilde{S} = \bigcup_{\alpha} \alpha \cdot S(1-\alpha)$$

where

$$S(1-\alpha) = \{\mathbf{x}^* \in X \ / \ \forall \mathbf{c} \in \mathbf{I}(\alpha), \ \mathbf{x} \in X \Leftrightarrow f(\mathbf{x}^*, \mathbf{c}) \geq f(\mathbf{x}, \mathbf{c})\}$$

Unfortunately, there are no much general-oriented solution methods to solve nonlinear parametric programming problems in the literature, although it deserves to mention the cases of linear programming problems in which data are continuously varied as a linear function of a single parameter. Therefore in order to theoretically solve (2) we shall try to find an approximate solution. It is patent that *Evolutionary Algorithms* (EA) [7] could be used to solve fuzzy nonlinear programming problems like the above one because of EA are solution methods potentially able of solving general nonlinear programming problems or, at least, of approaching theoretic solution ways that, each case, are to be specified according to the concrete problem to be solved. An evolutionary-parametric based approach to solve fuzzy transportation problems have been proposed in [9]. In [11], a fuzzy problem with fuzzy constraints is solved for a finite set of values of the parameter α by means of an EA for constrained nonlinear optimization problems. Final solution is constructed with numerical approximation techniques. The main disadvantage of this approach arises in the need for run an EA for each value of the parameter α. Moreover, numerical approximation does not ensures the feasibility of solutions. In [13], a multi-objective approach to solve nonlinear optimization problems with fuzzy constraints is described. In this paper we propose a multi-objective evolutionary approach to solve (2).

With this background, the paper have been organized as follows: in sect. 2 a multi-objective technique for fuzzy programming problems is approached and describes an ad-hoc Pareto-based multi-objective EA to solve the multi-objective problems connected with the fuzzy programming problems, in sect. 3 a nonlinear fuzzy problem is considered as case study and results of experiments are shown. Finally sect. 4 indicates the main conclusions.

2 A Multi-objective Approach and a Pareto-based Multi-objective Evolutionary Algorithm

We propose a multi-objective approach to solve the problem (2). The problem (5) can be transformed into a multi-objective nonlinear programming problem in which the parameter α is treated as a new decision variable. Besides the decision variable α, we also consider n new decision variables β_i, $i = 1, \ldots, n$

to transform the intervals $I_i(\alpha) = [h_{1i}^{-1}(1-\alpha), h_{2i}^{-1}(1-\alpha)]$ into functions of the form $z_i(\alpha, \beta_i) = h_{1i}^{-1}(1-\alpha) + \beta_i(h_{2i}^{-1}(1-\alpha) - h_{1i}^{-1}(1-\alpha))$.

Solution to (5) is composed by the solutions with maximum values for $f(\mathbf{x}, \mathbf{z}(\alpha, \boldsymbol{\beta}))$ for each value of the parameters α, β_i, $i = 1, \ldots, n$, with $\boldsymbol{\beta} = (\beta_1, \ldots, \beta_n)$. Then, by maximizing $f(\mathbf{x}, \mathbf{z}(\alpha, \boldsymbol{\beta}))$, and by maximizing and minimizing α, β_i simultaneously, we obtain a set of non-dominated solutions which represents the solution of (5).

The multi-objective problem is stated as follows:

$$\begin{array}{l} max\ f(\mathbf{x}, \mathbf{z}(\alpha, \boldsymbol{\beta})), \alpha, 1-\alpha, \beta_i, 1-\beta_i,\ i = 1, \ldots, n \\ s.t.: \\ \qquad g_j(\mathbf{x}) \leq b_j,\ j = 1, \ldots, m \end{array} \tag{6}$$

where $\mathbf{z}(\alpha, \boldsymbol{\beta}) = (z_1(\alpha, \beta_1), z_2(\alpha, \beta_2), \ldots, z_n(\alpha, \beta_n))$; $z_i(\alpha, \beta_i) = h_{1i}^{-1}(1-\alpha) + \beta_i(h_{2i}^{-1}(1-\alpha) - h_{1i}^{-1}(1-\alpha))$, $\alpha, \beta_i \in [0, 1]$, $i =, 1, \ldots, n$.

Multi-objective Pareto-based EA [3, 4, 10] are specially appropriated to solve multi-objective nonlinear optimization problems because they can capture a set of Pareto solutions in a single run of the algorithm. We propose an ad hoc multi-objective Pareto-based EA to solve the problem (6) with the following characteristics:

- Pareto-based multi-objective EA; it finds, in a single run, multiple non-dominated solutions.
- The EA has a real-coded representation. Each individual of a population contains $2n+1$ real parameters to represent the solution $(x_1, \ldots, x_n, \alpha, \beta_1, \ldots, \beta_n)$.
- The initial population is generated randomly with a uniform distribution within the boundaries of the search space $x_i \in [l_i, u_i]$, $\alpha, \beta_i \in [0, 1]$, $i = 1, \ldots, n$.
- The variation operators act on real numbers. It has been used two cross types, *uniform cross* and *arithmetical cross*, and three types of mutation, *uniform mutation, non-uniform mutation, minimal mutation* [10].
- Diversity among individuals is maintained by using an ad-hoc elitist generational replacement technique.
- It uses the $min - max$ formulation to handle constrains.

2.1 Constraint Handling

The populations generated by the algorithm are made up of both feasible and unfeasible individuals. Guided by the multi-objective optimization Pareto concept, the feasible individuals evolve towards optimality, while the non-feasible individuals evolve towards feasibility guided by an evaluation function based on the *min-max* formulation. See below for details. Thus the resulting algorithm is weakly dependent on the problem to be optimized since it is the evolutionary heuristics itself that is used to satisfy the constrains, unlike the repair, decoding or penalty techniques which tend to be heavily dependent on the problem.

2.2 Variation Operators

Bearing in mind that the EA uses a floating point representation and given the coexistence of feasible and unfeasible individuals within the EA populations, the variation operators therefore act on chains (sequences) of real numbers without any consideration regarding the feasibility of new descendants. After experimenting for real parameter optimization with different variation operators proposed in the literature and with others, it was finally decided to use two cross types, *uniform cross* and *arithmetical cross*, and three types of mutation, *uniform mutation*, *non-uniform mutation* and *minimal mutation*. The first four have been studied and described in depth by other authors [12]. Minimal mutation causes a minimal change in the descendant as compared to the father, and it is especially appropriate in fine tuning real parameters [10].

2.3 Generating a New Population

The algorithm performs the following steps in the generation of a new population:

1. Two random individuals are selected.
2. Two offspring are obtained by parent crossing, mutation and repair.
3. The offspring are inserted into the population.

The insertion of the offspring is the fundamental point for maintaining diversity. We use an ad hoc technique for insertion. Objectives space is distributed into $D = N$ slots, where N is the population size. We use $N = (nslots + 1)^{n+1}$ where $nslots$ is the number of slots (given by user) for each decision variable $\alpha, \beta_i \in [0, 1]$, $i = 1, \ldots, n$. The insertion of an individual $(x_1, \ldots, x_n, \alpha, \beta_1, \ldots, \beta_n)$ is performed as follows:

- Calculate the slot t the individual belongs to according with the following expression $t = \lceil \alpha D \rceil + \sum_{i=1}^{n} \lceil \beta_i D^{i+1} \rceil$.
- If individual is better than some individual in slot t, then replace the worse individual in slot t by the new individual.

In order to determine if an individual is better than another, the following criteria are established:

- A feasible individual is better than another unfeasible one.
- One unfeasible individual $\mathbf{x}$ is better than another one $\mathbf{x}'$ if its function:

$$\max_{j=1,\ldots,m} \{g_j(\mathbf{x}) - b_j\} \leq \max_{j=1,\ldots,m} \{g_j(\mathbf{x}') - b_j\}$$

  is smaller.

- One feasible individual is better than another one if the first dominates the second.

It should be observed that we are using the *min-max* formulation to satisfy the constrains. This method has been used in multi-objective optimization [2] to minimize the relative deviations of each objective function from its individual optimum, and the best compromise solution can be obtained when objectives of equal priority are optimized. Since constrains and objectives can be treated in a similar way, and it is assumed that all constrains have equal priority, the *min-max* formulation is appropriate for satisfying constrains and is, furthermore, a technique which is independent of the problem.

It should also be noted that insertion of the new individuals is not always carried out, but only in those cases in which the new individual is better than the individual replaced and the diversity is not worsened in any case. Thus the technique simultaneously permits optimization and conservation of the diversity. It is also an elitist technique, since an individual is only replaced by another individual which is better than itself.

3 Experiments and Results: Exporting Company

3.1 A Instance of a Problem of an Exporting Company

In this section we set out a nonlinear fuzzy optimization problem as case study which describes a possible situation in a exporting company. The problem is the following:

Two products for export A and B are to be produced by utilizing three different processes (p_1, p_2 and p_3). The production of units of product A (B) requires $10x_1^{0.975}$ ($6x_2^{0.975}$) minutes of processing time in the p_1 department, 5 (10) minutes in the p_2 department, and $7x_1^{0.95}$ ($10x_2^{0.95}$) minutes in the p_3 department, where x_1 and x_2 are the units of product A and B, respectively. The total time available for each production process is 2500 minutes for p_1, 2000 minutes for p_2 and 2050 minutes for p_3. When sold abroad, product A (B) yields a profit of 23 (32) per unit, although it is made a discount increasing of 4 (3) cent from each order. The manager wants to maximize the benefit.

The problem can be modeled as follows:

$$
\begin{aligned}
&Max\ \widetilde{23}x_1 + \widetilde{32}x_2 - \widetilde{0.04}x_1^2 - \widetilde{0.03}x_2^2 \\
&s.t.: \\
&\qquad 10x_1^{0.975} + 6x_2^{0.975} \leq 2500 \\
&\qquad 5x_1 + 10x_2 \leq 2000 \\
&\qquad 7x_1^{0.95} + 10x_2^{0.95} \leq 2050 \\
&\qquad x_i \geq 0, i = 1, 2
\end{aligned}
$$

The currencies change among the countries produces uncertainty in the costs which can be modeled with the following membership functions:

$$\mu_{23}(v) = \begin{cases} 0 & \text{if } v < 20.8 \text{ or } v > 25.3 \\ (\frac{v-20.8}{2.2})^{1/2} & \text{if } 20.8 \le v \le 23 \\ (\frac{25.3-v}{2.3})^2 & \text{if } 23 \le v \le 25.3 \end{cases}$$

$$\mu_{32}(v) = \begin{cases} 0 & \text{if } v < 28.04 \text{ or } v > 35.2 \\ (\frac{v-28.94}{3.06})^{1/2} & \text{if } 28.94 \le v \le 32 \\ (\frac{35.2-v}{3.2})^2 & \text{if } 32 \le v \le 35.2 \end{cases}$$

$$\mu_{0.04}(v) = \begin{cases} 0 & \text{if } v < 0.0362 \text{ or } v > 0.044 \\ (\frac{v-0.0362}{0.0038})^{1/2} & \text{if } 0.0362 \le v \le 0.04 \\ (\frac{0.044-v}{0.004})^2 & \text{if } 0.04 \le v \le 0.044 \end{cases}$$

$$\mu_{0.03}(v) = \begin{cases} 0 & \text{if } v < 0.0271 \text{ or } v > 0.033 \\ (\frac{v-0.0271}{0.0029})^{1/2} & \text{if } 0.0271 \le v \le 0.03 \\ (\frac{0.033-v}{0.003})^2 & \text{if } 0.03 \le v \le 0.033 \end{cases}$$

Figure 1 show the fuzzy sets defining extreme values for any crisp solution of the problem.

The problem can be transformed into the following parametric interval programming problem:

$$\begin{array}{ll} Max & \left[20.8 + 2.2(1-\alpha)^2, 25.3 - 2.3\sqrt{1-\alpha}\right] x_1 + \\ & \left[28.94 + 3.06(1-\alpha)^2, 35.2 - 3.2\sqrt{1-\alpha}\right] x_2 - \\ & \left[0.0362 + 0.0038(1-\alpha)^2, 0.044 - 0.004\sqrt{1-\alpha}\right] x_1^2 - \\ & \left[0.0271 + 0.0029(1-\alpha)^2, 0.033 - 0.003\sqrt{1-\alpha}\right] x_2^2 \\ s.t.: & \\ & 10x_1^{0.975} + 6x_2^{0.975} \le 2500 \\ & 5x_1 + 10x_2 \le 2000 \\ & 7x_1^{0.95} + 10x_2^{0.95} \le 2050 \\ & x_1, x_2 \ge 0, \alpha \in [0,1] \end{array}$$

In order to solve the nonlinear parametric programming problem we consider the following multi-objective nonlinear optimization problem according to (6):

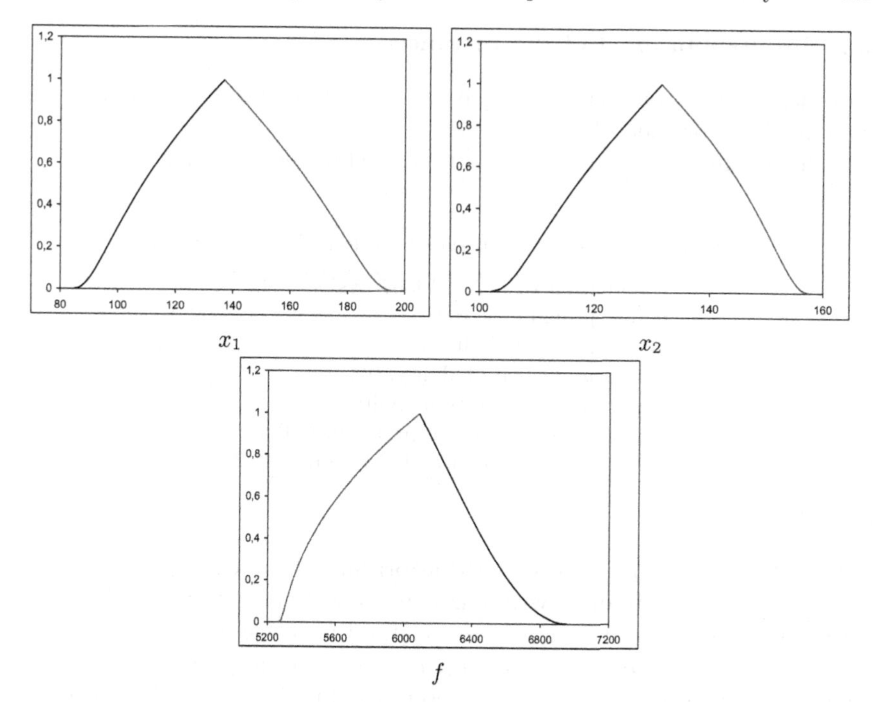

Fig. 1. Fuzzy solutions

$$
\begin{aligned}
&Max\ f_1 = [(20.8 + 2.2(1 - x_3)^2) + x_4(4.5 - 2.3\sqrt{1 - x_3} - 2.2(1 - x_3)^2)]x_1 + \\
&\quad [(28.94 + 3.06(1 - x_3)^2) + x_5(6.26 - 3.2\sqrt{1 - x_3} - 3.06(1 - x_3)^2)]x_2 - \\
&\quad \left[(0.0362 + 0.0038(1 - x_3)^2) + x_6(0.0078 - 0.004\sqrt{1 - x_3} - 0.0038(1 - x_3)^2)\right] x_1^2 - \\
&\quad \left[(0.0271 + 0.0029(1 - x_3)^2) + x_7(0.0059 - 0.003\sqrt{1 - x_3} - 0.0029(1 - x_3)^2)\right] x_2^2 \\
&Max\ f_2 = x_3 \\
&Max\ f_3 = 1 - x_3 \\
&Max\ f_4 = x_4 \\
&Max\ f_5 = 1 - x_4 \\
&Max\ f_6 = x_5 \\
&Max\ f_7 = 1 - x_5 \\
&Max\ f_8 = x_6 \\
&Max\ f_9 = 1 - x_6 \\
&Max\ f_{10} = x_7 \\
&Max\ f_{11} = 1 - x_7 \\
&s.t.: \\
&\quad 10x_1^{0.975} + 6x_2^{0.975} \leq 2500 \\
&\quad 5x_1 + 10x_2 \leq 2000 \\
&\quad 7x_1^{0.95} + 10x_2^{0.95} \leq 2050 \\
&\quad x_1, x_2 \geq 0, 0 \leq x_3, x_4, x_5, x_6, x_7 \leq 1
\end{aligned}
\tag{7}
$$

3.2 Results Obtained of the Instance

In order to check out our technique, 10 runs of the evolutionary algorithm were made on problem detailed in (7).

The parameters given in Table 1 were used in the executions.

Table 1. Parameters in the execution of the algorithm

Parameters
Population size (N): 16807 ($nslots = 6$)
Cross probability: 0.9
Mutation probability: 0.2
Uniform cross probability: 0.3
Uniform mutation probability: 0.1
Non uniform mutation probability: 0.4
Parameter c for non uniform mutation: 2.0

The best results obtained with the algorithm are shown in Fig. 2. We compare the solutions obtained with our multi-objective evolutionary algorithm with solutions obtained by a gradient method for constant values of the parameters α, β_1, β_2, β_3 and β_4 (x_3, x_4, x_5, x_6 and x_7 in problem (7)) showed in Table 2 and graphically in Fig. 2 (note that for constants values of x_3, x_4, x_5, x_6 and x_7 the problem is single-objective). It can be observed that non dominated points are obtained by the multi-objective evolutionary algorithm evenly distributed in the whole Pareto optimal front.

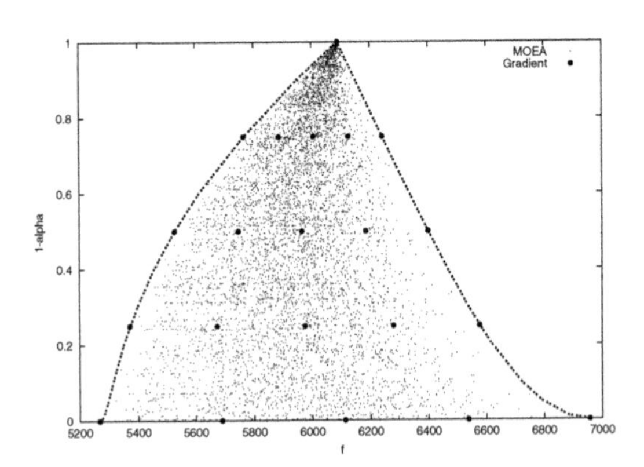

Fig. 2. Non dominated points obtained for the problem (7)

Various metrics for both convergence and diversity of the populations obtained have been proposed for a more exact evaluation of the effectiveness of the evolutionary algorithms. In his book, Deb [4] assembles a wide range of the metrics which figure in the literature. For this paper we propose the

Table 2. Results obtained with a gradient method for the problem (7)

x_1	x_2	x_3	x_4	x_5	x_6	x_7	f_1
123,732	138,134	1	0	0	1	1	5267,928
127,036	136,482	0,75	0	0	1	1	5371,144
129,606	135,197	0,5	0	0	1	1	5527,119
132,823	133,588	0,25	0	0	1	1	5765,396
136,842	131,579	0	0	0	1	1	6089,474
138,519	130,741	1	0,1	0,2	0	0,43	5689,670
138,011	130,994	0,75	0,1	0,2	0	0,43	5673,762
137,666	131,167	0,5	0,1	0,2	0	0,43	5746,888
137,277	131,361	0,25	0,1	0,2	0	0,43	5886,181
136,842	131,579	0	0,1	0,2	0	0,43	6089,474
138,253	130,873	1	0,5	0,5	0,435	0,5	6112,376
137,854	131,073	0,75	0,5	0,5	0,435	0,5	5975,567
137,569	131,216	0,5	0,5	0,5	0,435	0,5	5965,202
137,234	131,383	0,25	0,5	0,5	0,435	0,5	6005,566
136,842	131,579	0	0,5	0,5	0,435	0,5	6089,474
134,994	132,503	1	0,6	0,9	0,2	0,5	6537,814
135,492	132,254	0,75	0,6	0,9	0,2	0,5	6279,237
135,868	132,066	0,5	0,6	0,9	0,2	0,5	6184,834
136,317	131,841	0,25	0,6	0,9	0,2	0,5	6125,658
136,842	131,579	0	0,6	0,9	0,2	0,5	6089,474
152,647	123,677	1	1	1	0	0	6957,364
148,022	125,989	0,75	1	1	0	0	6576,556
144,764	127,618	0,5	1	1	0	0	6398,774
141,036	129,482	0,25	1	1	0	0	6241,949
136,842	131,579	0	1	1	0	0	6089,474

use of two metrics to evaluate the goodness of the algorithm. The first metric, the generational distance (Υ) proposed by Veldhuizen [16] evaluates the proximity of the population to the Pareto optimal front by calculating the average distance of the population Q from an ideal population P^* made up of solutions distributed uniformly along the Pareto front. This metric is shown in the following expression:

$$\Upsilon = \frac{\left(\sum_{i=1}^{|Q|} d_i^p\right)^{1/p}}{|Q|}$$

For $p = 2$, parameter d_i is the Euclidean distance (in the objective space) between the solution $i \in Q$ and the nearest solution in P^*:

$$d_i = \min_{k=1}^{|P^*|} \sqrt{\sum_{m=1}^{M} \left(f_m{}^{(i)} - f_m^*{}^{(k)}\right)^2}$$

where ${f_m^*}^{(k)}$ is the value of the m-th objective function for the k-th solution in P^*, and M is the number of objectives. For our problem, we use the 25 solutions showed in Table 2 as ideal population P^*.

To evaluate the diversity of the population we use the measurement put forward by Deb et al. [4]:

$$\Delta = \frac{\sum_{m=1}^{M} d_m^e + \sum_{i=1}^{|Q|} |d_i - \overline{d}|}{\sum_{m=1}^{M} d_m^e + |Q|\,\overline{d}}$$

where d_i may be any metric of the distance between adjacent solutions, and $\overline{d}$ is the mean value of such measurements. In our case, d_i has been calculated using the Euclidean distance. Parameter d_m^e is the distance between the extreme solutions in P^* and Q corresponding to the m-th objective function.

Table 3 shows the values for convergence and diversity metrics Υ and Δ respectively obtained with the proposed multi-objective algorithm for the problem (7).

Table 3. Convergence and diversity values

$\Upsilon = 0.582842$
$\Delta = 1.175131$

4 Conclusions

Nonlinear constrained optimization problems with fuzzy costs are, in general, difficult to solve. Parametric programming techniques have been shown in literature as suitable methods to approach these kinds of problems. However, parametric programming problems have been solved mainly for linear case. Multi-objective evolutionary computation provides a chance to solve nonlinear parametric programming problems. The set of points composing the parametric solutions can be capture in a single run of the algorithm besides the power of these techniques in solving hard problems. Obtained results show a real ability of the proposed approach to solve problems arising in exporting companies from South of Spain.

Acknowledgements

Research supported in part by MCyT/FEDER under project TIC2001-0245-CO2-01, TIC2002-04021-CO2-01 and TIC2002-04242-CO3-02.

References

1. Ali FM (1998) A differencial equation approach to fuzzy non-linear programming problems. Fuzzy Sets and Systems 93 (1):57–61
2. Chankong V, Haimes YY (1983) Multiobjective Decision Making: Theory and Methodology. In: Sage AP (ed) Series in Systems Science and Engineering. North-Holland
3. Coello CA, Veldhuizen DV, Lamont GB (2002) Evolutionary Algorithms for Solving Multi-Objective Problems. Kluwer Academic/Plenum publishers, New York
4. Deb K (2001) Multi-Objective Optimization using Evolutionary Algorithms. John Wiley and Sons, LTD
5. Delgado M, Verdegay JL, Vila MA (1987) Imprecise costs in mathematical programming problems. Control and Cybernet 16 (2): 113–121
6. Ekel P, Pedrycz W, Schinzinger R (1998) A general approach to solving a wide class of fuzzy optimization problems. Fuzzy Sets and Systems 97 (1):49–66
7. Goldberg DE (1989) Genetic Algorithms in Search, Optimization, and Machine Learning. Addison-Wesley
8. Gómez-Skarmeta AF (IP) et al. (2001) Un Sistema Inteligente para la Ayuda en la Toma de Decisiones en el Entorno Agrícola del Sudeste Español Aplicado a la Fertirrigación, el Control Fitosanitario y la Evaluación de Suelos para Uso Agrícola (1FD97-0255-C03-01). Comisión Interministerial de Ciencia y Tecnología - Feder. Participantes: Universidad de Murcia, Universidad de Granada, Universidad de Almería, 01/10/98-31/09/01 (in spanish)
9. Jiménez F, Verdegay JL (1999) Solving fuzzy solid transportation problems by an evolutionary algorithm based parametric approach. European Journal of Operational Research, 113 (3):688–715
10. Jiménez F, Gómez-Skarmeta AF, Sánchez G, Deb K (2002) An evolutionary algorithm for constrained multi-objective optimization. 2002 IEEE World Congress on Evolutionary Computation
11. Jiménez F, Cadenas JM, Verdegay JL, Sánchez G (2003) Solving fuzzy optimization problems by evolutionary algorithms. Information Science 152:303–311
12. Michalewicz Z, Schoenauer M (1996) Evolutionary Algorithms for constrained parameter optimization problems. Evolutionary Computation 4 (1):1–32
13. Sanchez G, Jiménez F, GómezSkarmeta AF (2003) Multiobjective evolutionary algorithm based fuzzy optimization. 2003 IEEE International Conference on Systems, Man, and Cybernetics
14. Ramik J, Vlach M (2002) Fuzzy Mathematical Programming: A unified Approach Based on Fuzzy Relations. Fuzzy Optimization and Decision Making 1:335–346
15. Tanaka H, Okuda T, Asai K (1974) On fuzzy mathematical programming. Journal of Cybernetics 3 (4):37–46
16. Veldhuizen DV, Lamont GB (1999) Multiobjective evolutionary algorithms: Classifications, Analyses, and New Innovations. Ph. D. Thesis Dayton, OH: Air Force Institute of Technology. Technical Report No. AFIT/DS/ENG/99-01
17. Verdegay JL (1982) Fuzzy mathematical programming. In: Gupta MM, Sánchez E (eds) Fuzzy Information and Decision Processes
18. Zimmermann HJ (1976) Description and optimization of fuzzy system. International Journal of General Systems 2:209–215

Part VII

Fuzzy Statistical Decision-making Techniques

Multi Attribute Performance Evaluation Using a Hierarchical Fuzzy TOPSIS Method

Nüfer Yasin Ateş, Sezi Çevik*, Cengiz Kahraman, Murat Gülbay, S. Ayça Erdoğan

Istanbul Technical University, Department of Industrial Engineering 34367 Maçka Istanbul Turkey

Summary. Performance of a faculty is vital both for students and school, and must be measured and evaluated for positive reinforcement to faculty. Faculty performance evaluation problem is a difficult and sensitive issue which has quantitative and qualitative aspects, complexity and imprecision. In literature many different approaches are proposed in order to evaluate faculty performance. To deal with imprecision and vagueness of evaluation measures, fuzzy multi-attribute evaluation techniques can be used. In this paper, a comprehensive hierarchical evaluation model with many main and sub-attributes is constructed and a new algorithm for fuzzy Technique for Order Preference by Similarity to Ideal Solution (TOPSIS) that enables taking into account the hierarchy in the evaluation model is proposed. The obtained results from this new fuzzy TOPSIS approach are compared with fuzzy Analytic Hierarchy Process (AHP) on an application in an engineering department of a university and some sensitivity analyses are presented.

Key words: Multi-attribute, AHP, TOPSIS, fuzzy sets, faculty performance, sensitivity.

1 Introduction

Performance evaluation is required to develop a performance management system which is a systematic and data oriented approach to managing people at work that relies on positive reinforcements as the major way to maximizing performance in organizations such as production facilities and service companies. Educational institutions are one of those service organizations which strongly need to have a performance management system for continuing to supply qualified graduates to today's competitive

* Corresponding author: cevikse@itu.edu.tr, fax: +90 212 240 72 60, tel: +90 212 293 13 00 Ext. 2669

N.Y. Ateş et al. : *Multi Attribute Performance Evaluation Using a Hierarchical Fuzzy TOPSIS Method*, StudFuzz **201**, 537–572 (2006)
www.springerlink.com

business environment and, in those organizations faculty of the school are the people who directly contact, educate and contribute to students' higher knowledge. Thus, performance of a faculty is vital both for students and school, and must be measured and evaluated for positive reinforcement to faculty. On the other hand, evaluation of a faculty is a difficult and sensitive issue which has quantitative and qualitative aspects, complexity and imprecision. In the professional literature this has been emphasized by several authors stated below.

Jauch and Glueck (1975) develop measures of research output -both objective and subjective- in order to identify which ones are effective for evaluation of research performance. Relative effectiveness of different measures for evaluating the research performance of professors is discussed. It is found that simply the number of publications is the best objective measure of research performance and effectiveness. It is the recommendation of the paper that the number of publications modified by weighting the number by a journal quality index would be the most effective method. Mesak and Jauch (1991) develop a model by which college and university administrators might evaluate performances of major components of a faculty work: research, teaching, and service. Overall faculty performance is determined by a function of teaching achievements, research achievements, and service achievements under sub-criteria related to those achievements.

Ellington and Ross (1994) propose a teaching evaluation scheme to assess the university teachers in the Robert Gordon University. This scheme is based on a teaching skills profile that enables academic staff to undertake self-rating in respect of a set of basic criteria for effective performance in teaching and related activities. It is stated that an evaluation based on self-assessment of the faculty would be more successful and effective. Agrell and Steuer (2000) develop a multi-criteria decision support system for the performance review of individual faculty. They propose a multi-criteria evaluation system for individual faculty member's performance, which consists of five criteria which are research output, teaching output, external service, internal service, and cost. The results of the system they present identify promotional candidates, reveal underlying problems in managerial consistency, and suggest categorizations for faculty groupings. Meho and Sonnenwald (2000) analyze the relationship between citation ranking and peer evaluation to asses faculty research performance with a multi-criteria approach. They use two sources of peer evaluation data: citation content analysis and book review content analysis. This study presents many subjective and objective criteria on the area of research performance and concludes that citation ranking can provide a valid indicator for comparative evaluation of senior faculty research performance.

Sproule (2002) mentions reporting errors, inadequate sample size (rather degrees of freedom), the presence of sample-selection bias, the presence of reverse causation, and no vetting for accuracy of data as the reasons of underdetermination of teaching performance by student evaluations. Sproule (2002) suggests that commissions in universities whose mission is the adjudication of matters related to reappointment, pay, merit pay, tenure, and promotion should consider non-trivial, incalculable, and systemic errors in student evaluation data. Otherwise their decision rules will be invalid, unreliable, and flawed. Paulsen (2002) says that the purpose of evaluating teaching effectiveness can be grouped as formative and summative. Formative evaluation tries to provide informative feedback to assist faculty in improving effectiveness but summative evaluation tries to help personnel decisions related to hiring, awarding etc. Weistroffer *et al.* (1999) propose a structured model for faculty performance evaluation that considers both quality and quantity of faculty outputs in the areas of teaching, scholarship, and service. They identify the criteria related to measuring the quantity of performance outputs and assigning quality weights. Huberty (2000) conducts a survey at the University of Georgia to assess and evaluate faculty performance. He considers instruction, research, service, and administration activities for evaluation of faculty. To analyze the faculty performance he uses a college education form, the assigned work load, self assessment, student feedbacks for sections, and faculty members' professional writings.

In the problem of faculty performance evaluation, the evaluation attributes are generally multiple and often structured in multilevel hierarchies. Additionally, since the judgments from experts are usually vague rather than crisp, a judgment should be expressed by using fuzzy sets which has the capability of representing vague data. Two multi-attribute evaluation methods, AHP and TOPSIS, can handle and solve this problem by integrating fuzzy set theory. There are few papers handling fuzzy sets and a multi-attribute method together for the evaluation of faculty performance. Deutsch and Malmborg (1985) show that a fuzzy representation can be used to compare alternative collections of performance measures. They provide insights to the value of these collections of measures for decision-making and control purposes. They present an example of the performance evaluation of university professors. For the same problem, Hon et al. (1996) propose a multi-attribute method based on fuzzy weighted average and Saaty's pairwise comparisons.

This paper is organized as follows: In the following section determination of performance attributes in a faculty is given. In the third section, two fuzzy multi-attribute evaluation techniques are given in detail: fuzzy AHP and fuzzy TOPSIS. The fourth section includes the development of fuzzy hierarchical TOPSIS. In the fifth section, the applications of fuzzy AHP and fuzzy hierarchical TOPSIS on the performance evaluation of a faculty together with comparison results are given. In that section some sensitivity analyses are also presented. Finally, the conclusion remarks are given.

2 Determination of Performance Attributes in a Faculty

Evaluation of a faculty is a difficult and sensitive issue which has quantitative and qualitative aspects, complexity and imprecision. In the literature, the faculty evaluation attributes are divided into main three attributes: teaching, research, and service. The main criterion *teaching* includes the sub-attributes like teaching skill, student learning performance, and student's appraisal to lecturer. The main criterion *research* includes the sub-attributes like number of papers published, gained patent, and periodical quality index. The main criterion *service* includes the sub-attributes like internal service, service to the profession, and community service. Table 1 gives the attributes for faculty evaluation, proposed by different researchers. Some of the items below are main attributes in some works whereas they are divided into some sub-attributes in some other works. The explanations of the numbers in Table 1 are listed in the following:

1. Journal papers published
2. Books
3. Technical reports published
4. Conference papers published
5. Journal Quality Index
6. Citations to published materials
7. Success rate of proposals for research support
8. Referee or editor of scientific journal
9. Recognition-honors and awards from profession
10. Officer of national professional association
11. Invited papers and quest lectures
12. Dissertations supervised
13. Peer evaluations of research and publications
14. Self evaluations of research and publications
15. Teaching contents
16. Teaching methods
17. Teaching attitude
18. Service performance
19. Research projects
20. Conference participation
21. Courses taught by faculty member
22. Student evaluations
23. Curriculum reorganizations
24. Participation in administration and management functions
25. Contribution to academic field (organization of conferences, editorships, etc.)
26. Contribution to professional field (publications in professional journals, collaboration with industry)
27. Societal service (public lectures, media visibility, etc.)

28. Grade distribution
29. Thesis supervised
30. Presentations, writings and workshops
31. Grant work
32. Assigned load
33. Book chapters
34. Faculty governance
35. Organization support
36. Activate participation in academic and professional organizations
37. Technical advice
38. Recommendation letters
39. Sponsorship of visitors
40. Record keeping
41. Course preparations during year
42. Number of new courses taught during year
43. Course developments during year
44. Independent study projects at the undergraduate or graduate level
45. Thesis and dissertation committees on which the faculty member served
46. Articles in trade journals
47. Advising student organizations
48. Serving on academic editorial boards
49. Providing technical assistance to public and private organizations
50. Conducting public policy analysis for local, state, national, and international governmental agencies
51. Appearing on television and medial events
52. Testifying before legislative and confessional committees
53. Serving as an expert for the press and other media
54. Teaching plan
55. Teaching skill
56. Professional ability
57. Professional spirit
58. Teaching material preparation
59. Student learning performance
60. Student accepting degree
61. Organization ability
62. Teaching innovation
63. Student's appraisal to teacher
64. Publishing textbook
65. Developing experimental equipments and manual
66. Gained patent
67. Acting as the session chair and commentator of professional seminar
68. Research plans in charge
69. Directing various lectures
70. Commissioner of committee in school
71. Times and sum of holding continuing educational class
72. Striving practice and employment opportunity for students
73. Participating in social activities by the specialty
74. Professional lecture, advisor of consultation
75. Performance in professional academy
76. Published papers in newspapers and magazine
77. Communication
78. Course scheduling
79. Recruitment of students
80. Undergraduate courses taught during year,
81. Graduate courses taught during year

Table 1. The attributes for faculty evaluation in the literature

Source	Main Attributes		
	Teaching	Research	Service
Jauch and Glueck (1975)	-	1, 2, 3, 4, 5, 6, 7, 8, 9, 10, 11, 12, 13, 14	-
Hon, C.C. et al (1996)	15, 16, 17	1, 9, 20	18
Agrell, P.J. et al (2000)	21, 22, 23	1, 6, 12, 20	24, 25, 26, 27
Huberty C.J. (2000)	9, 12, 21, 22, 23, 28, 29, 30, 31, 32	1, 2, 3, 4, 9, 31, 32	9, 23, 31, 34, 35, 36, 37, 38, 39, 40, 41, 42, 43
Weistroffer, H.R. et al. (2001)	44, 45, 46, 47, 48, 49, 50	1, 2 , 4, 31, 33, 51	8, 24, 34, 52, 53, 54, 55, 56, 47, 58
Kuo, Y.F. and Chen, L.S., (2002)	15, 17, 32, 59, 60, 61, 62, 63, 64, 65, 66, 67, 68, 69, 70	1, 3, 4, 5, 8, 9, 71, 72, 73	34, 52, 74, 75, 76, 77, 78, 79, 80, 81

In this study, the hierarchy in Figure 1 is used after eliminating the sub-attributes having the same or very similar meanings.

3 Multi-Attribute Evaluation Techniques: Fuzzy AHP and Fuzzy TOPSIS

To deal with vagueness of human thought, Zadeh (1965) first introduced the fuzzy set theory, which was oriented to the rationality of uncertainty due to imprecision or vagueness. A major contribution of fuzzy set theory is its capability of representing vague data. The theory also allows mathematical operators and programming to apply to the fuzzy domain. To deal with imprecision and vagueness of evaluation measures, two fuzzy multi-attribute evaluation techniques are be used in this study.

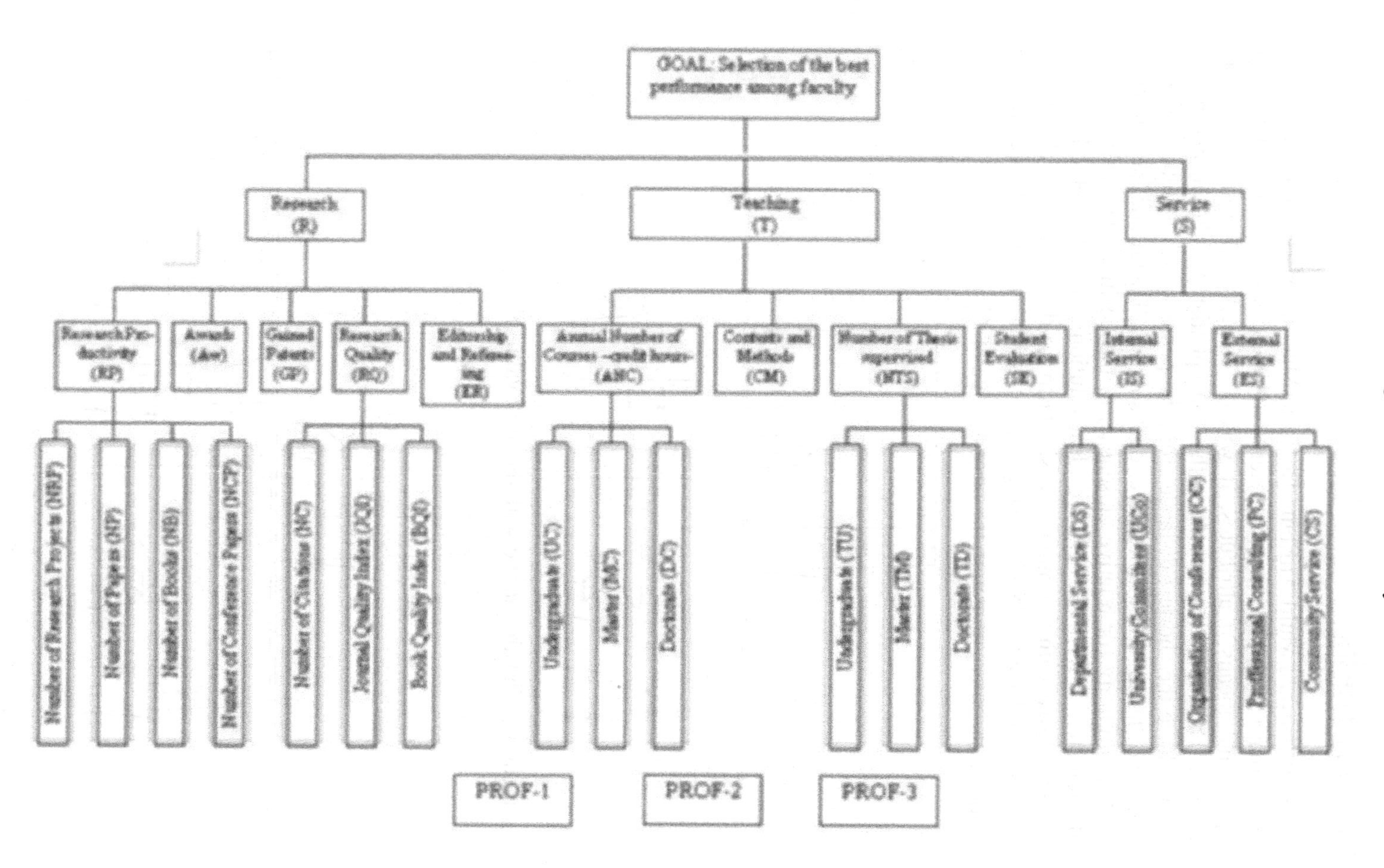
GOAL: Selection of the best performance among faculty
Research (R)
Teaching (T)
Service (S)
Research Productivity (RP)
Awards (Aw)
Gained Patents (GP)
Research Quality (RQ)
Editorship and Refereeing (ER)
Annual Number of Courses –credit hours- (ANC)
Contents and Methods (CM)
Number of Thesis supervised (NTS)
Student Evaluation (SE)
Internal Service (IS)
External Service (ES)
Number of Research Projects (NRP)
Number of Papers (NP)
Number of Books (NB)
Number of Conference Papers (NCP)
Number of Citations (NC)
Journal Quality Index (JQI)
Book Quality Index (BQI)
Undergraduate (UC)
Master (MC)
Doctorate (DC)
Undergraduate (TU)
Master (TM)
Doctorate (TD)
Departmental Service (DS)
University Committees (UCo)
Organization of Conferences (OC)
Professional Consulting (PC)
Community Service (CS)
PROF-1
PROF-2
PROF-3

3.1 Fuzzy TOPSIS

TOPSIS views a Multi Attribute Decision Making (MADM) problem with m alternatives as a geometric system with m points in the n-dimensional space that means n attributes. It was developed by Hwang and Yoon (1981). The method is based on the concept that the chosen alternative should have the shortest distance from the positive-ideal solution and the longest distance from the negative-ideal solution. TOPSIS defines an index called similarity (or relative closeness) to the positive-ideal solution and the remoteness from the negative-ideal solution. Then the method chooses an alternative with the maximum similarity to the positive-ideal solution (Yoon and Hwang, 1995). Using the vector normalization, the method chooses the alternative with the largest value of C_i^*:

$$C_i^* = \frac{\sqrt{\sum_{j=1}^{n}\left(w_j \frac{x_{ij}}{\sqrt{\sum_{i=1}^{m} x_{ij}^2}} - v_j^-\right)^2}}{\sqrt{\sum_{j=1}^{n}\left(w_j \frac{x_{ij}}{\sqrt{\sum_{i=1}^{m} x_{ij}^2}} - v_j^*\right)^2} + \sqrt{\sum_{j=1}^{n}\left(w_j \frac{x_{ij}}{\sqrt{\sum_{i=1}^{m} x_{ij}^2}} - v_j^-\right)^2}} \tag{1}$$

or it chooses the alternative with the least value of C_i^-:

$$C_i^- = \frac{\sqrt{\sum_{j=1}^{n}\left(w_j \frac{x_{ij}}{\sqrt{\sum_{i=1}^{m} x_{ij}^2}} - v_j^*\right)^2}}{\sqrt{\sum_{j=1}^{n}\left(w_j \frac{x_{ij}}{\sqrt{\sum_{i=1}^{m} x_{ij}^2}} - v_j^*\right)^2} + \sqrt{\sum_{j=1}^{n}\left(w_j \frac{x_{ij}}{\sqrt{\sum_{i=1}^{m} x_{ij}^2}} - v_j^-\right)^2}} \tag{2}$$

where i (i = 1,...,m) and j (j = 1,...,n) indicate number of alternatives and attributes respectively; w_j is the weight of the jth attribute; x_{ij} is the attribute rating for ith alternative's jth attribute; v_j^* is the positive-ideal value for jth attribute, where it is a maximum for benefit attributes and a minimum for cost attributes; v_j^- is the negative-ideal value for the jth attribute, where it is a minimum for benefit attributes and a maximum for cost attributes.

In the last fifteen years, some fuzzy TOPSIS methods were developed in the literature: Chen and Hwang (1992) transform Hwang and Yoon's (1981) method to the fuzzy case. Liang (1999) presents a fuzzy multi-criteria decision-making based on the concepts of ideal and anti-ideal points. The concepts of fuzzy set theory and hierarchical structure analysis are used to develop a weighted suitability decision matrix to evaluate the weighted suitability of different alternatives versus criteria. Chen (2000) describes the rating of each alternative and the weight of each criterion by linguistic terms which can be expressed in triangular fuzzy numbers. Then, a vertex method for TOPSIS is proposed to calculate the distance between two triangular fuzzy numbers. Tsaur et al. (2002) apply the fuzzy set theory to evaluate the service quality of airline. By applying AHP in obtaining criteria weight and TOPSIS in ranking, they find the most concerned aspects of service quality. Chu (2002) presents a fuzzy TOPSIS model for solving the facility location selection problem. Chu and Lin (2003) propose a fuzzy TOPSIS approach for robot selection where the ratings of various alternatives under different subjective attributes and the importance weights of all attributes are assessed in linguistic terms represented by fuzzy numbers. Zhang and Lu (2003) present an integrated fuzzy group decision-making method in order to deal with the fuzziness of preferences of the decision-makers. In Zhang and Lu's (2003) paper, the weights of the criteria are crisp values gathered by pairwise comparisons where the preferences of the decision-makers are represented by fuzzy triangular numbers.

Table 2 gives the comparison of the fuzzy TOPSIS methods in the literature. The comparison includes the computational differences among the methods. In this paper, we prefer Chen and Hwang's (1992) fuzzy TOPSIS method since the other fuzzy TOPSIS methods are derived from this method with minor differences.

Table 2. Comparison of Fuzzy TOPSIS Methods

Source	Attribute Weights	Type of Fuzzy Numbers	Ranking Method	Normalization Method
Chen and Hwang (1992)	Fuzzy Numbers	Trapezoidal	Lee and Li's (1988) generalized mean method	Linear Normalization
Liang (1999)	Fuzzy Numbers	Trapezoidal	Chen's (1985) ranking with maximizing set and minimizing set	Manhattan distance
Chen (2000)	Fuzzy Numbers	Triangular	Chen (2000) assumes the fuzzy positive and negative ideal solutions as (1, 1, 1) and (0, 0, 0) respectively.	Linear Normalization
Chu (2002)	Fuzzy Numbers	Triangular	Liou and Wang's (1992) ranking method of total integral value with =1/2	Modified Manhattan distance
Tsaur et al. (2002)	Crisp Values	Triangular	Zhao and Govind's (1991) center of area method	Vector Normalization
Zhang and Lu (2003)	Crisp Values	Triangular	Chen's (2000) fuzzy positive and negative ideal solutions: as (1, 1, 1) and (0, 0, 0) respectively.	Manhattan distance
Chu and Lin (2003)	Fuzzy Numbers	Triangular	Kaufmann and Gupta's (1988) mean of the removals method	Linear Normalization

In the following, the steps of fuzzy TOPSIS developed by Chen et al. (1992) are given. First, a decision matrix, D, of $m \times n$ dimension is defined:

$$D = \begin{matrix} & \begin{matrix} X_1 & \cdots & X_j & \cdots & X_n \end{matrix} \\ \begin{matrix} A_1 \\ \vdots \\ A_i \\ \vdots \\ A_m \end{matrix} & \begin{bmatrix} x_{11} & \cdots & x_{1j} & \cdots & x_{1n} \\ \vdots & & \vdots & & \vdots \\ x_{i1} & \cdots & x_{ij} & \cdots & x_{in} \\ \vdots & & \vdots & & \vdots \\ x_{m1} & \cdots & x_{mj} & \cdots & x_{mn} \end{bmatrix} \end{matrix} \tag{3}$$

where $x_{ij}, \forall i, j$ may be crisp or fuzzy. If x_{ij} is fuzzy, it is represented by a trapezoidal number as $x_{ij} = (a_{ij}, b_{ij}, c_{ij}, d_{ij})$. The fuzzy weights shall be described by $w_j = (\alpha_j, \beta_j, \chi_j, \delta_j)$

Algorithm

The problem is solved using the following steps.

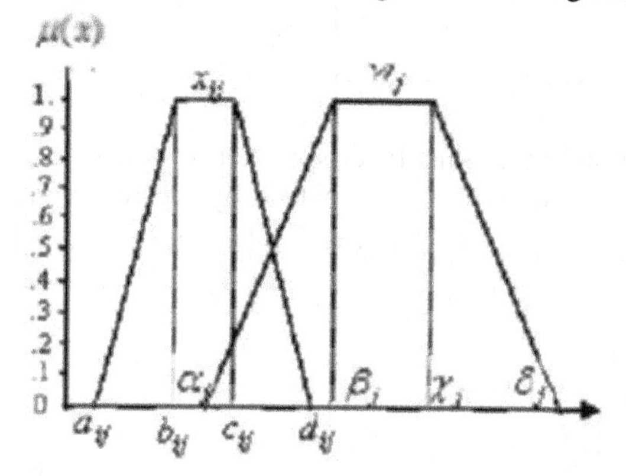

Fig. 1. Trapezoidal fuzzy numbers

Step 1. Normalize the Decision Matrix. The decision matrix must first be normalized so thet the elements are unit-free. Here, the r_{ij} values are normalized values. x_j^* and x_j^- are the maximum and minimum values of the columns in the decision matrix, respectively. To avoid the complicated normalization formula used in classical TOPSIS, we use linear scale transformation as follows:

$$r_{ij} = \begin{cases} x_{ij} / x_j^*, \forall j, x_j \text{ is a benefit attribute} \\ x_j^- / x_{ij}, \forall j, x_j \text{ is a cost attribute} \end{cases} \quad (4)$$

By applying Eq. (4), we can rewrite the decision matrix as:

$$D' = \begin{array}{c} \\ A_1 \\ \vdots \\ A_i \\ \vdots \\ A_m \end{array} \begin{array}{c} \begin{array}{ccccc} X_1 & \cdots & X_j & \cdots & X_n \end{array} \\ \left[\begin{array}{ccccc} r_{11} & \cdots & r_{1j} & \cdots & r_{1n} \\ \vdots & & \vdots & & \vdots \\ r_{i1} & \cdots & r_{ij} & \cdots & r_{in} \\ \vdots & & \vdots & & \vdots \\ r_{m1} & \cdots & r_{mj} & \cdots & r_{mn} \end{array} \right] \end{array} \tag{5}$$

When x_{ij} is crisp, its corresponding r_{ij} must be crisp; when x_{ij} is fuzzy, its corresponding r_{ij} must be fuzzy. Eq.(4) is then replaced by the following fuzzy operations: Let $x_{ij} = (a_{ij}, b_{ij}, c_{ij}, d_{ij})$. and $x_j^* = (a_j^*, b_j^*, c_j^*, d_j^*)$ we have

$$r_{ij} = \begin{cases} x_{ij}(+)x_j^* = \left(\dfrac{a_{ij}}{d_j^*}, \dfrac{b_{ij}}{c_j^*}, \dfrac{c_{ij}}{b_j^*}, \dfrac{d_{ij}}{a_j^*} \right) \\ x_j^-(+)x_{ij} = \left(\dfrac{a_i^-}{d_{ij}}, \dfrac{b_i^-}{c_{ij}}, \dfrac{c_i^-}{b_{ij}}, \dfrac{d_i^-}{a_{ij}} \right) \end{cases} \tag{6}$$

In the formula above x_j^* and x_j^- represent the largest and the lowest scores respectively.

Step 2. Obtain the Weighted Normalized Decision Matrix. This matrix is obtained using

$$v_{ij} = r_{ij} w_j, \forall_j, j \tag{7}$$

When both r_{ij} and w_{ij} are crisp, v_{ij} is crisp; while when either r_{ij} or w_{ij}, (or both) are fuzzy, Eq.(7) may be replaced by the following fuzzy operations:

$$v_{ij} = r_{ij}(.)w_j = \left(\frac{a_{ij}}{d_j^*} \alpha_j, \frac{b_{ij}}{c_j^*} \beta_j, \frac{c_{ij}}{b_j^*} \chi_j, \frac{d_{ij}}{a_j^*} \delta_j \right) \tag{8}$$

$$v_{ij} = r_{ij}(.)w_j = \left(\frac{a_i^-}{d_{ij}} \alpha_j, \frac{b_i^-}{c_{ij}} \beta_j, \frac{c_i^-}{b_{ij}} \chi_j, \frac{d_i^-}{a_{ij}} \delta_j \right) \tag{9}$$

Equation (8) is used when the *j*th attribute is a benefit attribute. Equation (9) is used when the *j*th attribute is a cost attribute. The result of Eqs (8) and (9) can be summarized as:

$$v = \begin{array}{c} \\ A_1 \\ \vdots \\ A_i \\ \vdots \\ A_m \end{array} \begin{array}{c} \begin{array}{ccccc} X_1 & \cdots & X_j & \cdots & X_n \end{array} \\ \begin{bmatrix} v_{11} & \cdots & v_{1j} & \cdots & v_{1n} \\ \vdots & & \vdots & & \vdots \\ v_{i1} & \cdots & v_{ij} & \cdots & v_{in} \\ \vdots & & \vdots & & \vdots \\ v_{m1} & \cdots & v_{mj} & \cdots & v_{mn} \end{bmatrix} \end{array} \tag{10}$$

Step 3. Obtain the Positive Ideal Solution (PIS), A^*, and the Negative Ideal Solution (NIS). PIS and NIS are defined as:

$$A^* = \left[v_1^*, \ldots, v_n^*\right], \tag{11}$$

$$A^- = \left[v_1^-, \ldots, v_n^-\right], \tag{12}$$

where $v_j^* = \max_i v_{ij}\, and\, v_j^- = \min_i v_{ij}$.

For crisp data, v_j^* and v_j^- are obtained straight forward. For fuzzy data, v_j^* and v_j^- may be obtained through some ranking procedures. The authors used Lee and Li's ranking method for comparison of fuzzy numbers. The v_j^* and v_j^- are the fuzzy numbers with the largest generalized mean and the smallest generalized mean, respectively. The generalized mean for fuzzy number $v_{ij}, \forall_j, j,$ is defined as:

$$M(v_{ij}) = \frac{-a_{ij}^2 - b_{ij}^2 + c_{ij}^2 + d_{ij}^2 - a_{ij}b_{ij} + c_{ij}d_{ij}}{\left[3\left(-a_{ij} - b_{ij} + c_{ij} + d_{ij}\right)\right]} \tag{13}$$

For each column j, we find a v_{ij} which has the greatest mean as the v_j^* and which has the lowest mean as the v_j^-.

Step 4. Obtain the Separation Measures S_i^* and S_i^-. In the classical case, separation measures are defined as:

$$S_i^* = \sum_{j=1}^{n} D_{ij}^*, \quad i = 1,...,n \tag{14}$$

and

$$S_i^- = \sum_{j=1}^{n} D_{ij}^-, \quad i = 1,...,n \tag{15}$$

For crisp data, the difference measures D_{ij}^* and D_{ij}^- are given as:

$$D_{ij}^* = \left| v_{ij} - v_j^* \right| \tag{16}$$

$$D_{ij}^- = \left| v_{ij} - v_j^- \right| \tag{17}$$

The computation is straight forward. For fuzzy data, the difference between two fuzzy numbers $\mu_{v_{ij}}(x)$ and $\mu_{v_j^*}(x)$ (based on Zadeh) as:

$$D_{ij}^* = 1 - \left\{ \sup_x \left[\mu_{v_{ij}}(x) \wedge \mu_{v_j^*}(x) \right] \right\} = 1 - L_{ij}, \forall i, j \tag{18}$$

where L_{ij} is the highest degree of similarity of v_{ij} and v_j^*. The value of L_{ij} is depicted in Fig. 2.

Similarly, the difference between $\mu_{v_{ij}}(x)$ and $\mu_{v_j^-}(x)$ is defined as:

$$D_{ij}^- = 1 - \left\{ \sup_x \left[\mu_{v_{ij}}(x) \wedge \mu_{v_j^-}(x) \right] \right\} = 1 - L_{ij}, \forall i, j \tag{19}$$

Note that both D_{ij}^*, D_{ij}^- are crisp numbers.

Step 5. Compute the Relative Closeness to Ideals. This index is used to combine S_i^* and S_i^- indices calculated in Step 4. Since S_i^* and S_i^- are crisp numbers, they can be combined:

$$C_i = S_i^- / \left(S_i^* + S_i^- \right) \tag{20}$$

The alternatives are ranked in descending order of the C_i index.

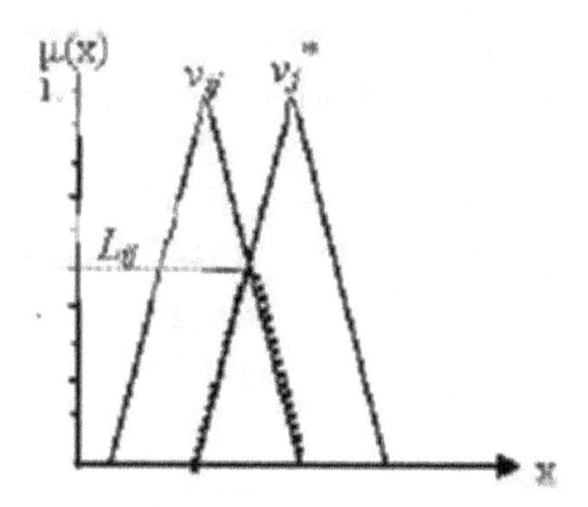

Fig. 2. The derivation of L_{ij}

3.2 Fuzzy AHP

There are many papers both on theory and application of fuzzy AHP in the literature. A vast literature review about the techniques can be find in Kahraman *et al*, (2004). In this paper, we prefer Chang's (1992; 1996) extent analysis method since the steps of this approach are relatively easier than the other fuzzy AHP approaches and similar to the conventional AHP.

Let $X = \{x_1, x_2, ..., x_n\}$ be an object set, and $U = \{u_1, u_2, ..., u_m\}$ be a goal set. According to the method of Chang's (1992) extent analysis, each object is taken and extent analysis for each goal, g_i, is performed respectively. Therefore, *m* extent analysis values for each object can be obtained, with the following signs:

$$M_{g_i}^1, M_{g_i}^2, ..., M_{g_i}^m, i = 1,2,..., n \tag{21}$$

where all the $M_{g_i}^j \ (j = 1,2,..., m)$ are triangular fuzzy numbers (TFNs) whose parameters are *l*, *m*, and *u*. They are the least possible value, the most possible value, and the largest possible value respectively. The triangular fuzzy conversion scale given in Table 3 is used in our evaluation model. This scale is not the same as the original linguistic scale of 1-9 but suitable for the usage of fuzzy AHP. A different scale in fuzzy AHP can be found in the literature as in (Abdel-Kader and Dugdale, 2001).

Table 3. Triangular fuzzy conversion scale

Linguistic scale	Triangular fuzzy scale	Triangular fuzzy reciprocal scale
Just equal	(1, 1, 1)	(1, 1, 1)
Equally important	(1/2, 1, 3/2)	(2/3, 1, 2)
Weakly important	(1, 3/2, 2)	(1/2, 2/3, 1)
Strongly more important	(3/2, 2, 5/2)	(2/5, 1/2, 2/3)
Very strong more important	(2, 5/2, 3)	(1/3, 2/5, 1/2)
Absolutely more important	(5/2, 3, 7/2)	(2/7, 1/3, 2/5)

The steps of Chang's extent analysis can be given as in the following:

Step 1. The value of fuzzy synthetic extent with respect to the ith object is defined as

$$S_i = \sum_{j=1}^{m} M_{g_i}^{j} \otimes \left[\sum_{i=1}^{n} \sum_{j=1}^{m} M_{g_i}^{j} \right]^{-1} \tag{22}$$

$$S_i = \sum_{j=1}^{m} M_{g_i}^{j} \otimes \left[\sum_{i=1}^{n} \sum_{j=1}^{m} M_{g_i}^{j} \right]^{-1} \tag{23}$$

To obtain $\sum_{j=i}^{m} M_{g_i}^{j}$, perform the fuzzy addition operation of m extent analysis values for a particular matrix such that

$$\sum_{j=1}^{m} M_{g_i}^{j} = \left(\sum_{j=1}^{m} l_j, \sum_{j=1}^{m} m_j, \sum_{j=1}^{m} u_j \right), \qquad \text{i} = 1,\ldots,\text{n} \tag{24}$$

and to obtain $\left[\sum_{i=1}^{n} \sum_{j=1}^{m} M_{g_i}^{j} \right]^{-1}$, perform the fuzzy addition operation of $M_{g_i}^{j}$ $(j = 1,2,\ldots,m)$ values such that

$$\sum_{i=1}^{n} \sum_{j=1}^{m} M_{g_i}^{j} = \left(\sum_{i=1}^{n} l_i, \sum_{i=1}^{n} m_i, \sum_{i=1}^{n} u_i \right) \tag{25}$$

and then compute the inverse of the vector in Eq. (25) such that

$$\left[\sum_{i=1}^{n}\sum_{j=1}^{m} M_{g_i}^{j}\right]^{-1} = \left(\frac{1}{\sum_{i=1}^{n} u_i}, \frac{1}{\sum_{i=1}^{n} m_i}, \frac{1}{\sum_{i=1}^{n} l_i}\right) \tag{26}$$

Step 2. The degree of possibility of $M_2=(l_1, m_1, u_1) \geq M_1=(l_2, m_2, u_2)$ is defined as

$$V(M_2 \geq M_1) = \sup_{y \geq x}\left[\min(\mu_{M_1}(x), \mu_{M_2}(y)\right] \tag{27}$$

and can be equivalently expressed as follows

$$V(M_2 \geq M_1) = hgt(M_1 \cap M_2) = \mu_{M_2}(d) = \begin{cases} 1, if\ m_2 \geq m_1 \\ 1, if\ I_1 \geq u_2 \\ \dfrac{l_1 - u_2}{(m_2 - u_2) - (m_1 - l_1)}, otherwise \end{cases} \tag{28}$$

where d is the ordinate of the highest intersection point D between μ_{M_1} and μ_{M_2} (see Fig. 4).

To compare M_1 and M_2, we need both the values of $V(M_1 \geq M_2)$ and $V(M_2 \geq M_1)$.

Step 3. The degree possibility for a convex fuzzy number to be greater than k convex fuzzy numbers $M_i (i = 1, 2, \ldots k)$ can be defined by

$$\begin{aligned} V(M \geq M_1, M_2, \ldots, M_k &= V\left[(M \geq M_1) and \ldots (M \geq M_k)\right] \\ &= \min V(M \geq M_i), i = 1,2,\ldots,k \end{aligned} \tag{29}$$

Assume that

$$d'(A_i) = \min V(S_i \geq S_k) \tag{30}$$

For $k = 1, 2, \ldots, n$; $k \neq i$. Then the weight vector is given by

$$W' = \left(d'(A_1), d'(A_2), \ldots, d'(A_n)\right)^T \tag{31}$$

where A_i $(i = 1, 2, \ldots, n)$ are n elements.

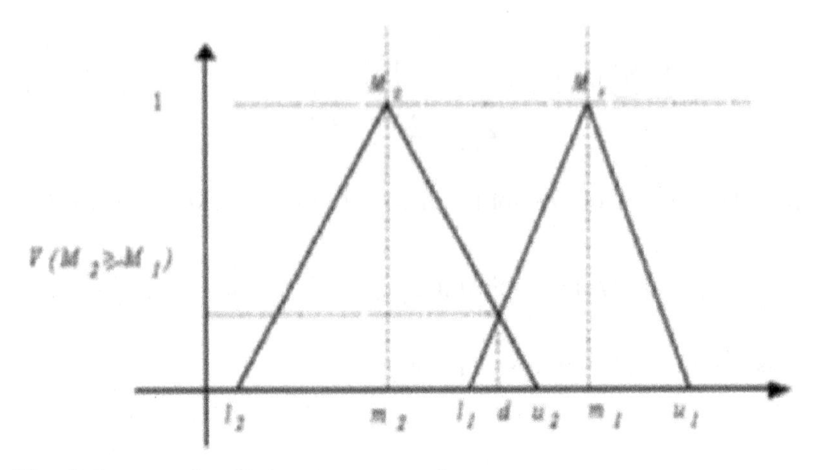

Fig. 3. Intersection Between μ_{M_1} and μ_{M_2}

Step 4. Via normalization, the normalized weight vectors are,

$$W = (d(A_1), d(A_2), ..., d(A_n))^T \tag{32}$$

where W is a nonfuzzy number.

4 Development of Hierarchical Fuzzy TOPSIS

Since this study aims at comparing two fuzzy multi-attribute methods, fuzzy AHP and fuzzy TOPSIS, a hierarchical fuzzy TOPSIS algorithm is needed. This algorithm is developed below. The hierarchy in Figure 4 will be considered.

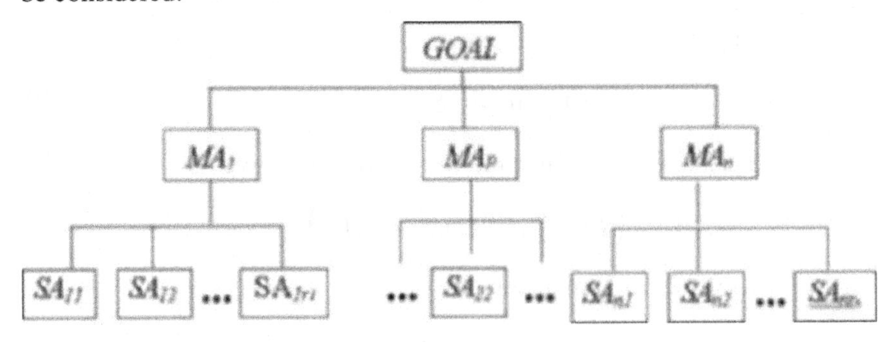

Fig. 4. The hierarchy considered in fuzzy TOPSIS algorithm

Assume that we have n main attributes, m sub-attributes, k alternatives, and s respondents. Each main attribute has r_i sub-attributes where the total number of sub-attributes m is equal to $\sum_{i=1}^{n} r_i$.

The first matrix ($\widetilde{I}_{MA}$), given by Eq. 33, is constructed from the weights of the main attributes with respect to the goal.

$$\widetilde{I}_{MA} = \begin{array}{c} \\ MA_1 \\ MA_2 \\ \vdots \\ MA_p \\ \vdots \\ MA_n \end{array} \begin{array}{c} Goal \\ \begin{bmatrix} \widetilde{w}_1 \\ \widetilde{w}_2 \\ \vdots \\ \widetilde{w}_p \\ \vdots \\ \widetilde{w}_n \end{bmatrix} \end{array} \tag{33}$$

where $\widetilde{w}_p$ is the arithmetic mean of the weights assigned by the respondents and is calculated by Eq. 34

$$\widetilde{w}_p = \frac{\sum_{i=1}^{s} \widetilde{q}_{pi}}{s}, \quad p = 1,2,...,n \tag{34}$$

where $\widetilde{q}_{pi}$ denotes the fuzzy evaluation score of pth main attribute with respect to goal assessed by the ith respondent. The second matrix ($\widetilde{I}_{SA}$) represents the weights of the sub-attributes with respect to the main attributes. The weights vector obtained from $\widetilde{I}_{MA}$ are written above this $\widetilde{I}_{SA}$ as illustrated in Eq. 35.

$$\tilde{I}_{SA} = \begin{array}{c} \\ \\ SA_{11} \\ SA_{12} \\ \vdots \\ SA_{1r_1} \\ SA_{21} \\ SA_{22} \\ \vdots \\ SA_{2r_2} \\ \vdots \\ SA_{pl} \\ \vdots \\ SA_{n1} \\ SA_{n2} \\ \vdots \\ SA_{nr_n} \end{array} \begin{array}{c} \begin{array}{cccccc} \tilde{w}_1 & \tilde{w}_2 & \cdots & \tilde{w}_p & \cdots & \tilde{w}_n \\ MA_1 & MA_2 & \cdots & MA_p & \cdots & MA_n \end{array} \\ \begin{bmatrix} \tilde{w}_{11} & 0 & \cdots & 0 & \cdots & 0 \\ \tilde{w}_{12} & 0 & \cdots & 0 & \cdots & 0 \\ \vdots & \vdots & & \vdots & & \vdots \\ \tilde{w}_{1r_1} & 0 & \cdots & 0 & \cdots & 0 \\ 0 & \tilde{w}_{21} & \cdots & 0 & \cdots & 0 \\ 0 & \tilde{w}_{22} & \cdots & 0 & \cdots & 0 \\ \vdots & \vdots & & \vdots & & \vdots \\ 0 & \tilde{w}_{2r_2} & \cdots & 0 & \cdots & 0 \\ \vdots & 0 & & & & \vdots \\ 0 & 0 & \cdots & \tilde{w}_{pl} & \cdots & 0 \\ \vdots & \vdots & & & & 0 \\ 0 & 0 & \cdots & 0 & \cdots & \tilde{w}_{n1} \\ 0 & 0 & \cdots & 0 & \cdots & \tilde{w}_{n2} \\ \vdots & \vdots & & \vdots & & \vdots \\ 0 & 0 & \cdots & 0 & \cdots & \tilde{w}_{nr_n} \end{bmatrix} \end{array} \tag{35}$$

where $\tilde{w}_{pl}$ is the arithmetic mean of the weights assigned by the respondents and it is calculated by Eq. 36.

$$\tilde{w}_{pl} = \frac{\sum_{i=1}^{s} \tilde{q}_{pli}}{s} \tag{36}$$

where $\tilde{q}_{pli}$ is the weight of lth sub-attribute with respect to p^{th} main attribute assessed by the ith respondent.

The third matrix ($\tilde{I}_A$) is formed by the scores of the alternatives with respect to the sub-attributes. The weights vector obtained from $\tilde{I}_{SA}$ are written above this $\tilde{I}_A$ as in Eq. 37.

$$\tilde{I}_A = \begin{array}{c} \\ \\ A_1 \\ A_2 \\ \vdots \\ A_q \\ \vdots \\ A_k \end{array} \begin{array}{c} \tilde{W}_{11} \quad \tilde{W}_{12} \quad \cdots \quad \tilde{W}_{1r_1} \quad \cdots \quad \tilde{W}_{pl} \quad \cdots \quad \tilde{W}_{nr_n} \\ SA_{11} \quad SA_{12} \quad \cdots \quad SA_{1r_1} \quad \cdots \quad SA_{pl} \quad \cdots \quad SA_{nr_n} \\ \begin{bmatrix} \tilde{c}_{111} & \tilde{c}_{112} & \cdots & \tilde{c}_{11r_1} & \cdots & \tilde{c}_{1pl} & \cdots & \tilde{c}_{1nr_n} \\ \tilde{c}_{211} & \tilde{c}_{212} & \cdots & \tilde{c}_{21r_1} & \cdots & \tilde{c}_{2pl} & \cdots & \tilde{c}_{2nr_n} \\ \vdots & \vdots & & \vdots & & \vdots & & \vdots \\ \tilde{c}_{q11} & \tilde{c}_{q12} & \cdots & \tilde{c}_{q1r_1} & \cdots & \tilde{c}_{qpl} & \cdots & \tilde{c}_{qnr_n} \\ \vdots & \vdots & & \vdots & & \vdots & & \vdots \\ \tilde{c}_{k11} & \tilde{c}_{k12} & \cdots & \tilde{c}_{k1r_1} & \cdots & \tilde{c}_{kpl} & \cdots & \tilde{c}_{knr_n} \end{bmatrix} \end{array} \quad (37)$$

where,

$$\tilde{W}_{pl} = \sum_{j=1}^{n} \tilde{w}_p \tilde{w}_{pj} \quad (38)$$

Since $w_{pj} = 0$ for $j \neq l$, we can use Eq. 39 instead of Eq. 38

$$\tilde{W}_{pl} = \tilde{w}_p \tilde{w}_{pl} \quad (39)$$

In $\tilde{I}_A$, $\tilde{c}_{qpl}$ is the arithmetic mean of the scores assigned by the respondents and it is calculated by Eq. 40

$$\tilde{c}_{qpl} = \frac{\sum_{i=1}^{s} \tilde{q}_{qpli}}{s} \quad (40)$$

where $\tilde{q}_{qpli}$ is the fuzzy evaluation score of qth alternative with respect to l^{th} sub-attribute under p^{th} main attribute assessed by i^{th} respondent.

After obtaining $\tilde{c}_{qpl}$ s hierarchical structure is ready to be included to the fuzzy TOPSIS algorithm described previously.

Table 4. The importance degrees

Very low	(0, 0, 0.2)
Low	(0, 0.2, 0.4)
Medium	(0.3, 0.5, 0.7)
High	(0.6, 0.8, 1)
Very high	(0.8, 1, 1)

Table 5. The scores

Very poor	(0, 0, 20)
Poor	(0, 20, 40)
Fair	(30, 50, 70)
Good	(60, 80, 100)
Very good	(80, 100, 100)

To determine the importance degree of each main attribute with respect to the goal and each sub-attribute with respect to the main-attributes, Table 4 will be used. The linguistic terms represented by TFNs for scoring the alternatives under the sub-attributes are given in Table 5.

5 Application in an Engineering Department

Three professors in the Industrial Engineering Department of Istanbul Technical University will be evaluated using the multi-attribute decision-making techniques given above. Taking into consideration the hierarchy in Figure 1, two sets of questionnaire, one for fuzzy AHP and the other for fuzzy TOPSIS, were prepared to receive the individual weights of main and sub-attributes and mailed to 150 academicians from different engineering departments in different universities in Turkey. The response rate was 42% that is 63 over 150. Samples from both sets of questionnaire are given in Appendix.

5.1 Fuzzy AHP Application Results

Using Chang's (1992) extent analysis, we obtained one eigenvector for the main attributes with respect to the goal, three eigenvectors for the sub-attributes with respect to the main attributes, six eigenvectors for the sub-sub-attributes with respect to the sub-attributes, and 23 eigenvectors for the alternatives with respect to the sub and sub-sub-attributes. In the following, one sample of pairwise comparisons for each level is given (Tables 6-9). The eigenvectors of other pairwise comparisons can be seen in Table 10. The fuzzy numbers in these matrices are the geometric means of the fuzzy weights assigned by the respondents.

Table 6. The fuzzy evaluation matrix with respect to the goal.

GOAL	R	T	S
R	(1, 1, 1)	(0.63, 1.145, 1.651)	(1.651, 1.817, 2.321)
T	(0.606, 0.874,1.587)	(1, 1, 1)	(1.31, 1.587, 1.842)
S	(0.606, 0.874,1.587)	(0.543, 0.63, 0.763)	(1, 1, 1)

$$S_R = (3.28,3.96,4,97) \otimes (1/12.75,1/9.93,1/8.35) = (0.257,0.399,0.596)$$
$$S_T = (2.92,3.46,4.43) \otimes (1/12.75,1/9.93,1/8.35) = (0.229,0.349,0.531)$$

and

$$S_S = (2.15,2.50,3.35) \otimes (1/12.75,1/9.93,1/8.35) = (0.168,0.252,0.401)$$

are obtained. Using these vectors, $V(S_R \geq S_T) = 1.00$, $V(S_R \geq S_S) = 1.00$, $V(S_T \geq S_R) = 0.84$, $V(S_T \geq S_S) = 1.00$, $V(S_S \geq S_R) = 0.50$, and $V(S_S \geq S_T) = 0.64$ are obtained. Thus, the weight vector from Table 6 is calculated as $W_G = (0.43,0.36,0.21)^T$.

Table 7. The fuzzy evaluation of sub attributes with respect to relevant main attribute.

R	RP	Aw	GP	RQ	ER
RP	(1, 1, 1)	(0.9, 1.44, 1.95)	(0.63, 1, 1.31)	(0.79, 1, 1.14)	(1.14, 1.65, 2.15)
Aw	(0.51, 0.69,1.1)	(1, 1, 1)	(0.63, 1, 1.31)	(0.69, 0.87, 1.26)	(0.63, 1, 1.31)
GP	(0.76, 1, 1.58)	(0.76, 1, 1.58)	(1, 1, 1)	(0.87, 1.31, 2)	(1.14, 1.71, 2.24)
RQ	(0.87, 1, 1.26)	(0.79, 1.14, 1.44)	(0.5, 0.76, 1.14)	(1, 1, 1)	(0.79, 1.31, 1.81)
ER	(0.76, 1, 1.58)	(0.76, 1, 1.58)	(0.44, 0.58, 0.87)	(0.55, 0.76, 1.26)	(1, 1, 1)

The weight vector from Table 7 is calculated as $W_R = (0.23,0.17,0.23,0.20,0.17)^T$.

Table 8. The fuzzy evaluation of sub-sub attributes with respect to relevant sub-attribute.

RP	NRP	NP	NB	NCP
NRP	(1, 1, 1)	(0.6, 0.87, 1.58)	(0.69, 1, 1.58)	(1.18, 1.65, 2.4)
NP	(0.6, 1.14,1.65)	(1, 1, 1)	(0.43, 0.56, 0.79)	(1.81, 2.32, 2.82)
NB	(0.63, 1, 1.44)	(1.26, 1.77, 2.28)	(1, 1, 1)	(1.14, 1.71, 2.24)
NCP	(0.41, 0.6, 0.84)	(0.35, 0.43, 0.5)	(0.44, 0.58, 0.87)	(1, 1, 1)

The weight vector from Table 8 is calculated as $W_{RP} = (0.28,0.30,0.33,0.09)^T$.

Table 9. The fuzzy evaluations of alternatives with respect to a sub attribute.

NRP	PROF-1	PROF-2	PROF-3
PROF-1	(1, 1, 1)	(1.145, 1.71, 2.241)	(1.145, 1.442, 1.71)
PROF-2	(0.446, 0.585, 0.874)	(1, 1, 1)	(1.357, 1.71, 2.19)
PROF-3	(0.585, 0.693,0.874)	(0.457, 0.585, 0.737)	(1, 1, 1)

The weight vector from Table 9 is calculated as $W_{NRP} = (0.53, 0.38, 0.09)^T$.

The weights of all main and sub-attributes and scores of alternatives are summarized in Table 10, and the priority weights obtained from fuzzy AHP are presented in Table 11.

Table 10. Weights of attributes and scores of alternatives – Summary Table

	R 0.43									
	RP 0.23				Aw 0.17	GP 0.23	RQ 0.2			ER 0.17
	NRP 0.28	NP 0.3	NB 0.33	NCP 0.09			NC 0.34	JQI 0.36	BQI 0.3	
PROF-1	0.53	0.55	0.44	0.45	0.43	0.33	0.48	0.41	0.41	0.62
PROF-2	0.38	0.29	0.37	0.22	0.38	0.33	0.27	0.29	0.33	0.20
PROF-3	0.09	0.16	0.19	0.33	0.19	0.33	0.25	0.30	0.26	0.18

Table 10. Weights of attributes and scores of alternatives (continued)

	T 0.36							
	ANC 0.38			CM 0.22	NTS 0.24			SE 0.16
	CU 0.27	CM 0.34	CD 0.39		TU 0.17	TM 0.25	TD 0.58	
PROF-1	0.27	0.35	0.39	0.42	0.37	0.36	0.39	0.39
PROF-2	0.33	0.27	0.30	0.31	0.33	0.32	0.38	0.32
PROF-3	0.39	0.38	0.30	0.27	0.30	0.31	0.23	0.29

Table 10. Weights of attributes and scores of alternatives (continued)

	S				
	0.21				
	IS		ES		
	0.46		0.54		
	DS	UC	OC	PC	CS
	0.34	0.66	0.32	0.33	0.35
PROF-1	0.24	0.38	0.43	0.24	0.42
PROF-2	0.37	0.34	0.22	0.48	0.35
PROF-3	0.39	0.28	0.35	0.28	0.23

Table 11. Results of Fuzzy AHP

	Priority Weights
PROF-1	0.41
PROF-2	0.32
PROF-3	0.27

The results show that Prof-1 has the highest performance among all professors where prof-3 is the third with corresponding priority weights in Table 11.

5.2 Fuzzy TOPSIS Application Results

First, equations in the fuzzy TOPSIS algorithm for trapezoidal fuzzy numbers given in Fuzzy TOPSIS section will be regenerated for TFNs which are considered in this application. Since a TFN *(a, b, c)* can be represented in trapezoidal form as *(a, b, b, c)*, it can be easily seen that Equation (6) can be expressed as follows

$$r_{ij} = \begin{cases} x_{ij}(+)x_j^* = \left(\dfrac{a_{ij}}{d_j^*}, \dfrac{b_{ij}}{b_j^*}, \dfrac{d_{ij}}{a_j^*} \right) \\ x_j^-(+)x_{ij} = \left(\dfrac{a_i^-}{d_{ij}}, \dfrac{b_i^-}{b_{ij}}, \dfrac{d_i^-}{a_{ij}} \right) \end{cases} \tag{41}$$

Equation (13) is then reduced to

$$M(v_{ij}) = \frac{-a_{ij}^2 + d_{ij}^2 - a_{ij}b_{ij} + b_{ij}d_{ij}}{[3(-a_{ij} + d_{ij})]} \tag{42}$$

D_{ij}^* and D_{ij}^- is calculated as shown below, where $v_j^* = (a^*, b^*, c^*)$ and $v_j^- = (a^-, b^-, c^-)$ are the fuzzy numbers with the largest generalized mean and the smallest generalized mean, respectively.

$$D_{ij}^* = \begin{cases} 1 - \dfrac{c_{ij} - a^*}{b^* + c_{ij} - a^* - b_{ij}} & for\ b_{ij} < b^* \\ 1 - \dfrac{c^* - a_{ij}}{b_{ij} + c^* - a_{ij} - b^*} & for\ b^* < b_{ij} \end{cases} \quad \forall i, j \tag{43}$$

$$D_{ij}^- = \begin{cases} 1 - \dfrac{c^- - a_{ij}}{b_{ij} + c^- - a_{ij} - b^-} & for\ b^- < b_{ij} \\ 1 - \dfrac{c_{ij} - a^-}{b^- + c_{ij} - a^- - b_{ij}} & for\ b_{ij} < b^- \end{cases} \quad \forall i, j \tag{44}$$

Then, the pre-defined hierarchical fuzzy TOPSIS algorithm steps are executed. In our application, we have 3 main attributes, 11 sub attributes, 18 sub sub-attributes and 3 alternatives. Evaluations from all 63 respondents are taken and $\tilde{I}_{MA}$, $\tilde{I}_{SA}$, $\tilde{I}_{SSA}$ and $\tilde{I}_A$ are obtained. $\tilde{I}_{MA}$ and $\tilde{I}_{SA}$ are given in Tables 12 and 13.

Table 12. $\tilde{I}_{MA}$

	GOAL
R	(0.26, 0.42, 0.57)
T	(0.29, 0.38, 0.49)
S	(0.06, 0.23, 0.36)

Table 14 shows the distances from the ideal solution for each professor and the normalized values which will make the comparison of fuzzy AHP and fuzzy TOPSIS easy.

Table 13. $\widetilde{I}_{SA}$

	R	T	S
RP	(0.11, 0.22, 0.38)	0	0
Aw	(0.09, 0.18, 0.29)	0	0
GP	(0.10, 0.26, 0.37)	0	0
RQ	(0.08, 0.18, 0.3)	0	0
ER	(0.07, 0.15, 0.27)	0	0
ANC	0	(0.26, 0.35, 0.43)	0
CM	0	(0.12, 0.23, 0.35)	0
NTS	0	(0.05, 0.25, 0.39)	0
SE	0	(0.09, 0.22, 0.41)	0
IS	0	0	(0.28, 0.45, 0.56)
ES	0	0	(0.25, 0.55, 0.80)

Table 14. S_i^*, S_i^- *and* C_i

	S_i^*	S_i^-	C_i	Normalized C_i
PROF-1	0.178056365	1.933416045	0.915671943	0.63
PROF-2	1.075008442	0.879005405	0.449846047	0.31
PROF-3	1.807534264	0.1793032	0.09024553	0.06

The results in Table 14 indicate that Prof-1 achieves the highest performance whereas prof-3 has the lowest.

5.3 Comparison of two Methods and Sensitivity Analyses

The main theoretical differences between two techniques can be counted in that way: 1. AHP considers pairwise comparisons but TOPSIS does not. This causes AHP to take into consideration more information. 2. AHP makes a consistency check but TOPSIS does not, 3. TOPSIS is easy to implement, has less number of questions to consider, and its calculations are less tedious and faster, 4. Classical TOPSIS is inadequate to consider the hierarchy in the problem, 5. TOPSIS pre-defines negative and positive ideal solutions.

These two methods are superior to the other multi attribute methods such as additive weighting and weighted product methods since a hierarchy among goal, attributes, and alternatives is not considered. Those methods evaluate the alternatives with respect to only attributes with a single level. While TOPSIS uses positive and negative ideal solutions, AHP formalizes our intuitive understanding of a complex problem using a hierarchical structure. These characteristics provide superiority to AHP and TOPSIS against many other multi attribute decision making methods.

In our application the performance ranking is obtained equal to each other in both techniques where prof-1 is the best and prof-3 is the worst. When compared the performance scores (relative closeness to ideals and priority weights) of the professors in both techniques, prof-1 strongly dominates the others in fuzzy TOPSIS. The reason of the differences in the alternatives' weights is the pairwise comparisons in fuzzy AHP which provides this technique to reflect more information to results.

Figures 6, 7, and 8 show the results of sensitivity analyses of fuzzy AHP, obtained by changing the weights of main attributes, R, T, and S. In each figure, 11 different combinations of weights of main attributes are illustrated. The priority weights for each state can be seen in the figures.

States	R	T	S
1	1.00	0.00	0.00
2	0.90	0.05	0.05
3	0.80	0.10	0.10
4	0.70	0.15	0.15
5	0.60	0.20	0.20
6	0.50	0.25	0.25
7	0.40	0.30	0.30
8	0.30	0.35	0.35
9	0.20	0.40	0.40
10	0.10	0.45	0.45
11	0.00	0.50	0.50

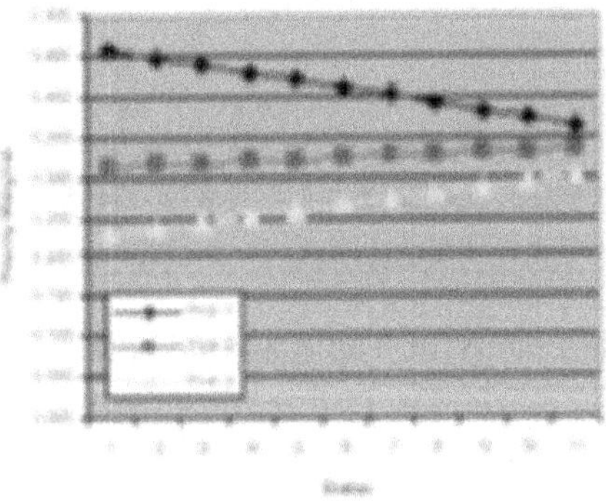

Fig. 5. Sensitivity analysis of fuzzy AHP: Case1

The sensitivity analyses indicate that Prof-1 is the best with respect to the criterion *research*. When the weights of the attributes *teaching* and *service* are increased, Prof-3's performance is more affected with respect to Prof-2's. This can be seen clearly from Fig. 6.: The slope of the curve of Prof-3 is larger than Prof-2's. In all figures, it can be seen that the performance of Prof-2 is not sensitive to the changes in the weights of the attributes. The changes in the weight of the criterion *research* affect the performance of Prof-1 more than the other professors' performances. Prof-1 and Prof-2 have the same performances only if the weight of the main attribute *service* equals to 1. Prof-1 dominates the others with the exception of this state. In all states Prof-3 never dominates the others.

States	R	T	S
1	0.00	0.00	1.00
2	0.05	0.05	0.90
3	0.10	0.10	0.80
4	0.15	0.15	0.70
5	0.20	0.20	0.60
6	0.25	0.25	0.50
7	0.30	0.30	0.40
8	0.35	0.35	0.30
9	0.40	0.40	0.20
10	0.45	0.45	0.10
11	0.50	0.50	0.00

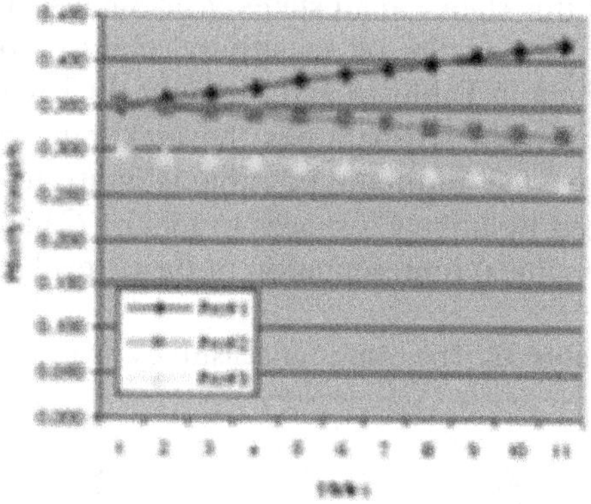

Fig. 6. Sensitivity analysis of fuzzy AHP: Case 2

States	R	T	S
1	0.00	1.00	0.00
2	0.05	0.90	0.05
3	0.10	0.80	0.10
4	0.15	0.70	0.15
5	0.20	0.60	0.20
6	0.25	0.50	0.25
7	0.30	0.40	0.30
8	0.35	0.30	0.35
9	0.40	0.20	0.40
10	0.45	0.10	0.45
11	0.50	0.00	0.50

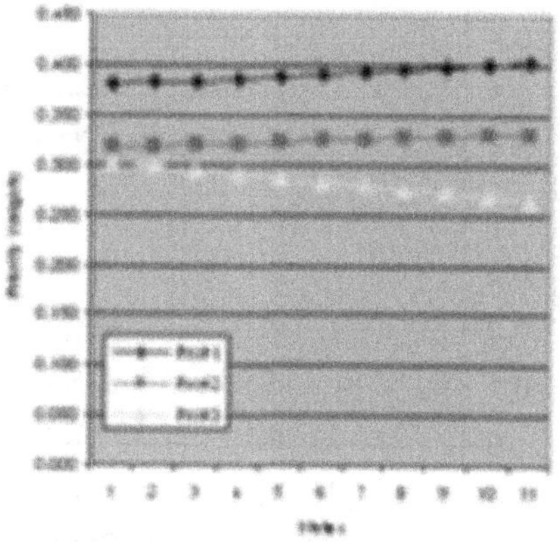

Fig. 7. Sensitivity analysis of fuzzy AHP: Case 3

The sensitivity analyses indicate that Prof-1 is the best with respect to the criterion *research.* When the weights of the attributes *teaching* and *service* are increased, Prof-3's performance is more affected with respect to Prof-2's. This can be seen clearly from Fig. 6.: The slope of the curve of Prof-3 is larger than Prof-2's. In all figures, it can be seen that the performance of Prof-2 is not sensitive to the changes in the weights of the attributes. The changes in the weight of the criterion *research* affect the performance of Prof-1 more than the other professors' performances. Prof-1 and Prof-2 have the same performances only if the weight of the main attribute *service* equals to 1. Prof-1 dominates the others with the exception of this state. In all states Prof-3 never dominates the others.

The results of sensitivity analyses of Hierarchical Fuzzy TOPSIS are given in Table 15 and Figure 9. In Table 15, the extreme states where only one main attribute has the maximum possible weight whereas the others have the minimum possible weights are examined. For each state, Normalized Relative Closeness to Ideals (C_i) are computed. Figure 9 illustrates the graphical representation of these results.

Table 15. The results of sensitivity analyses of Hierarchical Fuzzy TOPSIS

				Normalized C_i		
R	**T**	**S**	**States**	**Prof-1**	**Prof-2**	**Prof-3**
0.8, 1.0, 1.0	0.0, 0.0, 0.2	0.0, 0.0, 0.2	1	0.6742	0.2953	0.0305
0.0, 0.0, 0.2	0.8, 1.0, 1.0	0.0, 0.0, 0.2	2	0.5338	0.3129	0.1533
0.0, 0.0, 0.2	0.0, 0.0, 0.2	0.8, 1.0, 1.0	3	0.4453	0.3750	0.1797

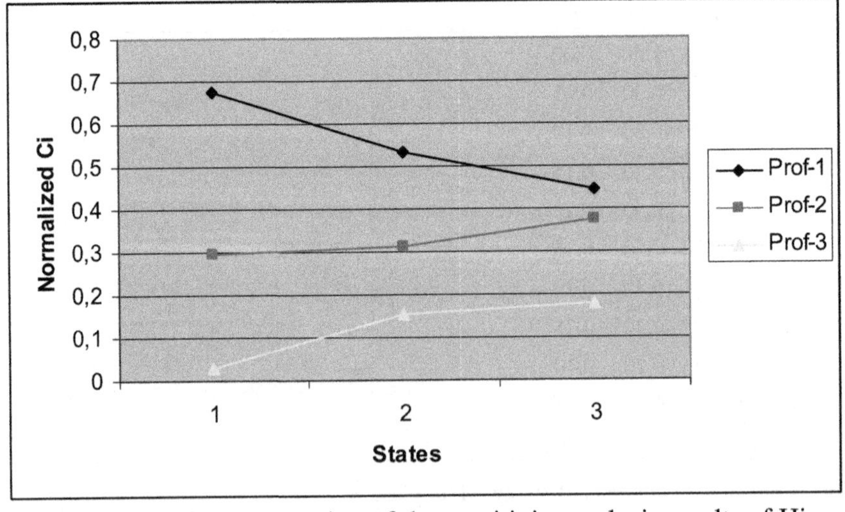

Fig. 8. Graphical representation of the sensitivity analysis results of Hierarchical Fuzzy TOPSIS

It can be seen from Figure 9 that Prof-1 is the best alternative for all states. Prof-1 is obviously superior to the other professors with respect to the attribute *research*. Prof-1 and Prof-3 are more sensitive to the changes in the weights of the attribute *teaching*. As the weight of this attribute increases, the performance of Prof-1 decreases and the performance of Prof-3 increases. When the weight of the attribute *service* has the highest value as in state 3, the performances of the professors become the closest to each other. Prof-2 is insensitive to the changes in the weights of attributes *research* and *teaching* when the weight of *service* remains constant. Prof-3 is insensitive to the changes in the weights of attributes *service* and *teaching* when the weight of *research* remains constant.

6 Conclusion

In this paper a model for evaluating faculty performance in the areas of teaching, research, and service has been presented. The model is based on the premise that faculty performance should be viewed as a product of quality and quantity.

Performance evaluation of a faculty is a complex problem in which many qualitative attributes must be considered. These kinds of attributes make the evaluation process hard and vague. Hierarchical structure is a good approach to describe complicated system. The judgments from experts are always in vague rather than in crisp numbers. It is suitable and flexible to express the judgments of experts in fuzzy number instead of in crisp number. Fuzzy AHP has the capability of taking pairwise comparisons of these attributes into account with a hierarchical structure. Many fuzzy TOPSIS methods have been proposed without considering these pairwise comparisons between attributes and a hierarchical structure by today. To be able to compare these methods, a fuzzy TOPSIS method has to take this hierarchy into account. The hierarchical fuzzy TOPSIS method developed in this paper has this ability without making pairwise comparisons.

The results of the application made in a faculty of a university show that both fuzzy AHP and hierarchical fuzzy TOPSIS methods give the same ranking. The reasons for the significant differences in the alternative weights may be the differences in input data and the additional information through pairwise comparisons in fuzzy AHP. Furthermore, under the assumption of consistency fuzzy hierarchical TOPSIS outperforms fuzzy AHP from a practitioner's point of view in terms of easiness. It is clear that the evaluation of a faculty is a difficult and sensitive issue which has quantitative and qualitative aspects, complexity and imprecision. However, these two fuzzy methods seem to be usable for the solution of this problem. These methods can be used for the performance evaluation of a faculty in any department of any university.

For further research, a hierarchical fuzzy TOPSIS method that can take pairwise comparisons between main and sub-attributes into account with a different manner from AHP may be developed. Furthermore, different levels of consistencies would be searched. Especially, it is critical to find the break-even consistency level where TOPSIS outperforms AHP.

References

Abdel-Kader, M.G., Dugdale, D., Evaluating investments in advanced manufacturing technology: A fuzzy set theory approach, British Journal of Accounting, Vol. 33, pp. 455-489, 2001.

Agrell, P., J., Steuer, R., E., ACADEA—A decision support system for faculty performance reviews, Journal of Multi-Criteria Decision Analysis 9, 191-204, 2000.

Chang, D-Y., Applications of the Extent Analysis Method on Fuzzy AHP, European Journal of Operational Research, Vol. 95, pp. 649-655, 1996.

Chang, D-Y., Extent Analysis and Synthetic Decision, Optimization Techniques and Applications, Vol. 1, World Scientific, Singapore, p. 352, 1992.

Chen, S.-H., Ranking fuzzy numbers with maximizing set and minimizing set. Fuzzy Sets and Systems, 17, 113–129, 1985.

Chen, S.-J., Hwang, C.-L., Fuzzy Multiple Attribute Decision Making Methods and Applications, Springer-Verlag, Berlin, 1992.

Chen, T.-C., Extensions of the TOPSIS for group decision-making under fuzzy environment, Fuzzy Sets and Systems, Vol. 114, pp. 1-9, 2000.

Chu, T.-C, Facility location selection using fuzzy topsis under group decisions, International Journal of Uncertainty, Fuzziness and Knowledge-Based Systems, Vol. 10, No. 6, pp. 687-701, 2002.

Chu, T.-C., Lin, Y.-C., A Fuzzy TOPSIS Method for Robot Selection, International Journal of Advanced Manufacturing Technology, Vol. 21, pp. 284-290, 2003.

Deutsch, S., J., Malmborg, C., J., Evaluating organizational performance using fuzzy subsets, European Journal of Operational Research 22(2), 1985, 234-242, 2003.

Ellington, H., Ross, G., Evaluating Teaching Quality throughout a University A Practical Scheme Based on Self-assessment, Quality Assurance in Education, Vol. 2, No. 2, pp. 4-9,1994.

Hon, C.-C., Guh, Y.- Y., Wang, K.-M., Fuzzy Multiiple Attributes and Multiple Hierarchical Decision Making, Computers Math. Applic., Vol. 32, No. 12, pp. 109-119, 1996.

Huberty, C., J., An approach to annual assesment and evaluation of university faculty, Journal of Personal Evaluation in Education 14(3), 241-251, 2000.

Hwang, C.-L., Yoon, K., Multiple Attribute Decision Making Methods and Applications, Springer-Verlag, New York, 1981.

Jauch, L.R., Glueck, W.F., Evaluation of university professors' research performance, Management Science, Vol. 22, No. 1, pp. 66-75, 1975.

Kahraman, C., Cebeci, U., Ruan, D., Multi-attribute comparison of catering service companies using fuzzy AHP: The case of Turkey, International Journal of Production Economics, Vol. 87, Issue 2, pp. 171-184, 2004.

Kaufmann, A., Gupta, M, M, Fuzzy Mathematical Models in Engineering and Management Science, North Holland, 1988.

Kuo, Y., Chen, L., Using the fuzzy synthetic decision approach to assess the performance of university teachers in Taiwan, International Journal of Management 19 (4), 593-604, 2002.

Lee, E.S, Li, R.L., Comparison of fuzzy numbers based on the probability measure of fuzzy events, Computer and Mathematics with Applications, Vol. 15, pp 887-896, 1998.

Liang, G.-S., Fuzzy MCDM based on ideal and anti-ideal concepts, European Journal of Operational Research, Vol. 112, pp.682-691, 1999.

Liou, T., S., Wang, M., J., J., Ranking fuzzy numbers with integral value, Fuzzy Sets and Systems, 50, 247, 1992.

Meho, L., Sonnenwald, D., H., Citation ranking versus peer evaluation of senior faculty research performance: A case study of Kurdish Scholarship, Journal of American Society for Information Science 51(2), 123-138, 2000.

Mesak, H.I., Jauch, L.R., Faculty Performance Evaluation: Modeling to Improve Personnel Decisions, Decision Sciences, Vol-22, pp```` 1142-1157

Paulsen, M.B., Evaluating teaching performance, New Directions for Institutional Research, 14, 5-18, 2002.

Sproule, R., The under determination of instructor performance by data from the student evaluation of teaching, Economics of Education Review 21, 287-294, 2002.

Tsaur, S.-H, Chang, T.-Y, Yen, C.-H, The evaluation of airline service quality by fuzzy MCDM, Tourism Management, Vol. 23, pp. 107-115, 2002.

Weistroffer, H.R., Spinelli, M.A., Canavos, G.C., Fuhs, F.P., A merit pay allocation model for college faculty based on performance quality and quantity, Economics of Education Review 20, 41-49, 2001.

Zadeh, L., Fuzzy sets, Information Control, Vol. 8, pp. 338-353., 1965.

Zhang, G., Lu, J., An Integrated Group Decision-Making Method Dealing with Fuzzy Preferences for Alternatives and Individual Judgments for Selection Criteria, Group Decision and Negotiation, Vol. 12, pp. 501-515, 2003.

Zhao, R., Govind, R., Algebraic Characteristics of Extended Fuzzy Numbers, Information Sciences, 54(1-2), 103-130, 1991.

APPENDIX

A Part from the Questionnaire of Fuzzy AHP

Read the following questions and put check marks on the pairwise comparison matrices. If an attribute on the left is more important than the one matching on the right, put your check mark to the left of the importance "*Equal*" under the importance level you prefer. If an attribute on the left is less important than the one matching on the right, put your check mark to the right of the importance '*Equal*' under the importance level you prefer.

With respect to the overall goal "*selection of the best performance among faculty*",

Q1. How important is Research (R) when it is compared with Teaching (T)?

Q2. How important is Research (R) when it is compared with Service (S)?

Q2. How important is Teaching (T) when it is compared with Service (S)?

With respect to the selection of the best performance among faculty	Importance (or preference) of one main-attribute over another										
Questions	Attributes	Absolutely More Important	Very Strongly More Important	Strongly More Important	Weakly Important	Equally Important	Just Equal	Absolutely More Important	Very Strongly More Important	Strongly More	Attributes
Q1	R					✔					T
Q2	R				✔						S
Q3	T			✔							S

Fig. 9. Questionnaire form used to facilitate comparisons of main attributes

With respect to the sub-sub-attribute "Community Service (CS)"

Q106. How important is Prof-1 when it is compared with Prof-2?

Q107. How important is Prof-1 when it is compared with Prof-3?

Q108. How important is Prof-2 when it is compared with Prof-3?

With respect to Community service		Importance (or preference) of one alternative over another									
Questions	Attributes	Absolutely More Important	Very Strongly More Important	Strongly More Important	Weakly Important	Equally Important	Just Equal	Absolutely More Important	Very Strongly More Important	Strongly More Important	Attributes
Q106	Prof-1									✔	Prof-2
Q107	Prof-1				✔						Prof-3
Q108	Prof-2			✔							Prof-3

Fig. 10. Two of the 23 questionnaire forms used to facilitate comparisons of alternatives Continued

A Part from the Questionnaire of Fuzzy TOPSIS

With respect to the overall goal "Selection of the best performance among faculty"

Q1. What degree of importance do you assign to the main attribute Research (R)?

Q2. What degree of importance do you assign to the main attribute Teaching (T)?

Q3. What degree of importance do you assign to the main attribute Service (S)?

With respect to Overall goal		Importance of one attribute with respect to overall goal				
Ques-	Attrib-	(0, 0, 0.2) Very Low	(0, 0.2, 0.4) Low	(0.3, 0.5, 0.7) Medium	(0.6, 0.8, 1) High	(0.8, 1 1) Very High
Q1	R		✔			
Q2	T			✔		
Q3	S					✔

Fig. 11. Questionnaire form used to facilitate importance degrees of main attributes wrt. goal

Q22. What scores do you assign to each professor with respect to the sub-sub attribute Number of Research Projects (NRP)?

Q23. What scores do you assign to each professor with respect to the sub-sub attribute Number of Papers (NP)?

Q44. What scores do you assign to each professor with respect to the sub-sub attribute Community Service (CS)?

Questions	Attributes	Alternatives	Very Poor (0, 0, 20)	Poor (0, 20, 40)	Fair (30, 50, 70)	Good (60, 80, 100)	Very Good (80, 100, 100)
		PROF-1		✔			
Q22	NRP	PROF-2	✔				
		PROF-3			✔		
		PROF-1				✔	
Q23	NP	PROF-2		✔			
		PROF-3			✔		
				…			
		PROF-1			✔		
Q44	CS	PROF-2		✔			
		PROF-3				✔	

Fig. 12. Questionnaire form used to facilitate scores of alternatives wrt. sub and sub-sub attributes

Fuzzy Quantitative Association Rules and Its Applications

Peng Yan and Guoqing Chen[1]

School of Economics and Management, Tsinghua University, Beijing 100084, China

Summary: In recent years, association rules from large databases have received considerable attention and have been applied to various areas such as marketing, retail and finance, et al. While conventional approaches usually deal with databases with binary values, this chapter introduces an approach to discovering association rules from quantitative datasets. To remedy possible boundary problems due to sharp partitioning and to represent linguistic knowledge, fuzzy logic is used to "discretize" quantitative domains. A method of finding fuzzy sets for each quantitative attribute by using clustering is proposed based on different overlapping degrees. This proposed method is then applied to two real datasets *housing* and *credit*.

Key words: Fuzzy quantitative association rules, clustering, housing data, credit data

1 Introduction

The past few decades have witnessed an explosion in the amount of electronically stored data due to advances in information technology and massive applications in business and scientific domains. Data mining, sometimes also referred to as knowledge discovery in database (KDD), is concerned with the nontrivial extraction of implicit, previously unknown, and potentially useful knowledge from data (Piatetsky-Shapiro et al. 1991). Association rules mining has become a focal point of research and applications since Agrawal (Agrawal et al., 1993) proposed an approach to discovering boolean association rules for the so-called "market-basket problem". Given a set of transactions where each transaction is a set of items, an association rule is an expression of the form $X \Rightarrow Y$, where X and Y are two sets of items, $X \cap Y = \emptyset$. Usually, a rule $X \Rightarrow Y$ can

[1] Corresponding author: Tel.: 86-10-62772940, Fax: 86-10-62785876, E-mail: chengq@em.tsinghua.edu.cn.

P. Yan and G. Chen: *Fuzzy Quantitative Association Rules and Its Applications*, StudFuzz **201**, 573–587 (2006)
www.springerlink.com

be obtained if its degree of support, $Dsupp(X \Rightarrow Y) = ||XY||/|D|$, is greater than or equal to the pre-specified threshold (*Min_supp*), and its degree of confidence, $Dconf(X \Rightarrow Y) = ||XY||/||X||$, is greater than or equal to the pre-specified threshold (*Min_conf*), where $||X||$ is the number of transactions that contain X, and $|D|$ is the total number of transactions in dataset D.

In recent years, there have been many attempts (Chen et al. 2002; Mannila et al. 1994; Roberto et al. 1999; Srikant and Agrawal 1994; Srikant and Agrawal 1995) to improve or extend Agrawal's work (hereafter referred to as the classical approach). Since real-world applications usually consist of quantitative values, mining quantitative association rules has been regarded meaningful and important, and carried out, for instance, by partitioning attribute domains and then transforming the quantitative values into binary ones in order to apply the classical mining algorithm (Srikant and Agrawal 1996). Since discovered rules are highly dependant on the partition methods, several efforts have been proposed to build "good" partitions for quantitative association rules (Aumann and Lindell 2003; Fukuda et al. 2001; Miller and Yang 1997; Rastogi and Shim 2001).

Fuzzy logic has been used widely in intelligent systems owing to its similarity to human reasoning (Chen 1998). Since fuzzy logic often represents rules in a more interesting manner and top management decision makers often refer to decision rules in terms of linguistic words, various efforts have been made to incorporate the notion of fuzziness into association rules mining. Fuzzy association rules (FAR) with fuzzy taxonomies and FAR with linguistic hedges were introduced in (Chen and Wei 2002). Meanwhile, to deal with the "sharp boundary" problem in partitioning quantitative data domains, fuzzy quantitative association rules were discussed in (Au and Chan 1997; Chien et al. 2001; Gyenesei 2000; Hong et al. 2001; Zhang 1999). Furthermore, from a more logic-oriented viewpoint, the notion of fuzzy implication was introduced in fuzzy association rules in (Chen et al. 2004; Hullermeier 2001). Moreover, different fuzziness-related interestingness measures were proposed in (Graff et al. 2001) to extend the notions of Dsupp-Dconf, including so-called weighted association rules applied to distinguish the importance of different items (Cai et al. 1998).

For the given interval width of a quantitative attribute, crisp partition will influence the algorithm's mining results in two ways: sharp boundaries and split point selections. As an example, for a company with 100 employees with age distribution shown in Figure 1, given *Min_supp* = 0.4 and interval width 10, the results of two partition methods are shown in Table 1.

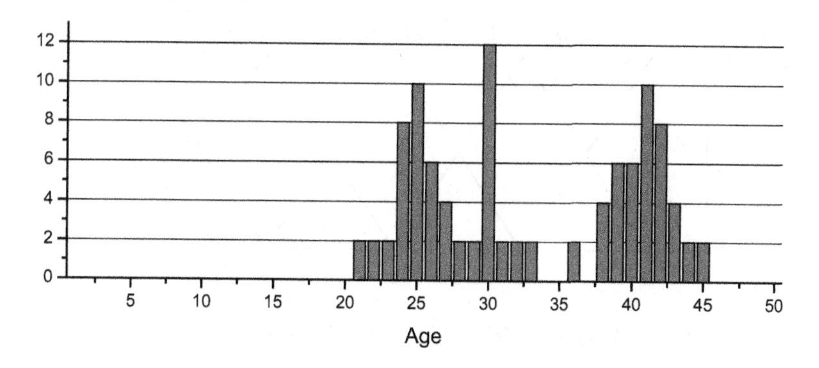

Fig. 1. Age distribution of a company

Table 1. *Dsupp* of two partition methods

Partition I		Partition II	
Age	*Dsupp*	Age	*Dsupp*
[1, 10)	0	[5, 15)	0
[10, 20)	0	[15, 25)	0.14
[20, 30)	0.38	[25, 35)	0.42
[30, 40)	0.3	[35, 45)	0.42
[40, 50)	0.32	[45, 55)	0.02

It can be seen that none of the intervals of partition I is frequent. However, the non-frequent interval [20, 30) with support 0.38 should be interesting if we consider the values near its right boundary with age=30, that is, *Dsupp*(Age[20, 31))= 0.50. That is, crisp partition may under-emphasize or over-emphasize the elements near the boundaries of intervals in the mining process. Furthermore, for different split point selections of partition I and partition II, partition II finds two frequent intervals ([25, 35) and [35, 45)) whereas partition I finds none. It means that the split points will greatly influence the mining results.

Fuzzy logic therefore can be employed to naturally deal with such a sharp boundary problem and weaken the influence of split point selections. In our example, for the partition I, if we define fuzzy sets on the domain of age, age of 30 will partly belongs to the interval [20, 30), which will make the interval frequent. For the fuzzy sets shown in Figure 2, *Dsupp*(about 25) equal to 0.404 and is frequent. In addition, since fuzzy logic does not clearly distinguish elements around split points, it can efficiently reduce the influence of arbitrary split point selections.

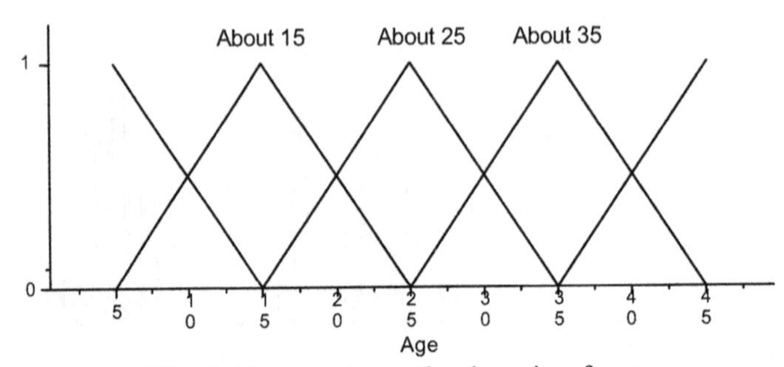

Fig. 2. Fuzzy sets on the domain of age

In dealing with quantitative datasets for association rule mining, the original quantitative dataset D can be transformed into a fuzzy dataset (note D_f) with attribute values in [0, 1]. In this chapter, clustering methods are used to find suitable partitions to reflect data distribution for each attribute, based on which the fuzzy sets can be built in different overlapping degrees. For the sake of easy comprehension and presentation, the number of clusters in our algorithm is 3. That is, linguistic terms can be used to represent quantitative attributes (e.g., young, middle, old) by "discretizing" attribute domains such that any numeric value for an attribute in the original dataset D is transformed into a value in [0,1] for the new attributes in D_f. Then, the definition of fuzzy quantitative association rules is proposed and classical Apriori algorithm is extended. The proposed algorithm is applied into real datasets to show its effectiveness.

The remaining part of this chapter is organized as follows. Clustering based fuzzy sets generation, fuzzy quantitative datasets and fuzzy quantitative association rules are introduced in Section 2. Several optimization strategies are discussed and the algorithm is described in detail in Section 3. In Section 4, the results of the algorithm applied on housing and credit are discussed.

2 Fuzzy Quantitative Association Rules

2.1 Clustering Based Fuzzy Sets Generation

Based on the notion that resemble transactions may belong to the same partition, clustering algorithm has been introduced into quantitative association rules in (Miller and Yang 1997) to find "good" intervals. In this chapter, the well-known *k* mean algorithm is used to built clustering center, and then to derive fuzzy sets in difference overlapping degrees.

2.1.1 Iterative Distance-based Clustering

K-means is one of the simplest clustering algorithms and has been widely used in real applications. First, *k* clusters centroids are specified in advance. Then k points are chosen as cluster centers, which should be placed as much as possible far away from each other. Instances are assigned to their closest clusters centroids according to the ordinary Euclidean distance function. Next the centroid of all instances in each cluster is calculated, which are taken to be new center values for their respective clusters. The whole process is repeated with the new cluster centroids until the cluster centroid have stabilized and will remain the same thereafter (Yu and Chen 2005).

2.1.2 Determining Fuzzy Sets by Using the Discovered Clustering

Let *k* be equal to 3, two split points can be found using k-means algorithm. Given the overlapping degree σ, three fuzzy sets of the attribute can be defined as follows (Figure 3 and formulas (1), (2), (3)):

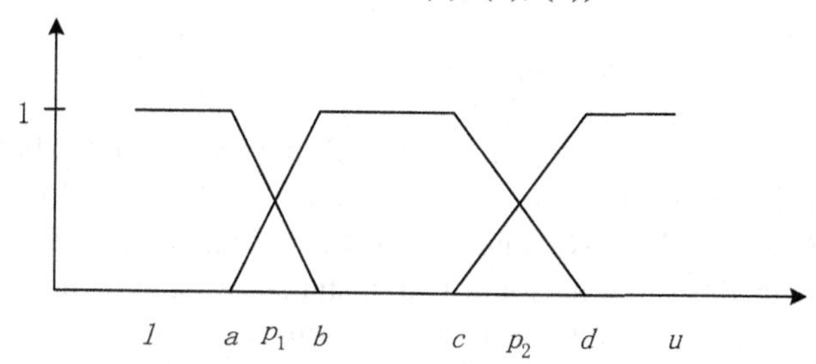

Fig. 3. Generation of fuzzy sets

$$f_1(x)=\begin{cases}1 & l\le x\le a\\ \dfrac{x}{a-b}+\dfrac{b}{b-a} & a\le x\le b\end{cases} \tag{1}$$

$$f_2(x)=\begin{cases}\dfrac{x}{b-a}+\dfrac{a}{a-b} & a\le x\le b\\ 1 & b\le x\le c\\ \dfrac{x}{c-d}+\dfrac{d}{d-c} & c\le x\le d\end{cases} \tag{2}$$

$$f_3(x)=\begin{cases}\dfrac{x}{d-c}+\dfrac{c}{c-d} & c\le x\le d\\ 1 & d\le x\le u\end{cases} \tag{3}$$

where p_1 and p_2 are two split points of the attributes, u and l are the maximal value and minimal value of the attribute respectively, a=(1-σ)p_1+σl, b=p_1+σ (p_2-p_1)/2, c=p_2-σ (p_2-p_1)/2, d=(1-σ)p_2+σu.

2.2 Data Transformation

Classically, for any given quantitative dataset D, sharp partitioning is used to discretize each of attribute domains into a number of distinct intervals. To deal with "the boundary problem" and to naturally facilitate knowledge representation, a number of fuzzy sets, usually labeled by linguistic terms, could be defined upon each domain as values of the attribute concerned. Concretely, let $I = \{I_1, I_2, \ldots, I_m\}$ be the item set where each I_j ($1 \leq j \leq m$) is an attribute of the original quantitative dataset D with domain Δ_j and several fuzzy sets on Δ_j associated with I_j. Then the original quantitative dataset D can be transformed into an extended dataset D_f with all the attribute values in [0, 1].

2.3 Fuzzy Quantitative Association Rules

Given fuzzy dataset $D_f = \{t_1, t_2, \ldots, t_n\}$ with the extended attribute set I_f, the discovered rules are of the form: $X \Rightarrow Y$, where t_i ($i=1, 2, \ldots, n$) is a transaction in D_f, $X \subset I_f$, $Y \subset I_f$, $X \cap Y = \varnothing$, and $X \cup Y$ do not contain any two items that are associated with the same attribute (for instance, $X \cup Y$ will not contain "age young and age old"). Like boolean association rules mining, X is called antecedent of the rule and Y is called consequent of the rule. An example of such rules is: "age young ⇒ monthly-income high".

Usually, in the mining process, degrees of support of itemsets (i.e., sets of items) are calculated first and those itemsets with their degrees of support greater than or equal to a user specified threshold are called "*frequent itemsets*". In fact, one will generally be interested only in rules with high degrees of support, otherwise, the rules may not be worth being considered.

For any itemset $X = \{x_1, x_2, \ldots, x_p\} \subset I_f$, the degree of support for X with respect to the ith transaction, $Dsupp_i(X)$, can be obtained as follows:

$$Dsupp_i(X) = T(x_{1i}, x_{2i}, \ldots, x_{pi}) \tag{4}$$

where x_{ji} is the value of x_j for ith transaction, $x_{ji} \in [0, 1]$, $i = 1, 2, \ldots, n$, $j = 1, 2, \ldots, p$, and T represents a t-norm. T-norms are mappings from $[0, 1]^2$ to [0, 1] and the commonly used t-norms include min (T(a, b) = min (a, b)), product (T(a, b)=a*b), and so on (Gupta and Qi, 1991).

Furthermore, in terms of how many transactions in D_f supporting X, $Dsupp$ of itemset X for the whole dataset D_f is defined as:

$$Dsupp(X) = \frac{\sum_{i=1}^{n} Dsupp_i(X)}{|D_f|} \tag{5}$$

where $|D_f|$ is the total number of transactions in D_f, which is equal to the number of transactions in D.

Note that *Dsupp* of a rule $X \Rightarrow Y$ ($Dsupp(X \Rightarrow Y)$) equals *Dsupp* of the itemset $X \cup Y$. Thus, $Dsupp(X \Rightarrow Y)$ can be obtained as:

$$Dsupp(X \Rightarrow Y) = \frac{\sum_{i=1}^{n} Dsupp_i(X \cup Y)}{|D_f|} \tag{6}$$

where $\cup$ is a union operator for fuzzy sets.
Similar to definition of confidence in classical association rules, degree of confidence, Dconf is defined as:

$$Dconf(X \Rightarrow Y) = \frac{Dsupp(X \cup Y)}{Dsupp(X)} \tag{7}$$

Next, based upon the notions of *Dsupp* and *Dconf*, fuzzy quantitative association rule (FQAR) is defined as follows.

Definition 1: The rule $X \Rightarrow Y$ is a FQAR if:
$Dsupp(X \Rightarrow Y) \geq Min\_supp$
$Dimp(X \Rightarrow Y) \geq Min\_conf$

where *Min_supp* and *Min_conf* are two thresholds defined by experts or decision makers.

3 The Proposed Algorithm

Since the rules in consideration are those whose degrees of support and of confidence are greater than or equal to the pre-specified *Min_supp* and *Min_conf*, a straightforward extended mining algorithm of (Chen et al. 2004) could be obtained easily. It is basically composed of three phases: generating fuzzy datasets D_f, generating all frequent itemsets from D_f based on the classical approach, then generating all rules than satisfy confidence constraint from frequent itemsets. The input of the algorithm includes original dataset D, *Min_supp* and *Min_conf*. The output of the algorithm will be the fuzzy quantitative association rules.

Notations used in the algorithm are as follows:

D: original quantitative dataset
D_f: fuzzy dataset
t: a transaction in D_f
L_k: set of frequent k-itemsets
C_k: set of candidate k-itemsets
$|D_f|$: the number of transactions in D_f
Ψ: set of fuzzy quantitative association rules
f_j^k: the kth fuzzy membership function associated with the attribute I_j
u_{jt}: item I_j's value of tth transaction of D_f
v_{jt}: item I_j's value of tth transaction of D
T: t-norm operator
σ: overlapping degree

3.1 Generating Fuzzy Datasets (D_f)

Given the original quantitative datasets and number of the clustering algorithm, the fuzzy sets associated with each attribute can be built as follows:

Algorithm building fuzzy sets

for all attribute $I_j \in I$ **do**
  generating 3 clusters and 2 split points (p_1, p_2) using k-mean algorithm
  generating 3 fuzzy sets according to (1), (2), (3)
end for

Using fuzzy sets associated with attributes, the algorithm transforms original dataset D into D_f. In this step, each attribute is associated with several fuzzy sets and an attribute in D is transformed into several new attributes in D_f corresponding to these fuzzy sets. Concretely, we have the following algorithm transform.

Algorithm transform

for all transaction $t \in D$ **do**
  for all attributes $I_j \in I$ **do**
    for all fuzzy membership functions f associated with attribute I_j **do**
      $u_{jt} = f_j^k(v_{jt})$
      insert u_{jt} **into** D_f,
    end for
  end for
end for

3.2 Discovering Frequent Itemsets

Once D_f is generated, the next step is to discover frequent itemsets from D_f. Recall the well-known Apriori algorithm (Srikant and Agrawal 1994), our extended Apriori algorithm is given as follows:

Extended Apriori algorithm

L_1 = {frequent 1-itemsets}
for $\{k = 2; L_{k-1} \neq \emptyset; k{+}{+}\}$ **do**
C_k = Apriori-Gen(L_{k-1}); // Generating candidates from L_{k-1}//
forall transactions $t \in D_f$ **do**
C_t = subset(C_k, t); // Generating candidate subsets of t //
forall candidates $c \in C_t$ **do**
c.support = c.support + $T(u_{1t}, u_{2t}, \ldots, u_{kt})$
endfor
endfor
$L_k = \{c \in C_k \mid \text{c.support} \geq Min\_supp \times |D_f|\}$
endfor
All frequent itemsets = $\cup_k L_k$

where Apriori-Gen is a procedure to generate the set of all candidate itemsets and represented as follows.

Candidate Itemsets Generation: Assuming that the items in each itemset are kept stored in lexicographic order. First, join L_{k-1} with L_{k-1}:

insert into C_k
select $p.item_1, p.item_2, \ldots, p.item_{k-1}, q.item_{k-1}$
from L_{k-1} p , L_{k-1} q
where $p.item_1 = q.item_1, \ldots, p.item_{k-1} = q.item_{k-2}, p.item_{k-1} < q.item_{k-1}$,

$q.item_{k-1}$ and $p.item_{k-1}$ associated different attributes in D

In this process, items associated with the same attribute will not join together. Based on the fact that any subset of a frequent itemset is also frequent, delete all itemsets $c \in C_k$ such that any (k-1)-subset of c is not in L_{k-1}. This can be a pruning strategy.

Pruning

forall candidates $c \in C_k$ **do**
forall (k-1)-itemsets $d \subseteq c$ and $c \in C_k$ **do**
if $d \notin L_{k-1}$ **then**

delete c **from** C_k

endif

endfor

endfor

Note that the major differences between the above extension and the Apriori algorithm are that the values of the dataset are in domain [0, 1] and t-norms are used to calculate rules' degrees of support, and that the items in D_f resulting from those fuzzy sets associated with the same attribute in D will not appear in the same candidate itemset.

3.3 Generating FQAR

The third phase is to generate all qualified FQAR from frequent itemsets using the similar method of classical Apriori. Concretely:

Algorithm generating FQAR

$\forall$ frequent k itemset $X \in L_k$

Generating(X, m) //Generating rules from X with m-1 partition antecedent

Generating(X, m)

if confidence($X_{(m-1)} \Rightarrow (X - X_{(m-1)})$) $\geq$ Min_conf // $X_{(m-1)}$ is m-1 subset of X

Then insert $X_{(m-1)} \Rightarrow (X - X_{(m-1)})$ into Ψ

endif

Generating (X, m-1)

4 Applications

Experiments on two real datasets are carried out to show effectiveness of the proposed algorithm. One is housing with 506 instances obtained from the well-known databanks at UCI KDD (http://kdd.ics.uci.edu/) that concerns housing values in suburbs of Boston. The other is credit rating for Industrial and Commercial Bank of China.

4.1 Housing

The housing dataset consists of 13 continuous attributes, 1 binary-valued attribute, which are shown in Table 2.

Table 2. Data description of housing

Name and Description		Type
1. CRIM	per capita crime rate by town	Quantitative
2. ZN	proportion of residential land zoned for lots over 25,000 sq.ft.	Quantitative
3. INDUS	proportion of non-retail business acres per town	Quantitative
4. CHAS	Charles River dummy variable (= 1 if tract bounds river; 0 otherwise)	Binary
5. NOX	nitric oxides concentration (parts per 10 million)	Quantitative
6. RM	average number of rooms per dwelling	Quantitative
7. AGE	proportion of owner-occupied units built prior to 1940	Quantitative
8. DIS	weighted distances to five Boston employment centres	Quantitative
9. RAD	index of accessibility to radial highways	Quantitative
10. TAX	full-value property-tax rate per $10,000	Quantitative
11. PTRATIO	pupil-teacher ratio by town	Quantitative
12. B	1000(Bk - 0.63)^2 where B_k is the proportion of blacks by town	Quantitative
13. LSTAT	% lower status of the population	Quantitative
14. MEDV	Median value of owner-occupied homes in $1000's	Quantitative

Given σ=20%, *Min_supp* = 0.5 and *Min_conf* = 0.95, some selected discovered rules are shown in Table 3.

Table 3. Some discovered FQAR from housing when σ=20%

	Rules	*Dsupp*	*Dconf*
r_1	RAD=easy ⇒ CRIM=low	0.5742	1.0000
r_2	MEDV=middle ⇒ CRIM=low	0.5953	0.9960
r_3	CHAS =1, RM=middle ⇒ CRIM=low	0.7525	0.9676
r_4	AGE=high ⇒ DIS=near	0.5573	0.9527
r_5	DIS=near ⇒ ZN=low	0.5782	0.9690
r_6	RAD=easy ⇒ B=high	0.5563	0.9689

4.2 Credit

The data are about 254 companies that gained loans from Industrial and Commercial Bank of China between 2000 and 2002, which are mainly the clients from Bank of Communications, Bank of China and CITIC Industrial Bank. According to the expertise from credit bureau of Industrial and Commercial Bank in Shanghai, the credit dataset contains 18 quantitative and 1 binary attributes. The indexes summarized the financial structure, debt paying ability, management ability and operation profitability, which are listed in Table 4 (Yu and Chen 2005).

Table 4. Financial indexes for Credit Rating

Financial Structure	1. Net Asset/Loan ratio	Quantitative
	2. Asset/Liability ratio	Quantitative
	3. Net fix asset/fix asset book value ratio	Quantitative
	4. Long-term asset/shareholder equity ratio	Quantitative
Debt Paying Ability	5. Current Ratio	Quantitative
	6. Quick Ratio	Quantitative
	7. Non-financing cash inflow/liquidity liability ratio	Quantitative
	8. Operating cash inflow/liquidity liability	Quantitative
	9. Interest Coverage	Quantitative
	10. Contingent debt/net asset ratio	Quantitative
Management Ability	11. Cash revenue/operating revenue ratio	Quantitative
	12. Account Receivable Turnover Ratio	Quantitative
	13. Inventory Turnover Ratio	Quantitative
	14. Fix Asset Turnover Ratio	Quantitative
Operation Profitability	15. Gross Profit Margin	Quantitative
	16. Operating Profit Ratio	Quantitative
	17. Return on Equity	Quantitative
	18. Return on Assets	Quantitative
	19. Credit Rating	Binary

Some discovered FQAR from credit when σ=20%, *Min_supp* = 0.4 and *Min_conf* = 0.90 are given in Table 5.

Table 5. Some discovered FQAR from credit when σ=20%, *Min_supp* = 0.4 and *Min_conf* = 0.90

	Rules	*Dsupp*	*Dconf*
r_1	Current Ratio=low, Fix Asset Turnover Ratio=low⇒Operating cash inflow/liquidity liability=low	0.4050	0.9083
r_2	Account Receivable Turnover Ratio=low, Inventory Turnover Ratio=low⇒Operating cash inflow/liquidity liability=low	0.4532	0.9161
r_3	Net Asset/Loan ratio=low⇒Operating cash inflow/liquidity liability=low	0.8122	0.9172
r_4	Cash revenue/operating revenue ratio=middle, Return on Assets=middle⇒Return on Equity=middle	0.5102	0.9075
r_5	Quick Ratio=low⇒Net Asset/Loan ratio=low, Operating cash inflow/liquidity liability=middle	0.5460	0.9591
r_6	Account Receivable Turnover Ratio=low, Fix Asset Turnover Ratio=low⇒Net Asset/Loan ratio=low	0.5348	0.9282

	Rules	*Dsupp*	*Dconf*
r_7	Quick Ratio=low⇒Net Asset/Loan ratio=low	0.5583	0.9806
r_8	Current Ratio=low, Fix Asset Turnover Ratio=low⇒Net Asset/Loan ratio=low	0.4449	0.9978
r_9	Current Ratio=low⇒Quick Ratio=low	0.4724	0.9307
r_{10}	Operating Profit Ratio=high⇒credit rating=low	0.7511	0.9320
r_{11}	Net Asset/Loan ratio=low, Long-term asset/shareholder equity ratio=low⇒credit rating=low	0.4864	0.9673
r_{12}	Long-term asset/shareholder equity ratio=low⇒Operating Profit Ratio=high, credit rating=low	0.5657	0.9169
r_{13}	Long-term asset/shareholder equity ratio=low, Account Receivable Turnover Ratio=low⇒Operating Profit Ratio=high	0.4417	0.9143
r_{14}	Net Asset/Loan ratio=low, Long-term asset/shareholder equity ratio=low⇒Operating Profit Ratio=high	0.4620	0.9187
r_{15}	Long-term asset/shareholder equity ratio=low, Return on Equity=middle⇒Operating Profit Ratio=high	0.4083	0.9206
r_{16}	Asset/Liability ratio =middle, Cash revenue/operating revenue ratio=middle⇒Operating Profit Ratio=high	0.4736	0.9213
r_{17}	Long-term asset/shareholder equity ratio=low, Cash revenue/operating revenue ratio=middle⇒Operating Profit Ratio=high	0.5072	0.9321
r_{18}	Long-term asset/shareholder equity ratio=low, Current Ratio=middle⇒Operating Profit Ratio=high	0.4109	0.9442
r_{19}	Gross Profit Margin =middle, credit rating=low⇒Cash revenue/operating revenue ratio=middle	0.4490	0.9076
r_{20}	Current Ratio=middle, Operating Profit Ratio=high⇒Long-term asset/shareholder equity ratio=low	0.4109	0.9344

As shown in table 5, there are some rules such as current ratio and quick ratio that are highly related (r_9), operating cash inflow/liquidity liability is dependant on account receivable turnover and inventory turnover (r_2). Consider the rules r_{12}, r_{13}, r_{15}, r_{17} and r_{18}, they reflect that operating profit ratio of a company will be influenced by long-term asset/shareholder equity ratio, account receivable Turnover, return on equity cash revenue/operating revenue and current ratio. Some novel rules are also discovered, e.g., rules r_{16} and r_{20}. r_{16} means that operating profit ratio of a stable company with middle asset/liability ratio and middle cash revenue/operating revenue is high, and r_{20} means that a company's long-term asset/shareholder equity ratio is low if its current ratio is middle and operating profit ratio is high. There are also some redundant rules, for example, r_5 is redundant because r_7 gives a more specific kind of knowledge, r_1 is redundant because it can be derived from r_3 and r_8, and r_{11} can be derived from r_{14} and r_{10}.

5 Concluding Remarks

To avoid sharp boundary and split point selection problems, fuzzy logic has been introduced into the quantitative association rules mining, in which clustering algorithm is used to build fuzzy sets. The fuzzy quantitative association rules have been defined and the classical Apriori algorithm has been extended to extract more linguistic rules. Real applications on the data set housing and credit have been shown.

Acknowledgments

The work was partly supported by National Natural Science Foundation of China (70231010/70321001). We would like to thank Xing Zhang and Ling Lin for their contributions to the experiments of the work.

References

Agrawal R, Imielinski T, Swarmi A (1993) Mining Association Rules between Sets of Items in Large Databases. In: Proceeding of the ACM-SIGMOD 1993, pp 207-216

Au W, Chan K (1999) FARM: A Data Mining System for Discovering Fuzzy Association Rules. In: Proceedings of 1999 IEEE International Fuzzy Systems Conference (Seoul, Korea), pp 1217-1222

Aumann Y, Lindell Y (2003) A Statistical Theory for Quantitative Association Rules. Jouranl of Intelligent Information systems 20(3): 255-283

Cai CH, Fu AW, Cheng CH, Kwong WW (1998) Mining association rules with weighted items. In: Proceedings of 1998 Intl. Database Engineering and Applications Symposium, pp 68-77

Chen GQ (1998) Fuzzy Logic in Data Modeling: semantics, constraints and database design. Kluwer Academic Publishers, Boston

Chen GQ, Wei Q (2002) Fuzzy Association Rules and the Extended Mining Algorithms. Information Sciences 147: 201-228

Chen GQ, Wei Q, Liu D, Wets G (2002) Simple association rules (SAR) and the SAR-based rule discovery. Computer & Industrial Engineering 43: 721-733

Chen GQ, Yan P, Kerre EE (2004) Computationally efficient mining for fuzzy implication- based association rules in quantitative databases. International Journal of General Systems 33(2-3): 163-182

Chien BC, Lin ZL, Hong TP (2001) An Efficient Clustering Algorithm for Mining Fuzzy Quantitative Association Rules. In: Proceedings of the 9th International Fuzzy Systems Association World Congress, pp 1306-1311

Fukuda T, Morimoto Y, Morishita S, Tokuyama T (2001) Data Mining with Optimized Two-Dimensional Association Rules. ACM Transactions on Database Systems 26 (2): 179-213

Graff JM, Kosters WA, Witteman JJW (2001) Interesting Fuzzy Association Rules in Quantitative Databases. Lecture Notes in Computer Science 2168: 140-151

Gupta MM, Qi J (1991) Theory of T-norms and Fuzzy Inference Methods. Fuzzy Sets and Systems 40(3): 431-450

Gyenesei A (2000) A fuzzy approach for mining quantitative association rules. TUCS technical reports 336.

Hong T, Kuo C, Chi S, (2001) Trade-off between Computation Time and Number of Rules for Fuzzy Mining from Quantitative data. International Journal of Uncertainty, Fuzziness and Knowledge-Based Systems 9 (5): 587-604

Hullermeier E (2001) Implication-Based Fuzzy Association Rules. In: Proceedings of ECML/PKDD 2001, pp 241-252

Mannila H, Toivonen H, Verkamo I (1994) Efficient Algorithms for Discovering Association Rules. In: Proceedings of AAAI Workshop on Knowledge Discovery in Databases, pp 181-192

Miller RJ, Yang Y (1997) Association Rules over Interval Data. ACM SIGMOD 26(2): 452-461

Piatetsky-Shapiro G, Frawley WJ (1991) Knowledge Discovery in Databases . AAAI Press/The MIT Press, Menlo Park, California

Rastogi R, Shim K (2001) Mining Optimized Support Rules for Numeric Attributes. Information Systems 26: 425–444

Roberto J, Bayardo J, Agrawal R (1999) Mining the Most Interesting Rules. In: Proceeding of the Fifth ACM-SIGKDD International Conference on Knowledge Discovery and Data Mining, pp 145-154

Srikant R, Agrawal R, (1994) Fast Algorithms for Mining Association Rules. In: Proceedings of VLDB Conference, pp 487-499

Srikant R, Agrawal R, (1995) Mining Generalized Association Rules. In: Proceedings of the 21st VLDB Conference, pp 407-419

Srikant R, Agrawal R (1996) Mining Quantitative Association Rules in Large Relational Tables. In: Proceeding of 1996 ACM-SIGMOD International Conference Management of Data, pp 1-12

Witten IH, Frank E (1996) Data mining practical machine learning tools and techniques with Java implementations. Morgan Kaufmann Publishers

Yu L, Chen GQ (2005) Application and Comparison of Classification Techniques in Controlling Credit Risk. submitted

Zhang W (1999) Mining fuzzy quantitative association rules. In: Proceedings of 11th IEEE International Conference on Tools with Artificial Intelligence, (Chicago, Illinois), pp 99-102

Fuzzy Regression Approaches and Applications

Cengiz Kahraman[1] Ahmet Beşkese[2] F. Tunç Bozbura[2]

[1] Istanbul Technical University, Department of Industrial Engineering, 34367 Maçka Istanbul Turkey
[2] Bahçeşehir University, Engineering Faculty, Department of Industrial Engineering, Bahçeşehir Istanbul Turkey

Summary: Fuzzy regression is a fuzzy variation of classical regression analysis. It has been studied and applied to various areas. Two types of fuzzy regression models are Tanaka's linear programming approach and the fuzzy least-squares approach. In this chapter, a wide literature review including both theoretical and application papers on fuzzy regression has been given. Fuzzy regression models for nonfuzzy input/nonfuzzy output, nonfuzzy input/fuzzy output, and possibilistic regression model have been summarized. An illustrative example has been given. Fuzzy hypothesis testing for the coefficients of a linear regression function has been explained with two numerical examples.

Key words: fuzzy, linear regression, non-linear regression, alpha-cut, multi-variate regression, hypothesis testing, least squares

1 Introduction: Basics of Fuzzy Linear Regression

Tanaka et al. (1982) introduced fuzzy linear regression as a means to model casual relationships in systems when ambiguity or human judgment inhibits a crisp measure of the dependent variable. Unlike conventional regression analysis, where deviations between observed and predicted values reflect measurement error, deviations in fuzzy regression reflect the vagueness of the system structure expressed by the fuzzy parameters of the regression model. The fuzzy parameters of the model are considered to be possibility distributions, which corresponds to the fuzziness of the system. The fuzzy parameters are determined by a linear programming procedure, which minimizes the fuzzy deviations subject to constraints of the degree of membership fit.

Tanaka et al.'s (1982) model took the general form:

$$\widetilde{Y} = \widetilde{A}_0 + \widetilde{A}_1 x_1 + \ldots + \widetilde{A}_n x_n \tag{1}$$

C. Kahraman et al.: *Fuzzy Regression Approaches and Applications*, StudFuzz **201**, 589–615 (2006)
www.springerlink.com

where $\widetilde{Y}$ is the fuzzy output, $\widetilde{A}_i$, j=1,2,..., n, is a fuzzy coefficient, and $x = (x_1,...,x_n)$ is an n-dimensional non-fuzzy input vector. The fuzzy components were assumed to be triangular fuzzy numbers (TFNs). Consequently, the coefficients, for example, can be characterized by a membership function, $\mu_A(a)$, a representation of which is shown in Figure 1.

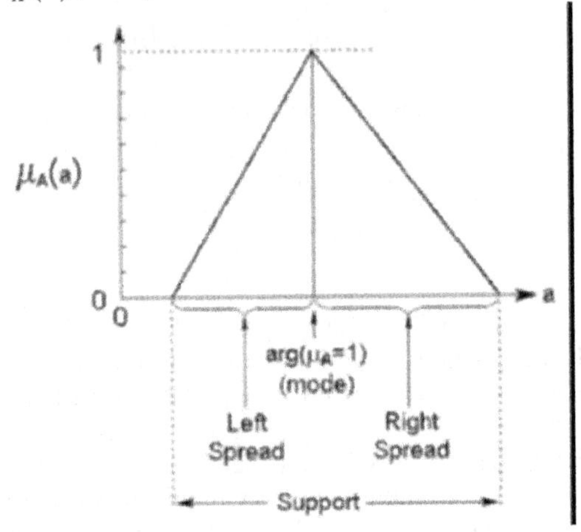

Fig. 1. Fuzzy Coefficient

The basic idea of the Tanaka et al.'s (1982) approach, often referred to as possibilistic regression, was to minimize the fuzziness of the model by minimizing the total spread of the fuzzy coefficients, subject to including all the given data.

Fuzzy set theory has been used to develop quantitative forecasting models such as time series analysis and regression analysis, and in qualitative models such as the Delphi method. In these applications, fuzzy set theory provides a language by which indefinite and imprecise demand factors can be captured. The structure of fuzzy forecasting models is often simpler yet more realistic than non-fuzzy models, which tend to add layers of complexity when attempting to formulate an imprecise underlying demand structure. When demand is definable only in linguistic terms, fuzzy forecasting models must be used.

The components of fuzzy regression

Crisp statistical linear regression takes the form

$$y_i = \beta_0 + \beta_1 x_{i1} + ... + \beta_k x_{ik} + \varepsilon_i, \qquad i = 1,2,...,m \tag{2}$$

where the dependent variable, y_i, the independent variables, x_{ij}, and the coefficients (parameters), β_j, are crisp values, and ε_i is a crisp random error term with $E(\varepsilon_i)=0$, variance $\sigma^2(\varepsilon_i)=\sigma^2$, and covariance $\sigma(\varepsilon_i,\varepsilon_j)=0, \quad \forall i,j,i\neq j$.

Although statistical regression has many applications, problems can occur in the following situations (Shapiro, 2004):

- Number of observations is inadequate (Small data set),
- Difficulties verifying distribution assumptions,
- Vagueness in the relationship between input and output variables,
- Ambiguity of events or degree to which they occur,
- Inaccuracy and distortion introduced by linearalization.

Thus, statistical regression is problematic if the data set is too small, or there is difficulty verifying that the error is normally distributed, or if there is vagueness in the relationship between the independent and dependent variables, or if there is ambiguity associated with the event or if the linearity assumption is inappropriate. These are the very situations fuzzy regression was meant to address. There are two general ways (not mutually exclusive) to develop a fuzzy regression model: (1) models where the relationship of the variables is fuzzy; and (2) models where the variables themselves are fuzzy.

For any given data pair, (x_i, y_i), the fuzzy regression interval $[Y_i^L, Y_i^U]$ is shown in Figure 2:

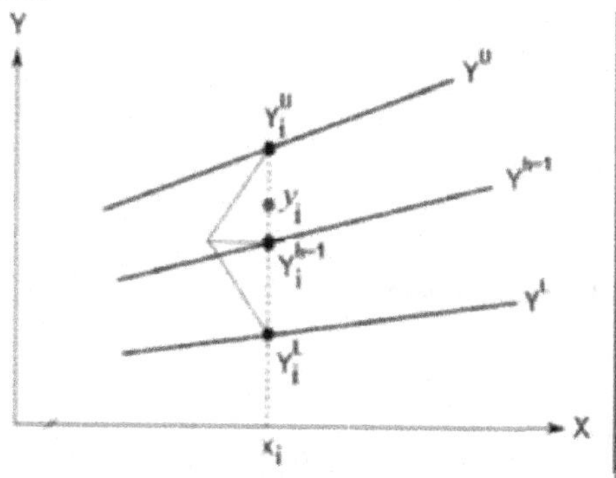

Fig. 2. Fuzzy Regression Interval

$Y_i^{h=1}$ is the mode of the membership function and if a symmetrical TFN is assumed, $Y_i^{h=1}=\overline{Y}_i=(Y_i^U+Y_i^L)/2$. Given the parameters, $(Y^U, Y^L, Y^{h=1})$, which characterize the fuzzy regression model, the *i*th data pair (x_i, y_i), is associated with the model parameters $Y_i^U, Y_i^L, Y_i^{h=1}$. From a regression perspective, it is relevant to note that $(Y_i^U - y_i)$and $(y_i - Y_i^L)$ are components of the SST, $(y_i - Y_i^{h=1})$ a component of the SSE, and $(Y_i^U - Y_i^{h=1})$ and $Y_i^{h=1} - Y_i^L$ are components of the SSR.

The fuzzy coefficients

Combining Equation (1) and Figure 1, the membership function of the j-th coefficient, may be defined as:

$$\mu_{A_i}(a) = \max\left\{1 - \frac{|a - a_j|}{c_j}, 0\right\} \tag{3}$$

where a_j is the mode and c_j is the spread, and represented as shown in Figure 3.

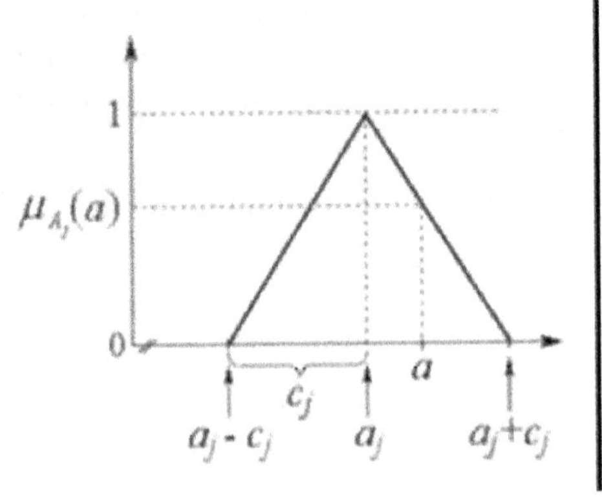

Fig. 3. Symmetrical fuzzy parameters

Defining

$$\widetilde{A}_j = \{a_j, c_j\}_L = \{\widetilde{A}_j : a_j - c_j \leq \widetilde{A}_j \leq a_j + c_j\}_L, \qquad j = 0,1,\ldots,n \tag{4}$$

and restricting consideration to the case where only the coefficients are fuzzy, we can write

$$\widetilde{Y}_i = \widetilde{A}_0 + \sum_{j=1}^{n} \widetilde{A}_1 x_{ij} = (a_0, c_0)_L + \sum_{j=1}^{n} (a_j, c_j)_L x_{ij} \tag{5}$$

This is a useful formulation because it explicitly portrays the mode and spreads of the fuzzy parameters.

2 Literature Review

Many authors have studied fuzzy regression and applied it to many areas. Many different fuzzy regression approaches have been proposed by different researchers since fuzzy linear regression was proposed by Tanaka et al. in 1982. Many of these approaches can be classified depending on the data and the treatment of the variables. Nonfuzzy input, nonfuzzy output, and fuzzy system structure is one of these approaches. The other two are "nonfuzzy input, fuzzy output, and fuzzy system structure" and "fuzzy input, fuzzy output, and fuzzy system structure". In the following, the studies on fuzzy regression are divided into two categories: The theoretical papers and the application papers on fuzzy regression.

2.1 Theoretical Research Outputs

Chang (1997) presents a fuzzy forecasting technique for seasonality in the time-series data. First, with the fuzzy regression analysis the fuzzy trend of a time-series is analyzed. Then the fuzzy seasonality is defined by realizing the membership grades of the seasons to the fuzzy regression model. Both making fuzzy forecast and crisp forecast are investigated. Seasonal fuzziness and trends are analyzed. The method is applied to the sales forecasting problem of a food distribution company.

Kim and Bishu (1998) propose a modification of fuzzy linear regression analysis based on a criterion of minimizing the difference of the fuzzy membership values between the observed and estimated fuzzy numbers.

Cheng and Lee (1999a) formulate a fuzzy inference system based on the Sugeno inference model for fuzzy regression analysis. This system is then represented by a fuzzy adaptive network. This approach combines the power of representation of fuzzy inference system with the ability of learning of the neural network. Cheng and Lee (1999b) investigate fuzzy regression without predefined functional form, or nonparametric fuzzy regression. The two most basic nonparametric regression techniques in statistics, namely, k-nearest neighbor smoothing and kernel smoothing, are fuzzified and analyzed. Algorithms are proposed to obtain the best smoothing parameters based on the minimization of cross-validation criteria. Chen (1999) uses three different ranking methods, namely, the overall existence ranking index (OERI), the approach proposed by Diamond (1988) and a new two-step method based on OERI, to estimate the distance between two fuzzy numbers. This distance parameter is then used in the least square or quadratic regression. Nonlinear programming is used to solve the resulting quadratic regression equations with constraints, and simulation is used to evaluate the performance of the approaches. The criterion used to evaluate the performance is the average of the absolute difference between the estimated and the observed values. Yen et al. (1999) extend the results of a fuzzy linear regression model that uses symmetric triangular coefficient to one with non-symmetric fuzzy triangular coefficients. This work eradicates the inflexibility of existing fuzzy linear regression models.

Özelkan and Duckstein (2000) develop several multi-objective fuzzy regression (MOFR) techniques to overcome the problems of fuzzy regression by enabling the decision maker to select a non-dominated solution based on the tradeoff between data outliers and prediction vagueness. It is shown that MOFR models provide superior results to existing fuzzy regression techniques; furthermore the existing fuzzy regression approaches and classical least-squares regression are specific cases of the MOFR framework. The methodology is illustrated with rainfall-runoff modeling examples; more specifically, fuzzy linear conceptual rainfall-runoff relationships, which are essential components of hydrologic system models, are analyzed. Buckley and Feuring's (2000) evolutionary algorithm searches the library of fuzzy functions (which includes linear, polynomial, exponential and logarithmic) for a fuzzy function which best fits the data when given some data consisting of pairs of fuzzy numbers. Dunyak and Wunsch (2000) describe a method for nonlinear fuzzy regression using neural network models. In earlier work, strong assumptions were made on the form of the fuzzy number parameters: symmetric triangular, asymmetric triangular, quadratic, trapezoidal, and so on. Their goal is to substantially generalize both linear and nonlinear fuzzy regression using models with general fuzzy number inputs, weights, biases, and outputs. This is accomplished through a special training technique for fuzzy number neural networks. The technique is demonstrated with data from an industrial quality control problem. D'Urso and Gastaldi (2000) deal with a new approach to fuzzy linear regression analysis. A doubly linear adaptive fuzzy regression model is proposed, based on two linear models: a core regression model and a spread regression model. The first one "explains" the centers of the fuzzy observations, while the second one is for their spreads. As dependence between centers and spreads is often encountered in real world applications, their model is defined in such a way as to take into account a possible linear relationship among centers and spreads. Wang and Tsaur (2000a) provide an insight into regression intervals so that regression interval analysis, data type analysis and variable selections can be analytically performed. Wang and Tsaur (2000b) consider a type of problem with crisp input and fuzzy output described by Tanaka of which a modified fuzzy least square method was proposed for solution. It shows that with such an approach the predictability in the new model is better than Tanaka's and its computation efficiency is better than the conventional fuzzy least square method. Wang and Tsaur (2000c) propose a variable selection method for a fuzzy regression equation with crisp-input and fuzzy-output by considering two criteria of minimum total sum of vagueness and minimum total sum of squares in estimation. A branch-and-bound algorithm is designed and "the least resistance principle" is adopted to determine the set of compromised solutions.

Ishibuchi and Nii (2001) explain several versions of fuzzy regression methods based on linear fuzzy models with symmetric triangular fuzzy coefficients. They point out some limitations of such fuzzy regression methods. Then they extend the symmetric triangular fuzzy coefficients to asymmetric triangular and trapezoidal fuzzy numbers. They show that the limitations of the fuzzy regression methods with the symmetric triangular fuzzy coefficients are remedied by such extension. Several formulations of linear programming problems are proposed for determining asymmetric fuzzy coefficients from numerical data. Finally, they show how fuzzified neural networks can be utilized as nonlinear fuzzy models in fuzzy regression. Yu et al. (2001) devises general fuzzy piecewise regression analysis with automatic change-point detection to simultaneously obtain the fuzzy regression model and the positions of change-points. Fuzzy piecewise possibility and necessity regression models are employed when the function behaves differently in different parts of the range of crisp input variables. Cheng and Lee (2001) use radial basis function network in fuzzy regression analysis without predefined functional relationship between the input and the output. The proposed approach is a fuzzification of the connection weights between the hidden and the output layers. This fuzzy network is trained by a hybrid learning algorithm, where self-organized learning is used for training the parameters of the hidden units and supervised learning is used for updating the weights between the hidden and the output layers. The c-mean clustering method and the k-nearest-neighbor heuristics are used for the self-organized learning. The supervised learning is carried out by solving a linear possibilistic programming problem. Chang (2001) develop a method for hybrid fuzzy least-squares regression. The method uses the new definition of weighted fuzzy-arithmetic and the well-accepted least-squares fitting criterion. First, a bivariate regression model using asymmetrical triangular fuzzy variables is derived. Then, the method is extended to multiple regression analysis. Chang and Ayyub (2001) identify the fundamental differences between fuzzy regression and ordinary regression. Through a comprehensive literature review, three approaches of fuzzy regression are summarized. The first approach of fuzzy regression is based on minimizing fuzziness as an optimal criterion. The second approach uses least-squares of errors as a fitting criterion, and two methods are summarized in this paper. The third approach can be described as an interval regression analysis. For each fuzzy regression method, numerical examples and graphical presentations are used to evaluate their characteristic and differences with ordinary least-squares regression. Based on the comparative assessment, the fundamental differences between ordinary least-squares regression and conventional fuzzy regression are concluded. In order to integrate both randomness and fuzziness types of uncertainty into one regression model, the concept of hybrid fuzzy least-squares regression analysis is proposed. Chen (2001) focuses on nonfuzzy input and fuzzy output data type and proposes approaches to handle the outlier problem. The proposed approach keeps most of the normal data in the estimated interval and avoids the influence of outliers.

Wu and Tseng (2002) construct a fuzzy regression model by fuzzy parameters estimation using the fuzzy samples. It deals with imprecise measurement of observed variables, fuzzy least square estimation and nonparametric methods. Wünsche and Nather (2002) present some contributions to the theoretical regression problem which consists in finding the best approximation of a given fuzzy random variable $\widetilde{Y}$ by another fuzzy random variable $\widetilde{X}$. They find some results structurally similar to the well known classical results. Tran and Duckstein (2002) develop a multiobjective fuzzy regression model. This model combines central tendency and possibilistic properties of statistical and fuzzy regressions and overcomes several shortcomings of these two approaches. Kao and Chyu (2002) propose a two-stage approach to construct the fuzzy linear regression model: In the first stage, the fuzzy observations are defuzzified so that the traditional least-squares method can be applied to find a crisp regression line showing the general trend of the data. In the second stage, the error term of the fuzzy regression model that represents the fuzziness of the data in a general sense is determined to give the regression model the best explanatory power for the data. Tseng and Tzeng (2002) propose a fuzzy seasonal ARIMA forecasting model that combines the advantages of the seasonal time series ARIMA model and the fuzzy regression model. It is used to forecast two seasonal time series data of the total production value of the Taiwan machinery industry and the soft drink time series. Yang and Lin (2002) use a fuzzy regression model to evaluate the functional relationship between the independent and dependent variables in a fuzzy environment. They propose alternative fuzzy least-squares methods for fuzzy linear regression models with fuzzy input-output data.

Höppner and Klawonn (2003) study the fuzzy partitions for fuzzy regression models. They modify the objective function used in fuzzy clustering and obtain different membership functions that better suit these purposes. They show that the modification can be interpreted as standard FCM using distances to the Voronoi cell of the cluster rather than using distances to the cluster prototypes. Kao and Chyu (2003) propose an idea stemmed from the classical least squares to handle fuzzy observations in regression analysis. Based on the extension principle, the membership function of the sum of squared errors is constructed. The fuzzy sum of squared errors is a function of the regression coefficients to be determined, which can be minimized via a nonlinear program formulated under the structure of the Chen–Klein method for ranking fuzzy numbers. To illustrate how the proposed method is applied, three cases, one crisp input-fuzzy output, one fuzzy input-fuzzy output, and one non-triangular fuzzy observation, are exemplified. The results show that the least-squares method of this paper is able to determine the regression coefficients with better explanatory power. Yang and Liu (2003) propose new types of fuzzy least-squares algorithms with a noise cluster for interactive fuzzy linear regression models. These algorithms are robust for the estimation of fuzzy linear regression models, especially when outliers are present. D'Urso (2003) suggests regression models with crisp or fuzzy inputs and crisp or fuzzy output by taking into account a least-squares approach. In particular, for these fuzzy regression models, unconstrained and constrained (with inequality restrictions) least-squares estimation procedures are developed.

Nasrabadi and Nasrabadi (2004) consider fuzzy linear regression models with fuzzy/crisp output, fuzzy/crisp input, and an estimated method along with a mathematical-programming-based approach is proposed. The advantages of the proposed approach are simplicity in programming and computation, and minimum difference of total spread between observed and estimated values. Hong and Hwang (2004) extend the models of Diamond (1988) to fuzzy multiple linear regression models and to extend the model F3 with numerical input of Diamond (1988) to fuzzy nonlinear model. Here, all extensions are based on regularization technique. Regularization approach to regression can be easily found in Statistics and Information Science literatures. The technique of regularization was introduced as a way of controlling the smoothness properties of regression function.

Hojati et al. (2005) propose a new method for computation of fuzzy regression that is simple and gives good solutions. They consider two cases: First, when only the dependent variable is fuzzy, their approach is given and is compared with those suggested in the literature. Secondly, when both dependent and independent variables are fuzzy, their approach is extended and compared with those given in the literature. Nasrabadi et al. (2005) criticize fuzzy regression because it is sensitive to outliers, it does not allow all data points to influence the estimated parameters, and the spread of the estimated values become wider as more data are included in the model. They develop a multi-objective fuzzy linear regression model to overcome these shortcomings.

2.2 Works Reporting Fuzzy Regression Applications

Heshmaty and Kandel (1985) utilize fuzzy regression analysis to build forecasting models for predicting sales of computers and peripheral equipment. The independent variables are: user population expansion, microcomputer sales, minicomputer sales, and the price of microcomputers. The forecasts for the sales of computers and peripheral equipment are given as fuzzy sets with triangular membership functions. The decision-maker then selects the forecast figure from within the interval of the fuzzy set.

Bell and Wang (1997) build fuzzy linear regression models to reveal the relationship of cumulative trauma disorders (CTD) risk factors, to predict the injuries, and to evaluate risk levels of individuals. Twenty-seven keyboard users (twenty-two for model building and five for model validation) and three CTD experts participated in the model building. Four fuzzy models are built corresponding to four risk categories, and a final fuzzy linear model is established using AHP pairwise comparisons. Multicolinearity effects are addressed and a partial standard deviation scaling method is used to eliminate the effects. From a methodological point of view, fuzzy linear regression models provide useful insight into risk factors-CTD relationship.

Wen and Lee (1999) perform a fuzzy linear regression analysis to study the cost function of wastewater that is developed by using a pool of data from 26 municipal wastewater treatment systems in Taiwan. The choice of a primary or higher wastewater treatment process is intimately related to financial, land availability, effluent standards and managerial factors. Especially, financial consideration plays an important role in the planning stage. In addition, cost estimation is difficult to precisely evaluate in an uncertain environment. In the conventional regression model, deviations between the observed values and the estimated values are supposed to be due to measurement errors. Here, taking a different perspective, these deviations are regarded as the fuzziness of the system's parameters. Thus, these deviations are reflected in a linear function with fuzzy parameters. Using linear programming algorithm, this fuzzy linear regression model might be very convenient and useful for finding a fuzzy structure in an evaluation system. In this paper, the details of the fuzzy linear regression concept and its applications to the cost function in an uncertain environment are shown and discussed.

Based on the seasonal time series ARIMA(p,d,q)(P,D,Q)s model (SARIMA) and fuzzy regression model, Tseng et al. (1999) combine the advantages of two methods to propose a procedure of fuzzy seasonal time series and apply this method to forecasting the production value of the mechanical industry in Taiwan. The intention is to provide the enterprises, in this era of diversified management, with a fresh method to conduct short-term prediction for the future in the hope that these enterprises can perform more accurate planning. This method includes interval models with interval parameters and provides the possibility distribution of future value. From the results of practical application to the mechanical industry, this method makes it possible for decision makers to forecast the possible situations based on fewer observations than the SARIMA model and has the basis of pre-procedure for fuzzy time series.

Considering the time-series ARIMA(p, d, q) model and fuzzy regression model, Tseng et al. (2001) develop a fuzzy ARIMA (FARIMA) model and applies it to forecasting the exchange rate of NT dollars to US dollars. This model includes interval models with interval parameters and the possibility distribution of future values is provided by FARIMA. This model makes it possible for decision makers to forecast the best- and worst-possible situations based on fewer observations than the ARIMA model.

In modeling a fuzzy system with fuzzy linear functions, the vagueness of the fuzzy output data may be caused by both the indefiniteness of model parameters and the vagueness of the input data. This situation occurs as the input data are envisaged as facts or events of an observation which are uncontrollable or uninuenced by the observer rather than as the controllable levels of factors in an experiment. Lee and Chen (2001) concentrate on such a situation and refer to it as a generalized fuzzy linear function. Using this generalized fuzzy linear function, a generalized fuzzy regression model is formulated. A nonlinear programming model is proposed to identify the fuzzy parameters and their vagueness for the generalized regression model. A manpower forecasting problem is used to demonstrate the use of the proposed model.

Wu and Tseng (2002) construct a fuzzy regression model by fuzzy parameters estimation using the fuzzy samples. It deals with imprecise measurement of observed variables, fuzzy least square estimation and nonparametric methods. They test their model using the monthly Taiwan Business Index from November 1987 to February 1997.

Soliman et al. (2003) propose a new technique for frequency and harmonic evaluation in power networks. This technique is based on fuzzy linear regression and uses the digitized voltage samples, which are fuzzy numbers, to estimate the frequency and harmonic contents of the voltage signal. In this technique they formulate a linear optimization problem, where the objective is to minimize the spread of the voltage samples at the relay location subject to satisfying two inequality constraints on each voltage sample. Effects of sampling frequency, data window size, and the degree of fuzziness on the estimated parameters are investigated. The performance of the proposed technique is illustrated using simulated data. Sanchez and Gomez (2003, 2004) attempt to provide a solution to the problem of interest rate uncertainty with a method for adjusting the temporal structure of interest rates (TSIR) that is based on fuzzy regression techniques.

Xue et al. (2005) present the possibilities of the fuzzy regression method in modeling the bead width in the robotic arc-welding process. Fuzzy regression is a well-known method to deal with the problems with a high degree of fuzziness so that the approach is employed to build the relationship between four process variables and the quality characteristic, such as bead width. Using the developed model, the proper prediction of the process variables for obtaining the optimal bead width can be determined.

3 Some Approaches on Fuzzy Regression

3.1 Nonfuzzy Input and Nonfuzzy Output

Consider the set of observed values (x_i, y_i); i=1; 2; ... ; n; with the expression $Y^* = f(X, A)$, where Y^* is the estimated output, $X = (x_1, x_2, ..., x_n)^t$ is a vector of nonfuzzy inputs, and $A = (A_1, A_2, ..., A_n)$ is a vector of fuzzy parameters with the following symmetrical triangular membership functions (Chen, 2001):

$$\mu_{A_j}(a_j) = \begin{cases} 1 - \dfrac{|\alpha_j - a_j|}{c_j}, \text{if} |\alpha_j - a_j| \le c_j \\ 0, \quad otherwise \end{cases} \tag{6}$$

where α_j is the center (mode) value of Aj and cj is the spread or width (variability) around the mode value. In vector notation, the fuzzy parameter A can be written as

$$A = (\alpha, c) \tag{7}$$

with $\alpha = (\alpha_1,...,\alpha_n)^t$ and $c = (c_1,...,c_n)^t$. Obviously, when c_j =0; A_j reduces to an ordinary number. The estimated output Y^* can be obtained by using the extension principle. The membership function of Y^* is

$$\mu_{Y^*}(y) = \begin{cases} 1 - \dfrac{|y - \alpha^t x|}{c^t|x|}, x \neq 0 \\ 1, \quad x = 0, y = 0 \\ 0, \quad x = 0, y \neq 0 \end{cases} \tag{8}$$

where $|x| = (|x_1|,...,|x_n|)^t$. The center of Y^* is $\alpha^t x$ and the spread (range) of Y^* is $c^t|x|$ with $c^t|x| \geq 0$. The value of Y^* can be obtained by the estimation equation

$$Y_i^* = (\alpha_0, c_0)x_{0i} + (\alpha_1, c_1)x_{1i} + ... + (\alpha_n, c_n)x_{ni}, \quad i = 1,2,...,N \tag{9}$$

A typical representation of Y^* is illustrated in Fig. 4.

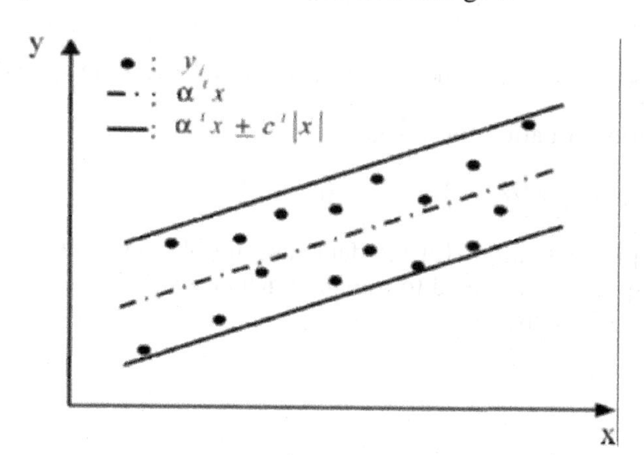

Fig. 4. Estimated values for (N, N, F) case

3.2 Nonfuzzy Input and Fuzzy Output

The membership function for the observed output fuzzy data $\widetilde{Y}_i = (y_i, e_i)$ can be represented by the following symmetric triangular function:

$$\mu_{\widetilde{Y}_i}(y) = \begin{cases} 1 - \dfrac{|y_i - y|}{e_i}, & |y_i - y| \le e_i \\ 0, & otherwise \end{cases} \tag{10}$$

where y_i is the center of the output data and e_i is the spread of the fuzzy data.

Applying the extension principle, the membership function, $\mu_{\widetilde{Y}_i}(y)$, for the estimated output variable $\widetilde{Y}^*$ can be expressed as

$$\mu_{\widetilde{Y}^*}(y) = \begin{cases} 1 - \dfrac{|y - \alpha^t x|}{c^t|x|}, & if\ x \ne 0 \\ 1, & if\ x = 0, y = 0 \\ 0, & if\ x = 0, y \ne 0 \end{cases} \tag{11}$$

where $|x| = (|x_1, ..., |x_n||)^t$, the center of $\widetilde{Y}^*$ is $\alpha^t x$ and the spread is $c^t|x|$. The estimation equation for $\widetilde{Y}^*$ is

$$\widetilde{Y}_i^* = (\alpha_0, c_0)x_{0i} + (\alpha_1, c_1)x_{1i} + ... + (\alpha_n, c_n)x_{ni}, \quad i = 1,2,..., N \tag{12}$$

and two typical situations for the data case are plotted in Figs. 5 and 6. In Fig.5, the spread is assumed to remain constant while in Fig. 6, the spread increases as x increases.

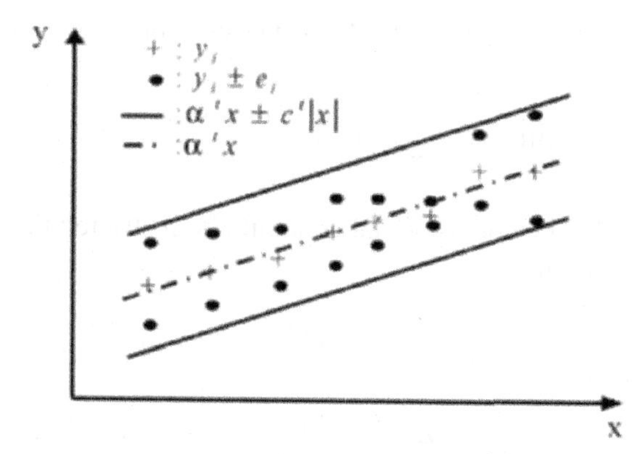

Fig. 5. Estimated value for the (N, F, F) case with constant spread

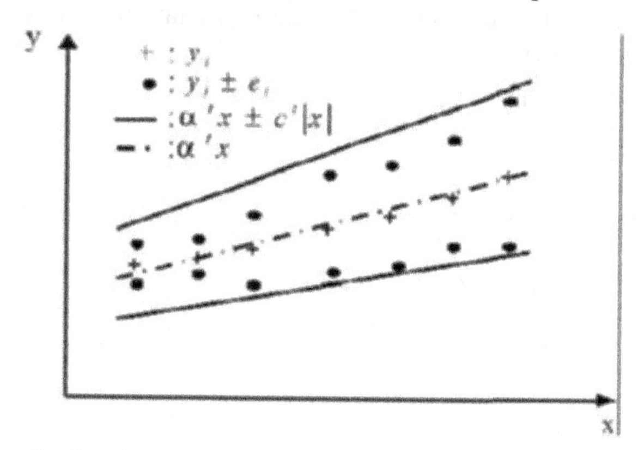

Fig. 6. Estimated value for the (N, F, F) case with increasing spread as x increases

3.3 The Possibilistic Regression Model

The possibilistic regression model was first proposed by Tanaka and Guo (1999) to reflect the fuzzy relationship between the dependent and the independent variables. The upper and the lower regression boundaries are used in the possibilistic regression to reflect the possibilistic distribution of the output values. By solving the linear programming (LP) problem, the coefficients of the possibilistic regression can easily be obtained.

The general form of a possibilistic regression can be expressed as

$$y = A_0 + A_1 x_1 + \ldots + A_n x_n = A' x \tag{13}$$

where A_i is a asymmetrical possibilistic regression coefficient which is denoted as $(a_i - c_{iL}, a_i, a_i + c_{iR})$. In order to obtain the minimum degree of uncertainty, the fitness function of the possibilistic regression can be defined as

$$\min_{a,c} J = \sum_{j=1,\ldots,m} c_L^{'}\left|x_j\right| + c_R^{'}\left|x_j\right| \tag{14}$$

In addition, the dependent variable should be restricted to satisfy the following two equations:

$$y_j \geq a'x_j - c_L^{'}\left|x_j\right| \tag{15}$$

$$y_j \leq a'x_j + c_R^{'}\left|x_j\right| \tag{16}$$

On the basis of the concepts above the formulation of a possibilistic regression model is as follows:

$$\min_{a,c} J = \sum_{j=1,\ldots,m} c_L^{'}\left|x_j\right| + c_R^{'}\left|x_j\right| \tag{17}$$

s.t.

$$y_j \geq a'x_j - c_L^{'}\left|x_j\right|$$

$$y_j \leq a'x_j + c_R^{'}\left|x_j\right|, \quad j = 1,\ldots,m$$

$$c_L, c_R \geq 0$$

4 Fuzzy Hypothesis Testing in Regression

In this section, the hypothesis testing for the coefficients of the linear function $y = a + bx$ will be considered.

4.1 $H_0 : a = a_0$, $H_1 : a \succ a_0$

In the crisp case, when the hypotheses $H_0 : a = a_0$ and $H_1 : a \succ a_0$ are considered, the following statistic is used:

$$t_0 = \frac{\hat{a} - a_0}{\sqrt{\hat{\sigma}^2/(n-2)}} \tag{18}$$

H_0 is rejected if $t_0 \geq t_\gamma$; and H_0 is accepted when $t_0 \prec t_\gamma$ where γ is the significance level of the test.

Now assume that a is a triangular shaped fuzzy number, $\tilde{a}$. Let the fuzzy estimator for σ^2 be $\tilde{\sigma}^2$. Then the fuzzy statistic is (Buckley, 2004)

$$\tilde{T} = \frac{\tilde{a} - a_0}{\sqrt{\tilde{\sigma}^2/(n-2)}} \tag{19}$$

All fuzzy calculations will be performed via α-cuts and interval arithmetic. After substituting the intervals for an α-cut of $\tilde{a}$ and $\tilde{\sigma}^2$ into the expression for $\tilde{T}$, using the interval arithmetic, and simplification, the following equation is obtained:

$$\tilde{T}[\alpha] = \left[\sqrt{R(\lambda)/n}\left(t_0 - t_{\alpha/2}\right), \sqrt{L(\lambda)/n}\left(t_0 + t_{\alpha/2}\right)\right] \tag{20}$$

where

$$L(\lambda) = [1-\lambda]\chi^2_{R,\gamma/2} + \lambda n \tag{21}$$

and

$$R(\lambda) = [1-\lambda]\chi^2_{L,\gamma/2} + \lambda n \tag{22}$$

where $0 \leq \lambda \leq 1$ and it is started with a 99% confidence interval when $\lambda = 0$ and ended up with a 0% confidence interval for $\lambda = 1$.

It is assumed that all intervals are positive in the derivation of Equation (20). The interval for an alpha-cut of $\tilde{a}$ may be positive or negative, but the interval for an alpha-cut of $\tilde{\sigma}^2$ is always positive. When the left end point (or both end points) of the interval for an alpha-cut of $\tilde{a}$ is negative, some changes in Equation (20) have to be made.

Alpha-cuts of the fuzzy critical value $C\tilde{V}$ for this one-sided test can now be found since the alpha-cuts of the fuzzy statistic are known. The fuzzy critical value can be calculated using the following equation:

$$C\tilde{V}[\alpha] = \left[\sqrt{R(\lambda)/n}\left(t_\gamma - t_{\alpha/2}\right), \sqrt{L(\lambda)/n}\left(t_\gamma + t_{\alpha/2}\right)\right] \tag{23}$$

In the above equation, γ is fixed and alpha varies in the interval $[0.01, 1]$.

The final decision will depend on the relationship between $\widetilde{T}$ and $C\widetilde{V}$: H_0 is rejected if $\widetilde{T} \succ C\widetilde{V}$ and H_0 is accepted if $\widetilde{T} \prec C\widetilde{V}$ and there is no decision on H_0 if $\widetilde{T} \approx C\widetilde{V}$.

4.2 $H_0 : b = 0$, $H_1 : b \neq 0$

This is a two-sided test and the statistic in the crisp case is

$$t_0 = \frac{\hat{b} - 0}{\sqrt{\dfrac{n\hat{\sigma}^2}{\sum_{i=1}^{n}(x_i - \overline{x})^2} \Big/ n-2}} \tag{24}$$

H_0 is rejected if $t_0 \geq t_{\gamma/2}$ or $t_0 \leq -t_{\gamma/2}$. Otherwise, H_0 is accepted.

In the fuzzy case, b is a triangular shaped fuzzy number: $\widetilde{b}$. The fuzzy estimator $\widetilde{\sigma}^2$ of σ is also a fuzzy number. Then the fuzzy statistic is

$$\widetilde{T} = \frac{\widetilde{b} - 0}{\sqrt{\dfrac{n\widetilde{\sigma}^2}{\sum_{i=1}^{n}(x_i - \overline{x})^2} \Big/ n-2}} \tag{25}$$

All fuzzy calculations will be performed via α-cuts and interval arithmetic. After substituting the intervals for an α-cut of $\widetilde{b}$ and $\widetilde{\sigma}^2$ into the expression for $\widetilde{T}$, using the interval arithmetic, assuming all intervals are positive, the following equation is obtained:

$$\widetilde{T}[\alpha] = \left[\sqrt{R(\lambda)/n}\left(t_0 - t_{\alpha/2}\right), \sqrt{L(\lambda)/n}\left(t_0 + t_{\alpha/2}\right)\right] \tag{26}$$

where

$$L(\lambda) = [1 - \lambda]\chi^2_{R,\gamma/2} + \lambda n \tag{27}$$

and

$$R(\lambda) = [1 - \lambda]\chi^2_{L,\gamma/2} + \lambda n \tag{28}$$

Alpha-cuts of the fuzzy critical value $\tilde{CV}_i$ for this one-sided test can now be found since the alpha-cuts of the fuzzy statistic are known. The fuzzy critical value can be calculated using the following equation:

$$\tilde{CV}_1[\alpha] = \left[\sqrt{L(\lambda)/n}\left(-t_{\gamma/2} - t_{\alpha/2}\right), \sqrt{R(\lambda)/n}\left(-t_{\gamma/2} + t_{\alpha/2}\right)\right] \tag{29}$$

and because $\tilde{CV}_2 = -\tilde{CV}_1$

$$\tilde{CV}_2[\alpha] = \left[\sqrt{R(\lambda)/n}\left(t_{\gamma/2} - t_{\alpha/2}\right), \sqrt{L(\lambda)/n}\left(-t_{\gamma/2} + t_{\alpha/2}\right)\right] \tag{30}$$

In the above equation, γ is fixed and alpha varies in the interval $[0.01, 1]$.

Given the fuzzy numbers $\tilde{T}$ and the $\tilde{CV}_i$, i=1,2, $\tilde{T}$ can be compared with $\tilde{CV}_1$ and $\tilde{CV}_2$ to determine the final decision on H_0.

5 An Application of Fuzzy Regression Models to Transfer Moulding of Electronic Packages (Ip et al., 2003)

Transfer moulding is one of the popular processes to perform microchip encapsulation. In this process, the thermoset moulding compound (typically a solid epoxy reform) is preheated and then placed into the pot of the moulding tool. A transfer cylinder, or plunger, is used to inject the moulding compound into the runner system and gates of the mould. The moulding compound then flows over the chip, wire-bonds and leadframes, encapsulating the microelectronic device. In this study, a comparatively new method for microelectronic encapsulation, liquid epoxy moulding (LEM), is investigated. LEM is classified as a type of transfer moulding. The moulding process is very similar with that of conventional transfer moulding, but the epoxy moulding compound used is liquid in nature. The relatively low viscosity of liquid epoxy moulding compound allows low operating pressures and minimises problems associated with wire sweep. The in-mould cure time for the liquid epoxy resin is no longer than that for current transfer moulding compounds. LEM is a highly complex moulding process, which involves more than 10 influential material properties, mould design parameters and process parameters in process design. Although modelling the LEM process is critical to understand the process behaviour and optimise the process, the process is much difficult to characterise due to the complex behaviour of epoxy encapsulant and the inherent fuzziness of the moulding system such as inconsistent properties of epoxy moulding compound and environmental effect on moulding temperature. In current practice, a time-consuming and expensive trial-and-error approach is adopted to qualify a mould design or to optimise the process conditions for a given mould. In the following, development of a process model of LEM process for microchip encapsulation using fuzzy regression is described.

To develop a fuzzy regression model for relating process parameters and quality characteristics of LEM, significant process parameters and quality characteristics have to be identified first. Eight significant process parameters, which are named as independent variables were identified and their operating values are shown below:

- Top mould temperature (150, 180 ° C), x_1.
- Bottom mould temperature (150, 180 ° C), x_2.
- Filling time (30, 60 s), x_3.
- Transfer force (90, 100 kgf), x_4.
- Curing time (30, 50 s), x_5.
- 1st injection profile (− 1 mm/2 s, − 1 mm/4 s), x_6.
- 2nd injection profile (− 7 mm/2 s, − 7 mm/4 s), x_7.
- 3rd injection profile (− 8 mm/2 s, − 8 mm/4 s), x_8.

Three quality characteristics (named as dependent variables) were also identified as shown below:

- Wire sweep (%), y.
- Void (mm^2), z.
- Flash (mm^2), w.

Thirty-two experiments were carried out in which 30 sets of the data and results were used to develop the fuzzy regression models of LEM and two sets of them were used to validate the developed models. The fuzzy linear regression model for dependent variable y (wire sweep) can be represented as follows:

$$y\psi = A_0 + A_1x_1 + A_2x_2 + A_3x_3 + A_4x_4 + A_5x_5 + A_6x_6 + A_7x_7 + A_8x_8$$

where $A_i\psi(i\psi = 0–8)$ are fuzzy parameters.ψ

The LP model can be formulated as follows:

$$Minimize\ J = \sum_{j=0}^{8}\left(c_j \sum_{i=1}^{30}\left|x_{ij}\right|\right)$$

subject to

$$\sum_{j=0}^{8}\alpha_j x_{ij} + (1-h)\sum_{j=0}^{8} c_j\left|x_{ij}\right| \geq y_i$$

$$\sum_{j=0}^{8}\alpha_j x_{ij} - (1-h)\sum_{j=0}^{8} c_j\left|x_{ij}\right| \leq y_i$$

$c_j \geq 0$, $\alpha_j \in R$, $j = 0, 1, 2, \ldots, 8$, $x_{i0} = 1$, $i = 1, 2, \ldots, 30$, $0 \leq h \leq 1$

With the threshold level $h\psi$= 0(1, the above LP model was solved using MATLAB and the central value and width of each fuzzy parameter were obtained as shown in Table 1. The fuzzy linear regression model for the dependent variable y (wire sweep) is shown below:

Table 1. The centre and width values of fuzzy parameters for wire sweep

Fuzzy parameters	Centre	Width
A_0	9.8907	0.0000
A_1	-0.0267	0.0218
A_2	0.0094	0.0134
A_3	0.0623	0.0000
A_4	-0.1481	0.0000
A_5	0.1447	0.0000
A_6	-0.8301	0.0000
A_7	-0.1084	0.0000
A_8	1.3133	0.0000

$y = (9.8907, 0.0000) + (-\ 0.0267, 0.0218)x_1 + (0.0094, 0.0134)x_2 + (0.0623, 0.0000)x_3 + (-\ 0.1481, 0.0000)x_4 + (0.1447, 0.0000)x_5 + (-\ 0.8301, 0.0000)x_6 + (-\ 0.1084, 0.0000)x_7 + (1.3133, 0.0000)x_8$

With the same calculation, the fuzzy linear regression models for dependent variables z (void) and $w\psi$(flash) can be obtained, respectively:

$z = (0.0392, 0.0000) + (0.0005, 0.0003)x_1 + (-0.0003, 0.0000)x_2 + (0.0003, 0.0000)x_3 + (-0.0011, 0.0000)x_4 + (0.0005, 0.0000)x_5 + (0.0048, 0.0000)x_6 + (0.0053, 0.0000)x_7 + (-0.0062, 0.0000)x_8$

$w = (0.0221, 0.0000) + (0.0004, 0.0003)x_1 + (0.0003, 0.0002)x_2 + (0.0000, 0.0000)x_3 + (-0.0005, 0.0000)x_4 + (-0.0013, 0.0000)x_5 + (0.0004, 0.0000)x_6 + (-0.0080, 0.0000)x_7 + (0.0053, 0.0000)x_8$

In order to evaluate the developed models, two validation tests were carried out, where the experimental setting is shown in Table 2. As can be seen from the results (Tables 3–5), all actual values of measurements are found to fall within their corresponding prediction ranges. The average prediction error of flash is found to be the greatest, whereas the prediction error of void is the lowest.

Table 2. Experimental settings for the validation tests

Setting	Test 1	Test 2
Top mould temperature, x_1 (°C)	150	150
Bottom mould temperature, x_2 (°C)	180	180
Filling time, x_3 (s)	60	60
Transfer force, x_4 (kgf)	90	100
Curing time, x_5 (s)	30	30
1st injection profile, x_6 (mm/s)	1/2.00	1/2.00
2nd injection profile, x_7 (mm/s)	7/2.00	7/2.00
3rd injection profile, x_8 (mm/s)	8/2.00	8/2.00

Table 3. Results of validation tests for wire sweep

Test	Fuzzy regression			Actual value (%)	Error (%)
	Spread	Centre	Range		
1	5.6787	2.8523	-2.8264 to 8.5310	2.7000	5.6400
2	5.6787	1.5884	-4.0903 to 7.2671	1.6000	0.7250
				Average error	3.1825

Table 4. Results of validation tests for void

Test	Fuzzy regression			Actual value (mm^2)	Error (%)
	Spread	Centre	Range		
1	0.0464	0.0063	0.0000–0.0527	0.0065	3.0800
2	0.0557	0.0099	0.0000–0.0656	0.0103	0.8800
				Average error	3.4800

Table 5. Results of validation tests for flash

Test	Fuzzy regression			Actual value (mm^2)	Error (%)
	Spread	Centre	Range		
1	0.0689	0.0312	0.0000–0.1001	0.0305	2.2951
2	0.0689	0.0423	0.0000–0.1112	0.0454	6.8282
				Average error	4.5617

6 An Application of Fuzzy Hypothesis Tests for Regression Coefficients

Consider the crisp data in the following table (Buckley, 2004).

Table 6. Crisp data

x	70	74	72	68	58	54	82	64	80	61
y	77	94	88	80	71	76	88	80	90	69

6.1 Test for *a*

From this data, we compute $\hat{a} = 81.3$, $\hat{b} = 0.742$ and $\hat{\sigma}^2 = 21.7709$ with n=10. Let $\gamma = 0.05$, $a_0 = 80$ and determine $t_0 = 0.7880$ and $t_{0.05} = 1.860$ with 8 degrees of freedom. We compute

$$L(\lambda) = 21.955 - 11.955\lambda$$

$$R(\lambda) = 1.344 + 8.656\lambda$$

$$\sqrt{R(\lambda)/n} = \sqrt{0.1344 + 0.8656\lambda}$$

$$\sqrt{L(\lambda)/n} = \sqrt{2.1955 - 1.1955\lambda}$$

From these results we may get the graphs of $\tilde{T}$ and $C\tilde{V}$ and they are shown in Figure 7. From this figure, we see that $\tilde{T} \prec C\tilde{V}$ since the height of the intersection is less than 0.8. We therefore conclude: do not reject H_0. Of course, the crisp test would have the same result.

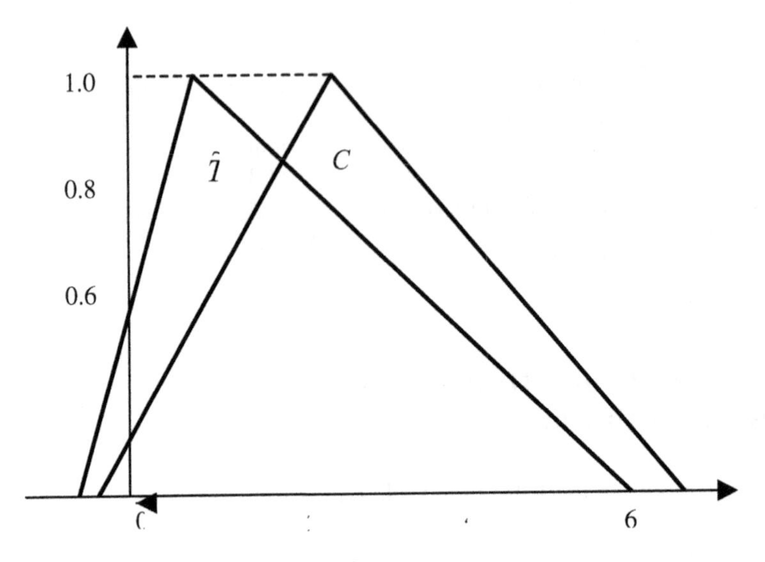

Fig. 7. Fuzzy test $\widetilde{T}$ verses $C\widetilde{V}$

6.2 Test for *b*

Let $\gamma = 0.05$, and compute $t_0 = 3.9111$ and $t_{0.025} = 2.306$ with 8 degrees of freedom. The values of $L(\lambda), R(\lambda), \sqrt{L(\lambda)/n}$, and $\sqrt{R(\lambda)/n}$ are all the same as in the test for *a*. All that has changed is the value of t_0 and that now we use both $C\widetilde{V}_1$ and $C\widetilde{V}_2$ for a two-sided test. The graphs of $\widetilde{T}$ and $C\widetilde{V}_i$ are shown in Figure 8. It is evident that $C\widetilde{V}_2 \prec \widetilde{T}$, because the height of the intersection is less than 0.8. Hence we reject H_0, the same as in the crisp case.

7 Conclusion

In conventional regression analysis, deviations between the observed values and the estimates are assumed to be due to random errors. The deviations are sometimes due to the indefiniteness of the sturucture of the system or imprecise observations. The uncertainty in this type of regression model becmes fuzziness, not randomness. Unlike statistical

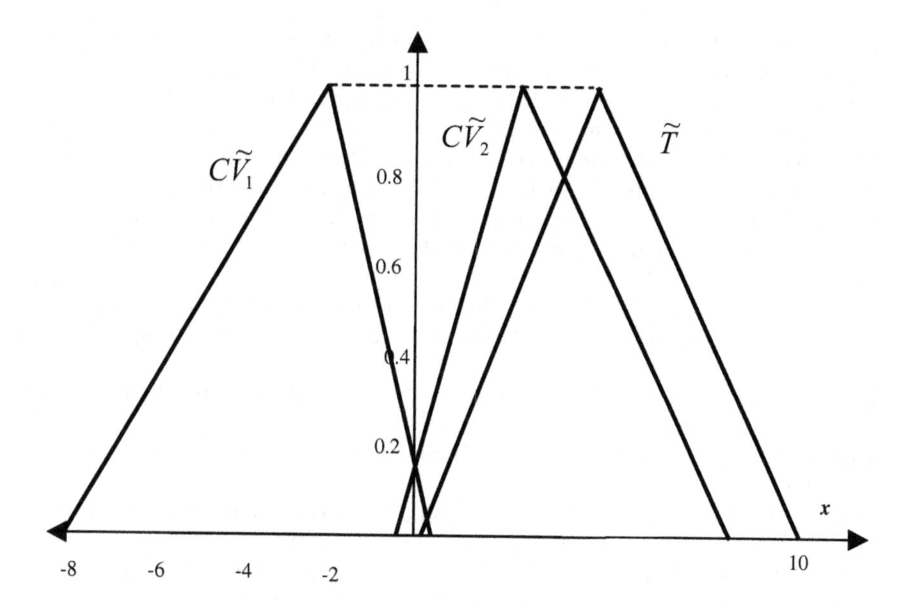

Fig. 8. Fuzzy test $\tilde{T}$ verses $C\tilde{V}_i$

regression modelling that is based on probability theory, fuzzy regression is based on possibility theory and fuzzy set theory. Deviations between observed values and estimated values are assumed to be due to system fuzziness. Fuzzy regression could be more effective than statistical regression when the degree of system fuzziness is high. Fuzzy regression has many directions and potential to make improvements in the future.

References

Bell, P.M., Wang, H., Fuzzy linear regression models for assessing risks of cumulative trauma disorders, Fuzzy Sets and Systems, Vol. 92, pp. 317-340, 1997.

Buckley, J.J., Fuzzy Statistics, Springer-Verlag, Berlin, 2004.

Buckley, J.J., Feuring, T., Linear and non-linear fuzzy regression: Evolutionary algorithm solutions, Fuzzy Sets and Systems, Vol. 112, pp. 381-394, 2000.

Chang, P-T., Fuzzy seasonality forecasting, Fuzzy Sets and Systems, Vol. 90, pp. 1-10, 1997.

Chang, Y-H. O., Hybrid fuzzy least-squares regression analysis and its reliability measures, Fuzzy Sets and Systems, Vol. 119, pp. 225-246, 2001.

Chang, Y-H. O., Ayyub, B.M., Fuzzy regression methods – a comparative assessment, Fuzzy Sets and Systems, Vol. 119, pp. 187-203, 2001.

Chen, Y-S., Fuzzy ranking and quadratic fuzzy regression, Computers & Mathematics with Applications, Vol. 38, pp. 265-279, 1999.

Chen, Y-S., Outliers detection and confidence interval modification in fuzzy regression, Fuzzy Sets and Systems, Vol. 119, pp. 259-272, 2001.

Cheng, C.-B., Lee, E.S., Nonparametric fuzzy regression—k-NN and kernel smoothing techniques, Computers & Mathematics with Applications, Vol. 38, pp. 239-251, 1999a.

Cheng, C.-B., Lee, E.S., Applying fuzzy adaptive network to fuzzy regression analysis, Computers & Mathematics with Applications, Vol. 38, pp. 123-140, 1999b.

Cheng, C-B., Lee, E.S., Fuzzy regression with radial basis function network, Fuzzy Sets and Systems, Vol. 119, pp. 291-301, 2001.

Diamond, P., Fuzzy least squares, Information Sciences, Vol. 46, pp. 141-157, 1988.

Dunyak, J.P., Wunsch, D., Fuzzy regression by fuzzy number neural networks, Fuzzy Sets and Systems, Vol. 112, pp. 371-380, 2000.

D'Urso, P., Gastaldi, T., A least-squares approach to fuzzy linear regression analysis, Computational Statistics & Data Analysis, Vol. 34, pp. 427-440, 2000.

D'Urso, P., Linear regression analysis for fuzzy/crisp input and fuzzy/crisp output data, Computational Statistics & Data Analysis, Vol. 42, pp. 47-72 , 2003.

Hestmaty, B., Kandel, A., Fuzzy linear regression and its applications to forecasting in uncertain environment, Fuzzy Sets and Systems, Vol.15, pp. 159-191, 1985.

Hojati, M., Bector, C.R., Smimou, K., A simple method for computation of fuzzy linear regression, European Journal of Operational Research, Volume 166, pp. 172-184 , 2005.

Hong, D.H., Hwang, C., Extended fuzzy regression models using regularization method, Information Sciences, Vol. 164, 2004, pp. 31-46, 2004.

Höppner, F. , Klawonn, F., Improved fuzzy partitions for fuzzy regression models, International journal of approximate reasoning, Vol. 32, pp. 85-102, 2003.

Ishibuchi, H., Nii, M., Fuzzy regression using asymmetric fuzzy coefficients and fuzzified neural networks, Fuzzy Sets and Systems, Vol. 119, pp. 273-290, 2001.

Kao, C., Chyu, C-L., A fuzzy linear regression model with better explanatory power, Fuzzy Sets and Systems, Vol. 126, pp. 401-409, 2002.

Kao, C., Chyu, C-L., Least-squares estimates in fuzzy regression analysis, European Journal of Operational Research, Vol. 148, pp. 426-435, 2003.

Kim, B., Bishu, R.R., Evaluation of fuzzy linear regression models by comparing membership functions, Fuzzy Sets and Systems, Vol. 100, pp. 343-352, 1998.

Ip, K.W., Kwong, C.K., Wong, Y.W., Fuzzy regression approach to modelling transfer moulding for microchip encapsulation, Journal of Materials Processing Technology, Vol. 140, pp. 147–151, 2003.

Lee, H.T., Chen, S.H., Fuzzy regression model with fuzzy input and output data for manpower forecasting, Fuzzy Sets and Systems, Vol. 119, pp. 205-213, 2001.

Nasrabadi, M.M., Nasrabadi, E., A mathematical-programming approach to fuzzy linear regression analysis, Applied Mathematics and Computation, Vol. 155, pp. 873-881, 2004.

Nasrabadi, M.M., Nasrabadi, E., Nasrabady, A.R., Fuzzy linear regression analysis: a multi-objective programming approach, Applied Mathematics and Computation, Vol. 163, pp. 245-251, 2005.

Özelkan, E.C., Duckstein, L., Multi-objective fuzzy regression: a general framework, Computers & Operations Research, Vol. 27, pp. 635-652, 2000.

Sánchez, J.A., Gómez, A.T., Estimating a term structure of interest rates for fuzzy financial pricing by using fuzzy regression methods, Fuzzy Sets and Systems, Vol. 139, pp. 313-331, 2003.

Sánchez, J.A., Gómez, A.T., Estimating a fuzzy term structure of interest rates using fuzzy regression techniques, European Journal of Operational Research, Vol. 154, pp. 804-818, 2004.

Shapiro, A. F., Fuzzy regression and the term structure of interest rates revisited, AFIR2004.

Soliman, S.A., Alammari, R.A., El-Hawary, M.E., Frequency and harmonics evaluation in power networks using fuzzy regression technique, Electric Power Systems Research, Vol. 66, pp. 171-177, 2003.

Tanaka, H. and Guo, P., Possibilistic Data Analysis for Operations Research, Physica-Verlag, 1999.

Tanaka, H., Uejima, S., Asai, K., Fuzzy linear regression model, IEEE Transactions on Systems Man Cybernetics, Vol. 12, pp. 903-907, 1982.

Tran, L., Duckstein, L., Multiobjective fuzzy regression with central tendecy and possibilistic properties, Fuzzy Sets and Systems, Vol. 130, pp. 21-31, 2002.

Tseng, F-M., Tzeng, G-H., Yu, H-C., Yuan, B.J.C.,Fuzzy ARIMA model for forecasting the foreign exchange market, Fuzzy Sets and Systems, Vol. 118, pp. 9-19, 2001.

Tseng, F-M., Tzeng, G-H., A fuzzy seasonal ARIMA model for forecasting, Fuzzy Sets and System, Vol. 126, pp. 367-376, 2002.

Tseng, F-M., Tzeng, G-H., Yu, H-C., Fuzzy seasonal time series for forecasting the production value of the mechanical industry in Taiwan, Technological Forecasting and Social Change, Vol. 60, pp. 263–273, 1999.

Yang, M-S., Lin, T-S., Fuzzy least-squares linear regression nalysis for fuzzy input-output data, Fuzzy Sets and Systems, Vol. 126, pp. 389-399, 2002.

Yang M-S., Liu, H-H., Fuzzy least-squares algorithms for interactive fuzzy linear regression models, Fuzzy Sets and Systems, Vol. 135, pp. 305-316, 2003.

Yen, K.K., Ghoshray, S., Roig, G., A linear regression model using triangular fuzzy number coefficients, Fuzzy Sets and Systems, Vol. 106, pp. 167-177, 1999.